Topics in Medicinal Chemistry

Volume 38

Topics in Medicinal Chemistry (TMC) covers all relevant aspects of medicinal chemistry research, e.g. pathobiochemistry of diseases, identification and validation of (emerging) drug targets, structural biology, drugability of targets, drug design approaches, chemogenomics, synthetic chemistry including combinatorial methods, bioorganic chemistry, natural compounds, high-throughput screening, pharmacological in vitro and in vivo investigations, drug-receptor interactions on the molecular level, structure-activity relationships, drug absorption, distribution, metabolism, elimination, toxicology and pharmacogenomics. Drug research requires interdisciplinary team-work at the interface between chemistry, biology and medicine. To fulfil this need, TMC is intended for researchers and experts working in academia and in the pharmaceutical industry, and also for graduates that look for a carefully selected collection of high quality review articles on their respective field of expertise.

Medicinal chemistry is both science and art. The science of medicinal chemistry offers mankind one of its best hopes for improving the quality of life. The art of medicinal chemistry continues to challenge its practitioners with the need for both intuition and experience to discover new drugs. Hence sharing the experience of drug research is uniquely beneficial to the field of medicinal chemistry.

All chapters from Topics in Medicinal Chemistry are published OnlineFirst with an individual DOI. In references, Topics in Medicinal Chemistry is abbreviated as Top Med Chem and cited as a journal.

More information about this series at http://www.springer.com/series/7355

Jesus Alcazar • Antonio de la Hoz •
Angel Díaz-Ortiz
Editors

Flow Chemistry in Drug Discovery

With contributions by

I. Abdiaj · J. Alcázar · E. Alza · L. Amenós · Y. Ashikari · M. L. Bolognesi · T. J. Brocksom · S. C. Cosgrove · F. F. de Assis · K. T. de Oliveira · A. A. N. de Souza · A. de la Hoz · Á. Díaz-Ortiz · K. Fox · S. Kobayashi · E. López · R. Luque · A. Nagaki · E. B. A. Paez · I. Peñafiel · M. A. Pericàs · T. H. Rehm · P. Richardson · C. R. Sagandira · Y. Saito · L. A. Soares Romeiro · M. Takumi · T. Tamaki · P. Watts · B. Winterson · T. Wirth

Springer

Editors
Jesus Alcazar
Discovery Chemistry, Janssen
Pharmaceutical Companies of J&J
Janssen-Cilag, S.A.
Toledo, Spain

Antonio de la Hoz
Faculty of Chemistry
University of Castilla-La Mancha
Ciudad Real, Ciudad Real, Spain

Angel Díaz-Ortiz
Faculty of Chemistry
University of Castilla-La Mancha
Ciudad Real, Spain

ISSN 1862-2461 ISSN 1862-247X (electronic)
Topics in Medicinal Chemistry
ISBN 978-3-030-85591-8 ISBN 978-3-030-85592-5 (eBook)
https://doi.org/10.1007/978-3-030-85592-5

This Springer imprint is published by the registered company Springer Nature Switzerland AG.
The registered company address is: Gewerbestrasse 11, 6330 Cham, Switzerland

Preface

The aim of the current book is filling a gap in terms of the application of flow chemistry for the discovery of new drugs, to boost medicinal chemist to use it as an additional tool that is opening new avenues in synthesis of biologically active compounds. Several books have reported the application of flow technology in organic chemistry highlighting advantages and limitations. They supported the spectacular development of flow chemistry tools in the last years as well as the development of new reactions and processes not achievable in batch, due to its capacity to use highly reactive and unstable reagents. In this regard, flow chemistry allows expand reaction parameters far beyond traditional batch processes can do. Flow reactions can be run in less than a second for temperatures and pressures far above solvent boiling point. However, none of them focused on how these advantages can be used to overcome the limitations of the current Drug Discovery process.

For this purpose, the book has been divided in several chapters, the first one covering an overview of the use of this technology in Drug Discovery. From the definition of a flow reaction and the fitting of the methodology in the discovery of new drugs, advantages, limitations, and outlook.

One of the advantages of flow is that covers all key principles of Green Chemistry, from the higher synthetic efficiency, the safe use of dangerous compounds and intermediates, reduction of residues due to telescoping reactions, reduction of activation and protecting groups and the employ of higher concentrations using even benign solvents (water, IL, supercritical fluids), to the higher energy transfer and the efficient synergy with non-conventional activation methods (microwave, ultrasound, electrochemistry, photochemistry,. . .). The efficient use of catalysis is another key principle of Green chemistry and is efficiently implemented in flow. Finally, on-line and in-line real-time analysis using spectroscopic methods pave the way to automatization and the use of artificial intelligence for optimization and scale-up of flow reactions.

One of the key features of flow chemistry has been its combination with other technologies that has been underused or overlooked by medicinal chemist due to technical limitations. In recent years, harvesting photons for running reactions has exploded as a controlled way to promote reactions not allowed by other procedures and has interested medicinal chemists all over the world. However, reproducibility of photochemical reactions has been a bottleneck for the development of drug candidates. To overcome this issue, flow chemistry has proved to be a tool for running reliable photochemical reactions at all scales.

Another way to activate molecules that has been underused in medicinal chemistry is electrochemistry. Adding or removing electrons from molecules using an electrical current has demonstrated over the years the capacity to break and make high energy bonds not accessible with chemical reagents, but reaction needs to take place at the surface of the electrode, limiting the capacity of the reaction at the size of the device. As surface to volume ratio is incredibly higher in flow than in batch, electrochemistry can be run more efficiently using a continuous process.

The fact that reaction parameters can be fully controlled in flow has also its application in enantioselective synthesis. Although this may not be a key step at the beginning of a drug discovery project, as the molecule evolve during the process stereochemical control becomes more relevant and essential to achieve the biochemical properties of the bioactive molecule. For this reason, a chapter dedicated to asymmetric catalysis in flow has been included to give the readers an overview on how enantioselective procedures in flow can be used efficiently for the synthesis of Active Pharmaceutical Ingredients (APIs) and useful building blocks in Drug Discovery.

Flash chemistry or chemistry that can be done in microseconds is a novel concept fully associated to the capability of flow to handle extremely reactive chemical species. This concept was originally defined by Prof. Yoshida and a chapter has been focused on this concept and how it can be useful to prepare amazing lithiated intermediates, impossible to be achieved by other methodologies, and how they can be used to prepare different drugs.

Organocatalysis is a methodology largely used in the pharma sector that is receiving much interest nowadays as a tool in drug discovery. Organocatalyst immobilization allows the recovery and reuse of this molecules and when combined with flow chemistry enables the improvement in the number of cycles that they can perform before deactivation. This has expanded the application to synthesis of bioactive products as it is overviewed in the corresponding chapter.

Similarly, biocatalysis has benefit from flow specially when the corresponding enzymes are immobilized in the corresponding support. This feature not only allows the preparation of singular molecules it also can be used for the preparation of libraries using enzymatic reaction, a key feature for medicinal chemists. Moreover, they are also used for the preparation of bioconjugates, a modern approach to achieve biobased medicines such as modified antibodies.

Beyond single step reaction, flow chemistry can be used for the total synthesis of biologically active molecules an important aspect from drug discovery as these highly reproducible protocols are essential to speed up the synthesis and evaluation of the drug candidate. The use of flow for the synthesis for APIs has been frequently associated to cost reductions and better efficiencies and it is becoming an essential element in their production.

We wanted to add two chapters that highlights new possibilities for flow chemistry. The first one is the use of biomass reagents for the preparation of bioactive compounds. This is an emerging field that clearly will benefit from flow for their production. The use of biomass for drug discovery is not only relevant at the level of recycling waste to reduce the carbon footprint in our planet, it also will allow to create novel building blocks for the synthesis of the molecules based on compounds of biological origin that will be more drug like than those coming from the petrochemical industry.

The second highlighted application of flow chemistry is the accessibility of drug discovery and APIs synthesis in developing countries. With this chapter we wanted to highlight how important is to consider that innovation can come from anywhere in the World nowadays when global solutions are required to build a more equitable and healthier planet.

The last chapter of the book is dedicated to library synthesis in flow and how important is this technology for automated synthesis, a hot topic in drug discovery as allows a much faster exploration and finding of drug candidates. This chapter is closing the book leave the door open to the future of the technology.

As editors of the book, we expect its reading will unlock the imagination to new possibilities for drug discovery and helps readers to pursue new challenges in the synthesis of bioactive products.

Active Pharmaceutical Ingredients (API) described in the book

	Chapter 2	Chapter 3	Chapter 4	Chapter 5	Chapter 6	Chapter 7	Chapter 8	Chapter 9	Chapter 10	Chapter 11	Chapter 12
(+)Femoxetine						X					
3-Demethoxyerythratidinone								X			
Actinomycin	X										
Acyclovir		X									
Adapelene					X						
Albendazole			X								
Artemisin	X							X			
Atorvastatin							X				
Atropine								X			
AZD1480							X				
AZD7594				X							
Baclofen		X		X		X					
Baricitinib								X			
Bassianolide								X			
Beauvericin								X			
Benoxaprofen			X								
Bis-demethoxycurcumin								X			
BMS-919373								X			
Boscalid					X						
Brivaracetam								X		X	
Buclizine (int)								X			
Bupivacaine										X	
Cantharidin									X		
Capecitabine								X		X	
Captopril	X										
Carfilzomib (int)								X			
Chloroxazone			X								
Chlorpromazine			X								

Ciprofloxacin hydrochloride								X			
Clitocybin			X								
Clozapine										X	
Copanlisib								X			
Crizotinib (int)								X			
Curcumine								X			
Cyclizine (int)								X			
Daclatasvir								X		X	
Dantrolene		X									
Darunavir										X	
Demethoxycurcumin								X			
D-erythro-CER[NDS]				X							
D-fagomine							X				
Diazepam	X							X			
Diclofenac			X								
Dihydroartemisinic acid								X			
Diphenhydramine	X							X			
Docloxacirin			X								
Dolutegravir										X	
Doravirine								X			
Efavirenz	X									X	
Eflornithine								X			
Elbasvir								X			
Emend							X				
Emtricitabine								X			
Enniantin C								X			
Ethambutol										X	
Feglymycin								X			
Finerenone	X										
Finrozole			X								

(continued)

	Chapter 2	Chapter 3	Chapter 4	Chapter 5	Chapter 6	Chapter 7	Chapter 8	Chapter 9	Chapter 10	Chapter 11	Chapter 12
Flucytosine								X			
Flunoxaprofen			X								
Fluorophenibut						X					
Fluoxetine	X							X			
Frentizole			X								
Fulvestrant (frg)								X			
Goniofufurone		X						X			
Goniothalamin								X			
Histone deacetylase (HDAC inhibitor)					X						
HNK-1							X				
Hydroxychloroquine (int)								X			
Idaxozan		X									
Imatinib											X
Ingenol-3-mebutate								X			
inhibitor derivatives for a liver X receptor		X									
Islatravir							X				
Isocryptolepine									X		
Isoniazid										X	
Kisqali								X			
Lamivudine								X		X	
Lesinurad								X			
Letrozole			X								
Levetiracetam			X							X	
Licarin A			X								
Lidocaine	X							X			
Lifitegrast (frg)								X			

Linezolid	X										
L-Norvaline							X				
Lomustine								X			
Losartan			X								
Lynparza		X									
Melitracen HCl								X			
Menadione *Vitamin K3*								X			
Mepivacaine										X	
Metaraminol	X										
Methoxamine	X										
Metronidazole (int)								X			
Mexiletine							X				
Miotine						X					
Monopril			X								
Naftopidil						X					
Neomarchantin A								X			
Neostigmine methylsulfate								X			
Nevirapine								X		X	
Nicardipine hydrochloride								X			
Norcantharimide									X		
Norephedrine	X										
NPS R-568						X					
Odanacatib		X									
O-Methoxyanacardate									X		
O-Methoxycardanol									X		
Oripavine	X										
Oseltamivir Phosphate										X	
Osimertinib								X			
Oxycodnone			X								

(continued)

	Chapter 2	Chapter 3	Chapter 4	Chapter 5	Chapter 6	Chapter 7	Chapter 8	Chapter 9	Chapter 10	Chapter 11	Chapter 12
Pagoclone			X								
Paroxetine				X		X		X			
Pauciflorol F					X						
Pazinaclone			X								
Pazopanib		X									
Phenibut		X				X					
Phortress			X								
Pregabalin		X				X	X				
Prexasertib	X							X			
Primidone			X								
Propanolol						X					
Remdesivir								X			
RHO kinase inhibitor						X					
Ribociclib								X			
Rivastigmine						X					
Rocaglamide								X			
Rocaglaol								X			
Rolipram				X							
Ropivacaine			X							X	
Rufinamide								X			X
Solanoeclepin A								X			
Sparteine					X						
TAC-101					X						
Tamiflu	X										
Tenofovir			X							X	
Tetracaine		X									
Thioflavin T			X								
Tolbutamide			X								

Tolmetin						X					
Trimethorprim		X									
Trioxanes	X										
Ursodeoxycholic Acid	X										
Vaborbactam								X			
Valacyclovir salt								X			
Valsartan	X										
Venlafaxine	X										
Verubecestat	X							X			
Vildagliptin (int)								X			
Vitamin D3	X										
Vorinostat									X		
Warfarin (S)						X					
WB-4101		X									
Yohimbine			X								
β-Artemether	X										

Int: intermediate. Frg: fragment

Contents

List of Abbreviations

$[Ir\{dF(CF_3)ppy\}_2(dtbbpy)]PF_6$	4,4′-bis(1,1-dimethylethyl)-2,2′-bipyridine-*N1, N1′*-bis3,5-difluoro-2-5-(trifluoromethyl)-2-pyridinyl-*N*-phenyl-*C*-Iridium(III) hexafluorophosphate
$[N_{14,\ 222}][Tf_2N]$	*N*-tetradecyl-triethylammonium bis (trifluoromethanesulfonyl)imide
$[N_{8,\ 222}][Tf_2N]$	*N*-octyl-triethylammonium bis (trifluoromethanesulfonyl)imide
1,4-DHP	1,4-dihydropyridine
2-MeTHF	2-methyl tetrahydrofuran
3TC	Lamivudine
4CzIPN	1,2,3,5-tetrakis(carbazol-9-yl)-4,6-dicyanobenzene
4-DPA-IPN	2,4,5,6-tetrakis(diphenylamino)isophthalonitrile
4-HTP	4-hydroxythiophenol
7DHC	7-dehydrocholesterol
7βHSDH-GDH	7β-hydroxysteroid dehydrogenase-glucose dehydrogenase
AADH	Amino acid dehydrogenase
AAS	Acetylacetoin synthase
ABO	6-amino-2,2,7-trifluoro-4*h*-benzo[1,4]oxazin-3-one
ABT	(2S,3R)-2-amino-1,3,4-butanetriol
acac	Acetylacetonate
AceK*Ms*	acetate kinase from *Mycobacterium smegmatis*
AcK	Acetate kinase
ACR	Agitated cell reactors
Ad	Adamantyl
AD	Alzheimer’s Disease
ADH	Alcohol dehydrogenase

ADMET	Absorption, distribution, metabolism, excretion, toxicity
ADP	Adenosine diphosphate
AE	Atom economy
aga	Agarose
AI	Artificial intelligence
AIBN	2,2′-azobisisobutyronitrile
ALK	Anaplastic lymphoma kinase
Alloc	Allyloxycarbonyl
ALP	Alkaline phosphatase
AmDH	Amine dehydrogenase
Amphos	di-*tert*-butyl(4-dimethylaminophenyl)phosphine
AP	Acidification potential
API	Active pharmaceutical ingredient
ARDP	Abiotic resource depletion potential
*Ar*TA	*Arthobacter* transaminase
ATA	Transaminase
*At*HNC	Celite supported hydroxynitrile lyase
AtHNL	Arabidopsis thaliana
*At*TA	*Aspergillus terrius* transaminase
AU	African union
AYS	Analysis of the lipase
B_2pin_2	bis(pinacolato)diboron
BACE1	β-site APP cleaving enzyme-1
BCR-Abl	Breakpoint cluster region Abelson tyrosine kinase
BDD	Boron-doped diamond
BDP	(R,R,R)-bisDiazaPhos-SPE
BH_3DMS	Borane dimethylsulfide
BHMF	2,5-bis(hydroxymethyl)furan
BINAP	2,2′-bis(diphenylphosphino)-1,1′-binaphthyl
BINOL	1,1′-binaphthyl-2,2′-diol
BLB lamp	Blacklight blue lamps
*Bm*GDH	Glucose dehydrogenase from *Bacillus megaterium*
$BMIm[BF_4]$	1,3-dimethylimidazolium tetrafluoroborate
Boc	*tert*-butyloxycarbonyl
Bpin	Pinacolato boron
BPR	Back pressure regulation
Bpy	Bipyridine
BSA	*N,O*-bis(trimethylsilyl)acetamide
*Bs*AlaDH	Alanine dehydrogenase from *Bacillus subtilis*
*Bs*GDH	Glucose dehydrogenase from *Bacillus subtilis*
*Bs*PAD	*Bacillus subtilis*
C/PVDF	Carbon-filled polyvinylidene fluoride

CA	Cellulose acetate
CALB	Candida antarctica lipase B
CAPIC	2-chloro-3-amino-4-picoline
CataCXium-A	Di(1-adamantyl)-*n*-butylphosphine
CB	Carbon beads
CBS	Corey, Bakshi and Shibata oxazaborolidine
CCE	Constant current electrolysis
CD	Cyclodextrin
CDCA	Chenodeoxycholic acid
CDK	Cyclin-dependent kinase
CED	Cumulative energy demand
CFD	Computational fluid dynamics
CFL	Compact fluorescent lamp
cGMP	Current good laboratory practice
CHEK	Serine/threonine protein kinase
ChiAmDH	Chimeric AmDH
CM	Continuous manufacturing
CML	Chronic myeloid leukemia
CMOs	Contract manufacturing organizations
CNSL	Cashew Nut Shell Liquid
Cod	Cyclooctadiene
Cox	Glucose oxidase
CP	Controlled potential electrolysis
Cp	Cyclopentadienyl
CPA	Chiral phosphoric acid
CPME	Cyclopentyl methyl ether
CSM	Catalytic static mixers
C-SOPS	Engineering research centre for structured organic particulate systems
CSS	Central switching station (reactor and inline analysis)
Cu-BOX	Cu-bis(oxazoline)
CV	Cyclic voltammogram
CVs	Collection vessels
*Cv*TA	*Chromobacterium violaceum* transaminase
DABCO	1,4-diazabicyclo[2.2.2]octane
DAD	Diode array detector
Dba	Dibenzylideneacetone
DBU	1,8-diazabicyclo[5.4.0]undec-7-ene
DCE	Dichloroethane
DERA	Deoxyribose 5-phosphate aldolase
D-erythro-CER[NDS]	*N*-((2S,3R)-1,3-dihydroxyoctadecan-2-yl) stearamide
DES	Deep eutectic solvents

dF(CF_3)ppy	2-(2,4-difluorophenyl)-5-(trifluoromethyl) pyridine
DFF	2,5-Diformylfuran
DFMO	α-difluoromethylornithine
DIAD	Diisopropyl azodicarboxylate
DIPEA	Diisopropylethylamine
DKR	Dynamic kinetic resolution
DMA	*N,N*-dimethylacetamide
DMAP	4-dimethylaminopyridine
DMC	Dimethyl carbonate
Dme	Dimethoxyethane
DMF	*N,N*-dimethylformamide
dmppy	Dimethylphenylpyridine
DMPS-Pd/AC-CP	Dimethylpolysilane-Pd/Activated carbon-Calcium phosphate
DMSO	Dimethylsulfoxide
DoE	Design of experiments
DOS	Diversity-oriented synthesis
DPF	Drug product formulation
DPP-4	Dipeptidyl peptidase-4
DPPA	Diphenylphosphoryl azide
dppf	1,1′-bis(diphenylphosphino)ferrocene
dpy	Dipyridine
dtbbpy	4,4′-di-*tert*-butyl-2,2′-dipyridyl
dtbpf	bis(di-*tert*-butylphosphino)ferrocene
DTG	Dolutegravir
DuPhos	1,2-bis[(2R,5R)-2,5-diRphospholano]benzene
DVB	*p*-divinylbenzene
EDC	1-ethyl-3-(3-dimethylaminopropy)carbodiimide
ee	Enantiomeric excess
EGFR	Epidermal growth factor receptor
ELSD	Evaporative light scattering detector
ESI	Electrospray ionization
FAETP	Fresh aquatic eco-oxicity potential
FDA	Food and drug administration
FDH	Formate dehydrogenase
FDM	Fuse deposition modeling
FDP	Fossil fuel depletion potential
FEP	Fluorinated ethylene propylene copolymer
FGE	Formylglycine generating enzymes
FPE	Flow Pickering emulsion
FR	Flow rate
FruA	Fructose aldolase
FTC	Emtricitabine

G3PD*Ec*	*E. coli* glycerol-3-phosphate dehydrogenase
GABA	γ-aminobutiric acid
GalK	Galactokinase
GalT	β1,4-galactosyltransferase
GC	Glassy carbon
GCIPR	Green chemistry institute pharmaceutical roundtable
GC-MS	Gas chromatography-mass spectrometry
GDH	Glucose 1-dehydrogenase
GlpK*Tk*	Glycerol kinase from *Thermococcus kodakarensis*
Glyme	1,2-dimethoxyethane
GMP	Good laboratory practice
GOase	Galactose oxidase
GPR	Gaussian process regression
$gpsA_{Ec}$	Guanidinopropylsuccinic acid
Gs-Lys6DH	*Geobacillus stearothermophilus*
GVL	γ-valerolactone
GWP	Global warming potential
HALEX	HALogene EXchange (reaction)
HAT	Hydrogen atom transfer (reaction)
HCV	Hepatitis C virus
HDAC	Histone deacetylase
HeP5C	Pyrroline-5-carboxylate reductase from *Halomonas elongata*
HEWT	Transaminase from *Halomonas elongata*
HFIP	1,1,1,3,3,3-hexafluoroisopropanol
H-Gly-Pro-AMC	(*S*)-*N*-(2-aminoacetyl)-1-(4-methyl-2-oxo-2H-chromen-7-yl)pyrrolidine-2-carboxamide
HIPE	High internal phase emulsion
HIV	Human inmunodeficient virus
HMF	5-Hydroxymethylfurfural
HNK-1	Human natural killer cell-1
HNL	Hydroxynitrile lyase
HOMO	Highest occupied molecular orbital
HPLC	High-performance liquid chromatography
HRP	Horseradish peroxidase
HTE	High throughput experimentation
HtL	Hit to lead
HTP	Human toxicity potential
HTS	High throughput screening
HyperBTM	Isothiourea derivative
I.D.	Inner diameter
IC_{50}	Half-maximal inhibitory concentration

IL	Ionic liquid
IMAC Ni-NTA	Immobilized metal affinity chromatography Nickel charged
IPA	Isopropyl alcohol
IRED	Imine reductase
JAK2	Janus kinase 2
JohnPhos	(2-biphenyl)di-*tert*-butylphosphine
KIT-6	Mesoporous silica
K-OMS-2	K-doped manganese oxide
KPi	Potassium phosphate buffer
KR	Kinetic resolution
KRED	Ketoreductase
LbADH	Ketoreductase LbADH
LCA	Life cycle assessment
LCI	Life cycle inventory
LCIA	Life cycle impact assessment
LC-MS	Liquid chromatography–mass spectrometry
LDH	Lactate dehydrogenase
LDH-7αHSDH	Lactate dehydrogenase-α-hydroxysteroid dehydrogenase
LED	Light-emitting diode
LiNp	Lithium naphthalenide
LLO	Late lead optimization
LO	Lead optimization
LOM	Laminated object manufacturing
LSC-PM	Luminescent solar concentrator based photomicroreactor
LSF	Late-stage functionalisation
LTF	Little things factory
LUMO	Lowest unoccupied molecular orbital
MAETP	Marine aquatic eco-toxicity potential
MAO	Monoamine oxidase
MAT540	Silicamicrosphere
MB	Methylene blue
MBA	α-methylbenzylamine
MCF	Mesostructured cellular foam
m-CPBA	*Meta*-chloroperbenzoic acid
MDN	Methyleneglutarodinitrile
MDP	Metal depletion potential
MEP	Marine eutrophication potential
Mes	Mesityl
Mes-Acr-4	9-mesityl-10-methylacridinium tetrafluoroborate
MFC	Mass flow controller
MIDA	*N*-methyliminodiacetic acid

MISER	Multiple injections in a single experimental run
MJM	Multijet modeling
ML	Machine learning
MM	Monolithic microreactor
MN	Monolith
MNU	*N*-methyl-*N*-nitrosourea
MOF	Metal organic framework
MPIR	Multi-point injection reactor
MPLC	Medium pressure liquid chromatography
MS	Mass spectroscopy
MS 4Å	Molecular sieves 4Å
MSC	Mesenchymal stem cell
MTBE	Methyltertbutylether
MTPS	Methyltriethyl ammonium methylsulfate
MW	Microwaves
MW	Molecular weight (chapter 12)
MWCNT	Multiwalled carbon nanotube
NADH	Nicotinamide adenine dinucleotide
NADPH/NADP+	Nicotinamide adenine dinucleotide phosphate
NaDT	Sodium decatungstate
NaPi	Sodium phosphate buffer
NDK	5-nitrononane-2,8-dione
NFSI	*N*-fluorobenzenesulfonimide
NHC	*N*-heterocyclic carbenes
NK1	Neurokinin 1
NLTP	Natural land transformation potential
NMP	*N*-methylpyrrolidone
NMR	Nuclear magnetic resonance
NMS	Nelder–Mead simplex
NNRTI	Non-nucleoside reverse transcriptase inhibitor
NOX*Ca*	NADH oxidase from *Clostridium aminovalericum*
NP	Nutrification potential
NPW	Novel process window
NS5A	Non-structural protein 5A
NTf_2	bis(trifluoromethane)sulfonimide
OCb	*N,N*-diisopropylcarbamoyloxy
ODP	Ozone depletion potential
PanK	Pantothenate kinase
PAT	Process analytical technologies
Pc	Phthalocyanine
PCC	Pyridinium chlorochromate
PCN	Porous carbon nanosheets
Pd/AMP-KG	Palladium on aminopropyl-grafted silica

PDMS	Polydimethylsiloxane
*Pe*AAOx	*Pleurotus eryngii*
PEEK	Polyether ether ketone
PEG	Polyethylene glycol
PEPPSI™-SIPr	Pyridine-enhanced precatalyst preparation stabilization and initiation (1,3-bis(2,6-diisopropylphenyl)imidazolidene) (3-chloropyridyl) palladium(ii) dichloride
PET	Positron emission tomography
PFA	Perfluoroalkoxyalkane
PG	Protecting group
*Pglu*KRED	*Pichia glucozyma*
Pin	Pinocamphenyl
Pip*(O)][BF_4]	(4-acetamido-2,2,6,6-tetramethyl-1-oxopiperidinium tetrafluoroborate
Piv	Pivaloyl
PK	Pharmacokinetic
PK/PD	Pharmacokinetic/pharmacodynamic relationships
PLP	Pyridoxal-5′-phosphate
PMC	Parallel Medicinal Chemistry
PMI	Process mass intensification
PMP	*p*-methoxyphenyl
PNP	Purine nucleoside phosphorylase
POCP	Photochemical ozone creation potential
POFP	Photochemical oxidant formation potential
PPM	Phosphopentamutase
ppy	2-phenylpyridine
PRV	Pressure relief valve
PS	Polystyrene
PS-BEMP	2-*tert*-butylimino-2-diethylamino-1,3-dimethylperhydro-1,3,2-diazaphosphorine on polystyrene
PS-DVB	Polystyrene-divinylbenzene
PS-SO2NIK	PS-*N*-iodo-*p*-toluenesulfonamide potassium salt
PS-TRIP	PS-4-hydroxy-2,6-bis(2,4,6-triisopropylphenyl) dinaphtho[2,1-d:1′,2′-f][1,3,2]dioxaphosphepine 4-oxide
PS-TU	PS-4-(3-((1R,2R)-2-(piperidin-1-yl)cyclohexyl) thioureido)-2-(trifluoromethyl)benzoic acid
PTA	Phase transfer agent
PTC	Phase transfer catalysis
PTFE	Polytetrafluoroethylene
PTSA	*p*-toluenesulphonic acid
py	Pyridine

Pybox	Pyridinebisoxazoline
QbD	Quality by design
QC	Quality control
RA	Rheumatoid arthritis
rAaeUPO	A. aegerita unspecified peroxygenase
RCM	Ring closing methatesis
RCPE	Research centre for pharmaceutical engineering
RCY	Radiochemical yield
RDS	Reagent delivery system
RedAms	Reductive aminases
RmaNOD	Rhodothermus marinus nitric oxide dioxygenase
RSM	Residual surface model
RSO1	C-ros oncogene 1
RSPU	Reaction segment preparation unit
S/C	$mol_{substrate}/mol_{catalyst}$
SAHA	Vorinostat
SAR	Structure-activity relationships
Sc	Supercritical
SC	SpyCatcher
SCE	Saturated calomel electrode
SDG 3	No. 3 United Nation's sustainable development goal
SEM	Scanning electron microscopy
SFC	Supercritical fluid chromatography
SFU	Single frequency ultrasound
SHS	Switchable-hydrophilicity solvent
SILLPg	Supported ionic liquid-like phases
SILP	Supported ionic liquid phase
SLAP	Silicone amine protocol
SLILs	Sponge-likeiionic liquids
SLS	Laser sintering
SM	Small molecules
SM	Standby module (12th chapter)
SMCR	Stacked multichannel reactor
S_N2	Bimolecular nucleophilic substitution
SP	Sucrose phosphorylase
S-Phos	2-Dicyclohexylphosphino-2′,6′-dimethoxybiphenyl
Sppy	Sulfonated 2-phenylpyridine
SPR	Structure-property relationships
SrtA*	Sortase A
SS	Stainless steel
SSPC	Synthesis and solid-state pharmaceutical centre
SSRI	Serotonin reuptake inhibitor

ST	SpyTag
STY	Space-time yield
TA	Transaminase
t-AmOH	*tert*-amylalcohol
TAP	Terrestrial acidification potential
TB	Tuberculosis
TBAA	Tetrabutylammonium azide
TBADT	Tetrabutylammonium decatungstate
TBAF	Tetrabutylammonium fluoride
TBME	*tert*-butyl methyl ether
TBS	*tert*-butyldimethylsilyl
Tc	*Thermomonospora curvata*
TEMPO	2,2,6,6-tetramethyl-1-piperidinyloxy
TES	Triethylsilyl
*Tes*ADH	*Thermoanerobacter ethanolicus*
TETP	Terrestrial eco-toxicity potential
TFA	Trifluoroacetic acid
TFE	2,2,2-trifluoroethanol
ThDP	Thiamine diphosphate
THF	Tetrahydrofuran
THIQs	1,2,3,4-tetrahydroisoquinolines
TK	Transketolase
TL IM	*Thermomyces lanuginosus*
TMDAM	bis(dimethylamino)methane
TMS	Trimethylsilyl
TMSA	Trimethylsilyl azide
TMSCN	Trimethylsilyl cyanide
TMSOTf	Trimethylsilyl trifluoromethanesulfonate
TOF	Turnover frequency
TON	Turnover number
TPFPP	5,10,15,20-tetrakis(pentafluorophenyl)porphyrin
TPP	Tetraphenylporphyrin
TPPy	2,4,6-triphenylpyrilium tetrafluoroborate
TPSA	Topological polar surface area
Tr	Tritylium
TRIP	3,3′-bis(2,4,6-triisopropylphenyl)-1,1′-binaphthyl-2,2′-diyl hydrogenphosphate
Tris	Tris HCl buffer pH 8.0
UDCA	Ursodeoxycholic acid
UDL	Undetected leaching
UDP	Uridine 5′-diphosphate
UGDH	UDP-glucose-dehydrogenase
UIB	Universal interface box

UNIDO	United nations industrial development organisation
UPLC	Ultra high performance liquid chromatography
UPO	Unspecific peroxygenase
USP	UDP-sugar pyrophosphoralase
UV	Ultraviolet
VfTA	*Vibrio fuvalis* transaminase
VTO	Volume-time output
Xantphos	(9,9-Dimethyl-*9H*-xanthene-4,5-diyl)bis (diphenylphosphane)
Xphos	2-dicyclohexylphosphino-2′,4′,6′-triisopropylbiphenyl

Top Med Chem (2021) 38: 1–22
https://doi.org/10.1007/7355_2021_110

Published online: 12 February 2021

Flow Chemistry in Drug Discovery: Challenges and Opportunities

Enol López and Jesús Alcázar

Contents

Abstract The spectacular development of new chemical reactions and processes under continuous flow has attracted particularly the attention of the pharmaceutical industry. The chance to carry out complex chemistry, the sustainability of the process, and the possibility of adapting new technologies to this technique have paved the way to the integration of flow chemistry into drug discovery. Thus, this book chapter covers essential aspects of flow chemistry and how a variety of technologies and catalytic methods can be used to enable new chemical space in drug discovery programs.

Keywords Drug discovery, Enabling technology, Flow chemistry, Integrated platforms, Novel chemical space

---

E. López
Facultad de Ciencias y Tecnologías Químicas, Universidad de Castilla-La Mancha, Ciudad Real, Spain

J. Alcázar (✉)
Lead Discovery, Janssen Research and Development, Toledo, Spain
e-mail: jalcazar@its.jnj.com

1 Introduction

1.1 Introduction to Flow Chemistry

During the last 20 years, flow chemistry has been positioned as a very effective technology in the chemistry community and has gained the attention of researchers from both academia and pharma companies, which have both contributed vast number of publications covered by peer-reviewed articles in journals and in patents. Among these contributions, new synthetic chemistry approaches and important improvements of known batch procedures have been reported [1–3]. However, this technology has not been yet implemented in every synthetic laboratory as it is a disruptive technology and therefore very different from the traditional batch approach all chemists have been trained to perform.

The interest in flow chemistry resides on the capability to conduct chemical transformations in a continuous way by constant pumping of one or more solutions containing the reagents. A mixing point merges the solutions, triggering the reaction, which will continue through the channel where the stream is subjected to the required conditions (i.e., thermal, microwave, light, current, etc.). Following that, the resulting solution is collected in a flask ready for quenching or to be mixed with another solution in order to achieve multistep sequential reactions. In contrast to the batch mode [4, 5], the transformation takes place spatially along a tube, so their dimensions as well as the flow rate determine the residence time.

A great number of variables can be modified in continuous flow to control the reaction outcome, like temperature, reaction time, concentration, or energy source. The accurate control of these parameters usually provides more selective transformations, as side reactions can be avoided. As a result, better yields and purities are obtained. Continuous flow chemistry can also be used to control reactions in terms of extreme reactivity or safety; for instance, when handling hazardous reagents, unstable intermediates, or non-safe or toxic gases. Thus, risks associated with the scale-up of these transformations are reduced or completely avoided because the potentially dangerous species are generated in a small reactor size and reacted in situ (make and use concept). Furthermore, the elimination of headspace in comparison with the batch mode favors the control of low-boiling point solvents and reagents under pressurized flow reactors.

Another important feature is the control of highly exothermic self-accelerated reactions ("runaway" reactions). The better heat and mix transfer processes due to the high surface area-to-volume ratio allow for these transformations to be carried out under precise conditions, which are unattainable by traditional batch chemistry. Likewise, high-pressure reactions can be performed, thereby speeding up reaction times. Cryogenic conditions are commonly avoided as the unstable species, for instance organometallic intermediates, are present in the reaction line only for a few seconds. Thus, continuous flow serves as a "filter of safety" when a problematic reaction must be carried out, bringing back to the bench disregarded reactions commonly known as "forgotten chemistries."

These facts overcome important limitations found in traditional batch approaches showcasing important benefits of the technology [6], and expanding the chemistry space available to the chemists (Fig. 1) [7], for instance, extremely fast reactions [8] that can take place in less than 1 s ("flash chemistry"). However, there are some common limitations that must be considered. One of the main concerns is the clogging of the flow system, which can happen at any point in the flow unit. This can be due to the accumulation of particles on certain parts of the channels, crystallization processes, chemical reaction fouling, or corrosion. In the last two cases, the problem can be solved using a compatible material. For the first two cases, a change in the solvent media, droplet reaction techniques, or an in-situ cleaning by using sonication are common approximations to overcome the formation of solids. If these approaches are not enough, a novel engineering design should be considered in order to create an appropriate flow unit. Pumping systems are crucial as they are the ones to keep the system continuously moving. Piston pumps offer the advantage that they can handle higher pressures; however, pulsation problems may lead to uneven mixing of reagent solutions. Syringe and peristaltic pumps can overcome this issue but only work at lower pressure. All in all, most of the common issues found in continuous flow are well-known and can be overcome by customizing the reactor according to reaction needs.

1.2 Flow Chemistry Setup

One of the most important benefits of continuous flow chemistry includes the possibility of adapting the reactor unit to the requirements of the reaction. As several pieces of equipment can be interchanged or replaced at different points, flow chemistry is a toolkit, which provides great versatility in terms of setup. Most parts of the flow system are connected by *tubing* whose composition depends on the pressure and nature of the reagents. For low and medium pressures (<30 bar), inert perfluorinated polymers such as PTFE, PEEK, PFA, and FEP are suitable. Instead, stainless steel or Hastelloy are common alternatives when high temperature and pressure are needed. A variety of pieces (nuts, ferrules) are also used to attach and fix the tubing properly (Fig. 2).

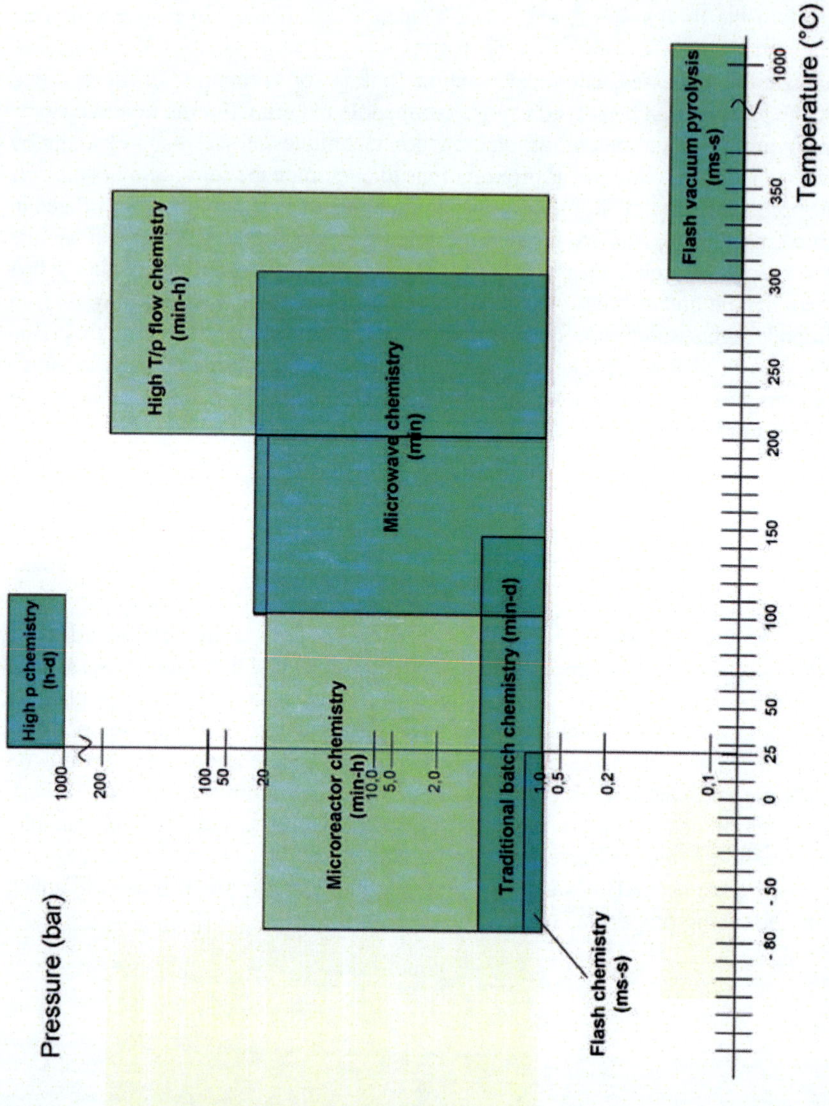

Fig. 1 Expansion of reaction parameters provided by flow chemistry. Reproduced with permission from Ref. [7]

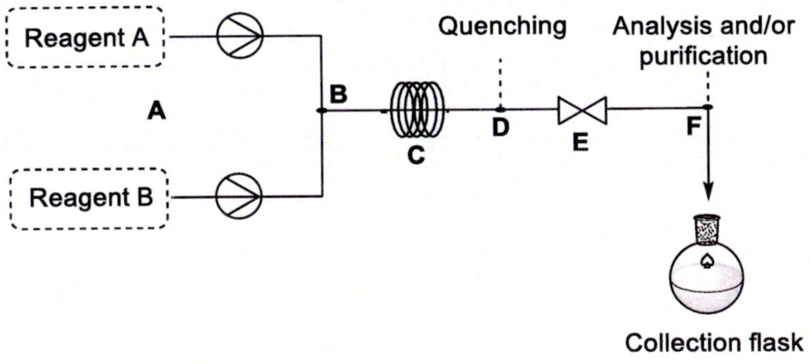

Fig. 2 General scheme of a flow system

(a) Very likely, *reagents delivery* is the first item to consider and a key element to run reactions in flow. Most commonly, reagents are introduced as solutions from a liquid delivery system in a sustained way to achieve the right mixing in the subsequent unit. To achieve that, three types of pumps can be envisioned depending on the flow rate, the nature of the solution, and the system pressure. Syringe pumps are the simplest and adequate for low flow rates, although they are limited to small-scale reactions and low pressurized systems. HPLC pumps work at high pressure, although pulsation issues and uneven delivery of solutions in terms of time can be observed specially at low flow rates preventing appropriate mixing. Recently, peristaltic pumps have appeared on the scene as a new solution using a central rotor that is constantly pushing the liquid through a flexible tube [9]. Deliveries at low flow rate are improved, but they can only work at limited system pressures. In the case of using gases, a gas bottle can be installed to the flow system by using a mass flow controller (MFC), which can control the amount of gas introduced into the reaction [10].

(b) The reaction starts at the *mixing point* where all reagents merge. Usually T or Y-shape connections are used to mix two or more streams before being introduced into the reaction unit. If a more efficient mixing is required, other special alternatives with optimal mixtures can be considered [11–13]. For instance, multilaminar mixers that separate both channels to micro streams to enhance the surface area and thereby facilitate diffusion [14].

In case of gas–liquid mixing, phases can be separated into two convergent lines separated by a gas permeable membrane: the "tube-in-tube" strategy developed by Ley and coworkers (Fig. 3). The gaseous reagent from the outer tube goes through the permeable membrane into the inlet tube to achieve the desired transformation [15].

(c) Once reagents are properly mixed the reaction continues in the *reactor unit*. Depending on the nature of the transformation, different reactors can be used. It usually consists of a coil, a chip, or a packed-bed reactor (Fig. 4). *Coils* are usually an easy access alternative as they are made up of a tube, which has been

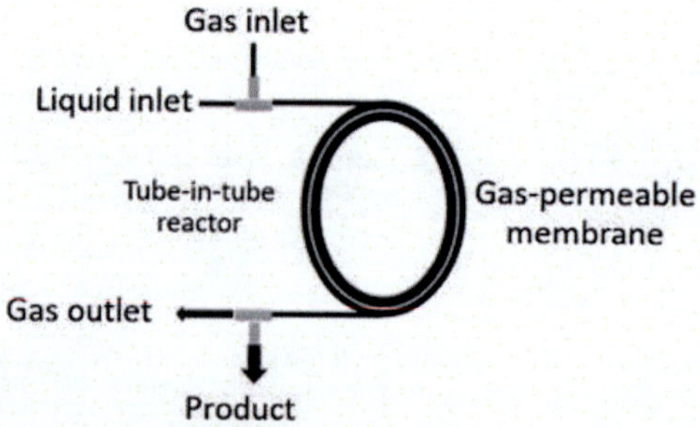

Fig. 3 General scheme of a tube-in-tube reactor

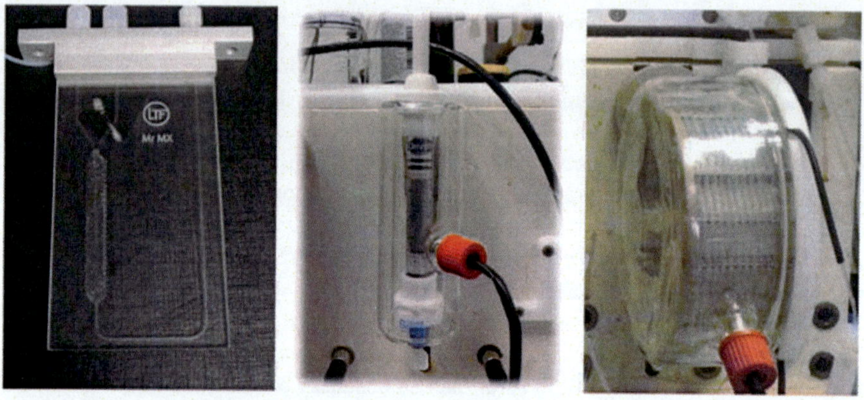

Fig. 4 Flow reactors from left to right: chip, packed-bed reactor, coil

wrapped around a circular support, and the material will depend on reagents and reaction conditions that will be used. *Chips* have been widely used in microfluidic thermal reactions due to the high surface area-to-volume ratio. Depending on the type of chemistry that is going to be developed, different materials can be used in these systems (e.g., glass, silicon, stainless steel). For instance, in photochemical reactions, the material needs to be permeable to light (glass). Recently, 3D printing technology has enabled the production of chips on-demand in order to adapt these devices into the flow unit [16, 17]. *Packed-bed reactors* are the best alternative when solid reagents or catalysts need to be used. It consists of a thicker tube that will contain the solid, and the other reagents as solution will flow through it. In the case of catalysts, they are usually immobilized [18].

Attending to electrochemical devices [19, 20], both cathode and anode electrodes are in close contact with the flow channel. While divided cell microreactors use a membrane to separate the cathode and anode channels, a sandwich disposition of the electrodes with the flow channel is commonly found in undivided cell microreactors (Fig. 5).

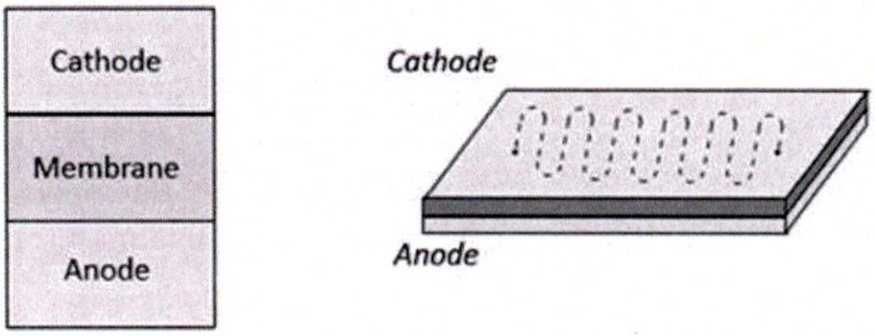

Fig. 5 Schemes of flow electrochemical reactors

One of the main drawbacks of flow chemistry is in-line precipitation of solids and the subsequent blockage of the system. Different solutions have been envisaged for this issue, i.e. the use of spiral tubing in combination with ultrasound irradiation [21] or the use of agitated cell reactors (ACR), which promote an efficient mixture through a lateral shaking of the reactor, thereby avoiding the precipitation of solids and preventing the separation of phases [22].

(d) Flow reactions are shut down at the *quenching point*. This can be done by a chemical agent, adding a quencher with an extra line containing or collecting over a quenching solution, depending on different parameters, such as solid formation or exothermicity. Another possibility is removing the effector; for instance: cooling the line in case of a thermal reaction, getting the solution out of the irradiation zone in the case of photochemistry or out of the electrical area in case of electrochemistry.
(e) *Back-pressure regulators* (BPR) are valves that maintain a constant pressure through the flow system. They are mainly used when gaseous reagents or intermediates participate in the transformation to keep the gas in solution and reduce possible residence time deviations. This gas can come also from the solvent when reactions are heated above their boiling point. Two kinds of devices can be found: either fixed when the regulator operates at predefined pressure or variable, which is able to adjust the system pressure to reaction needs.

Common aspects of flow chemistry equipment have been discussed above to provide a general idea of the different parts of the flow unit and show the versatility of the technology and how to adapt them to reaction requirements. In addition, the modularity of the technology affords new opportunities for attaching analytical equipment to follow the chemical transformation. It is possible to install different analytical techniques (LCMS, GC, NMR, etc.) to have a direct read out of the transformation [23–25]. A splitter device can sample the flow periodically and transfer them to the analytical equipment or spectroscopic methods can be attached in-line to collect the information as the reaction flows through. These techniques are non-destructive and allow real-time analysis of the reaction. Furthermore, purification techniques can also be added at the end of the flow unit, such as liquid–liquid separation, where the aqueous and organic layers are separated through a hydrophobic membrane [26]. Other alternatives are the use of scavengers to trap impurities at some points of the flow unit [27] or gas separators if the previously mentioned tube-in-tube reactor is connected to a vacuum line.

All the different modules described in this section can be interconnected and automated to accelerate the optimization of reaction parameters, exploration of the chemical space, and overcome chemist's routine operations. In other words, accelerating the synthesis of organic molecules in a multistep way [28].

1.3 Types of Transformations

A wide variety of transformations involving homo- or heterogeneous reactions can be carried out in continuous flow. In heterogeneous *gas–liquid reactions*, a large excess of the gas counterpart is necessary to favor the miscibility in the organic solvent. Microreactors are a convenient alternative as they eliminate the common headspaces found in batch chemistry, thereby increasing the interfacial area between the gas and liquid phases. Thus, for high pressures, manipulation of non-safe gases or scaling-up a reaction, flow chemistry provides a very convenient approach. Another type of heterogeneous transformations for which flow chemistry provides a better interaction of phases are *solid–liquid reactions* [29]. The solid material is immobilized in a packed-bed reactor and the liquid phase is pumped through it. In this manner, the large excess of solids remains in the flow system thereby allowing for subsequent reactions and facilitating work-up processes. If a gaseous reagent is participating in the process a *gas–liquid–solid reaction* is assumed, hydrogenation being the most common example [30].

Regarding homogeneous transformations in flow chemistry, the reactivity in *liquid–liquid reactions* is also enhanced due to the small reactor size. The challenge remains in maintaining a constant distribution of the fluid without affecting the residence times [31]. In this regard, flow rates should be carefully controlled to increase mass transfer efficiency.

Beyond conventional considerations of flow chemistry, the combination with photo-, mechano-, and electrochemistry is revolutionizing traditional chemistry approaches [32, 33].

2 The Drug Discovery Process in Pharma

Pharmaceutical companies invest large amounts of money and time to launch a new drug on the market, this take on average more than 12 years with an estimated cost of more than 1,000 million euros [34]. This process entails the preparation of a vast number of molecules, which therefore makes the drug discovery process both long and tedious. Moreover, drug discovery is a highly competitive field as different companies are working to achieve the best drug candidate for a limited number of protein targets or diseases [35, 36].

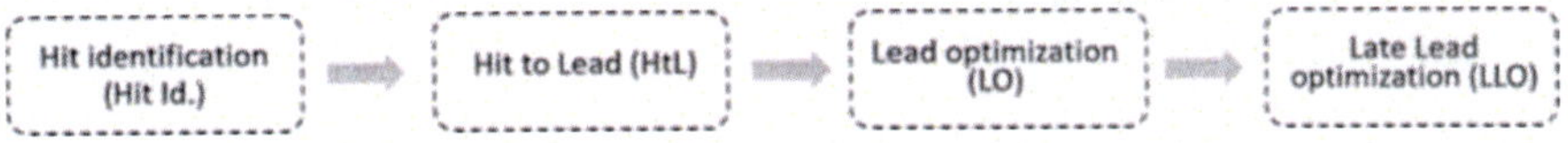

Fig. 6 Stages in the drug discovery process

The drug discovery process is divided into several stages from initial target identification to the selection of the clinical candidate (Fig. 6).

– Target validation-Hit identification

The first step of drug discovery is the identification and validation of a protein target, which is linked to a potential treatment of a certain disease. Then a first set of binders, also known as initial *hits*, needs to be identified as a starting point in order to develop the chemistry part of the program. Several strategies can be found to meet this end: high throughput screening (HTS) [37], chemical genomics [38], virtual screening [39], and fragment-based screening [40] (by using X-ray diffraction analysis or NMR techniques). The combination of these strategies with the most appropriate assays is necessary to select the best scaffolds to go through the next stage [41].

– Hit to Lead (HtL)

Once those hits are identified, which are usually molecules with micromolar activities at the target protein, further modifications of the structure are envisioned and carried out to establish structure–activity relationships (SAR) [42]. In parallel, their physicochemical properties are profiled to build the corresponding structure–property relationships (SPR). Typically, iterative design-synthesize-test cycles are used to evolve the chemistry program. First designing the molecules to be prepared, then they are synthetized and screened biologically. Data are collected and used in the next generation of molecules to be made. This iterative approach allows for both the improvement of binding affinity and pharmacokinetic (PK) properties of the molecules prepared. Optimally, compounds with nanomolar activity as well as suitable selectivity and pharmacokinetic profile (*leads*) are selected for subsequent test in in vivo models.

– Lead Optimization (LO)

At this stage, the optimization is focused on improving selectivity and ADMET properties (Absorption, Distribution, Metabolism, Excretion, Toxicity) in vivo of the compounds to establish their Pharmacokinetic/Pharmacodynamic relationships (PK/PD). This ratio will define the potential therapeutic window of suitable candidates that can be subjected to human testing.

– Late Lead Optimization (LLO)

The compounds with a suitable therapeutic window are prepared at multigram amounts to complete all preclinical studies needed in order to support their proposal as a new drug for human use to the regulatory agencies.

Based on the different stages of the drug discovery program, the synthesis of molecules varies from the single milligram scale in HtL to 50-gram scale for the selected candidate. The generation of a wide variety of molecular entities at different scales is a key feature in drug discovery, for this reason, the development of new technologies that can accelerate this process is therefore highly desirable.

3 Flow Chemistry as a Tool to Improve Drug Discovery

As we have seen so far, drug discovery is a high-cost process that is also associated with various risks that may end the course before delivering a drug candidate. Pharma companies are looking to diverse alternatives to reduce cycle times and attrition rates by accelerating the fast generation of the required data to stop unsustainable programs and focus efforts to progress the most promising ones.

Flow chemistry can be one of these alternatives. For instance, streamlining the preparation of a *Hit* in gram amounts to be tested in vivo without requiring a time-consuming re-designing of its synthesis. This simple advantage may reduce production time from weeks or months to just a few days. For the aforementioned reasons, flow reactor technology is continuously growing as a discovery technology and is being implemented in pharmaceutical companies [43–48] notwithstanding the fact it was initially primarily used in academia for the synthesis of Active Pharmaceutical Ingredient (API) manufacturing [49].

Flow has also been associated with the preparation of challenging molecules. For instance, it has proved its value as a tool to introduce C(sp^3) fragments and escape from flatland [50]. This abovementioned point is important in order to explore different dispositions of donor, acceptor, lipophilic or hydrophilic groups so as to find the most preferred interaction with the target protein, thereby improving their binding properties [51, 52].

3.1 *Green Components of Flow Chemistry*

Despite the well-established methodologies to prepare drug candidates [53–55], there is a recent tendency to develop chemical transformations from a more sustainable point of view that have indeed made the introduction of continuous flow very interesting from a green perspective [56–58]. Interestingly, the chances of combining continuous flow with catalytic methodologies or novel synthetic approaches (e.g., photochemistry, electrochemistry, ultrasounds, microwave irradiation, biomass reagents) open the doors to explore chemical space difficult-to-achieve by traditional methods.

Ultrasound irradiation has been widely explored in organic synthesis not only to favor heterogeneous processes, but also to induce new chemical reactivities [59]. This energy source facilitates the development of faster procedures under mild reaction conditions. Thus, its integration into continuous flow is very convenient to avoid clogging in microreactors, thereby broadening their applicability in drug discovery. However, scaling of these reactors is still a work in progress, as problems associated with uniform irradiation and low energy transfer efficiencies need to be overcome [60].

The ability of microwave heating (MW) to shorten reaction times allows the preparation of biologically active compounds in a much faster way, reducing cycle times in drug discovery. These applications have been widely described in literature [61, 62] and are particularly useful in many cross-coupling reactions involving transition metals as they usually have long reaction times. Despite the advantages shown in batch mode, there are some drawbacks concerning the scalability of such reactions mainly due to the limited penetration of the microwave irradiation. To overcome this, continuous flow setups have been investigated to exploit this technique in the drug discovery process, furnishing interesting transformations in organic synthesis [63, 64].

In addition to the previously mentioned methodologies, flow chemistry has also been considered for the synthesis of high-valued chemicals from renewable biomass [65]. This promising approach represents a sustainable alternative to limited fossil sources, as it is considered a virtually infinitive reservoir that can be used for the production of organic derivatives such as polyols, furanoids, or carboxylic acids. In this regard, microreactors provide better heat and mass transfer processes, which are mostly required for these multiphasic systems [66].

3.2 *Diversity Oriented Synthesis (DOS) in Flow*

The development of technologies, which can support the rapid generation of chemical entities, is a common goal to reduce cycle times and accelerate the finding of the potential clinical candidate. Thus, sequential synthesis is perfectly compatible with flow reactors and enables the preparation of chemical libraries for rapid screening vs biological targets. This is highly interesting for the pharmaceutical industry as it can be linked with automated compound synthesis. Thus, various reagents can be loaded into the flow system and the products are collected separately after multiple possible combinative processes. By using segmented flow, different micro-reactions are flowing continuously between an inert solvent, which separates the reaction droplets [67]. A subsequent analysis and purification of each reaction allows for the preparation of a diverse set of chemical entities.

Diversity Oriented Synthesis has appeared in recent years as an interesting tool to access novel chemical space [68]. It aims to generate a functionally diverse library based on a collection of compounds with known molecular structures of high complexities, which can interact with different biological targets [69]. In this manner, substantial chemical space can be covered by considering different small-molecular shapes and obtaining a diverse set of compounds with varying biological activities [70, 71]. For instance, unstable organometallic reagents can be prepared and reacted in situ with a variety of simple starting materials providing a set of diverse chemical analogues (Fig. 7). The generation of highly energetic intermediates can also be controlled in flow and it has been used to perform reactions in less

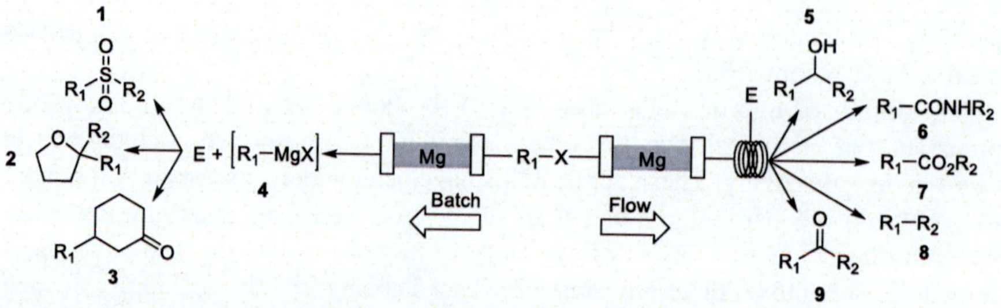

Fig. 7 Preparation of Grignard reagents in flow and reaction with electrophiles

than a second. This concept named as "flash chemistry" allows to carry out chemical transformations, which are unattainable by other methodologies [72].

3.3 *Catalysis in Flow*

The pharmaceutical industry is constantly requiring new chemical transformations to synthetize bioactive molecules in drug discovery programs. To achieve that end, new catalytic methodologies under flow conditions provide an opportunity to enable difficult chemistries or to improve well-established methodologies [73, 74].

3.3.1 Hetero- and Homogeneous Catalysis

Transition metal catalyzed transformations have clearly revolutionized synthetic chemistry. A convenient application of this synthetic field under continuous flow is the use of transition metal complexes as supported catalysts [29], by allowing the reuse of the catalyst in subsequent reactions. Among the variety of transition metals used for these purposes (gold, zirconium, iron), palladium-complexes have been most common due to their numerous synthetic applications, such as the Sukuki-Miyaura, Sonogashira, Heck or Negishi cross-couplings. These complexes have been immobilized on various solid supports, among which silica proved to be one of the most effective ones (Fig. 8) [75]. In addition, heterogeneous reactions can be carried out in flow with solid metal sources to prepare organometallic agents on-demand. For instance, Grignard reagents and organozinc species are synthetized as unstable intermediates by flowing the corresponding halo-derivative to a previously activated packed-metal [76].

The limitations associated with the applicability of supported catalysts, such as a possible leaching found in some palladium-catalyzed transformations [77], can be overcome under homogeneous catalysis [78], which present important advantages such as a better control of the catalyst loading. Although this approach is less friendly for the environment as recycling of the catalyst can be more challenging, some efforts have been done to overcome this issue. For instance, separation by

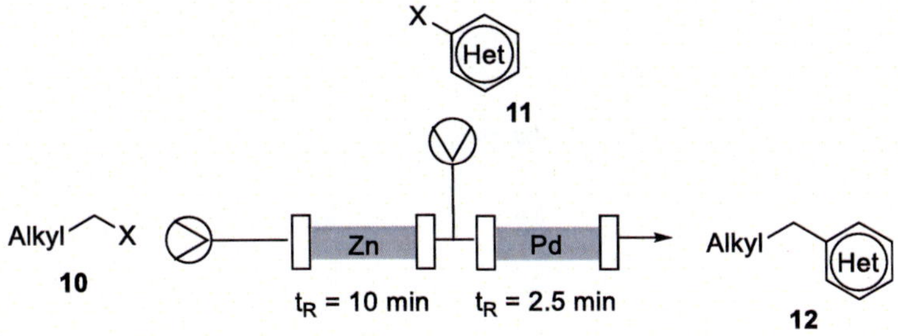

Fig. 8 Zinc insertion-Negishi coupling to introduce C(sp^3) motifs

liquid–liquid extraction, nanofiltration and recirculation were described as alternatives to reuse expensive and often toxic catalytic metals [79].

Organocatalysis and enantioselective catalysis have also been performed in flow using either homogeneous or heterogeneous catalyst to generate scaffolds in an enantioselective manner [80–82].

3.3.2 Biocatalysis

Biocatalytic methods under continuous flow have been developed to improve the interaction between the enzyme and the organic substrate [83]. Immobilizing the enzyme on solid supports and flowing the substrates through this bed allows an increase in the surface area thereby improving the efficiency and selectivity along with an increase of the overall turnover number (TON) of the protein. This is particularly useful because some enzymes are only available in small quantities and they can be recycled avoiding purification steps.

3.3.3 Photocatalysis

During the last years, photoredox catalysis has been positioned as an outstanding alternative for the synthesis of organic molecules [84, 85]. Despite the wide range of useful transformations reported in batch chemistry until now, some common limitations associated with the use of light are yet unsolved. For instance, scalability processes are limited due to the attenuation of light as a function of distance (Bouguer-Lambert-Beer law), which reduces the penetration of photons over the entire reaction mixture. On the other hand, over-irradiation can cause the formation of side products that complicates purification processes. These issues associated with the use of photochemistry in batch mode can be overcome with continuous flow photochemical microreactors that ensure a homogeneous and effective irradiation. In this manner, more selective transformations and an acceleration of the reaction times are achieved [86–90]. In this way, high value building blocks have been synthetized

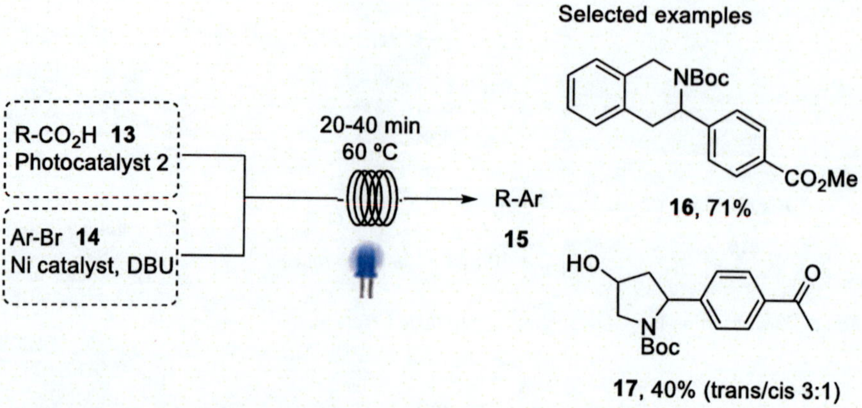

Fig. 9 Example of dual photoredox and nickel catalysis in flow

in drug discovery programs broadening chemical space, which cannot be obtained otherwise (Fig. 9) [91–93].

3.4 *Electrochemistry in Flow*

Electrochemistry has also been merged with organic synthesis as a disruptive technology for the development of sustainable methodologies [94–98]. The chance to carry out an organic reaction by using an electric current as a cheap reagent generates a greener alternative to strong oxidizing or reducing agents, by increasing safety and a smaller footprint. These transformations start with a mass transfer reaction of the substrate from the solution to the electrode surface. Then, the substrate is adsorbed onto the electrode and an electron transfer process occurs thereafter generating the product. Finally, the product is desorbed from the electrode and diffuses back to the liquid phase.

Electrochemical reactions are typically regarded as heterogeneous transformations in which both mass and charge transfer regimes must be considered. As a consequence, vigorous stirring is required in batch mode because only molecules in close contact with the electrodes are ready to react. The recent development of electrochemical machinery in continuous flow has overcome this problem as it offers a substantially enlarged surface area-to-volume ratio, thereby increasing the reproducibility and selectivity of the process [99]. In this regard, a great number of flow cells have been reported to date illustrating their versatility in organic synthesis [19, 20]. The green benefits of flow electrochemistry make it a valuable tool for the generation of new drug candidates in drug discovery programs [100, 101]. In particular, electrochemical transformations are extremely useful in late state functionalization as they avoid de novo synthesis of the derivatives, which may take weeks or months. The biological study of the new set of analogues obtained can

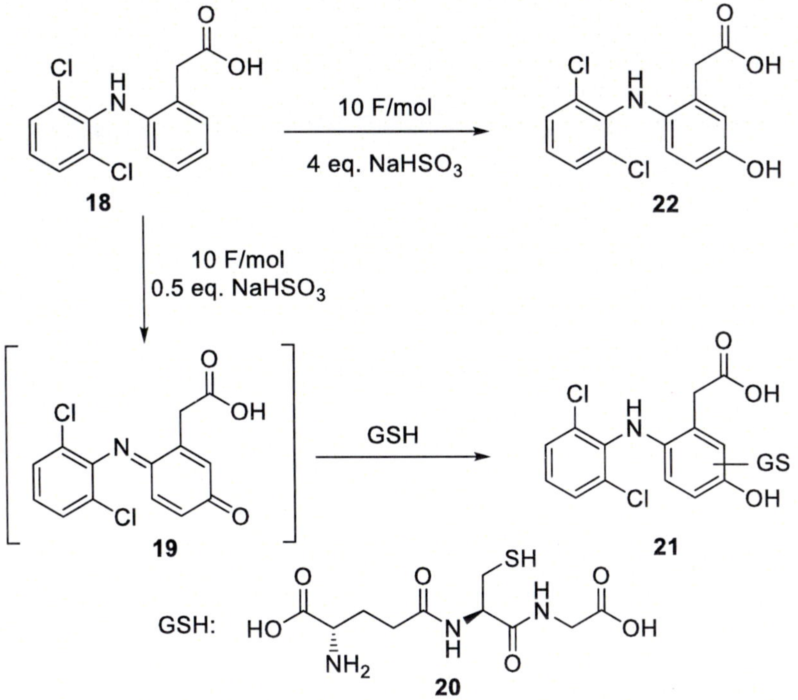

Fig. 10 Electrochemical preparation of metabolites in flow

provide valuable information in terms of metabolic stability or pharmacokinetics. For instance, oxidative processes can reveal the most susceptible positions to be oxidized in vivo, simulating CYP450 oxidation (Fig. 10). This late state electrochemical derivatization can drive decision making in drug discovery programs, accelerating the discovery of better drug candidates.

3.5 *Library Synthesis and Automation Using Flow*

Discovering a new drug is a slow and expensive process in which most attritions are due to different parameters related to the chemical structure of the molecule (e.g., toxicity, pharmacokinetics, cost of materials) [102]. One bottleneck is the synthesis of a great number of chemical entities, which can facilitate a better understanding of structure–activity relationships required for drug discovery program. In this regard, the corresponding cycle design-to-test usually takes various weeks, which is also an inconvenience for obtaining data rapidly. Furthermore, field competition makes the search of innovative technologies necessary to accelerate the drug discovery process.

Parallelization approaches using automated systems speed up the synthesis of drug candidates and can increase chemical diversity in a fast and effective manner, accelerating decision making in drug discovery programs while helping the scientist to maximize time, for instance in the exploration of new synthetic methodologies [28, 103–105]. They are generally based on an autosampler to load the reagents into the system and a fraction collector for reaction compilation. Thus, novel chemical series can be synthetized in a fast and efficient manner, by providing suitable information during early stages and expanding drug discovery capabilities. Typically, series of compounds in flow are made in a sequential way using plug flow approaches [106]. The optimization of synthetic routes can also be carried out in automated systems in combination with Design of Experiments (DoE) exploration [107]. A considerable reduction of cost, time, and human error is achieved because an array of reaction variables is optimized by using advanced algorithms.

Traditional methods for reaction optimization imply the generation of vast data, which require important amounts of starting materials. High throughput experimentation (HTE) techniques offer the opportunity to carry out the reaction optimization on smaller scale in order to identify the best reaction hotspots [108, 109]. The automated version of HTE has been envisioned in the pharmaceutical sector as an opportunity to catapult both reaction discovery and development due to the rapid generation of information [110]. However, the analysis of the automated reactions is relatively slow and it limits the benefits of the technology. The coupling with analytical techniques such as mass spectroscopy (MS) overcomes this issue [111, 112]. Through the continuous flow approach the screening datapoints can be scaled up and many issues can be resolved through the use of microfluidic cell-chips [113]. For instance, installation of proper equipment allows for carrying out 1,500 chemical combinations in a day with real-time analysis data, which speeds up drug discovery programs (Fig. 11) [114].

3.6 *Artificial Intelligence (AI) and Flow Chemistry*

Artificial Intelligence (AI) has also been implemented in continuous flow automated systems [115, 116]. These platforms are capable of selecting the best synthetic routes from scientific databases by utilizing a chemical programming language, which controls molecular assembly [117]. Thereafter, the system is able to test different reaction parameters and execute the synthesis of chemical libraries robotically. Ideally, the continuous flow approach should be compatible with different reaction conditions allowing for the translation of small-scale synthetic reactions to larger amounts if required. The integration of AI in continuous flow platforms provides a considerable advance in the preparation of drug candidates without human intervention (Fig. 12) [118].

The aforementioned area of automation and the variety of technologies that can be adapted to continuous flow made possible the development of fully integrated systems comprising design, synthesize, and test cycles [119]. Algorithmic predictions, purification steps, and biological screening can be combined in automated platforms to reduce the optimization time required for a drug candidate [120]. Typically, the experimentation is compartmentalized and several days are required from

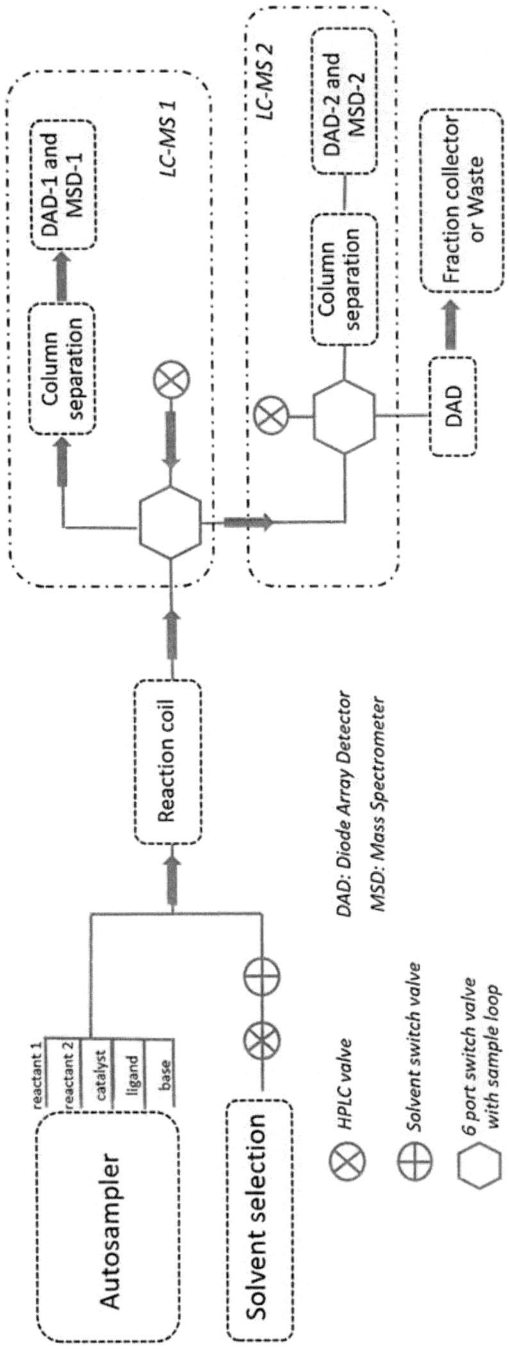

Fig. 11 Pfizer HTE platform

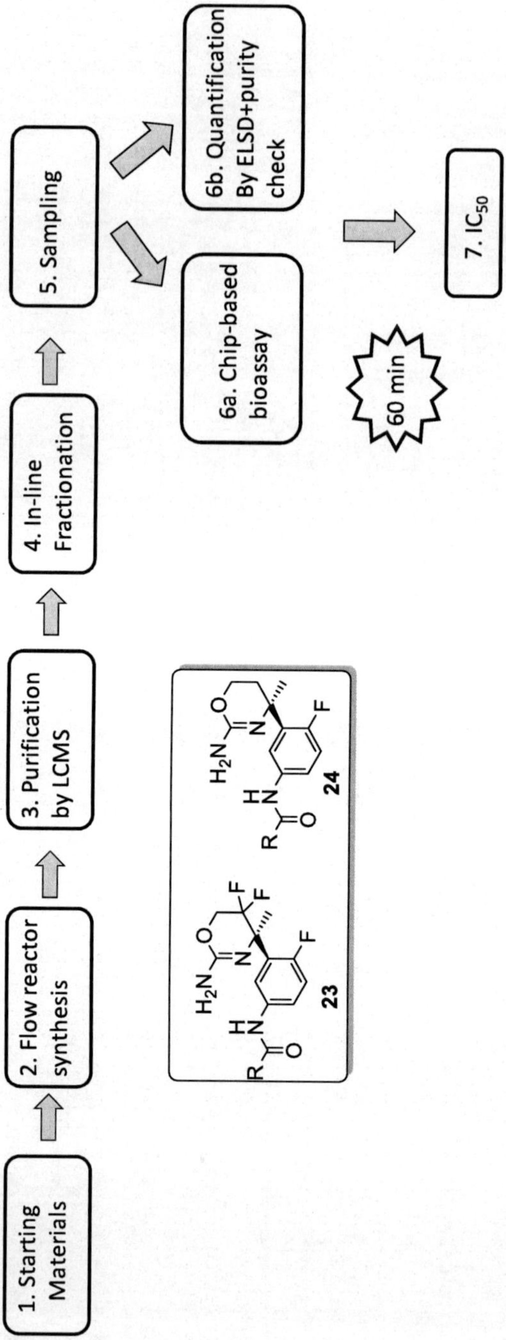

Fig. 12 Hoffman-La Roche platform for drug optimization

design to test. By using this methodology, a more rational approach is achieved, as the SAR data obtained from the drug candidates are used to design the next generation more efficiently [121]. These strategies will drastically reduce time, cost, and materials as well as accelerating library production and opening new avenues for understanding the drug discovery process.

4 Conclusions and Outlook

To sum up continuous flow chemistry offers unique opportunities for drug discovery. From expanding the medicinal chemist's toolbox to accelerating design-synthesize-test cycle through the integration of automation and AI.

Various aspects make flow chemistry extraordinarily attractive for the generation of bioactive target compounds such as drugs or natural products. First, flow chemistry enables forgotten and underused chemistries making them parallelizable for the rapid generation of organic molecule analogues expanding accessible chemical space. Then, an array of technologies can be installed to construct unique molecules, which could be problematic to synthetize otherwise. Altogether, the sustainable approach provided by flow chemistry has made this technology very convenient to obtain bioactive molecules during the drug discovery process and for reaction scale-up in industrial environments.

Nevertheless, there are still challenges where flow chemistry yet needs to demonstrate its value. Parallel batch and automated approaches are well established and can perform diverse chemistries regardless of the solubility of the mixture, a limitation in flow. Thus, chemists can run 1,536 reactions per plate in a well-plate format. Flow chemistry can achieve comparable numbers of reactions. Chemists should consider it as a complementary technology to other existing platforms. It is always an interesting approach for flow medicinal chemists to look for transformations that are not useful in batch to be translated in continuous flow and essentially make the best of it.

In the following chapters key authors will describe different applications in the drug discovery setting. This overview will provide the reader with a broad understanding of how flow chemistry can help medicinal chemists to improve the way in which clinical candidates can be discovered.

Compliance with Ethical Standards

Ethical Approval: This manuscript is a review of previously published accounts, as such, no animal or human studies were performed.

Informed Consent: No patients were studied in this chapter.

References

1. Gérardy R, Emmanuel N, Toupy T, Kassin V-E, Tshibalonza NN, Schmitz M, Monbaliu J-CM (2018) Eur J Org Chem:2301–2351
2. Akwi FM, Watts P (2018) Chem Commun 54:13894–13928
3. Trojanowicz M (2020) Molecules 25:1434–1486
4. Bana P, Orkenyi R, Lovei K, Lako A, Turos GI, Eles J, Faigl F, Greiner I (2017) Bioorg Med Chem 25:6180–6189
5. Cole KP, Johnson MD (2018) Expert Rev Clin Pharmacol 11:5–13
6. Plutschack MB, Pieber B, Gilmore K, Seeberger PH (2017) Chem Rev 117:11796–11893
7. Keserű GM, Soós T, Kappe CO (2014) Chem Soc Rev 43:5387–5399
8. Yoshida JI (ed) (2008) Flash chemistry: fast organic synthesis in microsystems. Wiley, Chichester
9. Murray PRD, Browne DL, Pastre JC, Butters C, Guthrie D, Ley SV (2013) Org Process Res Dev 17:1192–1208
10. Mallia CJ, Baxendale IR (2016) Org Process Res Dev 20:327–360
11. Hessel V, Löwe H, Schönfeld F (2005) Chem Eng Sci 60:2479–2501
12. Soleymani A, Yousefi H, Turunen I (2008) Chem Eng Sci 63:5291–5297
13. Ghanem A, Lemenand T, Valle DD, Peerhossaini H (2014) Chem Eng Res Des 92:205–228
14. Nagaki A, Togai M, Suga S, Aoki N, Mae K, Yoshida J-I (2005) J Am Chem Soc 127:11666–11675
15. Brzozowski M, O'Brien M, Ley SV, Polyzos A (2015) Acc Chem Res 48:349–362
16. Alimi OA, Bingwa N, Meijboom R (2019) Chem Eng Res Des 150:116–129
17. Symes MD, Kitson PJ, Yan J, Richmond CJ, Cooper GJT, Bowman RW, Vilbrandt T, Cronin L (2012) Nat Chem 4:349–354
18. Lumley EK, Dyer DE, Pamme N, Boyle RW (2012) Org Lett 14:5724–5727
19. Atobe M, Tateno H, Matsumura Y (2018) Chem Rev 118:4541–4572
20. Pletcher D, Green RA, Brown RCD (2018) Chem Rev 118:4573–4591
21. Noël T, Naber JR, Hartman RL, McMullen JP, Jensen KF, Buchwald SL (2011) Chem Sci 2:287–290
22. Browne DL, Deadman BJ, Ashe R, Baxendale IR, Ley SV (2011) Org Process Res Dev 15:693–697
23. Yue J, Schouten JC, Nijhuis TA (2012) Ind Eng Chem Res 51:14583–14609
24. Fabry DC, Sugiono E, Rueping M (2016) React Chem Eng 1:129–133
25. Reizman BJ, Jensen KF (2016) Acc Chem Res 49:1786–1796
26. Wang K, Luo G (2016) Chem Eng Sci 169:18–33
27. Ley SV (2012) Chem Rec 12:378–390
28. Pan JY (2019) ACS Med Chem Lett 10:703–707
29. Colella M, Carlucci C, Luisi R (2018) Top Curr Chem 376:46–93
30. Irfan M, Glasnov TN, Kappe CO (2011) ChemSusChem 4:300–316
31. Woitalka A, Kuhn S, Jensen KF (2014) Chem Eng Sci 116:1–8
32. Robertson JC, Coote ML, Bissember AC (2019) Nat Rev Chem 3:290–295
33. Verschueren RH, De Borggraeve WM (2019) Molecules 24:2122–2160
34. Edwards L, Fox A, Stonier P (2010) Principles and practice of pharmaceutical medicine, 3rd edn. Wiley-Blackwell, Oxford
35. Leeson PD, Empfield JR (2010) Annu Rep Med Chem 45:393–407
36. DiMasi JA, Faden LB (2011) Nat Rev Drug Discov 10:23–27
37. Hughes JP, Rees S, Kalindjian SB, Philpott KL (2011) Br J Pharmacol 162:1239–1249
38. Zanders JH, Bailey DS, Dean PM (2002) Drug Discov Today 7:711–718
39. Lavecchia A, di Giovani C (2013) Curr Med Chem 20:2839–2860
40. Erlanson DA, McDowell RS, O'Brien T (2004) J Med Chem 47:3463–3482
41. Keseru GM, Makara GM (2006) Drug Discov Today 11:741–748
42. Lipinski CA, Lombardo F, Dominy BW, Feeney PJ (2001) Adv Drug Deliv Rev 46:3–26
43. López E, Linares ML, Alcázar J (2020) Future Med Chem 12:1457–1563.

44. Bogdan AR, Dombrowski AW (2019) J Med Chem 62:6422–6468
45. Bogdan AR, Organ MG (2018) Sharman UK, Van der Eycken EV (eds) Flow chemistry for the synthesis of heterocycles. Springer, Cham, pp 319–341
46. Alcázar J, de la Hoz A, Díaz-Ortiz A (2019) Ballini R (ed) Green synthetic processes and procedures. RSC Publishing, Cambridge, pp 53–78
47. Hughes DL (2020) Org Process Res Dev 24:1850–1860
48. Baumann M, Moody TS, Smyth M, Wharry S (2020) Org Process Res Dev 24:1802–1813
49. Porta R, Benaglia M, Puglisi A (2016) Org Process Res Dev 20:2–25
50. Lovering F, Bikker J, Humblet C (2009) J Med Chem 52:6752
51. Alonso N, Miller LZ, Muñoz JM, Alcázar J, Mcquade T (2014) Adv Synth Catal 356:3737–3741
52. Huck L, Berton M, de la Hoz A, Díaz-Ortiz A, Alcázar J (2017) Green Chem 19:1420–1424
53. Boström J, Brown DG, Young RJ, Keserü GM (2018) Nat Rev Drug Discov 17:709–727
54. Roughley SD, Jordan AM (2011) J Med Chem 54:3451–3479
55. Brown GD, Boström J (2016) J Med Chem 59:4443–4458
56. Rogers L, Jensen KF (2019) Green Chem 21:3481–3498
57. Alcázar J (2017) Vaccaro L (ed) Sustainable flow chemistry in drug discovery. Wiley-VCH, Weinheim, pp 135–164
58. Watts P (2017) Vaccaro L (ed) Sustainable flow chemistry: methods and applications. Wiley-VCH, Weinheim, pp 193–217
59. Cravotto G, Cintas P (2006) Chem Soc Rev 35:180–196
60. Dong Z, Delacour C, Mc Carogher K, Udepurkar AP, Kuhn S (2020) Materials 13:344–369
61. Alcázar J, Oehlrich D (2010) Future Med Chem 2:169–176
62. Alcázar J, Muñoz JM (2013) In: de la Hoz A, Loupy A (eds) Microwaves in organic synthesis, 3rd edn. Wiley-VCH, Weinheim, pp 1173–1204
63. Barham JP, Koyama E, Norikane Y, Ohneda N, Yoshimura T (2018) Chem Rec 19:188–203
64. Sadler S, Moeller AR, Jones GB (2012) Expert Opin Drug Discovery 7:1107–1128
65. Gérardy R, Debecker DP, Estager J, Luis P, Monbaliu J-C (2020) Chem Rev 120:7219–7347
66. Hommes A, Heeres HJ, Yue J (2019) ChemCatChem 11:4671–4708
67. Köhler JM, Cahill BP (eds) (2014) Micro-segmented flow – applications in chemistry and biology. Springer, Berlín
68. Schreiber SL (2000) Science 287:1964–1969
69. Galloway WRJD, Isidro-Llobet A, Spring DR (2010) Nat Commun 1:80–93
70. Schawrz MK, Sauer WHB (2003) J Chem Inf Comput Sci 43:987–1003
71. Pavlinov I, Gerlach EM, Aldrich LN (2019) Org Biomol Chem 17:1608–1623
72. Yoshida JI, Takahashi Y, Nagaki A (2013) Chem Commun 49:9896–9904
73. Santoro S, Ferlin F, Ackermann L, Vaccaro L (2019) Chem Soc Rev 48:2767–2782
74. Palao E, Alcázar J (2019) Luis SV, Garcia-Verdugo E (eds) Flow chemistry: integrated approaches for practical applications. RSC Publishing, Cambridge, pp 86–128
75. Ciriminna R, Pandarus V, Fidalgo A, Ilharco LM, Béland F, Pagliaro M (2015) Org Process Res Dev 19:755–768
76. Berton M, Huck L, Alcázar J (2018) Nat Protoc 13:324–334
77. Thathagar MB, ten Elshof JE, Rothenberg G (2006) Angew Chem Int Ed 45:2886–2890
78. Noël T, Buchwald SL (2011) Chem Soc Rev 40:5010–5029
79. Gürsel IV, Noël T, Wanga Q, Hessel V (2015) Green Chem 17:2012–2026
80. Yu T, Ding Z, Nie W, Jiao J, Zhang H, Zhang Q, Xue C, Duan X, Yamada YMA, Li P (2020) Chem A Eur J 26:5729–5747
81. Atodiresei I, Vila C, Rueping M (2015) ACS Catal 5:1972–1985
82. de Risi C, Bortolini O, Brandolese A, Di Carmine G, Ragno D, Massi A (2020) React Chem Eng 5:1017–1052
83. Britton J, Majumdar S, Weiss GA (2018) Chem Soc Rev 47:5891–5918
84. Romero NA, Nicewicz DA (2016) Chem Rev 116:10075–10166
85. Zhu C, Yue H, Chu L, Rueping M (2020) Chem Sci 11:4051–4064
86. Cambié D, Bottecchia C, Straathof NJW, Hessel V, Nöel T (2016) Chem Rev 116:10276–10341

87. Yang C, Li R, Zhang KAI, Lin W, Landfester K, Wang X (2020) Nat Commun:1239–1247
88. Sambiagio C, Nöel T (2020) Trends Chem 2:92–106
89. Beeler AB, Corning SR (2016) Albini A, Fasani E (eds) Photochemistry. RSC Publishing, Cambridge, pp 173–190
90. Politano F, Oksdath-Mansilla G (2018) Org Process Res Dev 22:1045–1062
91. Abdiaj I, Alcázar J (2017) Biorg Med Chem 25:6190–6196
92. Hsieh H-W, Coley CW, Baumgartner LM, Jensen KF, Robinson RI (2018) Org Process Res Dev 22:542–550
93. Di Filippo M, Bracken C, Baumann M (2020) Molecules 25:356
94. Francke R, Little RD (2014) Chem Soc Rev 43:2492–2521
95. Wiebe A, Gieshoff T, Möhle S, Rodrigo E, Zirbes M, Waldvogel SR (2018) Angew Chem Int Ed 57:5594–5619
96. Möhle S, Zirbes M, Rodrigo E, Gieshoff T, Wiebe A, Waldvogel SR (2018) Angew Chem Int Ed 57:6018–6041
97. Yan M, Kawamata Y, Baran P (2018) Angew Chem Int Ed 57:4149–4155
98. Kingston C, Palkowitz MD, Takahira Y, Vantourout JC, Peters BK, Kawamata Y, Baran PS (2020) Acc Chem Res 53:72–83
99. Noël T, Cao Y, Laudadio G (2019) Acc Chem Res 52:2858–2869
100. Stalder R, Roth GP (2013) ACS Med Chem Lett 4:1119–1123
101. Gao L, Teng Y (2016) Future Med Chem 8:567–577
102. Desai B, Dixon K, Farrant E, Feng Q, Gibson KR, Van Hoorn WP, Mills J, Morgan T, Parry DM, Ramjee MK, Selway CN, Tarver GJ, Whitlock G, Wright AG (2013) J Med Chem 56:3033–3047
103. Urge L, Alcázar J, Huck L, Dorman D (2017) Annu Rep Med Chem 50:87–147
104. Godfrey AG, Masquelin T, Hemmerle H (2013) Drug Discov Today 18:795–802
105. Sanderson K (2019) Nature 568:577–579
106. Thompson CM, Poole JL, Cross JL, Akritopoulou-Zanze I, Djuric SW (2011) Molecules 16:9161–9177
107. Moore JS, Jensen KF (2013) Wirth T (ed) Microreactors in organic chemistry and catalysis. Wiley-VCH, Weinheim, pp 81–100
108. Jhoti H, Rees S, Solari R (2013) Expert Opin Drug Discovery 8:1449–1453
109. Hüser J (ed) (2006) High throughput-screening in drug discovery. Wiley-VCH, Weinheim
110. Selekman JA, Qiu J, Tran K, Stevens J, Rosso V, Simmons E, Xiao Y, Janey J (2017) Annu Rev Chem Biomol Eng 8:525–547
111. Wleklinski M, Falcone CE, Loren BP, Jaman Z, Iyer K, Ewan HS, Hyun S-H, Thompson DH, Cooks RG (2016) Eur J Org Chem:5480–5484
112. Troshin K, Hartwig JF (2017) Science 357:175–181
113. Chi C-W, Ahmed AR, Dereli-Korkut Z, Wang S (2016) Bioanalysis 8:921–937
114. Perera P, Tucker JW, Brahmbhatt S, Helal CJ, Chong A, Farrel W, Richardson P, Sach NW (2018) Science 359:429–434
115. Sellwood MA, Ahmed M, Segler MHS, Brown N (2018) Future Med Chem 10:2025–2028
116. Coley CW, Thomas III DA, Lummiss JAM, Jaworski JN, Breen CP, Schultz V, Hart T, Fishman JS, Rogers L, Gao H, Hicklin RW, Plehiers PP, Byington J, Piotti JS, Green WH, Hart AJ, Jamison TF, Jensen KF (2019) Science 365:557–566
117. Steiner S, Wolf J, Glatzel S, Andreou A, Granda JM, Keenan G, Hinkley T, Aragon-Camarasa G, Kitson PJ, Angelone D, Cronin L (2019) Science 363:144–152
118. Werner M, Kuratli C, Martin RE, Hochstrasser R, Wechsler D, Enderle T, Alanine AI, Vogel H (2014) Angew Chem Int Ed 53:1704–1708
119. Gioiello A, Piccinno A, Lozza AM, Cerra B (2020) J Med Chem 63:6624–6647
120. Rodrigues T, Schneider P, Schneider G (2014) Angew Chem Int Ed 53:5750–5758
121. Holenz J, Brown DG (2016) Holenz J, Mannhold R, Kubinyi H, Folkers G (eds) Lead generation: methods and strategies. Wiley-VCH, Weinheim, pp 13–34

Top Med Chem (2021) 38: 23–70
https://doi.org/10.1007/7355_2021_111

Published online: 13 February 2021

Green Aspects of Flow Chemistry for Drug Discovery

Ángel Díaz-Ortiz and Antonio de la Hoz

Contents

Abstract Flow chemistry is considered to be an enabling technology that covers many of the key principles of Green Chemistry. In this chapter, we review the main

Á. Díaz-Ortiz and A. de la Hoz (✉)
Facultad de Ciencias y Tecnologías Químicas, Universidad de Castilla-La Mancha, Ciudad Real, Spain
e-mail: Antonio.Hoz@uclm.es

issues related to Green Chemistry, the use of green solvents, synergy with other enabling and green technologies that have been revitalized under flow conditions, in situ production and handling of hazardous reagents, and the possibility of telescoping, screening, optimization, automatization, and scale-up. We will finish with the topic of quantification of sustainability in comparison with batch conditions and the most important contributions for the design of greener processes.

Keywords Green, Hazard, LCA, Methodology, Solvents

1 Introduction

Flow chemistry has been highlighted by IUPAC as one of the emerging technologies in chemistry with the potential to make our planet more sustainable [1]. Flow chemistry was selected as one of the "Ten Chemical Innovations that will Change our World," together with nanopesticides, enantioselective organocatalysis, solid-state batteries, reactive extrusion, MOFs (metal organic frameworks) and porous materials for water harvesting, directed evolution of selective enzymes, recovery of monomers from plastics, reversible-deactivation of radical polymerization, and 3D-bioprinting [1].

Pharmaceutical industries are constantly searching for ways to increase the efficiency of drug discovery, from research to the development of new drugs and also in terms of new regulations and costs associated with environmental protection, even in developing countries [2]. Shortening the time from research to production by reducing the requirements for reoptimization of the reaction conditions in each step is also a prerequisite. Finally, the application of Green Chemistry principles is a central pillar in order to improve security, reduce waste and, in particular, to improve efficiency [3].

The introduction of flow chemistry in the pharmaceutical industry has made a remarkable contribution to the fulfillment of these goals, both in efficiency and Green Chemistry [4].

In this respect, the FDA declared continuous manufacturing (CM) as one of the most important tools in the modernization of the pharmaceutical industry [5].

The increased surface area-to-volume ratio in relation to batch chemistry permits rapid and controlled heating and heat transfer. This allows precise temperature control with an improvement in reaction rates and yields and minimizes thermal runaway in exothermic reactions. Flow systems also enable better mixing, which is especially important in biphasic reactions, thus improving reactions that involve gases and liquids.

Another benefit is the simplicity when it is necessary to perform reactions above the boiling point of the solvent, which can be achieved by regulating the pressure with a BBR (back pressure regulator).

1. Prevention of production of waste
2. Atom Economy
3. Less Hazardous Chemical Syntheses
4. Designing Safer Chemicals
5. Safer Solvents and Auxiliaries
6. Design for Energy Efficiency
7. Use of Renewable Feedstocks
8. Reduces Derivatives
9. Catalysis
10. Design for Degradation
11. Real-time Analysis for Pollution Prevention
12. Inherently Safer Chemistry for Accident Prevention

Fig. 1 Green Chemistry principles [6]

The small scale used and the ability to generate unstable reagents that are formed and consumed immediately are also principal advantages over batch mode operation and this permits the safe use of reagents that are otherwise avoided.

Photochemical and electrochemical reactions have been revitalized. Reactions can be performed at higher concentrations than in batch and, in the case of photochemical reactions, limitations of light penetration can be reduced, thus allowing the use of more simple irradiation sources such as LEDs.

Finally, the possibility of telescoping multi-step sequences avoids purification steps, reduces waste, and improves yields. Flow methodologies have a great impact on most of the Green Chemistry principles and, in consequence, on the efficiency and economy of a given process (Fig. 1). The possible use of telescoped synthesis and higher concentrations have a marked impact on the reduction of waste (principle 1). Reactions show higher efficiency both synthetically (principle 2) and in energy transfer (principle 6). Hazardous compounds and intermediates can be used safely due to the reduction of scale and the possibility of in situ generation and use (principles 3 and 12). Flow chemistry opens the window to reactions that otherwise cannot be performed (New Process Windows) (principles 4, 7, and 10). Procedures for using green solvents (supercritical, recyclable solvents, etc.) under flow conditions have been successfully implemented (principle 5). Multiphase reactions can be efficiently performed, thus reducing the need for derivatives or activating groups (principle 8). Homogeneous and heterogeneous catalysis and biocatalysis can take advantage of this methodology (principle 9), thus reducing the amount of catalyst required by immobilization. Finally, flow methodologies are especially suitable for the implementation of real-time analysis with spectroscopic methods (principle 11) and, consequently, for the implementation of automatization and artificial intelligence for optimization and scale-up. Scale-up under flow conditions is, in principle, easier than in batch and three possibilities can be envisaged: (1) running the process for a longer time (scaling out), (2) running multireactors in parallel (numbering up), and (3) using larger reactors (scaling-up) [7].

Optimization of the reaction parameters can lead to green-by-design products [5]. Flow chemistry is also an ideal system for computer-controlled reactions since several parameters, e.g., reaction time, temperature, flow rate, pressure, amongst others, can be easily and accurately controlled [8].

Finally, Life Cycle Assessment (LCA) studies – as well as other parameters – used to quantify the sustainability, E-factor, Atom Economy, Process Mass Intensification (PMI), solvent rate consumption or energy resource consumption have confirmed the benefits of switching to flow methodologies [9].

In 2005, the Green Chemistry Institute Pharmaceutical Roundtable (GCIPR) was formed with the mission of catalyzing the implementation of Green Chemistry and Green Engineering in the global pharmaceutical industry [10]. The main objectives were to monitor and identify new research opportunities, influence technical agendas, and encourage external funding. Amongst the projects granted, the subjects that received the most funding were amide reduction, catalysis, greener solvents, and flow chemistry, including biocatalysis in flow.

In this chapter we will cover some issues related to flow methodologies and Green Chemistry such as solvents, coupling with enabling technologies, the use of hazardous reagents, analysis, optimization, automatization, and quantitative analysis of sustainability.

2 Solvents

Solvents play a central role in chemistry since most reactions in the laboratory and on an industrial scale are performed in solution. Solvents allow the mixing of reagents at the molecular level and at the appropriate concentrations to achieve a suitable reaction rate. Solvents also transport reagents or products and facilitate the dosage when introducing a reagent into the reactor. Solvents also provide control over the reaction temperature; in endothermic processes, the reaction temperature can be increased by heating the solvent with an external heat source, while in exothermic processes they absorb the heat released by boiling the solvent (reflux) with steam cooling and direct solvent cooling.

Solvents are not only used in reactions but also in the analysis and purification (crystallization and chromatography) of compounds and, since they are used in a large excess, they are responsible for most of the waste associated with a chemical process [11].

Considering environmental issues, solvents present problems of toxicity, flammability, or explosion hazards and they may cause issues of persistent contamination and occupational diseases due to long-term exposure. In addition, the volume of solvent is usually much larger than that of reagents and products, so the costs and risks of production are largely determined by those of the solvents.

LCA studies have shown the important impact of solvents on the green aspects of pharmaceutical processes [12]. Three approaches have been used to reduce the impact of solvents: reduction of solvent volumes, solvent replacement, and the use of neoteric and green solvents. All of these possibilities have been used and promoted with flow methodologies.

Pharmaceutical companies have designed solvent selection guides in order to reduce the environmental impact by replacing toxic or harmful solvents [13]. Indeed the ACS GCI Pharmaceutical Roundtable has developed a solvent selection tool in order to facilitate solvent replacement [14]. In this regard, Yaseneva et al. found by LCA studies that solvent recycling as well as replacement of THF by Me-THF reduced the environmental impact of the flow process [12].

Ley et al. observed a dramatic reduction in the use of solvents on employing telescoped multi-step flow sequences as compared with batch operations, since solvent switching, extractions, evaporations, chromatography, etc. can be minimized or avoided [15].

Wong and Cernak consider that reaction miniaturization, both in well plates and in flow, can accelerate medicinal chemistry development and minimize the environmental impact as solvents preferred in miniaturization have the characteristics of green solvents [16]. Three main factors should be considered when selecting a solvent: (1) high boiling point to avoid evaporation of the solvent, (2) good solubility of substrates to avoid slurries, and (3) clogging and plastic compatibility since many reactors contain plastic components [16].

2.1 Supercritical Fluids and Ionic Liquids

The possibility of accessing high temperatures and pressures and the small scale used in flow reactions makes the use of supercritical fluids very attractive since safety is improved and the scale can be smaller than in batch reactions. Moreover, the reaction can be analyzed in-line and optimization of parameters such as temperature and pressure is facilitated as they can be modified in situ [17]. The properties of the supercritical solvent, i.e., density, viscosity, diffusivity, and surface tension, are intermediate between those of the gas and the liquid and they can be modulated continuously from gas-like to liquid-like properties. $scCO_2$ has been the solvent of choice in most processes because it is a good solvent for organic compounds, it can be generated under mild conditions (Tc = 31.1°C, Pc = 73.8 bar) and products can be recovered by reducing the pressure to atmospheric pressure. From the Green Chemistry perspective, the recovery of a waste gas that produces the greenhouse effect and the simplicity of recycling make $scCO_2$ an interesting solvent of choice. scH_2O can be obtained at high temperatures and pressures (Tc = 374.2°C, Pc = 220.5 bar) and consequently it is mostly used for oxidation reactions.

Poliakoff et al. described a photo-oxidation reaction in one step for the synthesis of the antimalarial trioxanes **5**. They started with an allylic alcohol and used 5,10,15,20-tetrakis(pentafluorophenyl)porphyrin (TPFPP) as a photosensitizer and

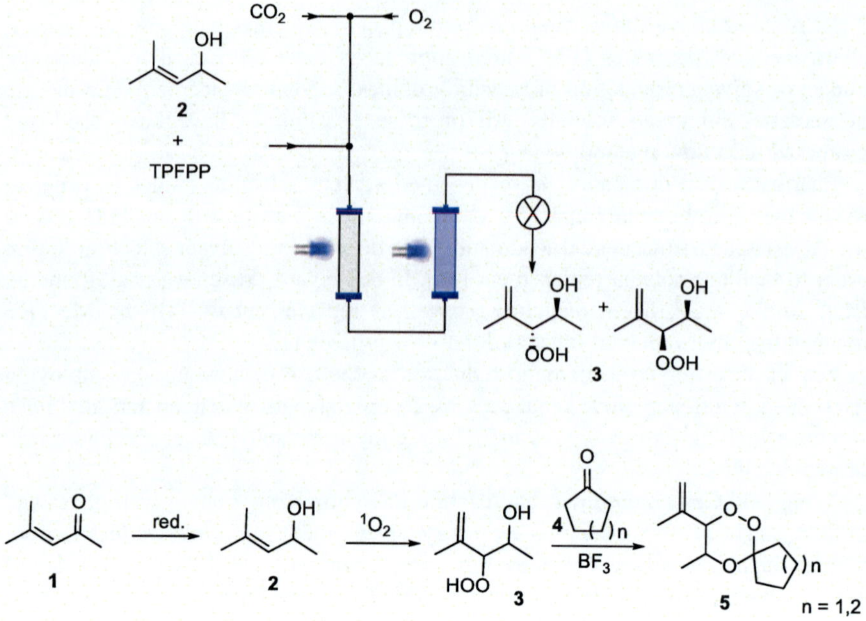

Fig. 2 Continuous flow photo-oxidation to produce antimalarial trioxanes **5** (in scCO_2)

scCO_2 as the solvent. **1** and **2** were obtained with a conversion >50%. Cyclic ketone **4** was used as a reagent and solvent in Lewis acid-catalyzed cyclization to **5** (Fig. 2) [18].

The application of supercritical fluids in flow conditions in hydrogenation, hydroformylation, and biocatalysis [19] has recently been reviewed [20]. In most cases immobilization of the catalyst gave improvements in productivity, higher stability of the catalytic system, and higher selectivity, even in stereoselective reactions [21]. Moreover, catalyst isolation and recovery were enhanced.

Ionic liquids (ILs) are a unique class of solvents. These materials consist of salts that are liquids at temperatures below 100°C. They are very polar solvents with high conductivity, negligible vapor pressure, high stability and they are non-flammable and show good solubility for many organic compounds. ILs are based on ammonium, phosphonium, and heterocyclic salts (imidazolium and pyridinium salts). The properties of the IL can be modulated by careful choice of the salt and the counterion and they can even be used as smart solvents that can act as a catalyst [22].

Enzymes, lipases, proteases, peroxidases, dehydrogenases, and glycosidases have been used in ILs because of their ability to dissolve both polar and non-polar compounds.

ILs and scCO_2 are immiscible and form biphasic systems under all conditions. Considering the solubility of hydrophobic compounds in scCO_2, this solvent can be used to extract such compounds from IL solutions. A new concept of biphasic

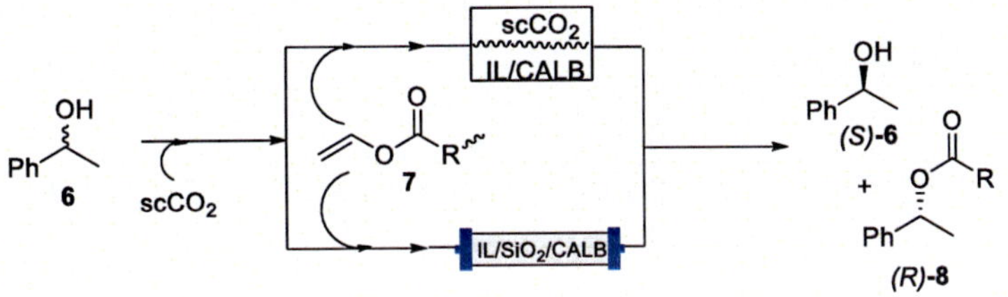

Fig. 3 CALB-catalyzed esterification in an IL/$scCO_2$ biphasic system

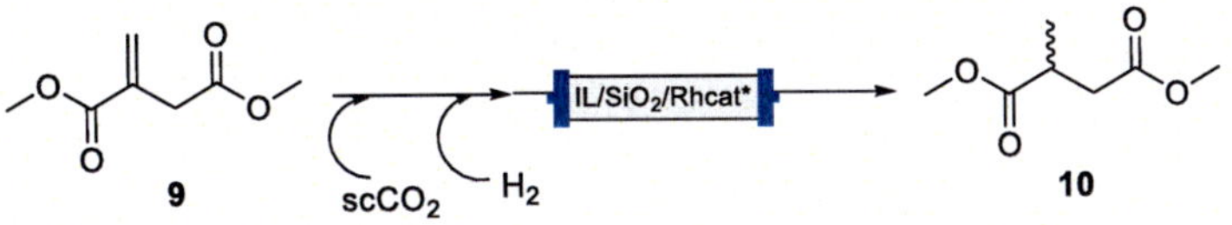

Fig. 4 Rh-catalyzed hydrogenation of methyl itaconate in a SILP system

biocatalysis has been described, where a biocatalyst is immobilized and coated in an IL (denoted the catalytic phase) and substrates and products reside in the $scCO_2$ phase (extractive phase). Products can be easily separated and recovered by evaporation of $scCO_2$ and this can be recycled by re-compression [23].

This approach was developed by Lozano and Leitner independently for the kinetic resolution of racemic phenylethanol (**6**) catalyzed by Candida Antarctica lipase B (CALB) suspended in the ionic liquid or immobilized in SiO_2 (Fig. 3) [24, 25].

The concept of supported ionic liquid phase (SILP) catalysis has subsequently been introduced as an attractive methodology in flow chemistry. In SILP, a molecular catalyst is dissolved in an immobilized ionic liquid (IL) anchored on a porous material (silica or alumina). $scCO_2$ has been used for organic synthesis in flow considering the low solubility of the supported IL in $scCO_2$. This methodology exploits the advantages of a homogeneous biphasic IL-$scCO_2$ system and that of an immobilized catalyst. $scCO_2$ is the mobile phase that brings the reactants to the IL and extracts the final product to the collector.

Poliakoff described the hydrogenation of dimethyl itaconate (**9**) using a chiral Rh catalyst supported on Al_2O_3 with a phosphotungstic linker (Fig. 4) [26]. Conversion, selectivity, and TON depended on the temperature, pressure, and flow rate, with the optimum conditions being 60°C, 10 MPa and 1.25 mL min^{-1} to give ee up to 83% and variable conversions.

A step further was the development of Supported Ionic Liquid-Like Phases (SILLPs), which are usually formed by immobilization of an IL moiety onto the surface of a polymer (polystyrene-divinylbenzene, PS-DVB). This approach permits a reduction in the amount of the IL used and facilitates its recovery and reuse. As an example, Luis et al. [27] described the preparation of citronellyl propionate (**15**) with an immobilized CALB-SILLP catalyst under continuous $scCo_2$ flow (Fig. 5).

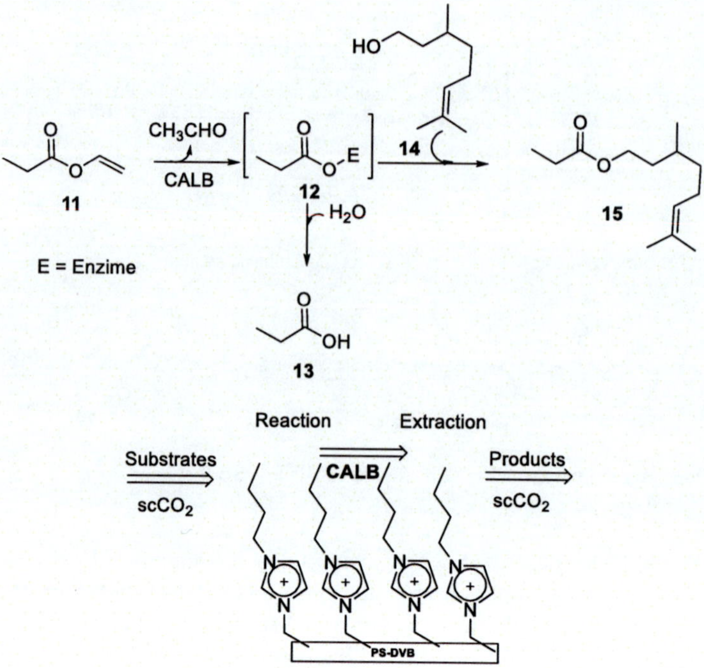

Fig. 5 CALB-catalyzed esterification of citronellol in a SILLPs system

Finally, the concept of Sponge-Like Ionic Liquids (SLILs) was introduced. This refers to hydrophobic ILs with long alkyl chains [28]. These materials have a very good ability to dissolve hydrophobic compounds and this results in a single phase liquid if they are heated above their melting point. The mixture becomes a solid phase upon cooling to room temperature and the solute is trapped in the IL net. On centrifugation at lower temperature, the ionic liquid (lower phase) behaves as a sponge and releases the liquid solute to the upper phase (sponge wrung out). Immobilized lipases have been used to catalyze the synthesis of flavor esters by esterification reactions in SLILs, thus simplifying the procedures described by this centrifugation procedure.

2.2 *Deep Eutectic Solvents*

Deep eutectic solvents (DES) have emerged as a green alternative to ionic liquids and they overcome many of the problems associated with the latter, i.e., toxicity and degradability. DES have been used for extraction, biocatalysis, and electrochemistry. DES are composed of two or three environmentally friendly components that can associate through hydrogen bonds to form a stable mixture with a melting point lower than that of the individual components. In a similar way to ILs, the properties

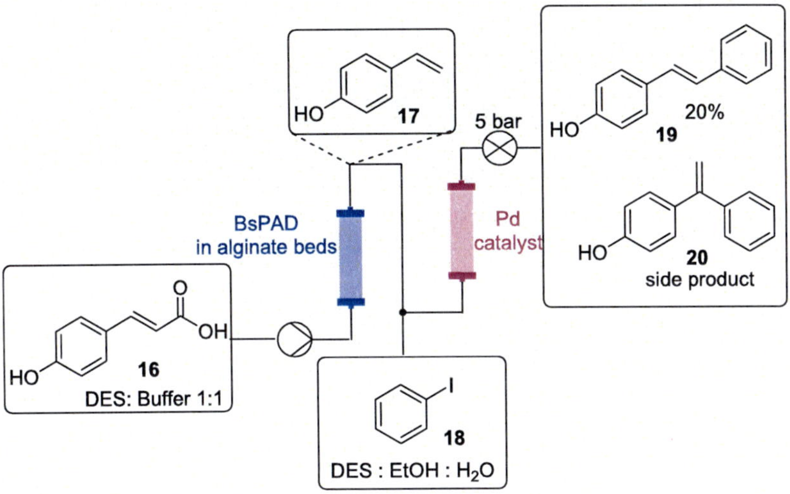

Fig. 6 Preparation of 4-hydroxystilbene from *p*-coumaric acid in DES

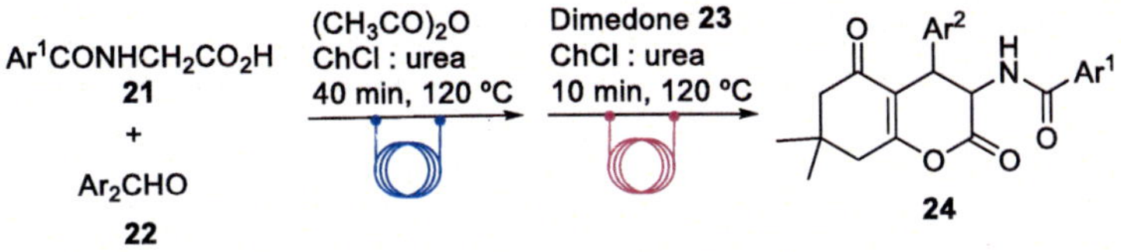

Fig. 7 Tandem synthesis of 3-aminohexahydrocoumarins **24**

of DES can be modulated by a careful choice of the components and of the anion. When these two components are primary metabolites, they are named natural deep eutectic solvents (NADES) [29]. The most popular choice is the use of choline chloride as the hydrogen bond acceptor and acids or alcohols as hydrogen bond donors.

The high viscosity of DES is a barrier for their use in flow chemistry; however, it is possible to reduce the viscosity by using water as a cosolvent. In this regard, Grabner et al. used low viscosity DES for chemoenzymatic reactions. They used immobilized enzymes for the tandem decarboxylation of *p*-coumaric acid (**16**) followed by a Pd-catalyzed Heck reaction to give 4-hydroxystilbene (**19**) in 20% overall yield and together with **20** as a side product in a ratio **19**: **20**, 3:1 (Fig. 6) [30].

Similarly, Guajardo et al. used this approach for lipase-catalyzed esterifications [31].

Finally, Zamani and Khosropour used choline chloride (ChCl) and urea as a DES for the tandem synthesis of 3-aminohexahydrocoumarins **24** from N-Arylamidoglycines **21** and aromatic aldehydes **22** followed by reaction with dimedone (**23**) in higher yields than the batch procedure (Fig. 7) [29].

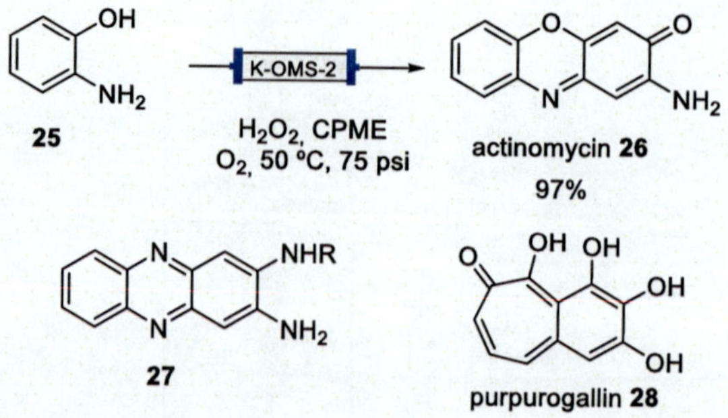

Fig. 8 Preparation of actinomycin (**26**) by oxidation of *o*-aminophenol (**25**) with K-OMS-2 in CPME and flow conditions

2.3 *Biomass-Derived Solvents*

Biomass opens up several possibilities in Green Chemistry from the design of new starting materials, for example glucose, to the extraction of valuable compounds like terpenes or proteins. The use of biomass-derived solvents is always a useful green alternative. Vaccaro et al. described the use of several biomass-derived solvents [11]. γ-Valerolactone (GVL), prepared by hydrogenation and cyclization of levulinic acid (from lignocellulosic material), can be used to replace dipolar aprotic solvents; cyclopentyl methyl ether (CPME), prepared from cyclopentene (from petrol), and methanol are used as replacements for THF; methyleneglutaronitrile (MDN), a waste product from the manufacture of Nylon-66; methyl lactate, obtained by esterification of lactic acid (from carbohydrates) [11], or *p*-cymene for esterification reactions [32].

As an example, Vaccaro et al. described the use of CPME in flow for the manganese-catalyzed synthesis of aminophenoxazinones **26**, diaminophenazines **27**, and purpurogallin (**28**). The procedure was optimized for the synthesis of a known API, actinomycin (**26**) (Fig. 8), and showed a good F-factor and increased TON and TOF values without metal contamination of the product [33].

There are several possibilities for the design of biomass-derived solvents for specific uses. Most of these require a comprehensive investigation into the characteristics and possibilities in synthetic chemistry and their impact in minimization of waste.

2.4 *Miscellaneous*

Other interesting types of green solvent are known as "switchable hydrophilicity solvents" (SHSs). An SHS is considered as a hydrophobic solvent that, in the presence of water and carbon dioxide (CO_2), can "switch" to a hydrophilic solvent (Fig. 9). These solvents have been used for extractions in flow and show improved extraction due to the enhanced interfacial area and lower solvent consumption, thus providing an interesting green approach [34].

$$NR_2 + H_2O + CO_2 \rightleftharpoons NR_2H^+ + HCO_3^-$$

Fig. 9 Switchable solvents

Fluorous solvents have been used as spacers in the segmented flow synthesis of a library of 20 pyrazole derivatives in 45 min, with a total consumption of 8 mL of DMF and complete recycling of the fluorous spacer. The authors showed that solvents like DMF increase the stability of the droplet and a minimal flow rate of 1.3 mL/min is required to obtain stable droplets [35].

3 Enabling Technologies

Drug Discovery can often be described as a "race." The enabling chemistry technologies currently used in the drug discovery process can provide a significant competitive advantage. The goal is not only to achieve a reduction in time or the cost of goods and an increase in the probability of success, but the development of safer, efficient, and green processes where flow chemistry in conjunction with other technologies enables chemistries that are difficult or dangerous to scale in batch mode. In a typical continuous process, the narrow diameters, increased surface areas, and the possibility of immobilized beds can be beneficial for photochemical, electrochemical, and biocatalyzed reactions [36, 37].

3.1 Photochemistry

Photochemical transformations are of great interest in drug discovery because the chemistry of excited singlet and triplet states can often lead to an increase in the complexity and stereochemistry of product structures. When performing photochemical reactions, several advantages over traditional batch processes can be found employing microfluidic technology, in-chip-based microreactors or transparent tubing offers. Another advantage is the uniform irradiation of the entire reaction solution and maximum penetration of light due to the short path lengths. A microfluidic system allows for improved temperature control, especially minimization of hot spots and heat dispersal. Finally, as an additional benefit photochemistry in flow offers the possibility to precisely control UV exposure time by the flow rate and reactor volume, thus avoiding the under- and over-irradiation problems found in batch reactors [36].

Simple flow reactors can be built from economical materials (polymer capillaries, low energy consumption and safer LEDs, amongst others). In parallel, many companies supply commercial equipment for continuous flow photochemical synthesis, including light sources.

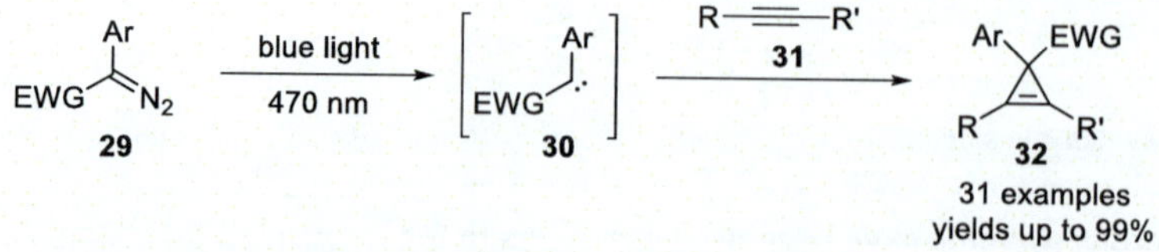

Fig. 10 Cyclopropenation reactions of carbenes under photochemical continuous flow conditions

Visible-light photocatalysis has received a great deal of attention recently. Well-known, but costly, iridium or ruthenium metal complexes have been widely used in this context. However, the use of metal-free sensitizers like porphyrin or riboflavin derivatives, or biological-derived catalysts of Rose Bengal broadens the application range of continuous flow photocatalysis and makes the process much greener [38]. An excellent review on heterogeneous photocatalysis in flow chemical reactors and its application in environmental remediation has recently been published [39].

Carbenes are very useful reactive intermediates to obtain a wide range of complex and valuable molecules of high interest in drug discovery. Typically, the chemistry of these systems is accessed by the use of transition metal catalysts. Koenigs et al. described the application of low-energy blue light for the photochemical generation of carbenes **30** from donor–acceptor diazoalkanes **29** (Fig. 10). This easy and catalyst-free approach permits highly efficient cyclopropenation reactions with alkynes **31**, which can be performed under continuous flow conditions. This protocol allows a 36-fold increase in productivity over conventional batch cyclopropenation and avoids the use of metal catalysts and the need to exclude moisture and air [40].

The sun is the most sustainable light source available on planet Earth. However, the limited applications of solar photochemistry can be associated with its broad spectral distribution, the fluctuations in solar irradiance, and the dilute energy content (up to 6.6 mol·m^{-2}·h^{-1}) [41]. Recently, Noël et al. [42] described, for the first time, a diverse set of photon-driven transformations that can be efficiently powered by solar irradiation using solvent-resistant and cheap luminescent solar concentrator-based photomicroreactors (LSC-PMs). In these devices, the LSC light guide embedded within microreactor channels can act as final absorbers of the sunlight-generated luminescent photons. As an application for the LSC-PM concept, the authors described the continuous solar-driven synthesis of artemisin (**35**), an antimalarial drug whose annual production is insufficient. A solar-based production plant could be very convenient to resolve the needs of areas where malaria is endemic. The semisynthetic approach to artemisin (**35**) involves the biomimetic photooxygenation of dihydroartemisinic acid **33** to the corresponding endoperoxide **34** as the crucial step. The authors adapted the reaction conditions described [43] for the use of methylene blue (MB) as a photocatalyst to perform the reaction on a window ledge under solar irradiation with a reaction control system (Fig. 11). Noël et al. obtained yields between 69 and 78%, with a significantly higher productivity (up to 21.2 mmol·m^{-2}·h^{-1}) during days of higher irradiance [42].

Cyclic amines are very interesting and useful compounds in drug discovery. MacMillan et al. described the photochemical decarboxylative coupling of amino

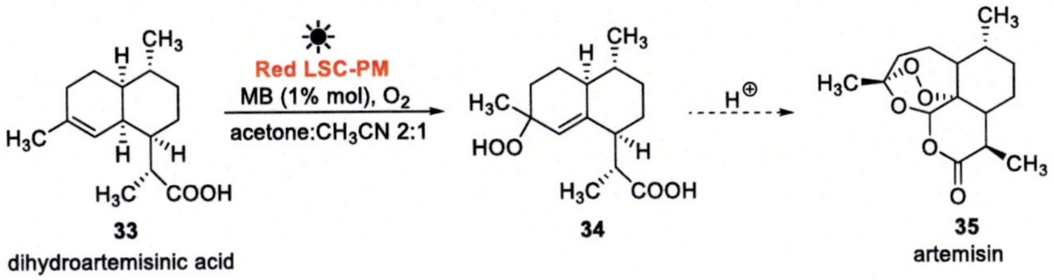

Fig. 11 Photooxygenation of dihydroartemisinic acid under irradiation with natural sunlight

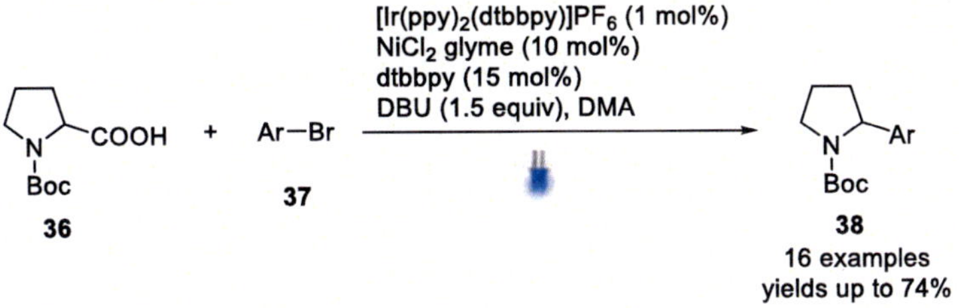

Fig. 12 Dual catalysis reaction in flow

acids with aryl halides as a novel C(sp^2)-C(sp^3) coupling process [44]. However, the scalability of photochemical transformations is limited by the attenuation effect of photon transport (Bouguer–Lambert–Beer law). To solve this issue, photoredox catalysis has been combined with continuous flow technology. For example, Alcazar [45] translated the dual photoredox and nickel catalysis for novel C(sp^2)-C(sp^3) coupling from batch to flow (Fig. 12). This protocol provides a huge time reduction in flow, from 3 days to 20–30 min, and it allows to increase the scale obtaining amounts of products that would be difficult to achieve in batch. Additionally, the use of flow demonstrates its sustainability, also markedly improving the space-time yield.

α-Arylation of carbonyl compounds is one of the most useful Negishi cross-coupling reactions for medicinal chemists. This functionality can be found in several pharmaceutical drugs, such as naproxen, ibuprofen, flurbiprofen, tolmetin, and fexofenadine. However, this reaction frequently fails when electron-rich heterocycles and chloro-derivatives are used. With the aim of solving this challenge to access products of great interest in drug discovery, Alcázar et al. designed a visible light-induced Negishi cross-coupling process, through activation of a Pd(0)-zinc complex, that allows the expansion of the scope of zinc enolates with deactivated aryl halides (Fig. 13) [46]. NMR experiments in the presence and absence of light confirmed that the formation of the palladium-zinc complex is key in accelerating the oxidative addition step.

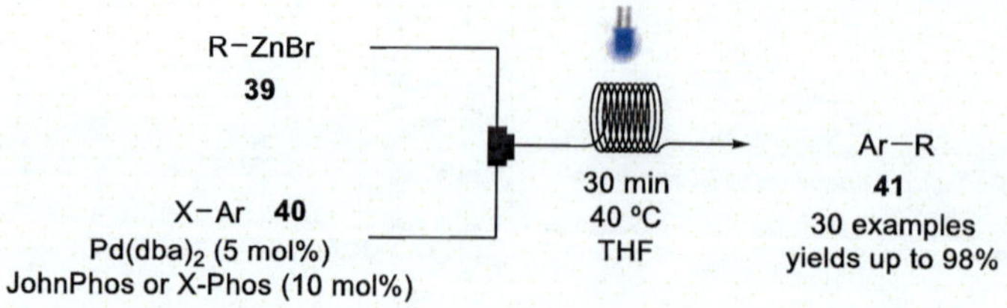

Fig. 13 Light-induced palladium-catalyzed Negishi reactions in flow conditions

3.2 Electrochemistry

Electrochemical processes replace dangerous and pollutant chemical oxidizing and reducing agents by electricity, which can be obtained from renewable sources to improve the overall ecological footprint of the chemical transformation. So, electrochemical processes are inherently green. However, the lack of standardized equipment, reproducibility, and protocols has prevented the participation of such techniques in conventional chemical research for many years [47]. Several requirements, such as very large electrodes to generate satisfactory electron flow, or inhomogeneous electric fields and energy loss owing to Joule heating, make it difficult to scale up electrochemical processes in batch mode [48].

The difficulties outlined above can be overcome by using electrochemical flow cells. Microreactors provide high surface/volume ratios and allow very accurate control over residence time, flow rate, temperature, and pressure. Additionally, the possibility of overoxidation is reduced under flow conditions as the reaction mixture flows continuously out of the reactor, in contrast to conventional batch electrolysis processes. In an electrochemical flow cell, the ohmic resistance is reduced by the short distance between the two electrodes and, in this way, electrolysis can be carried out with low concentrations of supporting electrolyte or even in its absence. As a result, the combination of the advantages of electroorganic synthesis and flow chemistry leads to more selective, safer, controllable, more efficient, economically viable, and ecofriendly chemical processes [49].

Finerenone (BAY-94-8862) (**42**) is a nonsteroidal mineralocorticoid receptor antagonist. This product, developed by Bayer to slow down kidney disease derived from type 2 diabetes mellitus, is prepared as a racemic mixture and purified by chiral column chromatography to obtain the active *S* enantiomer [50]. A recent patent application from Bayer described the recycling of the undesired *R* enantiomer by conversion back to racemic mixture through an oxidation/reduction flow electrochemical process (Fig. 14) [51]. The patent notes that 200 kg of the drug candidate has been generated for clinical trials using this process.

Hypervalent iodine reagents are mild, selective, versatile, and environmentally benign oxidants with wide applications in organic synthesis. These reagents can be prepared by anodic oxidation of iodoarenes, thus avoiding hazardous and frequently expensive chemical oxidants and improving the ecological impact of the process [52, 53]. Recently, Wirth et al. reported the first general synthetic approach to iodine

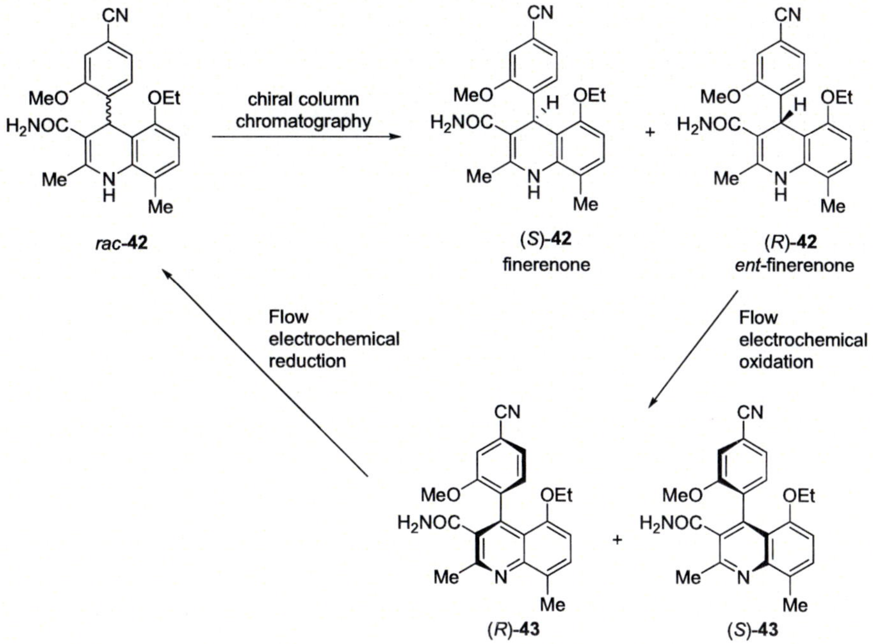

Fig. 14 Flow electrochemical oxidation/reduction of finerenone

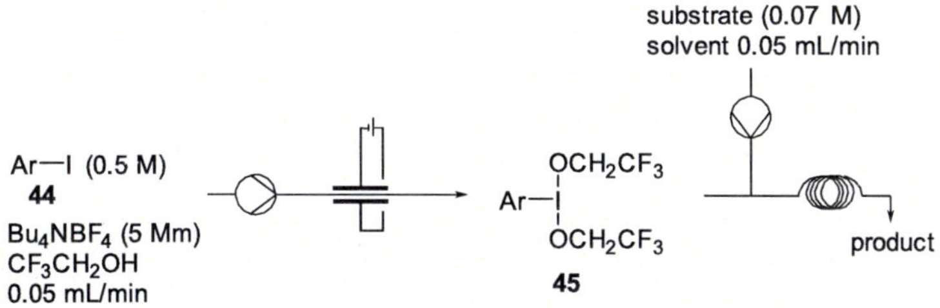

Fig. 15 Electrochemical generation and utilization of hypervalent iodine reagents in a coupled flow reaction system

(III) reagents **45** by anodic oxidation under continuous flow conditions using glassy carbon as the anode material and fluorinated alcohols [54]. Since the above iodine (III) reagents **45** are not sufficiently stable and decompose rapidly when the solvent is removed, the authors successfully used them directly in various oxidative transformations such as oxidation of sulfides, oxidative heterocyclization, phenol dearomatization, and α-functionalization of carbonyl compounds (Fig. 15). Additionally, Wirth proposed the transformation of these derivatives into stable hypervalent iodine reagents by quantitative ligand exchange through a treatment with acids in flow [54].

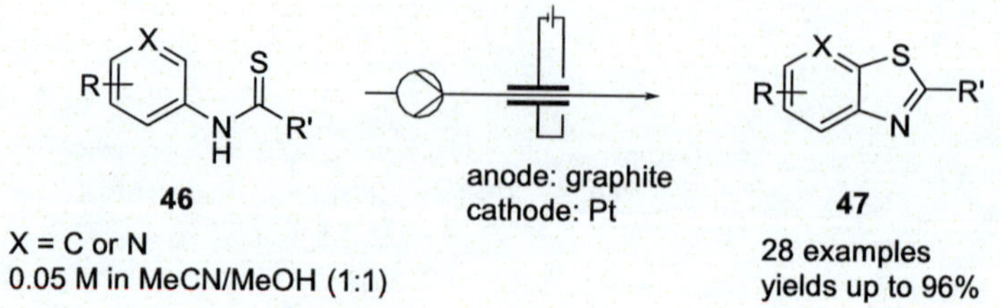

Fig. 16 Electrochemical flow synthesis of benzothiazoles and thiazolopyridines **47** by dehydrogenative C–S bond formation

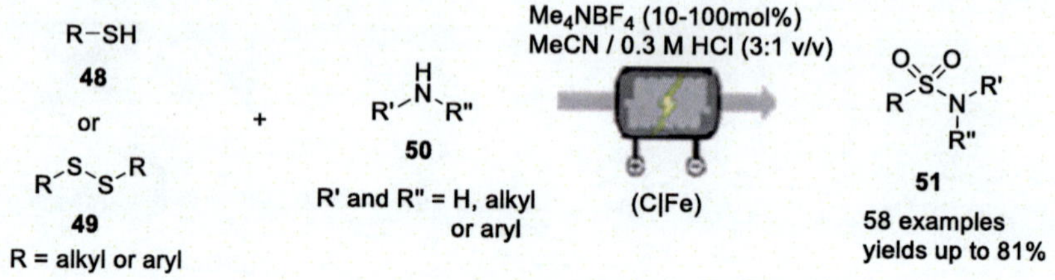

Fig. 17 Electrochemical flow synthesis of sulfonamides from sulfides or disulfides and amines

Benzothiadiazole is an important moiety present in many biologically active compounds that can be built by oxidative cyclization of *N*-arylthioamides **46**. Recently, an electrochemical synthesis of benzothiazoles and thiazolopyridines **47** in batch has been described that involves the use of supporting electrolytes, TEMPO as a mediator, and an inert atmosphere [55]. Xu and Wirth developed an efficient flow method without the need for an inert atmosphere, supporting electrolyte or catalyst and using laboratory grade solvents (Fig. 16) [56]. The process has been scaled up to gram scale, thus overcoming one of the most important limitations of the batch process.

It is well known that sulfonamides are key functionalities in APIs. Noël et al. developed an environmentally benign method that allows the oxidative coupling between thiols and amines to prepare a wide variety of sulfonamides (Fig. 17) [57]. The process is completely driven by electricity, can be performed in only 5 min, and does not need any sacrificial reagent or additional catalyst. These mild reaction conditions are compatible with a broad substrate scope and functional groups and only produce hydrogen as a benign by-product at the counter electrode.

In an interesting example, Atobe et al. described the electrochemical flow generation of *o*-quinones followed by a Diels–Alder cycloaddition in batch [58]. In many cases the oxidation potentials of an *o*-quinone precursor and dienophile are relatively close and thus competing oxidation processes can be expected. To overcome this impediment, researchers prepared the *o*-quinone in an electrochemical microreactor and then carried out the Diels–Alder cycloaddition by adding it to a batch vessel containing a solution of the dienophile. On using 4-*tert*-butylpyrocatechol as the *o*-quinone precursor and 6,6-dimethylfulvene as the dienophile, the researchers obtained the corresponding cycloadduct in 75% yield, while in batch only 13% was isolated.

3.3 Biocatalysis

Biocatalysis refers to the use of biological systems (mostly enzymes) as catalysts and the scope and application of this approach have been broadened, thanks to the extensive advances in protein and metabolic engineering together with biocatalyst immobilization. Whole cell catalysis uses the whole organism, such as *Escherichia coli*, for the transformation, while purified protein biocatalysis uses an extracted protein from the cell [19]. However, biocatalysis does suffer from several synthetic and operational problems.

The integration of biocatalyzed reactions with flow reactor technology leads to sustainable and highly productive continuous processes. The most important advantages of flow-based biocatalysis are: (1) limitation of substrate/product inhibition effects, (2) in-line purification with easy recovery of the product, and (3) no mechanical mixing. Flow processes accelerate biotransformations due to enhanced mass transfer, which in turn makes large-scale production more economically feasible in smaller equipment with a substantial decrease in reaction time, from hours to a few minutes, and in waste generation. Automated machines and devices for in-line product recovery are now available at relatively low cost, thus making flow-based biocatalysis an affordable technology [59].

Ursodeoxycholic acid (UDCA, **54**) is an effective drug for the treatment of hepatitis and it can be synthesized from cheap and more abundant chenodeoxycholic acid (CDCA, **52**). The preparation in batch involves the use of four enzymes in a two-step one-pot process. After the first step, the enzymes are inactivated by heat treatment to avoid the reverse reaction. This operation is not only costly and time-consuming but it also means that the enzymes cannot be reused [60]. Xu et al. transferred the process to flow to separate spatially the enzymes required for the first and second steps (Fig. 18) [61]. Thus, four enzymes were pairwise co-immobilized, LDH-7αHSDH and 7βHSDH-GDH, on an epoxy-functionalized resin by covalent bonding. The specific loading and activity recovery of the immobilized enzymes were significantly improved by 12- and 516-fold, respectively, and this resulted in a continuous process that avoided the heat treatment to denature the enzymes and allowed their reuse. The authors reported that with this biocatalyzed flow system UDCA (**54**) can be synthesized in approximately 100% yield and an excellent space-time yield of 88.5 $g{\cdot}L^{-1}{\cdot}d^{-1}$.

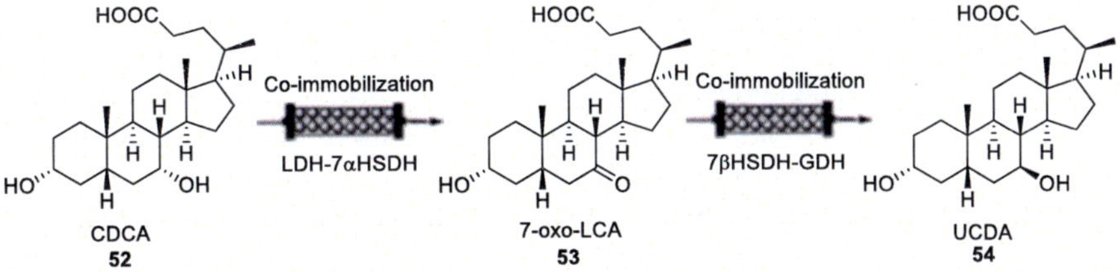

Fig. 18 The cascade enzyme reactions in the biocatalyzed flow synthesis of UDCA (**54**) from CDCA (**52**)

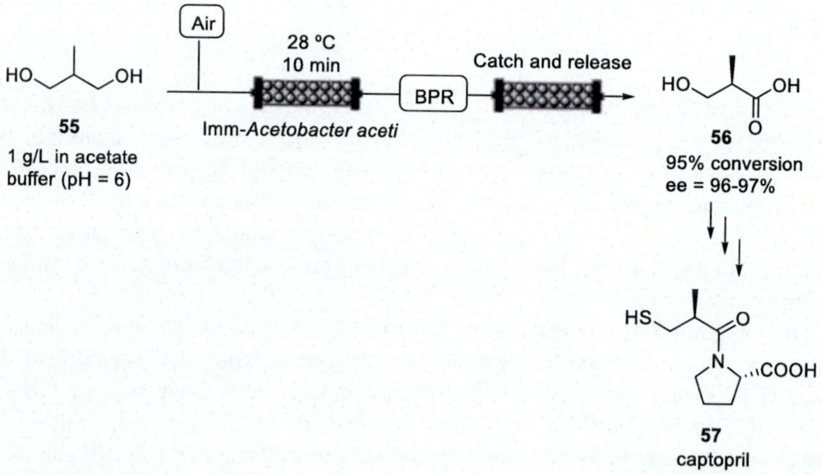

Fig. 19 Biocatalyzed heterogeneous flow oxidation of prochiral 2-methyl-1,3-propanediol (**55**) and in-line purification of the product through a catch-and-release protocol in the synthesis of captopril (**57**)

In 2017 a three-step flow chemical synthesis of captopril (**57**) was described. The first step is a heterogeneous biocatalyzed regio- and stereoselective oxidation of cheap prochiral 2-methyl-1,3-propanediol (**55**) (Fig. 19) [62]. This transformation was performed in flow using immobilized whole cells of *Acetobacter aceti*, thus avoiding aggressive and environmentally harmful chemical oxidants and the use of purified proteins, which would increase the cost of the process. The authors isolated the highly hydrophilic intermediate (*R*)-3-hydroxy-2-methylpropanoic acid (**56**) in-line by using a catch-and-release strategy. After this operation, captopril can be isolated by means of three sequential high-throughput chemical steps in only 75 min in an overall yield of 50% after crystallization.

3.4 Microwaves

Microwave radiation is an energy source widely employed with success in all fields of chemistry and its use has strongly modified the way in which chemical syntheses are performed. Microwave-assisted processes are characterized by short reaction times, simplified isolation procedures, higher product purities, increased yields and, in some cases, modifications in the selectivity. Furthermore, microwave irradiation can promote some reactions that do not occur under conventional heating [63].

Flow chemistry and microwaves (MW) have several aspects in common. Due to heat transfer is very rapid and overheating is possible, the translation of microwave irradiation to flow conditions has been described as a really easy process [64].

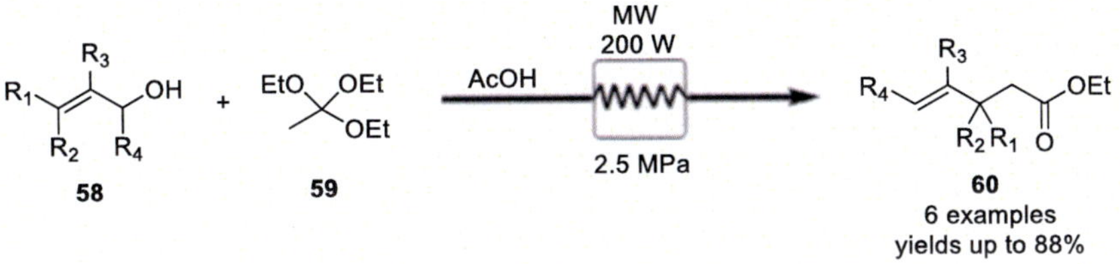

Fig. 20 Continuous Johnson–Claisen rearrangement under microwave irradiation

Microwave and flow processes can be easily performed under high temperatures and/or pressures and they are complementary when scale-up is considered. It is well known that one of the main advantages of flow reactions is the ease with which processes can be scaled without any reoptimization of the reaction conditions, even when operating with micro- and mesoreactors. The small diameters of flow reactors are ideal for the application of microwaves due to the low penetration depth of the waves, and this overcomes the problem encountered in scaling processes under microwave irradiation [65]. As a consequence of the current synthetic utility of microwave-assisted continuous flow processes, currently most of microwave manufacturers have systems that are adapted to continuous flow conditions [66].

γ,δ-Unsaturated esters have been employed in the synthesis of natural and unnatural bioactive compounds. These scaffolds can be prepared by conventional procedures through a [3,3]-sigmatropic rearrangement using harsh reaction conditions (high pressure and temperature, and often long reaction times) [67]. Hamashima et al. described the rapid and easy Johnson–Claisen rearrangement of allyl alcohols **58** and triethyl acetate **59** with a continuous flow system coupled to a microwave reactor (Fig. 20) [68]. The transformation was carried out without solvent using only a catalytic amount of acetic acid, thus avoiding the abovementioned harsh conditions. The productivity of the γ,δ-unsaturated esters **60** was 89.5 g/h under the optimal reaction conditions, which suggests that 2.1 kg of the target product could be obtained in 1 day.

Ley et al. described a single-mode bench-top resonator for the microwave-assisted flow generation of primary ketenes by thermal decomposition of α-diazoketones **61** at high temperature. The reaction of these compounds in situ with imines **62** by means of a [2 + 2] Staudinger cycloaddition afforded the corresponding *trans*-configured β-lactams **63** in good yields when aliphatic, aryl, or heteroaryl α-diazoketones were employed (Fig. 21) [69]. The authors also reported some insights into the mechanism of the reaction at high temperature, including the use of Augmented Reality.

In 2018 Cravotto et al. reported an efficient Pd-catalyzed microwave procedure for alkyne semihydrogenation in flow [70]. The authors employed alumina-sphere loaded Pd nanoparticles for the MW continuous flow semihydrogenation of 2-butyne-1,4-diol (**64**) in ethanol under H_2 flow (7.5 mL/min, counter pressure 4.5 bar). The corresponding alkene was obtained in good conversion (>90.5%) and very high selectivity to (Z)-2-butene-1,4-diol (**65**) (95.2%) (Fig. 22) [71].

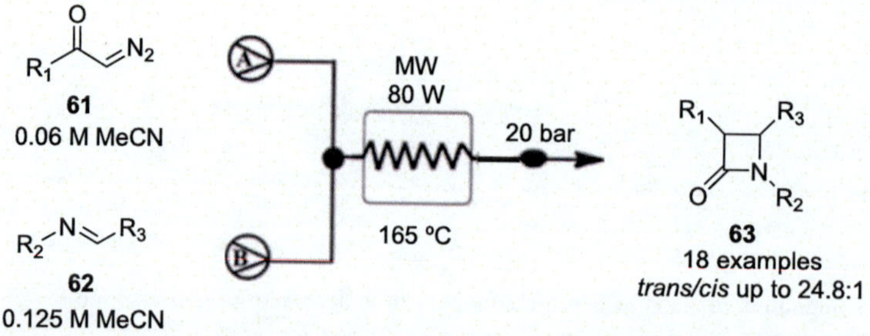

Fig. 21 Microwave-assisted Wolff–Staudinger flow strategy with the formation in situ of β-lactams **63**

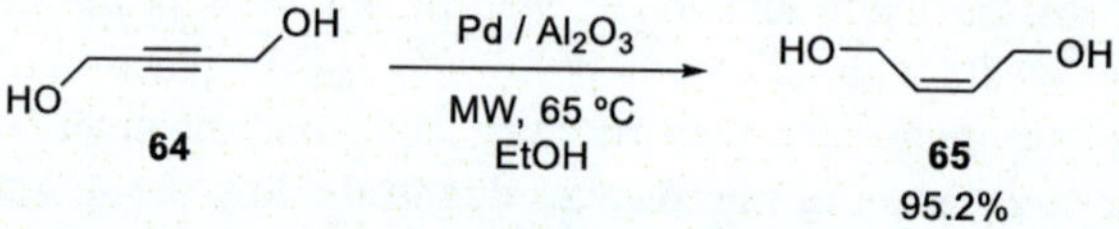

Fig. 22 MW flow semi-hydrogenation of alkyne **64** to (*Z*)-alkene **65** on Pd/Al2O3 catalyst

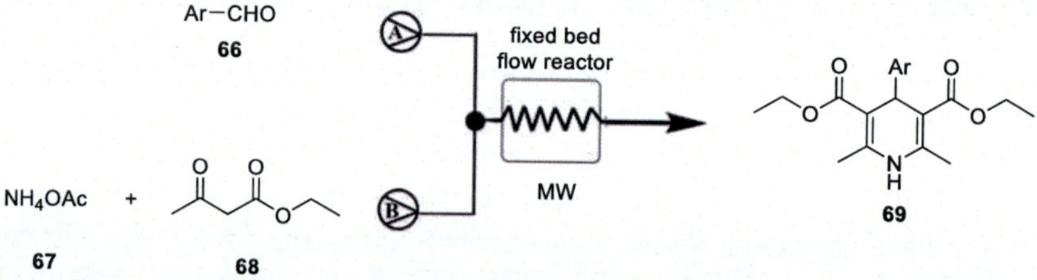

Fig. 23 Hantzsch microwave-assisted flow reaction under solvent-free conditions

Hantzsch 1,4-dihydropyridines (1,4-DHPs) have a wide variety of pharmacological activities, including the treatment of hypertension and angina pectoris, as well as antitrypanosomal, anticancer, antitubercular, antiplasmodial, antibacterial, antileishmanial, and antioxidant applications. Using a prototype 2.45 GHz MW-equipped MiniFlow 200SS apparatus designed and built by Sairem, Vanden Eynde described the preparation of 1,4-DHPs **69** [72]. The authors suggested the use of Raman spectroscopy as an aid to optimize better the wide range of reaction parameters. More recently, a 1,4-DHP **69** synthesis has been developed employing a fixed bed catalyst (γ-Fe_2O_3 nanoparticles) in an MW-assisted solvent-free flow reaction approach (Fig. 23) [73]. The convenient optimization of catalyst nanoparticle uniformity and size, temperature, and residence time can increase the reaction yield up to 98.7% avoiding by-products formation.

Bromine (Br_2) is probably the most typically employed reagent for bromination processes. However, Br_2 has several problems from the point of view of safety, atom economy, and selectivity. In fact, hydrogen bromide (HBr) is easy to handle and it achieves a high atom economy, so it is considered to be a better alternative for

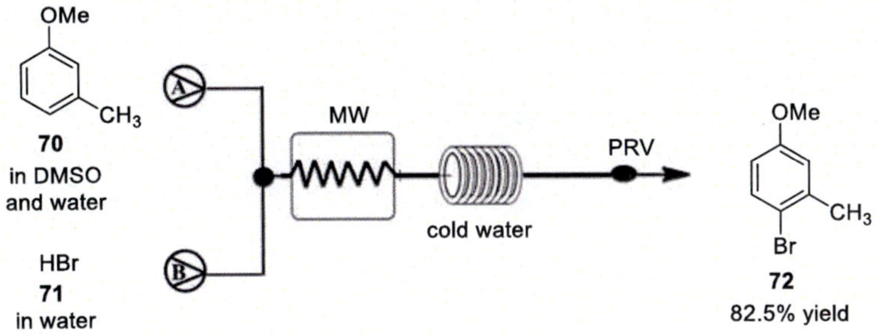

Fig. 24 Microwave-assisted continuous flow bromination with HBr-DMSO (PRV: pressure relief valve)

bromination processes. Among the chemical compounds that promote HBr oxidation during the bromination reaction, dimethyl sulfoxide (DMSO) is preferred because it is readily available, safe, and inexpensive. Thus, an efficient procedure has been reported for large-scale oxidative bromination with HBr-DMSO under continuous flow microwave-assisted conditions (Fig. 24) [74]. The reaction is performed using water as the solvent and in a metal-free manner to obtain a productivity of 60 g/min under the optimal conditions, suggesting a productivity of 8.6 Kg/day.

4 Hazardous Reagents

Various reagents and intermediates commonly used in drug discovery and in the preparation of APIs (Active Pharmaceutical Ingredients) are highly toxic, inflammable, explosive, corrosive, or carcinogenic. This has limited the use of such reagents in chemical synthesis and especially in the scaling of processes. Continuous flow processes are considered to be inherently safer for a variety of reasons, including lower reaction volumes, better temperature control, and ability to accommodate higher pressures without risk. Unstable, highly reactive or toxic intermediates can be generated in situ from benign, readily available and cheap precursors in a closed, pressurized system and converted directly into more stable, non-hazardous intermediates or products. Thus, flow processes offer the ability to prepare more hazardous and batch-inaccessible fine chemicals on a large scale, thereby unlocking access to products that would have been difficult to make competitively [75].

4.1 Azides

Azides are synthetically useful compounds, but the hazards associated with their preparation have limited their production and use. It is well known that organic

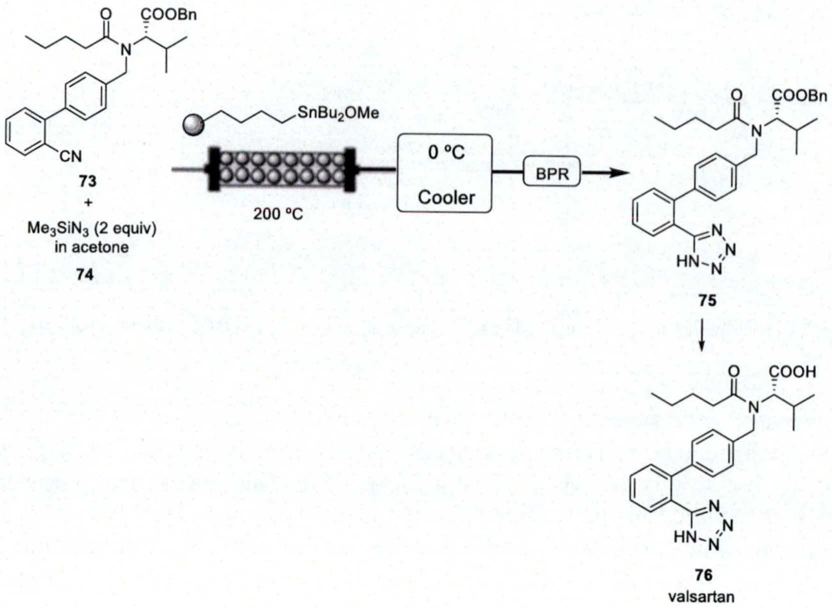

Fig. 25 Continuous flow synthesis of valsartan (BPR)

azides are very energetic and potentially explosive substances. In addition, toxic and explosive hydrazoic acid may be released in the synthetic transformation. For these reasons, azides are perfect candidates for use in continuous flow process.

A practical continuous method for the in situ generation of mesyl azide from sodium azide was reported for the preparation of a series of α-diazo-β-keto esters and α-diazo-β-ketosulfones in a safe process [76]. The use of an in-line liquid–liquid separator allowed the partition of the desired diazo compound in acetonitrile from the water-soluble by-products and the subsequent reaction was telescoped.

Recently, Le Grognec described an efficient continuous flow process that involved trimethylsilylazide (**74**) and a polystyrene-supported organotin reagent for the synthesis of 5-substituted 1*H*-tetrazoles from nitriles in a short time (7.5 min or 15 min for the less reactive nitriles) [77]. The products could be isolated with high purity and a low concentration of tin residues (less than 5 ppm). The authors applied this method to the synthesis of valsartan (**76**), an angiotensin II receptor antagonist (Fig. 25).

The acyl azide method is a very good synthetic approach for peptide bond formation while avoiding side reactions such as epimerization. However, the method is underutilized because acyl azides are considered to be highly unstable and potentially explosive intermediates. In 2020, Kappe and Hone developed a new protocol in which acyl azides were safely generated and reacted in situ within a continuous flow system [78]. The azide intermediate **79** was generated by using

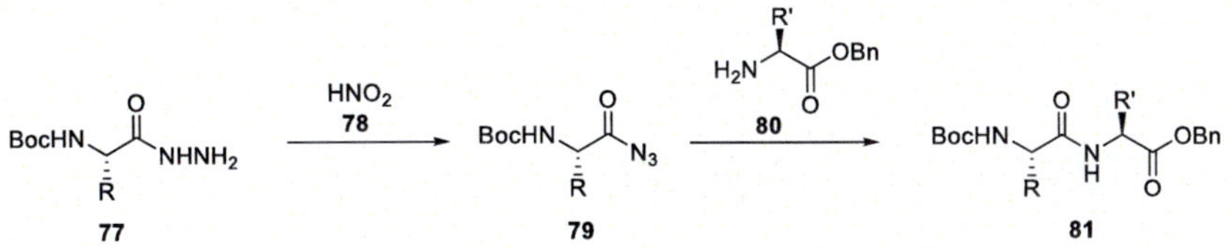

Fig. 26 Continuous flow acyl azide formation and subsequent amide coupling

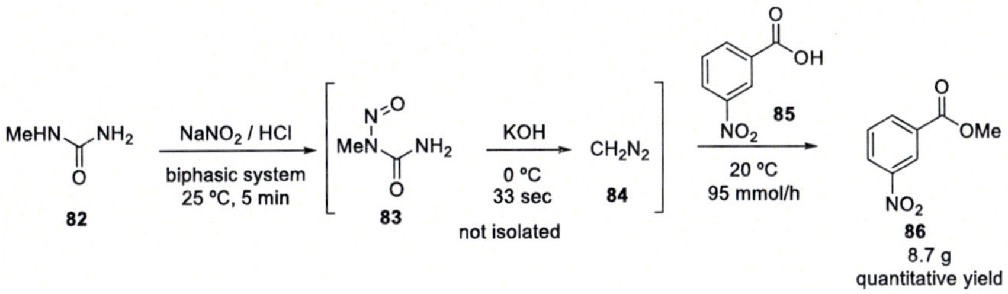

Fig. 27 Methylation of 3-nitrobenzoic acid (**85**) by in-line generated MNU (**83**) and diazomethane (**84**)

nitrous acid in water and extracted into the organic phase containing the amine nucleophile for peptide coupling without epimerization (>1%) (Fig. 26).

Isoxazoles are widely present in bioactive compounds and they serve as valuable synthetic intermediates for numerous drug candidates. In a recent application of azide chemistry, the synthesis of isoxazoles in a continuous flow system has been reported in a two-step process: the Friedel–Crafts acylation of alkynes and conjugate addition of the azide to the resulting β-chlorovinyl ketones followed by a photochemical-thermal reaction [79].

Tamiflu, one of the most effective anti-influenza drugs, is usually prepared from shikimic acid by employing potentially hazardous azide chemistry. Watts [80] developed a safe and easily scalable continuous flow route to Tamiflu.

4.2 *Diazomethane*

Diazomethane is a very versatile and useful C_1 building block that reacts selectively under mild conditions with improved atom efficiency (it produces only nitrogen as a by-product). However, the preparation and handling of diazomethane in organic synthesis is troublesome due to its toxicity, extreme instability, and volatility.

In 2017, Lehmann developed a two-step continuous process for the in-line on-demand generation of the highly toxic *N*-methyl-*N*-nitrosourea (MNU) (**83**) and diazomethane (**84**) from cheap and non-hazardous *N*-methylurea (**82**) (Fig. 27) [81]. This high-yielding and atom-efficient process allows both reagents to be

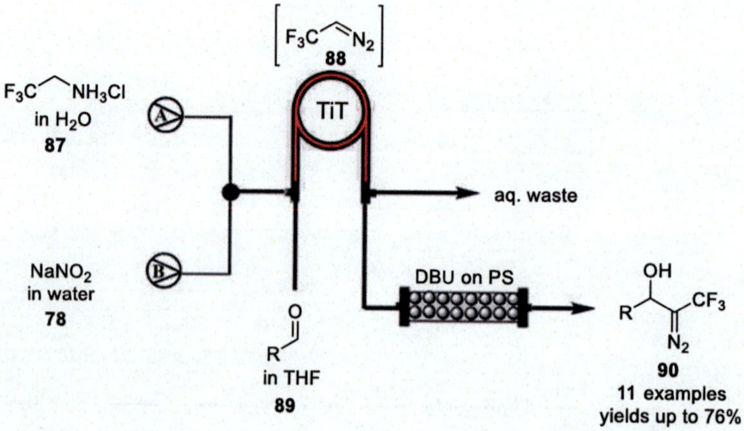

Fig. 28 Continuous flow preparation and separation of trifluoromethyl diazomethane (**88**) in a Tube-in-Tube (TiT) Reactor and its condensation with aldehydes **89**

generated and consumed directly without any isolation. Thus, the closed system and the small reaction volumes avoid human exposure and minimize the risk of explosive decomposition. The productivity of this continuous flow reactor reached up to 117 mmol·h^{-1}.

A recent review highlighted the developments in the synthesis of the diazomethane precursor, subsequent generation and purification of diazomethane, and the final transformation to afford the target product. This study focused on an industrial perspective where the process is classified by its capability to generate specific forms of diazomethane, e.g., as a neat gas, diluted gas, or in organic solution [82].

Kappe and Pieber reported a simple continuous flow technology protocol for the synthesis of anhydrous trifluoromethyl diazomethane (**88**) from the amine **87** and $NaNO_2$ (**78**) under acidic conditions. The diazo compound diffuses through a gas-permeable membrane into an organic stream and reacts with a carbonyl compound (**89**), thus resulting in highly functionalized building blocks (Fig. 28) [83]. The authors found that the success of the condensation required the use of an immobilized catalyst in a packed-bed reactor.

4.3 Hydrogenation

Hydrogenation is usually an exothermic molecular reduction. However, in a large-scale batch reactor the gas/liquid/solid multiphase mixing is problematic. In addition, hydrogen employed during the hydrogenation process is a highly flammable gas that readily forms explosive mixtures with air. In contrast, a flow-reactor approach provides numerous advantages, including the precision of reaction

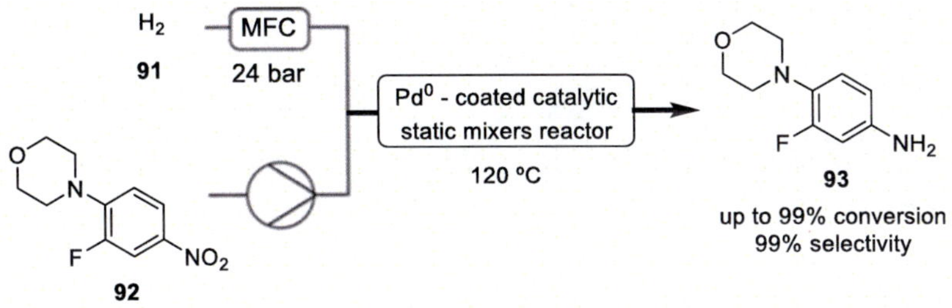

Fig. 29 Flow synthesis of the key intermediate of linezolid (**94**) using 3D-printed catalytic static mixers (MFC: Mass Flow Controller)

parameters, good multiphase mixing, automatic control, low carbon emissions, high reproducibility, and high safety.

To achieve safe, efficient, and sustainable continuous flow hydrogenations with reduced reaction times and increased selectivities, new catalyst types and immobilization methods as well as flow reactors have been developed in recent years. Heterogeneous catalysts are more stable and easier to recycle than homogeneous catalysts [84].

Primary amines are often present in bioactive compounds and they are employed as synthetic intermediates for APIs. Linezolid (**94**) (Zyvox) is a new class of antibiotic developed by Pfizer in 2000. Gardiner et al. reported a novel continuous flow hydrogenation reactor containing a series of 3D-printed catalytic static mixers (CSMs), which were coated with Pd catalysts by electroplating, for the synthesis of **93**, the key intermediate of linezolid (Fig. 29) [85]. This system reduced the manufacturing cost of the reactor, eliminated the need for catalyst filtration, and allowed an easy scale-up to produce 425 g of the product per day (with 99% conversion and 99% purity). In addition, a single set of twelve Pd-coated CSM inserts could be used almost continuously for hydrogenation reactions for over a year without significant loss of catalytic activity.

Recently, Smyth and Manyar developed a continuous flow packed-bed catalytic reactor for the hydrogenation of aromatic nitrobenzoic acids in water. These hydrogenations are green, more efficient, less consumptive, and safer than the conventional reduction process [86].

Catalytic hydrogenation of nitriles is also an attractive approach to primary amines from the viewpoint of sustainability. However, the process suffers from limited substrate scope and low selectivity, with secondary and tertiary amines frequently formed as by-products. Kobayashi et al. described a polysilane/SiO_2-supported Pd as an effective catalyst in the hydrogenation of nitriles to primary amine salts in almost quantitative yields under continuous flow conditions [87]. It is remarkable that a complex mixture was obtained in this reaction under batch conditions. The catalyst remained active for more than 300 h (TON >10,000) without loss of selectivity and no metal leaching. The authors applied this continuous flow reaction to the total synthesis of venlafaxine (**97**), which is a common

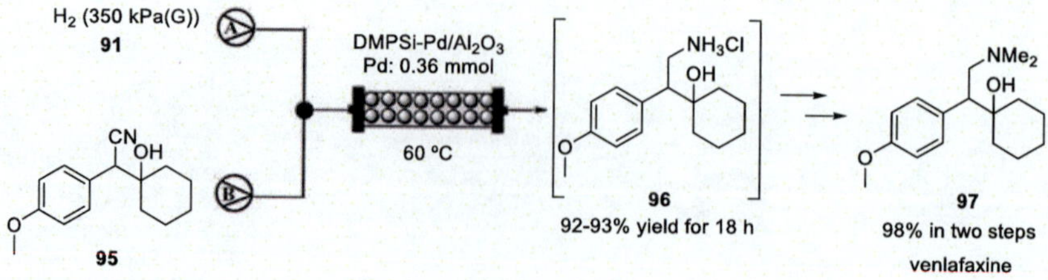

Fig. 30 Continuous flow synthesis of venlafaxine (66)

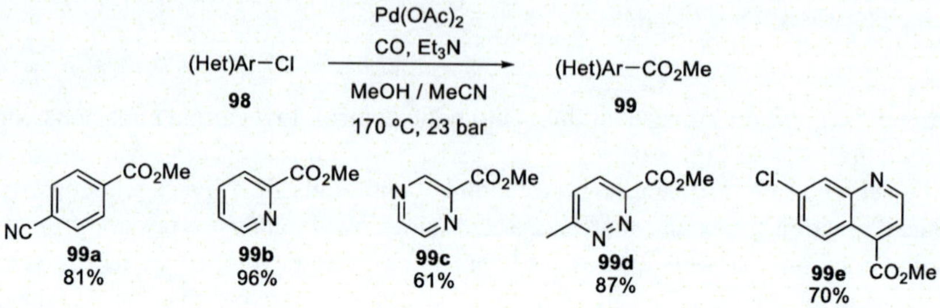

Fig. 31 Methoxycarbonylation of (hetero)aryl chlorides (98) under high temperature/pressure continuous flow conditions

antidepressant drug (Fig. 30). The hydrogenation step as the key reaction gave the desired compound **96** in 92–93% yield.

4.4 *Carbonylation*

Carbon monoxide, CO, is a cheap, atom efficient, and synthetically valuable C_1 building block for organic chemistry that is readily available in large quantities from the bulk chemical industry. However, there are significant risks in handling this substance due to its toxicity and flammability. Moreover, because CO gas has low solubility in many solvents, harsh reaction conditions (high temperatures and/or high pressures of CO) are usually required in conventional batch carbonylation reactions in flasks or autoclaves. All of these drawbacks can be safely solved by means of continuous flow processes [88].

Hetero/aryl chlorides are the least expensive of the aryl halides but they are underused in carbonylation reactions owing to their very poor reactivity. Kappe et al. described a continuous flow protocol for the Pd-catalyzed methoxycarbonylation of (hetero)aryl chlorides (**98**) using CO gas and methanol under high temperature and pressure conditions (Fig. 31) [89]. The protocol provides moderate to excellent product yields in a short 16 min residence time and the process is safe and potentially scalable. Likewise, in 2017 Kappe described a continuous flow Pd-catalyzed oxidative protocol employing CO and O_2 gas for the synthesis of

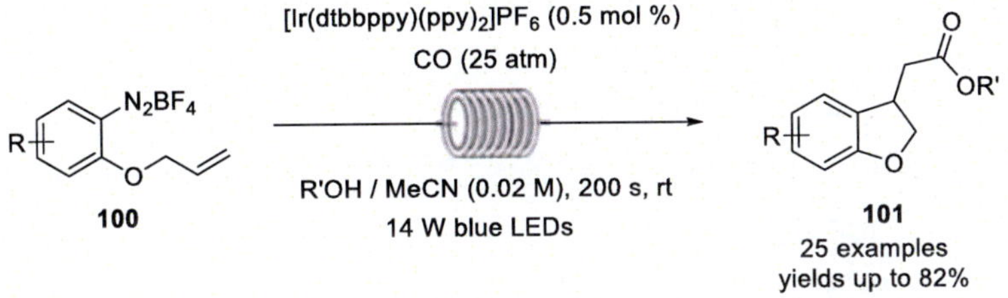

Fig. 32 Synthesis of acetate-functionalized 2,3-dihydrobenzofurans **101** by annulation/alkoxycarbonylation cascade under flow conditions

carbonylated heterocycles, such as benzoxazolones, 2-benzoxazolidinones, and other biologically important five- and six-membered carbonylated heterocycles [90]. The same author also described a similar protocol for the Pd-catalyzed reductive carbonylation of (hetero)aryl bromides to their corresponding (hetero)aryl aldehydes using syngas ($CO + H_2$) under flow conditions. This protocol proceeds with low catalyst and ligand loadings to afford products in good to excellent yields [91].

Polyzos et al. developed a continuous flow visible light photoredox catalytic approach for the annulative alkoxycarbonylation of alkenyl-tethered arenediazonium salts (**100**) [92]. This protocol involves room temperature, rapid reaction (200 s), moderate CO pressure (25 atm), facile and safe scale-up, and exclusive 5-exo cyclization. The authors applied the protocol to the synthesis of pharmaceutically relevant 2,3-dihydrobenzofurans (**101**) (Fig. 32).

More recently, Osako and Uozumi developed an aqueous continuous flow reaction system for the hydroxycarbonylation of aryl halides to the corresponding benzoic acids. This flow hydroxycarbonylation in aqueous solution proceeds efficiently in a flow reactor containing a palladium-diphenylphosphine complex immobilized on an amphiphilic polystyrene-poly(ethylene glycol) resin [93].

4.5 Miscellaneous

Concern over the utilization of hazardous materials is not only related to their handling but also to their transportation and on-site storage, especially on a large scale. The hazardous material must be a simple, low molecular weight and versatile compound and it must be produced from inexpensive, benign precursors. A complete review of this topic has recently been published [94].

Fluorination processes have become well-established methods in drug discovery to modify the metabolic stability and bioavailability of compounds. Nucleophilic fluorinations are hampered by the low solubility of metal fluorides, such as CsF, and the need for perfectly dry reaction conditions, so they are usually performed in the presence of phase transfer catalysts, at high temperatures and with long reaction times. In 2018, Lindhardt developed a CsF-CaF_2 packed-bed reactor to perform

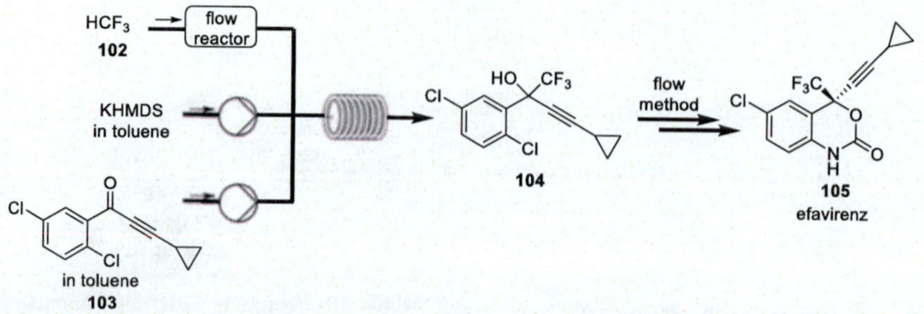

Fig. 33 Trifluoromethylation of **103** to afford efavirenz intermediate **104**

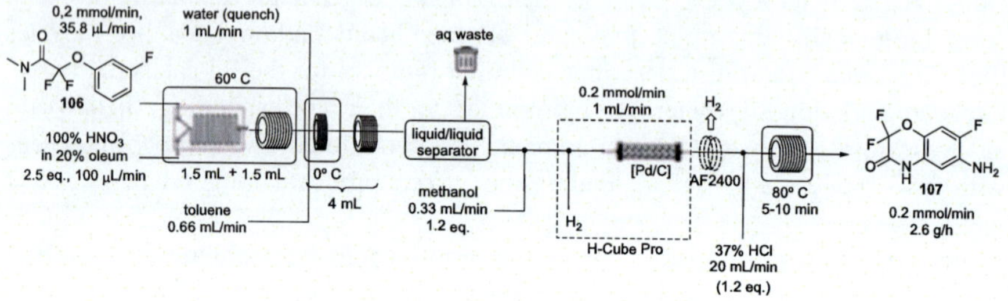

Fig. 34 Continuous flow setup developed for the integrated synthesis of ABO (**107**). Reproduced with permission of ACS (Ref. [97])

nucleophilic fluorinations in continuous flow [95]. The reactor material was obtained by evaporation of dissolved CsF onto appropriately sized CaF_2 particles, with efficient in-line drying achieved by passing superheated acetonitrile or toluene through the reactor bed. Nucleophilic trifluoromethylations are usually performed using the Ruppert–Prakash reagent (CF_3SiMe_3), which is not atom economical. Fluoroform (CF_3H) is a highly atom-economical gaseous source of CF_3 groups that is readily available (fluoroform is a chemical waste product that accumulates during the manufacture of Teflon®). In 2019, Shibata developed a gas/liquid phase micro-flow trifluoromethylation protocol for ketones, aldehydes, chalcones, and imines with excellent diastereoselectivity using the gaseous chemical waste fluoroform. The authors applied this protocol for the formal total micro-flow synthesis of the anti-HIV drug efavirenz (**105**) (Fig. 33) [96].

In 2017 Kappe et al. developed a continuous flow protocol for the synthesis of 6-amino-2,2,7-trifluoro-4*H*-benzo[1,4]oxazin-3-one (ABO, **107**) that employed a fully continuous, three-step sequential nitration/hydrogenation/cyclization process (Fig. 34) [97]. The sequence involves an exothermic nitration using 100% HNO_3 in oleum, a hydrogenation using heterogeneous Pd/C as catalyst at 45°C, and finally acidification in-line with 37% HCl to afford ABO (**107**) after an additional residence time at 80°C for 10 min. The overall yield for the continuous flow process was 83%, which represents a significant improvement on the yield obtained in the batch protocol.

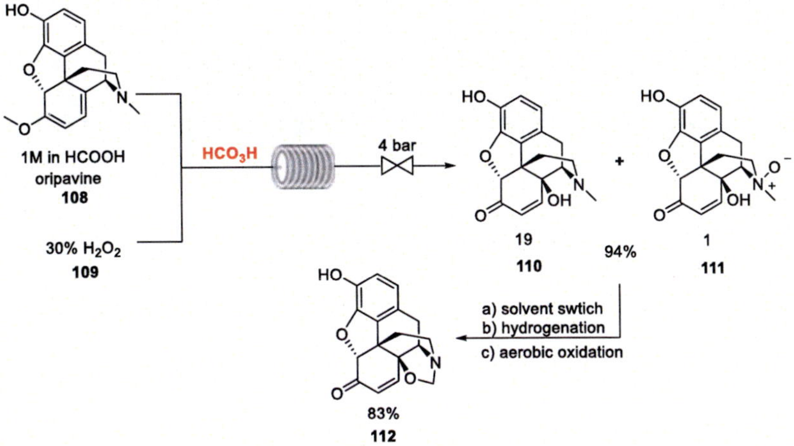

Fig. 35 HCO_3H generator and its use in the C14 hydroxylation of oripavine (**108**) for the telescoped synthesis of 1,3-oxazolidine **112**

Peracids are strong oxidizing agents that are employed for epoxidations, hydroxylations, and Baeyer–Villiger oxidations. It is well known that these compounds are also very unstable and prone to explosive decomposition. For example, performic acid (HCO_3H) at concentrations >50% is highly reactive and it decomposes upon heating and explodes when rapidly heated to 80–85°C. Kappe described the C14 hydroxylation of naturally occurring oripavine (**108**) in the telescoped continuous process to the noroxymorphone precursor **112** [98]. Performic acid was generated in situ from formic acid (HCO_2H) and this rapidly oxidized the diene moiety of oripavine at 100°C to provide 14-hydroxymorphinone **110**. A subsequent continuous solvent switch, hydrogenation in a packed-bed hydrogenator, and palladium-catalyzed *N*-methyl oxidation afforded the 1,3-oxazolidine derivative **112** (Fig. 35).

5 Monitoring, Optimization, and Scale-Up in the Pharmaceutical Industry

5.1 Monitoring

The Quality by Design (QbD) initiative launched by the pharmaceutical regulatory authorities encourages manufacturers to design and control processes based on having a thorough understanding of the reactions involved. Real-time acquisition of data for chemical process development and in-process monitoring by process analytical technology (PAT) is essential for this initiative [94].

Flow chemistry is very suitable for real-time monitoring (Green Chemistry principle 11). It enables very precise control over the chemical process, which in turn ensures high yields and selectivity and reduces waste. Real-time analysis can be

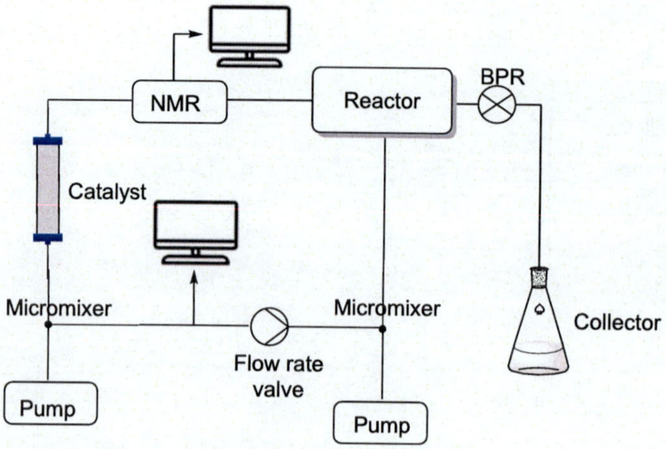

Fig. 36 General scheme for an NMR-monitored reaction in flow

used for the optimization of the reactor, for automatization and for scale-up of the process [4]. Real-time monitoring is also very important when hazardous reagents are used in flow [99].

IR [100], RAMAN [101], UV-vis [102], fluorescence [103], and NMR [104, 105] spectroscopy have been used to monitor flow reactions and, additionally, analytic techniques such as liquid and gas chromatography and MS [106] have been interfaced to flow systems to provide real-time reaction monitoring (Fig. 36).

In-line and on-line analyses have been widely used. For in-line analysis, the reaction and the monitoring system are connected in series and all of the reaction mixture passes through the analytical system and is continuously analyzed. In-line monitoring minimizes the time-lag between reaction and monitoring. For on-line analysis, the system is not directly connected to the stream and the sample is transferred as representative aliquots collected periodically during the reaction. This method is simpler and is used when direct connection is difficult [104].

Real-time analysis, particularly multireaction trace analysis by MS simplifies the hit identification and avoids the use of chromatographic procedures. As a result, the time required for hit identification is significantly reduced [106].

Another important point for automatization is the integration of in-line purification that can remove impurities and excess reagents from the main stream as these may not be compatible with the next reaction step [107]. Four methods have been commonly used: scavenger column, distillation, nanofiltration, and extraction. The use of scavenger resins is widespread due to its simplicity in separating the desired and undesired compounds. Distillation in flow chemistry has been performed by a membrane distillation unit, where a membrane layer is sandwiched between two channels and serves as liquid–vapor contact. Nanofiltration is based on molecular size and it has been possible to apply this technique to flow chemistry, thanks to the progress in materials for membranes and the development of organic solvent nanofiltration (OSN). Finally, liquid extraction is based on the different solubilities of liquids.

Flow chemistry is an ideal technology for computer-controlled optimization and the ability to monitor the results of a reaction also provides a unique opportunity for automation and scale-up. Several groups have developed methods for optimization as well as for obtaining useful kinetic data.

Information provided by the analytical instrument is sent to a computer, which in turn controls the reactor (temperature, pressure, flow rate, etc.). In this way stoichiometry, temperature, and time can be changed instantaneously by an algorithm that suggests new conditions. This process can be repeated in an iterative way until a maximum yield and/or selectivity is identified.

5.2 *Automatization*

Automatization involves the development of algorithms for the analysis of results (conversion, selectivity, etc.) based on the integration of analytical tools in real-time and protocols for the selection of better reaction conditions to obtain the optimal results.

Automation in flow synthesis can be categorized into three categories: auto-sampling and in-line analysis, optimization, and automation for control [108].

Important progress has been made in recent years towards the automatization of chemical synthesis. Two approaches have been used to achieve this goal. The first one is the automation of customized synthesis routes to different targets and this enables the use of many reactions and starting materials. The second approach is the automation of generalized platforms that can make many different targets using common coupling chemistry and building blocks [109].

Progress in the automated synthesis of small molecules has been made in two directions. Firstly, the design of machines for the customized synthesis of one specific small molecule and, secondly, the design of machines that can synthesize different types of small molecules.

From the synthetic point of view two possibilities have been envisaged: (1) automatization of customized synthesis routes and machines enables the use of different reactions and starting materials and (2) the design of platforms that make different targets using known building blocks. The first case involves the design and development of algorithms for retrosynthetic analysis. The building block approach enables the efficient, reproducible, and automated synthesis of a wide range of targets.

Customized routes to synthesize each small molecule require systematic planning known as retrosynthetic analysis. All chemical reactions should be considered in retrosynthesis by simplifying the target molecule until available starting materials are reached.

The selected synthesis should then be performed and the best route selected. The customized approach has several advantages for the preparation of very similar

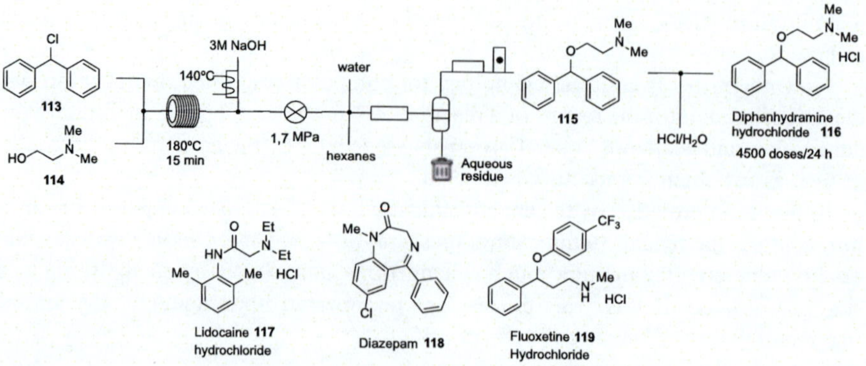

Fig. 37 Flow diagram of the on-demand flow synthesizer for the preparation of four APIs **116–119**

molecules since each synthetic step is highly optimized. There are some important disadvantages, especially related to automation, since each target requires a new design, development, and optimization of all of the synthetic steps involved and sometimes in different equipment.

Customized routes are generally slow and depend to a large extent on the design by the chemists.

Several machines have been designed for customized synthesis and most of these are discussed in the excellent review by Trobe and Burke [109].

It is worth highlighting the automated flow platform developed at MIT [110]. This platform consists of a modular system for reaction, separation, and purification and it can be easily reconfigured to produce a series of pharmaceuticals in the required doses on site if the drug is required in a crisis. The flexibility of this on-demand flow synthesizer was demonstrated by the production of diphenhydramine hydrochloride (**116**), lidocaine hydrochloride (**117**), diazepam (**118**), and fluoxetine hydrochloride (**119**) (Fig. 37). Other advantages of automatization are related to scalability, safety, speed, and reproducibility.

To solve the problems associated with customized synthesis, a second approach takes advantage of the use of building blocks. This methodology enables an efficient, reproducible, and flexible procedure for the preparation of a large variety of compounds. The simplicity of this methodology enables the synthesis to be performed even by non-specialists. In this strategy it is considered that many types of molecules are modular, for example natural products, and can be synthesized using discrete molecular building blocks. Once again, excellent reviews have been published and important examples of applications are described [109, 111].

A platform for the iterative synthesis of small-molecules requires a common type of reaction that can provide different types of C–C and C–X bonds by coupling reactions. These coupling reactions should be performed in tolerant conditions since functional groups and stereochemistry should be included in the building blocks and translated to the target molecule.

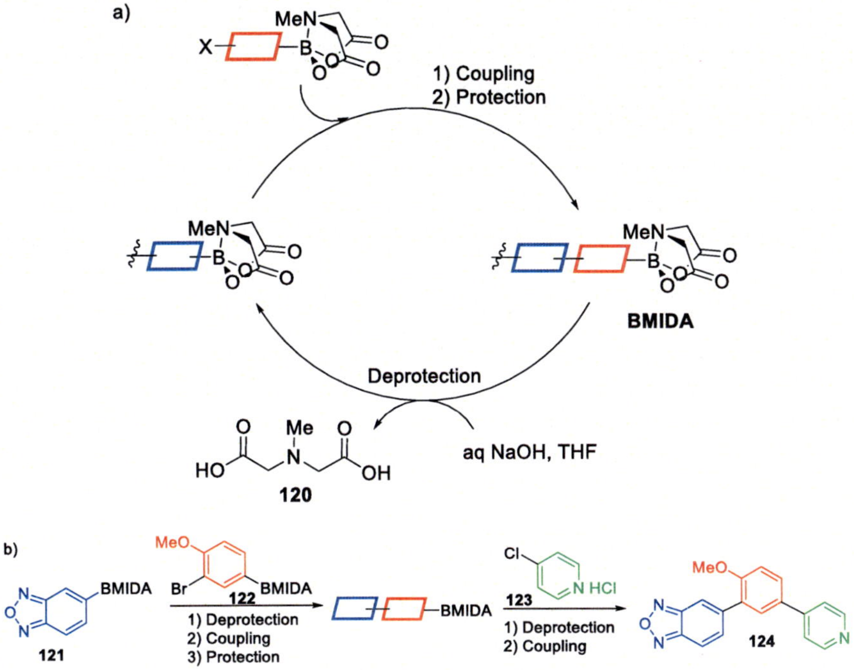

Fig. 38 (**a**) Protection, coupling and deprotection steps in MIDA building blocks and (**b**) automated synthesis of an active pharmaceutical compound **124** [112]

MIDA boronates are common building blocks for iterative synthesis since they are stable in coupling reactions, can be easily purified, and they are easily deprotected (Fig. 38) [112]. These compounds are compatible with a range of reaction conditions, for example, oxidations, reductions, nucleophilic displacements and addition to aldehydes, double and triple bonds, electrophilic substitution, transition metal-catalyzed reactions and they survive protection and deprotection steps. The protection coupling and deprotection steps for MIDA building blocks are shown in Fig. 38 along with an application in flow using an automatic synthesizer.

Automatization of the synthesis of useful molecules is common in industry and it has a positive influence on scalability. Safety can be another important advantage of automatization, for example in the preparation of radiolabeled compounds for PET imaging (Fig. 39) [113]. Automatization can address the problems of radiation and short half-life of the isotopes and allows the preparation of on-demand doses on site or reactions that require harsh conditions (temperature or pressure). Automated synthesis platforms can also increase the speed of a chemical process by avoiding tedious manual operations. Similarly, this aspect improves the reproducibility of chemical reactions since human error is eliminated and the whole process can be controlled perfectly.

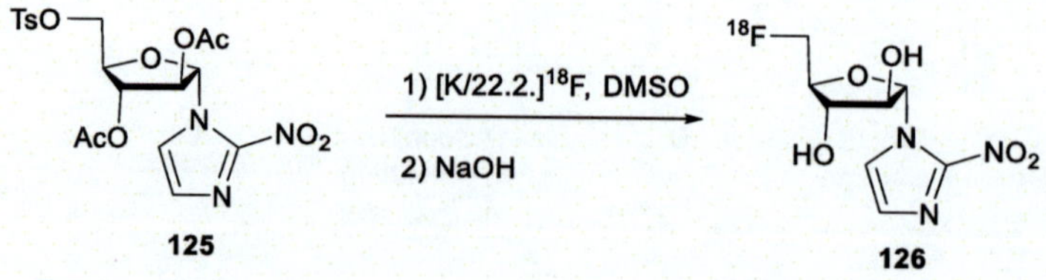

Fig. 39 Automated synthesis of an ^{18}F-labeled radiopharmaceutical

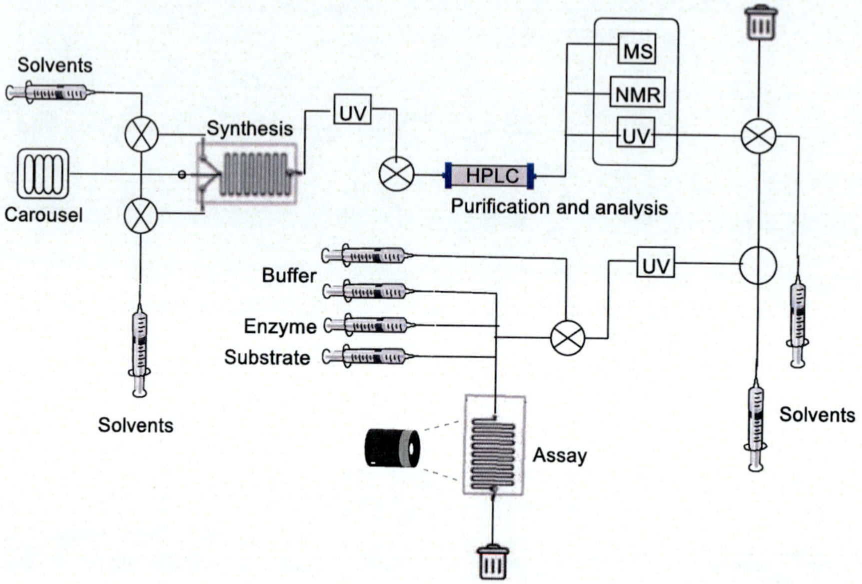

Fig. 40 General design of an automatic synthesis and assay platform

Finally, syntheses and biological assays have been automated to various degrees, thus closing the loop from automatic design, synthesis, and screening platforms (Fig. 40) [114].

5.3 3D Printing

Cronin et al. described a flexible synthesizer with a robotic platform and a 3D printer to design and print reaction systems optimized for different reactions and scales [115].

The development of 3D printing technology has permitted the design of flow reactors in the field of micro- and meso-fluidics and for the realization of chemical devices. 3D printing enables the fabrication of geometrically optimized reactors.

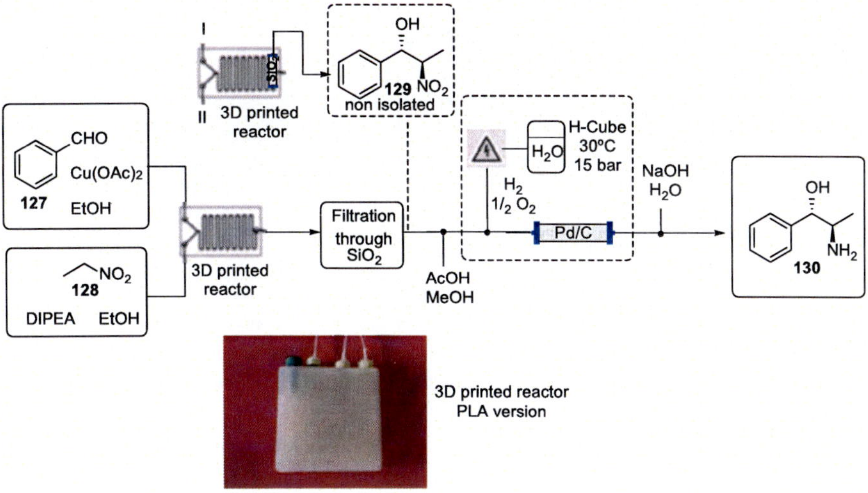

Fig. 41 Two-step flow synthesis of norephedrine (**130**) in a 3D printed reactor. Reproduced with permission of VCH (Ref. [119])

Flow reactors can be designed to accommodate specific reaction kinetics and thermodynamics, multicomponent reactions where each component can be added in a specific point during the reaction, and to design mixing chambers with specific geometries that are difficult to prepare by other methods. Some excellent reviews have been published in this area [116–118].

Benaglia et al. [119] described the stereoselective preparation of aminoalcohols (norephedrine (**130**), metaraminol (**131**), and methoxamine (**132**)) by a tandem Henry reaction followed by reduction. The authors used 3D printing with fused filament fabrication and different materials (PLA, HIPS, Nylon) to optimize the reactor size, shape, and channel dimensions (Fig. 41). High yields, moderate diastereoselectivities, and up to 90% ee were obtained with this system.

One limitation of 3D printing is the nature of the material used in the printing process, since its inertness is a prerequisite for any application. Photopolymers have poor resistance to the standard organic solvents; common thermoplastic polymers are sensitive to high temperatures and to acidic/basic conditions, while powder-based technologies require post-printing processes that can leave traces of unwanted materials in the printed device.

In this regard, Ferguson et al. described the use of PEEK (polyether ether ketone) as a high-performance plastic with excellent mechanical strength, a high melting point, and excellent chemical resistance. The printed reactors were used at pressures of 500 psi, high temperatures, and with superheated solvents [120].

Despite these limitations, many different materials could replace glass and silicon. Among the 3D printing processes only selective laser sintering (SLS), laminated object manufacturing (LOM), multiJet modeling (MJM), and fuse deposition modeling (FDM) have found application in this field.

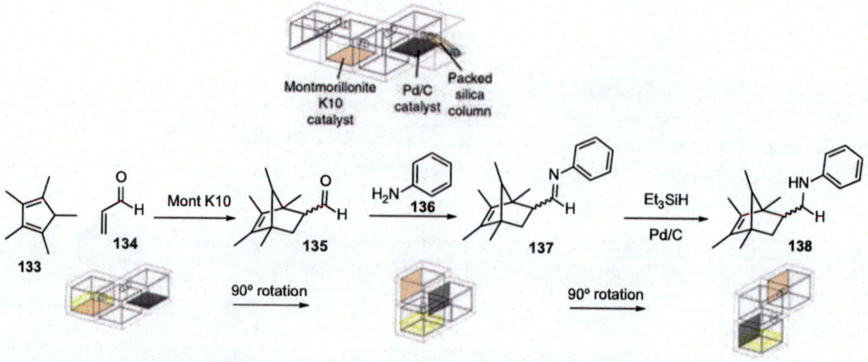

Fig. 42 3D printed reactionware for 1) Acid-catalyzed Diels–Alder reaction, 2) Imine formation, 3) Imine reduction. Catalyst, Montmorillonite K10, and Pd/C were printed in the reactor. Reproduced with permission of Springer (Ref. [115])

A very attractive application of 3D printing technology is undoubtedly the possibility of directly printing catalysts for chemical transformations.

Cronin et al. designed a multiple reactor for the sequence described in Fig. 42 [115]. Firstly, an acid-catalyzed Diels–Alder reaction of 1,2,3,4,5-pentamethylcyclopentadiene (**133**) with acrolein (**134**) was carried out to obtain a bicyclic compound **135**. In the second step the imine **137** was formed by reaction of the aldehyde function with aniline (**136**) and, thirdly, reduction of the imine with triethylsilane over Pd/C was carried out. Reactionware was designed with CAD and 3D printed with FDM. Catalyst, Montmorillonite K10, and Pd/C were also printed in the reactor. Once the Diels–Alder reaction was completed the reactor was rotated by 90° to react with the imine and again by 90° for the reduction reaction. At the end of the reaction it was possible to connect a silica gel cartridge to purify the final product.

3D printing is also a good option for the fabrication of droplet microfluidics flow systems. Complex and on-demand design structures can be fabricated in a simple and rapid manner [121]. 3D printing offers great freedom for the design and fabrication of devices for droplet formation and opens the door for new droplet manipulation strategies.

5.4 *Optimization*

Optimization is essential when developing new chemical reactions or processes with the desired yields, selectivity, and atom economy. In industrial processes a small increase in these parameters is of paramount importance to improve the economy of the process.

The objective is to obtain the highest possible yield and selectivity in the shortest time. These objectives are especially suited to microreactor technology. A general scheme of the optimization process is provided in Fig. 43.

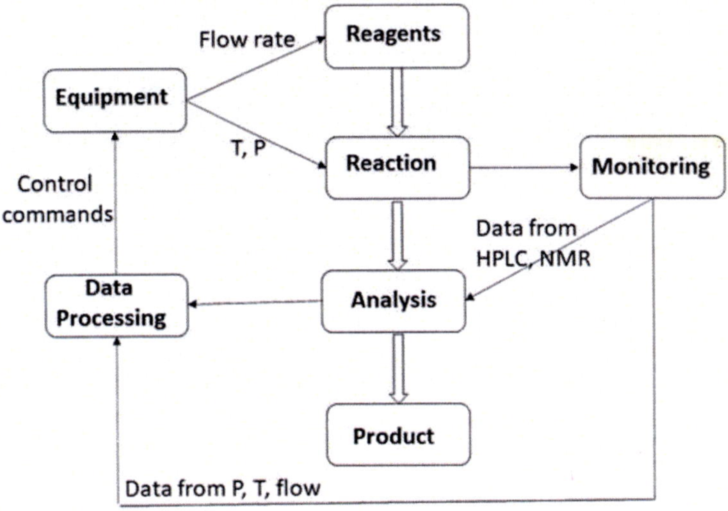

Fig. 43 General scheme of an optimization system including reaction, analysis, and data processing for automatic decision making

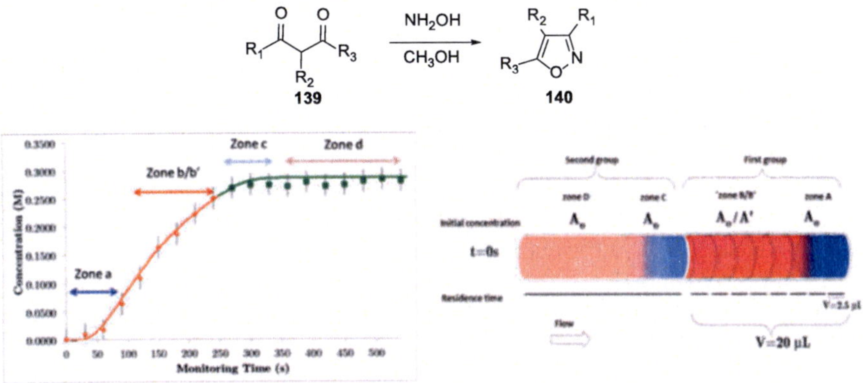

Fig. 44 On-line optimization of a series of isoxazoles with extraction of kinetic data

The first approach is the modification of one variable at a time. This approach requires the realization of one experiment for each variation when performing reactions in batch. However, in flow conditions variations can be done in the flow stream, especially if the results are analyzed in-line. In this way, optimization can be performed in a short time and with low consumption of solvents and reagents.

This approach has been described by Rodriguez et al. for the preparation of a series of isoxazoles. The experiments permitted the optimization of the procedure, the preparation of a library of isoxazoles **140**, the 100-fold scale-up of the procedure, and the extraction of kinetic data (Fig. 44) [122].

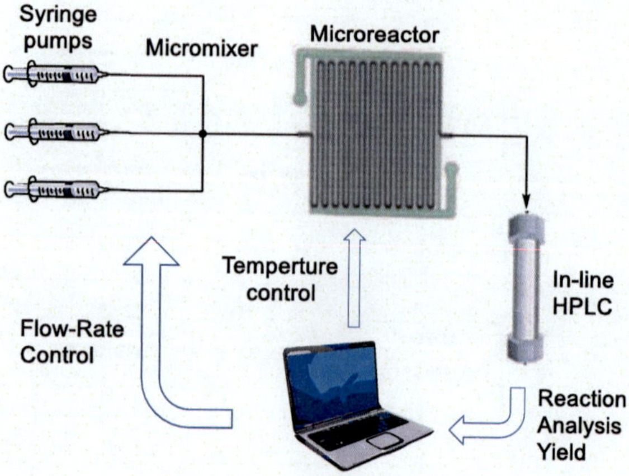

Fig. 45 Self-optimization system as described by McMullen et al. [124]

However, to achieve greater accuracy and efficiency in finding the optimal conditions, several designs and algorithms have been applied to optimization in continuous flow.

Semi-automated methods have been used for high-throughput screening and optimization. Design of Experiments (DoE) is a statistical approach in which a diverse set of experiments are selected before execution [123].

In DoE a matrix is created with the experimental parameters and the algorithm selects a series of experiments to cover the maximum parameter space. The results of the experiments are evaluated to design a model that can be confirmed or refined with a new set of experiments.

Another approach is the residual surface model (RSM), which simultaneously uses a symmetrical experimental design to optimize the reaction parameters.

In general, the optimal conditions found can even be translated to preparative scale production, with only minor adjustments required.

Incorporation of in-line and on-line analysis attached to computer-controlled algorithms has permitted the design of fully automatic systems. The combination with continuous production allows a rapid optimization that reduces waste, cost, and time. Several algorithms have been used for these optimizations and these include Steepest descent, Nelder–Mead simplex (NMS), and SNOBFIT [123].

As an example, McMullen et al. [124] designed a microreactor system for the self-optimization of a Heck reactor (Fig. 45). In this case, HPLC was used for analysis and NMS was used for optimization.

Cronin [125] and Mateos [8] published excellent reviews that cover all aspects of automatization and optimization from monitoring and algorithms to autonomous platforms.

5.5 *Process Intensification and Scale-Up*

Process intensification is described as a strategy to achieve tremendous reductions in plant size for a given plant volume. Continuous flow systems have also been exploited as tools to effect process intensification in chemical synthesis processes [126].

Continuous processing and process intensification have been designated as key green engineering research areas for sustainable manufacturing according to a 2011 roundtable study [127].

Process intensification is one of the main green aspects of flow chemistry aimed at reducing waste. Waste prevention for large-scale manufacturing in flow may improve the efficiency of mixing, reaction, separation, and purification modules over normal batch operations [126]. One main technique for increasing efficiency is through the implementation of recycling of solvents. Another example is liquid–liquid extractions that can be enhanced to batch extractions by running the extraction solvent counter-current to the process several times or by application of separation membranes [23].

Flow processes are ideal for rapid screening of the reaction conditions since many parameters can be evaluated in a short time and with low product consumption. Reaction scale in flow is easier than in batch and most of the problems encountered can be solved, e.g., heat and mass transfer or by-product formation. Moreover, this approach provides enhancements in safety and energy efficiency, and allows the possibility of developing small-scale manufacturing plants. However, many advantages of flow reactions are a consequence of the small dimensions inherent in the system scale and these can be lost on larger scales.

Flow reactors can be divided into three types: micro-, milli-, and macro-scale reactors [128].

Micro-scale reactors with microchannel diameter scales of hundreds of microns show short diffusion and excellent heat and mass transfer and they can produce products on the scale of kg/yr. These systems are suited to the synthesis of APIs, screening, and optimization. Milli-scale reactors, with diameter scales of 1–2 mm, are suited to the preparation of fine and speciality chemicals since they can produce hundreds of kg/yr. Macro-scale reactors have diameter scales greater than 5 mm and are capable of preparing products at industrial scales of ton/yr.

The performance across reaction scales is maintained if mass (Sherwood, Schmidt, Peclet, and Biot numbers), heat transfer (Nusselt, Prandtl, and Biot numbers), and mixing and reaction rates (Reynolds, Damköhler, and Thiele numbers) are similar.

Three different approaches can be used for scale-up in flow: [7]

- Scaling-out involves running the process for a longer time. This is the simplest method since none of the reaction parameters are changed. However, it is difficult to obtain the large quantities of product required on an industrial scale by this methodology.

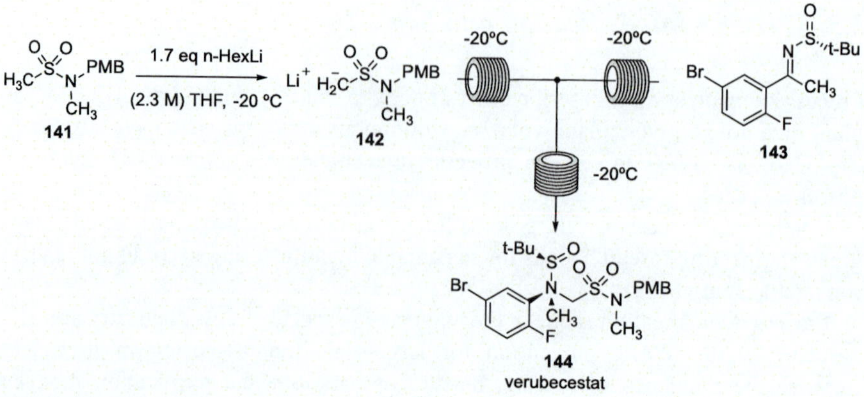

Fig. 46 Pilot plant preparation of verubecestat (**144**)

– Numbering-up involves running several reactions in parallel. Several channels or reactors are placed in parallel so that the reactions can be performed under identical conditions. Parallel numbering-up is the most common method in the scale-up of microreactors. However, this approach requires complex fluid flow distribution and control and is associated with extra energy consumption and may lower the reactor flow rate operating window.

Another possibility is consecutive numbering-up, in which several reactors or channels are used in series, with one reactant kept at a high flow rate and the second added stepwise in series. With this technique, care should be taken to ensure longer channels in downstream segments to ensure a consistent residence time when flow rates are increased [129].

– Scaling-up involves the use of larger continuous reactors. In this case it is very important to retain the advantages of the small scale, i.e., heat and mass transfer. Another advantage of this approach is that the possibility of clogging decreases.

An intermediate possibility is to increase the reactor scale in one dimension while maintaining the scale in the second dimension [129].

Some of the most important scale-up processes have been covered in several excellent reviews [129–131].

It is worth highlighting the preparation of verubecestat (**144**), an inhibitor of BACE1, at Pilot Plant Scale (>100 kg) in flow conditions [132]. A methodology was developed to evaluate the relationship between mass flow rate, temperature, and conversion to introduce modifications to enhance the process (Fig. 46).

The development of a kilogram-scale prexasertib monolactate monohydrate (**156**) synthesis under continuous flow CGMP conditions is also noteworthy. A total of 24 kg of the product was produced in CGMP conditions [133]. The benefits of CM included the ability to operate at high temperature in a low-boiling solvent, improved

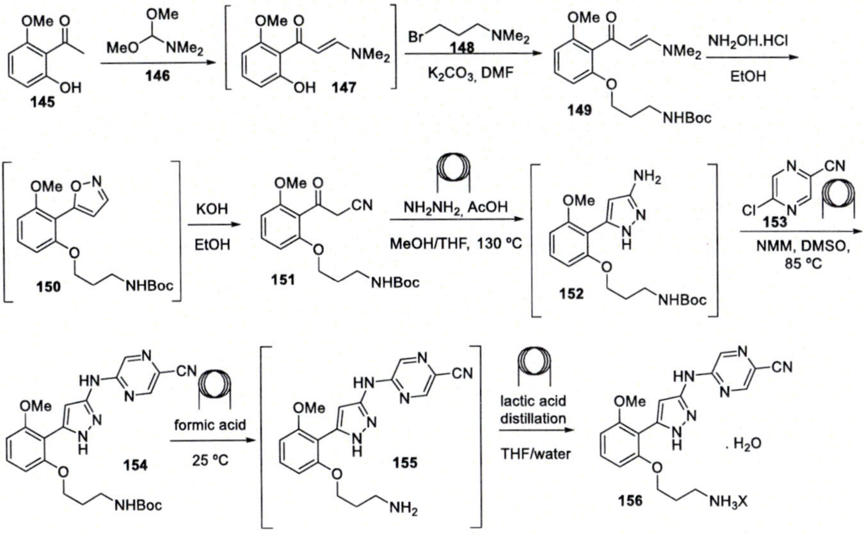

Fig. 47 Continuous kilogram-scale synthesis of prexasertib monolactate monohydrate (**156**)

safety for a hazardous reaction, better yield, improved purification procedures, and on-line PAT and process automation (Fig. 47).

6 Quantification of Sustainability (LCA)

Flow chemistry has always been considered a green methodology. It is presumed that this technique has a lower energy consumption due to the higher heat transfer, a reduced impact of solvents by replacement with green solvents, a reduction of solvent consumption, and the possibility of solvent recycling [12]. Finally, due to automatization there is a reduction in personnel costs and increased safety [8].

Quantification of sustainability is of paramount importance when translating a synthesis to the industrial scale. Several parameters have been used to evaluate this aspect. Atom economy (AE) (Eq. ch. 1) and E-factor (Eq. ch. 2) were the first parameters used but they are very limited since they do not consider many factors, especially toxicity of reagents and products and solvent consumption.

Process Mass Intensity (PMI) (Eq. 3) is defined as the mass of raw materials as a function of the mass of products and this is a key mass-based green metric. Solvent rate (Eq. 4), defined as the mass of solvent as a function of the mass of products, is another quantitative parameter to be considered. Finally, cumulative energy demand (CED) (Eq. 5) represents the energy demand over the entire process and considers both renewable and non-renewable energy [134].

$$\mathrm{AE} = \frac{\text{MW desired products}}{\text{MW of all products}} \tag{1}$$

$$\mathrm{E-factor} = \frac{\text{Weight of redisues}}{\text{weight of products}} \tag{2}$$

$$\mathrm{PMI} = \frac{\text{Mass of row materials}}{\text{Mass of products}} \tag{3}$$

$$\text{Solvent rate} = \frac{\text{Mass of solvent}}{\text{Mass of product}} \tag{4}$$

$$\mathrm{CED} = \frac{\text{Energy resource consumption (MJ)}}{\text{Mass of product (Kg)}} \tag{5}$$

LCA is an analysis of the impact of a chemical process from "cradle-to-grave" or more appropriately "cradle-to-gate." LCA is defined as "a technique for assessing the environmental aspects and potential impacts associated with a product by compiling an inventory of relevant inputs and outputs of a system; evaluating the potential environmental impacts associated with those inputs and outputs; and interpreting the results of the inventory and impact phases in relation to the objectives of the study" by the International Standards Organization (ISO) (ISO 14040, 1997).

The life cycle impact assessment (LCIA) includes the following impact categories to be evaluated in a Life Cycle Inventory (LCI): cumulative energy demand (CED), abiotic resource depletion potential (ADRP), global warming potential (GWP), ozone depletion potential (ODP), photochemical ozone creation potential (POCP), acidification potential (AP), and nutrification potential (NP). Additionally, an assessment of the human toxicity potential (HTP), fresh aquatic eco-toxicity potential (FAETP), marine aquatic eco-toxicity potential (MAETP), and the terrestrial eco-toxicity potential (TETP) was taken into account [135].

Eckelman evaluated the life cycle inherent toxicity and introduced the metrics of inherent hazard in chemical synthesis [136].

Lapkin and Yaseneva collected some important examples of LCA in flow conditions [9].

Lapkin et al. described the cradle-to-gate analysis of the antimalarial API artemether (**157**) (Fig. 48) [12]. The study emphasized the importance of solvents and the advantage of flow technology, which enabled small solvent inventories to be used. The most dominant impact in flow conditions is the solvent. In this case THF was replaced by 2-MeTHF, a biomass-derived and recyclable solvent, and Quadrasil-SA was used as catalyst. This produced a reduction of the CED.

The same authors described the LCA of a continuous flow Buchwald–Hartwig amination of a pharmaceutical intermediate (**160**) and the translation to a pilot plant [137]. The study confirmed the benefits of flow with respect to batch conditions. The consumption of solvent is higher in flow due to the higher dilution. However, in the pilot plant under flow conditions the concentrations are higher.

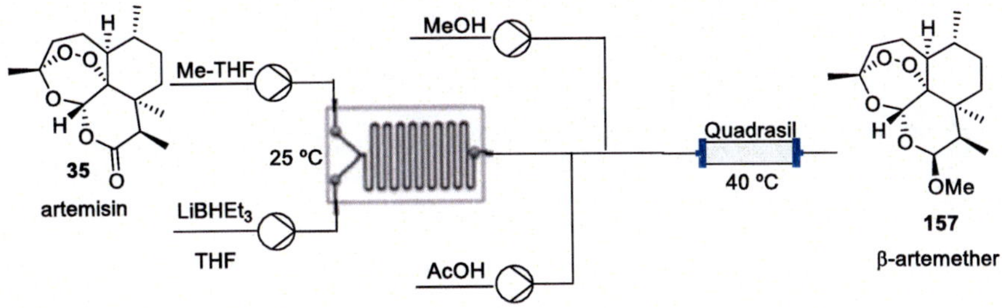

Fig. 48 Transformation of artemisin into β-artemether

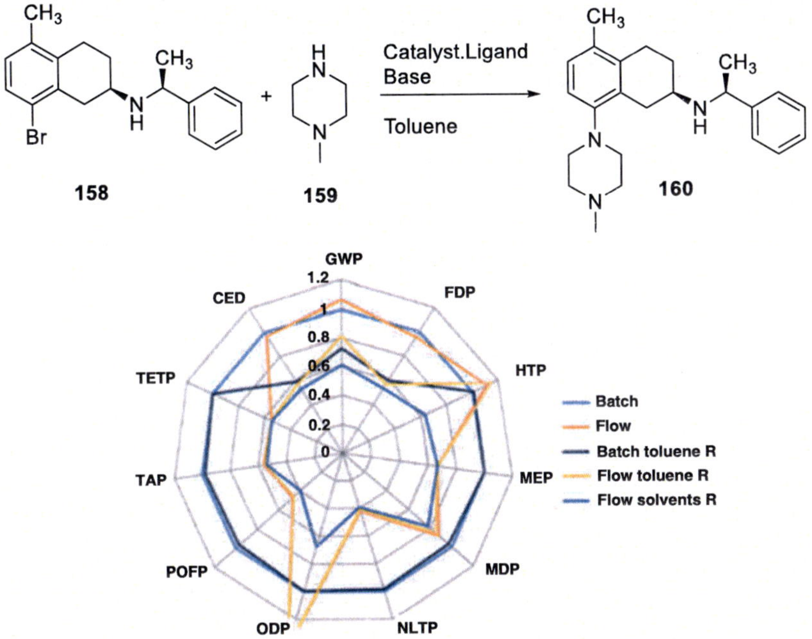

Fig. 49 LCA of a Buchwald–Hartwig amination of a pharmaceutical intermediate. GWP, Climate change. FDP, Fossil fuel depletion potential. HTP, Human toxicity potential. MEP, Marine eutrophication potential. MDP, Metal depletion potential. NLTP, Natural land transformation potential. ODP, Ozone depletion potential. POFP, Photochemical oxidant formation potential. TAP, Terrestrial acidification potential. TETP, Terrestrial eco-toxicity potential. CED, Cumulative energy demand. R, recycling. Published by The Royal Society of Chemistry (Ref. [137])

It was emphasized that the overall flow process with catalyst recycling has lower or comparable environmental impact scores for almost all impact categories (Fig. 49).

Hessel et al. reported the LCA of the photochemical synthesis of Vitamin D_3 (**163**) in flow conditions (Fig. 50) [138]. They used a cradle-to-gate approach that considered the materials, energy flows, and emission associated with the extraction

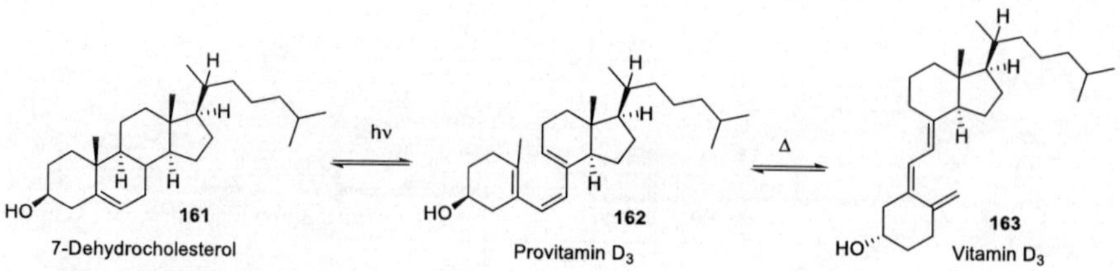

Fig. 50 Preparation of vitamin D_3 (**163**) from 7-DHC (**161**)

of the raw materials to the product. The authors included the formation of 7DHC (**161**), extraction of Vitamin D_3 (**163**), and purification by crystallization. The solvent selected was *tert*-butyl methyl ether (t-BME) instead of diethyl ether in order to avoid the potential formation of dangerous peroxides.

HUMBERTO NXT LCA software was used to obtain the environmental impacts and mass and energy balances. The results show that the continuous process has a lower environmental impact than any of the batch scenarios evaluated and that solvents are dominant in the LCA.

7 Conclusions

Flow chemistry can be classified as a green enabling technology due to the high impact shown on almost all of the green chemistry principles. In this chapter we have collected examples that emphasize these aspects: reduction of waste by the use of telescoped syntheses, minimization of purification steps and the use of solvent recycling; security in the use of hazardous compounds and the application of automatization and artificial intelligence; use of neoteric solvents, some of which are derived from biomass; higher synthesis efficiency by increased heat and mass transfer, which in turn reduces reaction times, the synergy with other enabling technologies and catalysis and the simplicity of optimization, screening and scale-up taking advantage of real-time analysis, which is especially suitable in flow conditions. Finally, the quantification of sustainability by LCA has demonstrated the impact of solvents on sustainability and the importance of recycling. Flow chemistry has shown a lower impact on almost all of the parameters studied.

As considered by IUPAC, flow chemistry is one of the emerging technologies with potential to make our planet more sustainable and should of great importance in the pharmaceutical industry.

Compliance with Ethical Standards

Ethical Approval: This manuscript is a review of previously published accounts, as such, no animal or human studies were performed.

Informed Consent: No patients were studied in this chapter.

References

1. Gomollón-Bel F (2019) Chem Int:12–17
2. Sharma S, Das J, Braje WM, Dash AK, Handa S (2020) ChemSusChem 13:2859–2875
3. Lummiss JAM, Morse PD, Beingessner RL, Jamison TF (2017) Chem Rec 17:667–680
4. Brandao P, Pineiro M, Pinho e Melo TMVD (2019) Eur J Org Chem:7188–7217
5. US Food and Drug Amministration. https://www.fda.gov/drugs/news-events-human-drugs/modernizing-way-drugs-are-made-transition-continuous-manufacturing. Accessed 15 Jul 2020
6. Anastas PT, Warner JC (1998) Green chemistry: theory and practice. Oxford University Press, Oxford
7. Porta R, Benaglia M, Puglisi A (2016) Org Process Res Dev 20:2–25
8. Mateos C, Nieves-Remacha MJ, Rincón JA (2019) Flow React Chem Eng 4:1536–1544
9. Lapkin AA, Yaseneva P (2017) Life cycle assessment of flow chemistry processes. In: Vaccaro L (ed) Sustainable flow chemistry. Wiley-VCH, Weinheim, pp 249–276
10. Koenig SG, Leahy DK, Wells AS (2018) Org Process Res Dev 22:1344–1359
11. Vaccaro L (2020) Eur J Org Chem:4273–4283
12. Yaseneva P, Plaza D, Fan X, Loponov K, Lapkin A (2015) Catal Today 239:90–96
13. Prat D, Hayler J, Wells A (2014) Green Chem 16:4546–4551
14. https://www.acs.org/content/acs/en/greenchemistry/research-innovation/tools-for-green-chemistry/solvent-selection-tool.html. Accessed 7 Jul 2020
15. Ley SV (2012) Chem Rec 12:378–390
16. Wong H, Cernak T (2018) Curr Opin Green Sus Chem 11:91–98
17. Newman SG, Jensen KF (2013) Green Chem 15:1456–1472
18. Hall JFB, Bourne RA, Han X, Earley JH, Poliakoff M, George MW (2013) Green Chem 15:177–180
19. Britton J, Majumdar S, Weiss GA (2018) Chem Soc Rev 47:5891–5918
20. Burguete MI, García-Verdugo E, Luis SV (2011) Beilstein J Org Chem 7:1347–1359
21. Kristofkov D, Modrock V, Meciarov M, Sebesta R (2020) ChemSusChem 13:2828–2858
22. Wasserscheid P, Welton T (2008) Ionic liquids in synthesis. Wiley-VCH, Weinheim
23. Lozano P, Alvarez E, Bernal JM, Nieto S, Gómez C, Sanchez-Gomez G (2017) Curr Green Chem 4:116–129
24. Reetz MT, Wiesenhöfer W, Franciò G, Leitner W (2002) Chem Commun:992–993
25. Lozano P, de Diego T, Carrié D, Vaultier M, Iborra JL (2002) Chem Commun:692–693
26. Stephenson P, Kondor B, Licence P, Scovell K, Ross SK, Poliakoff M (2006) Adv Synth Catal 348:1605–1610
27. Lozano P, García-Verdugo E, Piamtongkam R, Karbass N, De Diego T, Burguete MI, Luis SV, Iborra JL (2007) Adv Synth Catal 349:1077–1084
28. Villa R, Alvarez E, Porcar R, Garcia-Verdugo E, Luis SV, Lozano P (2019) Green Chem 21:6527–6544
29. Zamani P, Khosropour AR (2016) Green Chem 18:6450–6455
30. Grabner B, Schweiger AK, Gavric K, Kourist R, Gruber-Woelfler H (2020) React Chem Eng 5:263–269
31. Guajardo N, Schrebler RA, Domínguez de María P (2019) Bioresour Technol 273:320–325
32. Iemhoff A, Sherwood J, McElroy CR, Hunt AJ (2018) Green Chem 20:136–140
33. Ferlin F, Luque Navarro PM, Gu Y, Lanari D, Vaccaro L (2020) Green Chem 22:397–403
34. Han S, Raghuvanshi K, Abolhasani M (2020) ACS Sustain Chem Eng 8:3347–3356
35. Thompson CM, Poole JL, Cross JL, Akritopoulou-Zanze I, Djuric SW (2011) Molecules 16:9161–9177
36. Vasudevan A, Bogdan AR, Koolman HF, Wang Y, Djuric SW (2017) Prog Med Chem 56:1–35
37. Hughes DL (2020) Org Process Res Dev 24:1850–1860
38. Rehm TH (2020) Chem Eur J 71:16952–16974

39. Thomson CG, Lee A-L, Vilela F (2020) Beilstein J Org Chem 16:1495–1549
40. Hommelsheim R, Guo Y, Yang Z, Empel C, Koenigs RM (2019) Angew Chem Int Ed 58:1203–1207
41. Oelgemöller M (2016) Chem Rev 116:9664–9682
42. Cambié D, Dobbelaar J, Riente P, Vanderspikken J, Shen C, Seeberger PH, Gilmore K, Debije MG, Noël T (2019) Angew Chem Int Ed 58:14374–14378
43. Lévesque F, Seeberger PH (2012) Angew Chem Int Ed 51:1706–1709
44. Zuo Z, Ahneman DT, Chu L, Terret JA, Doyle AG, MacMillan DWC (2014) Science 345:437–440
45. Abdiaj I, Alcázar J (2017) Bioorg Med Chem 25:6190–6196
46. Abdiaj I, Huck L, Mateo JM, de la Hoz A, Gómez MV, Díaz-Ortiz A, Alcázar J (2018) Angew Chem Int Ed 57:13231–13236
47. Yan M, Kawamata Y, Baran PS (2018) Angew Chem Int Ed 57:4149–4155
48. Pletcher D, Green RA, Brown RCD (2018) Chem Rev 118:4573–4591
49. Elsherbini M, Wirth T (2019) Acc Chem Res 52:3287–3296
50. Bärfacker L, Kuhl A, Hillisch A, Grosser R, Figueroa-Pérez S, Heckroth H, Nitsche A, Ergüden J-K, Gielen-Haertwig H, Schlemmer K-H, Mittendorf J, Paulsen H, Platzek J, Kolkhof P (2012) ChemMedChem 7:1385–1403
51. Platzek KJ, Gottfried K, Assmann J, Lolli G (2017) WO 2017/032678 A1, March 2
52. Elsherbini M, Wirth T (2018) Chem Eur J 24:13399–13407
53. Francke R (2019) Curr Opin Electrochem 15:83–88
54. Elsherbini M, Winterson B, Alharbi H, Folgueiras-Amador AA, Génot C, Wirth T (2019) Angew Chem Int Ed 58:9811–9815
55. Qian X-Y, Li S-Q, Song J, Xu H-C (2017) ACS Catal 7:2730–2734
56. Folgueiras-Amador AA, Qian X-Y, Xu H-C, Wirth T (2018) Chem Eur J 24:487–491
57. Laudadio G, Barmpoutsis E, Schotten C, Struik L, Govaerts S, Browne DL, Noël T (2019) J Am Chem Soc 141:5664–5668
58. Tanaka K, Yoshizawa H, Atobe M (2019) Synlett 30:1194–1198
59. Tamborini L, Fernandes P, Paradisi F, Molinari F (2018) Trend Biotechnol 36:73–88
60. Zheng MM, Wang RF, Li CX, Xu JH (2015) Process Biochem 50:598–604
61. Zheng MM, Chen FF, Li H, Li CX, Xu JH (2018) Chembiochem 19:347–353
62. De Vitis V, Dall'Oglio F, Pinto A, De Micheli C, Molinari F, Conti P, Romano D, Tamborini L (2017) Chem Open 6:668–673
63. de la Hoz A, Loupy A (2012) Microwaves in organic chemistry.2nd edn. Wiley-VCH Verlag GmbH, Weinheim
64. Glasnov TN, Kappe CO (2011) Chem Eur J 17:11956–11968
65. de la Hoz A, Díaz-Ortiz A (2017) Vaccaro L (ed) Sustainable flow chemistry: methods and applications. Wiley-VCH, Weinheim, pp 219–248
66. Bálint E, Tajti Á, Keglevich G (2019) Materials 12:788
67. Fernandes RA, Chowdhury AK, Kattanguru P (2014) Eur J Org Chem 2014:2833–2871
68. Egami H, Tamaoki S, Abe M, Ohneda N, Yoshimura T, Okamoto T, Odajima H, Mase N, Takeda K, Hamashima Y (2018) Org Process Res Dev 22:1029–1033
69. Musio B, Mariani F, Śliwiński EP, Kabeshov MA, Odajima H, Ley SV (2016) Synthesis 48:3515–3526
70. Calcio Gaudino E, Manzoli M, Carnaroglio D, Wu Z, Grillo G, Rotolo L, Medlock J, Bonrath W, Cravotto G (2018) RSC Adv 8:7029–7039
71. Tagliapietra S, Calcio Gaudino E, Martina K, Barge A, Cravotto G (2019) Chem Rec 19:98–117
72. Christiaens S, Vantyghem X, Radoiu M, Vanden Eynde JJ (2014) Molecules 19:9986–9998
73. He W, Fang Z, Zhang K, Tu T, Lv N, Qiu C, Guo K (2018) Chem Eng J 331:161–168
74. Jin Y, Yang J, Feng X, Li J, Xu J, Chen X, Wang S, Lv Y, Yu J (2020) J Flow Chem 10:369–376
75. Baumann M, Moody TS, Smyth M, Wharry S (2020) Org Process Res Dev 24:1802–1813

76. O'Mahony RM, Lynch D, Hayes HLD, Ní Thuama E, Donnellan P, Jones RC, Glennon B, Collins SG, Maguire AR (2017) Eur J Org Chem:6533–6539
77. Carpentier F, Felpin F-X, Zammattio F, Le Grognec E (2020) Org Process Res Dev 24:752–761
78. Mata A, Weigl U, Flögel O, Baur P, Hone CA, Kappe CO (2020) React Chem Eng 5:645–650
79. Koo H, Kim HY, Oh K (2019) Org Lett 21:10063–10068
80. Sagandira CR, Watts P (2019) Beilstein J Org Chem 15:2577–2589
81. Lehmann H (2017) Green Chem 19:1449–1453
82. Yang H, Martin B, Schenkel B (2018) Org Process Res Dev 22:446–456
83. Pieber B, Kappe CO (2016) Org Lett 18:1076–1079
84. Yu T, Jiao J, Song P, Nie W, Yi C, Zhang Q, Li P (2020) ChemSusChem 13:2876–2893
85. Gardiner J, Nguyen X, Genet C, Horne MD, Hornung CH, Tsanaktsidis J (2018) Org Process Res Dev 22:1448–1452
86. Rahmana T, Wharry S, Smyth M, Manyar H, Moody TS (2020) Synlett 31:581–586
87. Saito Y, Ishitani H, Ueno M, Kobayashi S (2017) Chem Open 6:211–215
88. Hansen SVF, Wilson ZE, Ulven T, Ley SV (2016) React Chem Eng 1:280–287
89. Mata A, Hone CA, Gutmann B, Moens L, Kappe CO (2019) ChemCatChem 11:997–1001
90. Chen Y, Hone CA, Gutmann B, Kappe CO (2017) Org Process Res Dev 21:1080–1087
91. Hone CA, Lopatka P, Munday R, O'Kearney-McMullan A, Kappe CO (2019) ChemSusChem 12:326–337
92. Micic N, Polyzos A (2018) Org Lett 20:4663–4666
93. Osako T, Kaiser R, Torii K, Uozumi Y (2019) Synlett 30:961–966
94. Dallinger D, Gutmann B, Kappe CO (2020) Acc Chem Res 53(7):1330–1341
95. Johansen MB, Lindhardt AT (2018) Chem Commun 54:825–828
96. Hirano K, Gondo S, Punna N, Tokunaga E, Shibata N (2019) Chem Open 8:406–410
97. Cantillo D, Wolf B, Goetz R, Kappe CO (2017) Org Process Res Dev 21:125–132
98. Mata A, Cantillo D, Kappe CO (2017) Eur J Org Chem 2017:6505–6510
99. Sagmeister P, Williams JD, Hone CA, Kappe CO (2019) React Chem Eng 4:1571–1578
100. Brodmann T, Koos P, Metzger A, Knochel P, Ley SV (2012) Org Process Res Dev 16:1102–1113
101. Zapata F, Ortega-Ojeda F, García-Ruiz C, González-Herráez M (2018) Sensors 18:2196–2205
102. Yao X, Deng Q, Wang S, Wang W, Hou Y, Gao Z, Wu Y, Guo Z (2019) Chem Select 4:5116–5121
103. Shokoufi N, Vosough M, Rahimzadegan-Asl M, Abbasi-Ahd A, Khatibeghdami M (2020) Int J Anal Chem:2921417
104. Gomez MV, de la Hoz A (2017) Beilstein J Org Chem 13:285–300
105. Jacquemmoz C, Giraudb F, Dumez J-N (2020) Analyst 145:478–485
106. Wang Y, Lin W-Y, Liu K, Lin RJ, Selke M, Kolb HC, Zhang N, Zhao X-Z, Phelps ME, Shen CKF, Faull KF, Tseng H-R (2009) Lab Chip 9:2281–2285
107. Weeranoppanant N, Adamo A (2020) ACS Med Chem Lett 11:9–15
108. Shukla CA, Kulkarni AA (2017) Beilstein J Org Chem 13:960–987
109. Trobe M, Burke MD (2018) Angew Chem Int Ed 57:4192–4214
110. Adamo A, Beingessner RL, Behnam M, Chen J, Jamison TF, Jensen KF, Monbaliu J-CM, Myerson AS, Revalor EM, Snead DR, Stelzer T, Weeranoppanant N, Wong SY, Zhang P (2016) Science 352:61–67
111. Lehmann JW, Blair DJ, Burke MD (2018) Nat Rev 2:0115
112. Li J, Ballmer SG, Gillis EP, Fujii S, Schmidt MJ, Palazzolo AME, Lehmann JW, Morehouse GF, Burke MD (2015) Science 347:1221–1126
113. Reischl G, Ehrlichmann W, Bieg C, Solbach C, Kumar P, Wiebe LI, Machulla H-J (2005) Appl Radiat Isot 62:897–901
114. Parry DM (2019) ACS Med Chem Lett 10:848–856
115. Kitson PJ, Glatzel S, Chen W, Lin C-G, Song Y-F, Cronin L (2016) Nat Protoc 11:920
116. Rossi S, Puglisi A, Benaglia M (2018) ChemCatChem 10:1512–1525

117. Parra-Cabrera C, Achille C, Kuhn S, Ameloot R (2018) Chem Soc Rev 47:209
118. Hartings MR, Ahmed Z (2019) Nat Rev Chem 3:305–314
119. Rossi S, Porta R, Brenna D, Puglisi A, Benaglia M (2017) Angew Chem Int Ed 56:4290–4294
120. Harding MJ, Brady S, O'Connor H, Lopez-Rodriguez R, Edwards MD, Tracy S, Dowling D, Gibson G, Girardg KP, Ferguson S (2020) Chem Eng 5:728–735
121. Zhang JM, Ji Q, Duan H (2019) Micromachines 10:754–777
122. Rodriguez AM, Juan A, Gomez MV, Moreno A, de la Hoz A (2012) Synthesis 44:2527–2530
123. Lenstra DC, Rutjes FRJT (2017) Organic synthesis in flow: toward higher levels of sustainability. In: Vaccaro L (ed) Sustainable flow chemistry. Wiley-VCH, Weinheim, pp 103–134
124. McMullen JP, Stone MT, Buchwald SL, Jensen KF (2010) Angew Chem Int Ed 49:7076–7080
125. Sans V, Cronin L (2016) Chem Soc Rev 45:2032–2043
126. Akwi FM, Watts P (2018) Chem Commun 54:13894–13928
127. Kilcher E, Freymond S, Vanoli E, Marti R, Schmidt G, Abele S (2016) Org Process Res Dev 20:432–439
128. Bennett JA, Campbell ZS, Abolhasani M (2019) Curr Opin Chem Eng 26:9–19
129. Zhang J, Wang K, Teixeira AR, Jensen KF, Luo G (2017) Annu Rev Chem Biomol Eng 8:285–305
130. Rossetti I, Compagnoni M (2016) Chem Eng J 296:56–70
131. Pechlauer P, Skranc W (2017) Scale-up processes in the pharmaceutical industry. In: Vaccaro L (ed) Sustainable flow chemistry. Wiley-VCH, Weinheim, pp 73–94
132. Thaisrivongs DA, Naber JR, Rogus NJ, Spencer G (2018) Org Process Res Dev 22:403–408
133. Cole KP, Groh JM, Johnson MD, Burcham CL, Campbell BM, Diseroad WD, Heller MR, Howell JR, Kallman NJ, Koenig TM, May SA, Miller RD, Mitchell D, Myers DP, Myers SS, Phillips JL, Polster CS, White TD, Cashman J, Hurley D, Moylan R, Sheehan P, Spencer RD, Desmond K, Desmond P, Gowran O (2017) Science 356:1144–1150
134. Ott D, Borukhova S, Hessel V (2016) Green Chem 18:1096–1116
135. Kralisch D, Kreisel G (2007) Chem Eng Sci 62:1094–1100
136. Eckelman MJ (2016) Green Chem 18:3257–3264
137. Yaseneva P, Hodgson P, Zakrzewski J, Falß S, Meadows RE, Lapkin AA (2016) React Chem Eng 1:229–238
138. Morales-Gonzalez OM, Escribà-Gelonch M, Hessel V (2019) Int J Life Cycle Assess 24:2111–2127

Top Med Chem (2021) 38: 71–120
https://doi.org/10.1007/7355_2021_112

Published online: 13 February 2021

Photochemistry in Flow for Drug Discovery

Thomas H. Rehm

Contents

Abstract The discovery of new drug candidates and the development of new lead structures for future active pharmaceutical ingredients are continuous processes, which combine the expertise of many scientific disciplines. The race for novel molecular building blocks with potent biological activity triggered the development of new synthesis methodologies and forced the scientific communities to interact even stronger with each other. The result of one very fruitful interaction is the application of continuous flow chemistry and micro reaction technology to photochemistry and photocatalysis. The synergy of those research fields combined is an environment for mild and controlled reaction conditions, which allows new synthesis routes with higher conversion and selectivity, straightforward scale-up and possible integration in multi-step syntheses. In this chapter, several examples for molecular transformations are highlighted, which are important for the synthesis of complex molecular structures or for the integration of pharmaceutically active functional groups. This snapshot gives an overview of a vivid research field for

T. H. Rehm (✉)
Division Chemistry, Sustainable Chemical Syntheses Group, Fraunhofer Institute for Microengineering and Microsystems IMM, Mainz, Germany
e-mail: thomas.rehm@imm.fraunhofer.de; https://www.imm.fraunhofer.de; https://www.flowphotochemistry.com

drug discovery and illustrates the benefits for synthetic organic photochemistry by going to flow.

Keywords Continuous flow, Fine chemicals, LED, Microreactor, Process intensification, Reactive intermediates, Small molecules

Abbreviations

2-MeTHF	2-Methyltetrahydrofuran
4CzIPN	1,2,3,5-Tetrakis(carbazol-9-yl)-4,6-dicyanobenzene
4-DPA-IPN	2,4,5,6-Tetrakis(diphenylamino)isophthalonitrile
4-HTP	4-Hydroxythiophenol
acac	Acetylacetonate
ADP	Adenosine diphosphate
API	Active pharmaceutical ingredient
B_2pin_2	Bis(pinacolato)diboron
Boc	Tert-butyloxycarbonyl
bpy	Bipyridine
CFD	Computational fluid dynamics
CFL	Compact fluorescent lamp
cod	1,5-Cyclooctadien
Cp	Cyclopentadienyl
DABCO	1,4-Diazabicyclo[2.2.2]octane
dba	Dibenzylideneacetone
DBU	1,8-Diazabicyclo[5.4.0]undec-7-ene
DCE	Dichloroethane
$dF(CF_3)ppy_3$	3,5-Difluoro-2-[5-(trifluoromethyl)-2-pyridinyl-N]phenyl-C
DIPEA	Diisopropylethylamine
DMA	*N,N*-dimethylacetamide
DMAP	4-Dimethylaminopyridine
dme	Dimethoxyethane
DMF	*N,N*-dimethylformamide
dmppy	Dimethylphenylpyridine
DMSO	Dimethylsulfoxid
dtbbpy	4,4′-Di-tert-butyl-2,2′-dipyridyl
eq	Equivalent
FEP	Fluorinated ethylene propylene
GABA	γ-Aminobutyric acid
HALEX	HALogen EXchange (reaction)
HAT	Hydrogen atom transfer (reaction)
HFIP	1,1,1,3,3,3-Hexafluoro-propan-2-ol
$h\nu$	Photon energy
I.D.	Inner diameter
JohnPhos	(2-Biphenyl)di-tert-butylphosphine

KRED	Keto reductase
LC-MS	Liquid chromatography-mass spectrometry
LED	Light emitting diode
MAO	Monoamine oxidase
m-CPBA	Meta-chloroperbenzoic acid
Mes-Acr-4	9-Mesityl-10-methylacridinium tetrafluoroborate
NADPH/NADP+	Nicotinamide adenine dinucleotide
NaDT	Sodium decatungstate
NFSI	N-fluorobenzenesulfonimide
NHC	N-heterocyclic carbenes
NMR	Nuclear magnetic resonance
PDMS	Polydimethylsiloxane
PET	Positron emission tomography
PFA	Perfluoroalkoxy alkane
ppy	2-Phenylpyridine
r.t.	Room temperature
rAaeUPO	A. Aegerita unspecified peroxygenase
RCY	Radiochemical yield
SCE	Saturated calomel electrode
SLAP	Silicone amine protocol
sppy	Sulphonated 2-phenylpyridine
TBADT	Tetrabutylammonium decatungstate
TBAF	Tetrabutylammonium fluoride
TFA	Trifluoroacetic acid
THF	Tetrahydrofuran
TMDAM	*N,N,N′,N′*-tetramethyldiaminomethane
TMS	Trimethylsilyl
TMSCN	Trimethylsilyl cyanide
TMSOTf	Trimethylsilyl trifluoromethanesulfonate
TPP	2,4,6-triphenylpyrylium tetrafluoroborate
UPO	Unspecific peroxygenase
UV	Ultraviolet
W	Watt
Xphos	2-Dicyclohexylphosphino-2′,4′,6′-triisopropylbiphenyl

1 Introduction

The race for new active pharmaceutical ingredients (APIs) is triggered by the evolutionary development and adaptation of pathogens in nature and the increasingly competitive landscape of the pharmaceutical industry [1]. The central issue in medicinal chemistry, the search for new drug candidates with explicit biological activity, is inherently connected to the full spectrum of organic synthetic chemistry methods. Hence, progress can only be achieved in drug discovery by the application of novel synthesis concepts and technologies that result in improved or new lead structures and APIs accessed under mild, selective and low-cost conditions. The scientific community responded to these challenges with the application of photochemistry and photocatalysis as one possible route with high potential. The tremendously increased knowledge about photocatalysis with visible light opened the door for mild reaction conditions and pathways, which are not or not easily accessible under classic thermal conditions [2–19]. Various noble or non-precious metal complexes [20–22] or organic dyes [23–25] as well as inorganic semiconductor materials [26, 27] have been widely applied to visible light photocatalysis for the selective construction of molecules. Concepts for dual catalysis with photocatalysis jointly performed with classic metal complexes [28–37] or organocatalysts even enhanced the landscape to, e.g., chiral products. Selective irradiation of electron donor-acceptor complexes allows the photochemical conversions without a photocatalyst [38–53]. And the concept of consecutive photocatalyst excitation can result in reactive species with high reduction potential for the cleavage of, e.g., less reactive C–Cl bonds in aryl chlorides [54–56].

On the technology side, flow chemistry concepts and micro reaction technology offer novel process windows, which have excellent compatibility with photocatalysis [57–60]. In general, flow chemistry has the goal to overcome restrictions of chemical batch processing by providing a defined technological environment in which physical phases can be mixed and reacted in a small and defined volume under strictly controlled conditions [61–63]. With this definition, flow chemistry offers several advantages compared to reactions performed in batch mode: (1) High pressure and high temperature reactions are possible due to quick and efficient heating and cooling of a small reaction volume for advanced process control with higher conversion; (2) Higher selectivity (less by-products) can be achieved by improved mixing of substrates with reagents or quenching of reaction solutions on a small volume scale; (3) Exact time control of phase contacting is possible under defined physical conditions by the control of the flow rates; (4) Less complicated work-up and product separation is possible with heterogeneous catalysts immobilized inside the small reactor volume; (5) Photochemical reactions can be performed in small reaction volumes under full irradiation and defined residence time in the reactor resulting in higher conversion and selectivity due to less over-irradiation and consecutive degradation [64, 65]; (6) Modern LED technology can be used for the construction of light sources for flow photoreactors offering less energy consumption and less safety issues compared to mercury or xenon lamps [66].

The possibilities of photocatalysis concepts for organic synthesis as well as the available technology for flow chemistry clearly show synergy, if both research fields join forces for a complex mission like drug discovery. The following examples in this chapter have been selected in order to demonstrate important photochemical transformations for drug discovery and API synthesis. Hence, carbon–carbon and carbon–heteroatom coupling reactions as well as photochemical cyclization and rearrangement reactions will be discussed. The incorporation of fluorine and fluorine-containing groups will be presented, since fluorine has a very prominent role in medicinal chemistry. Although not yet available in continuous flow mode, photochemical-assisted biocatalysis will be discussed at the end of this chapter, as it is a current trend with very promising results for complex and chiral molecule synthesis.

2 Carbon–Carbon and Carbon–Heteroatom Bond Formation

2.1 Diazonium Salts and Diazo Compounds for C–C and C–X Bond Formation

Amines are the precursors of diazonium salts, which can be regarded as reactive intermediates towards the formation of new chemical bonds with carbon or other (hetero)atoms [67, 68]. Such coupling reactions are well-known for noble metal catalysis, but light-induced synthesis routes have been developed as well in recent years [69]. For instance, Ackermann et al. published their work on arene C-H activation in flow using a photocatalyst based on manganese, an earth abundant metal [70]. Their approach was applied to the synthesis of the precursor molecule of dantrolene, a drug for muscle relaxation in the case of neuroleptic malignant syndrome (Fig. 1) [71]. Here, the conversion of 4-nitrobenzenediazonium tetrafluoroborate **1** (1.6 mol L^{-1}) was performed with furfural **2** as heterocyclic substrate (15 eq.) for arylation and $CpMn(CO)_3$ as photocatalyst precursor (10 mol%) in DMSO as solvent. The flow setup consisted of commercially available photoreactor system (10 mL) with blue light irradiation from LEDs (450 nm). The reaction solution was pumped with a flow rate of 330 μL min^{-1} into the reactor, the temperature was set to 25°C. Within a residence time of 30 min, a yield of 65% could be obtained for the desired product molecule. Subsequent condensation of **3** with 1-aminohydantoin hydrochloride **4** yielded dantrolene (**5**) in 94% yield.

Fig. 1 Photocatalysed arylation of furfural in the synthesis route yielding dantrolene (**5**)

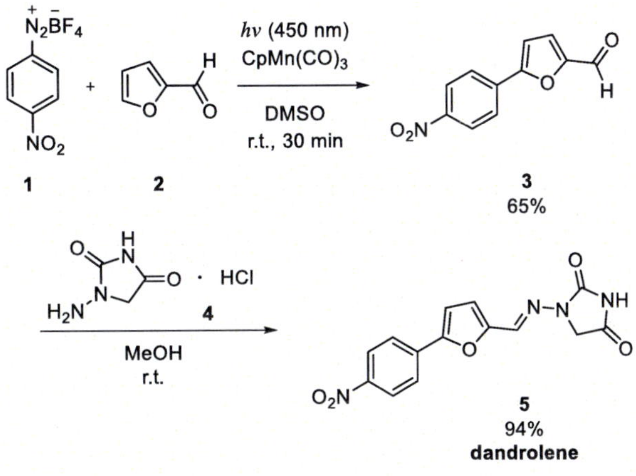

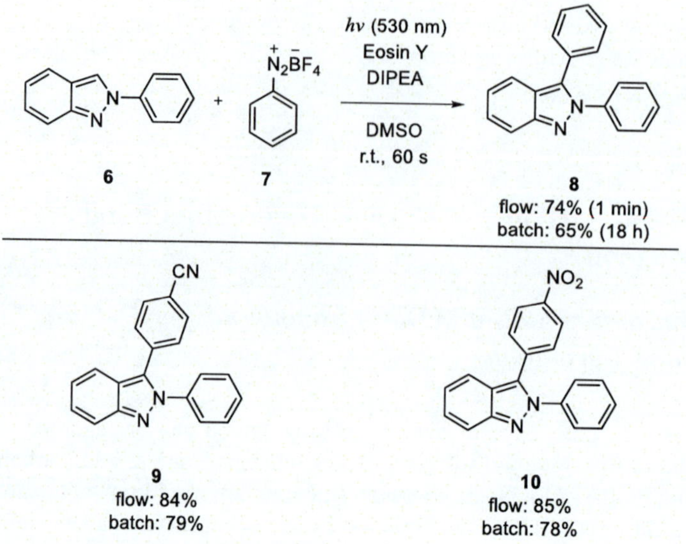

Fig. 2 Photocatalysed arylation of 2*H*-indazoles at C3-position with Eosin Y as sensitizer

The nitrogen-containing aromatic compound class of indazoles play an important role as key structural element in natural compounds and bioactive molecules for pharmaceutical applications. Indazoles are used as basic scaffold in, e.g., antimicrobial, anti-inflammatory or commercially available drugs like pazopanib [72–78]. To overcome the use of metal catalysts with expensive ligands and long reaction times for the functionalization of the rather less reactive C3 position of 2*H*-indazoles, Kim et al. developed a straightforward methodology via the photocatalytic activation of aryl diazonium ions (Fig. 2 top) [79]. Batch screening for optimal reaction conditions was started with 2-phenyl-2*H*-indazole (**6**) and phenyl diazonium tetrafluoroborate (**7**) as model substrates with Rose Bengal (3 mol%) as sensitizer in DMSO as solvent. Visible light irradiation was done with high power LEDs (70 W) emitting green light at 530 nm. Within 24 h reaction time, a moderate yield of 43% of the desired product was obtained. Addition of 1 eq. diisopropylethylamine (Hünig's base) as electron donor and the use of Eosin Y as sensitizer resulted in an increased yield of 65% for 2,3-phenyl-2*H*-indazole (**8**) within 24 h. The transfer from batch to continuous flow mode was done by using a PFA capillary (I.D. = 250 μm, length: 2 m) as coil, which was surrounded by a stripe of green light emitting LEDs cooled with air. The reaction solution was pumped through the capillary with a flow rate of 0.2 mL min^{-1} corresponding to a residence time of 30 s. Due to the improved light–matter interaction inside the capillary, the same yield of 64% could be obtained for the benchmark reaction within this dramatically shortened reaction time. An intensification factor of approx. 2,880 clearly proves the advancement of photochemical processing by the switch to flow mode [80]. Finally, Kim et al. decreased the flow rate to 0.1 mL min^{-1} resulting in an even more improved yield of 74% within 1 min residence time. Variation of the substituents at the indazole scaffold resulted in no conversion for, e.g., alkyl and benzyl groups in the 2-position or low reactivity for bromo substituents at the phenyl ring (<5%). Moderate to good yields were obtained for electron-donating groups at the phenyl ring (65–74%) as well as for fluorine- and bromine-substitution at the arene part of the indazole (63 and 68%, respectively). Different substitution patterns at the phenyl ring of the aryl diazonium salt resulted throughout in good yields, except for very bulky structures as in the case of mesitylene. Highest yields were obtained for electron-withdrawing substituents in *para*-position like –CN in **9** (84%) and $-NO_2$ in **10** (85%).

Kim et al. applied their novel photochemical approach also to the synthesis of inhibitor derivatives for a liver X receptor. The methoxy-substituted indazole **11** was arylated with *p*-chloro or *p*-cyanoaniline via the described diazonium salt route. The desired indazole-based inhibitor molecules could be obtained in good yields with 78% for **14** (-Cl) and with 80% for **15** (-CN) within a residence time of 1 min. Compared to the very long reaction time of 18 h in batch mode for only 65 and 72% yield, respectively, clearly shows the potential for industrial production of such indazole APIs in continuous flow mode under photochemical conditions.

In order to foster such progress and increase the acceptance of flow photochemistry for larger scale production, Kim et al. also improved their reactor concept [81]. The new approach foresees both the immobilization of the sensitizer on the inner walls of the PFA capillary and the parallelization of several capillaries within one reactor module (Fig. 3 bottom). The interior of a PFA capillary (I.D. = 400 μm) was coated with allylhydridopolycarbosilane, which was hydrolysed afterwards to generate reactive hydroxyl groups for further functionalization with aminopropyltriethylsilane. Eosin Y was coupled via amide bond formation to the surface with direct contact to the reaction solution and perfect irradiation from the outside through the PFA capillary wall. The second advancement was done by using 3D printing techniques to produce a polymer-based inlet and outlet for 10 PFA capillaries, each 1 m long. Computational fluid dynamics (CFD) studies were performed upfront to design inlet and outlet architectures for uniform flow distribution through all capillaries. A quartz tube was used as reactor housing carrying the capillaries in straight fashion from the bottom inlet to the top outlet. High-power LED stripes were placed around the quartz tube for equidistant irradiation of the capillaries. Finally, downstream processing was performed with a droplet-based liquid–liquid extraction of the DMSO solution mixed with water and diethyl ether. With this reactor concept and work-up process at hand, Kim et al. performed the synthesis of the liver X receptor inhibitor derivatives **14** and **15**. Within all-over 2.2 min process time, consisting of 0.63 min reaction time and 1.57 min for extraction, 63% yield was obtained for the desired product. This efficiency allows the production of 3.75 g h^{-1}, which is high enough for a daily amount to be used in first clinical trials.

Molecules with thioether as functional structure can also have bioactive properties. Hence, such molecules are interesting for drug research as well, but need mild reaction conditions due to the versatile reactivity at the sulphur atom [82]. One mild and biocompatible, i.e. metal-free arylation method was developed by Noël et al. for

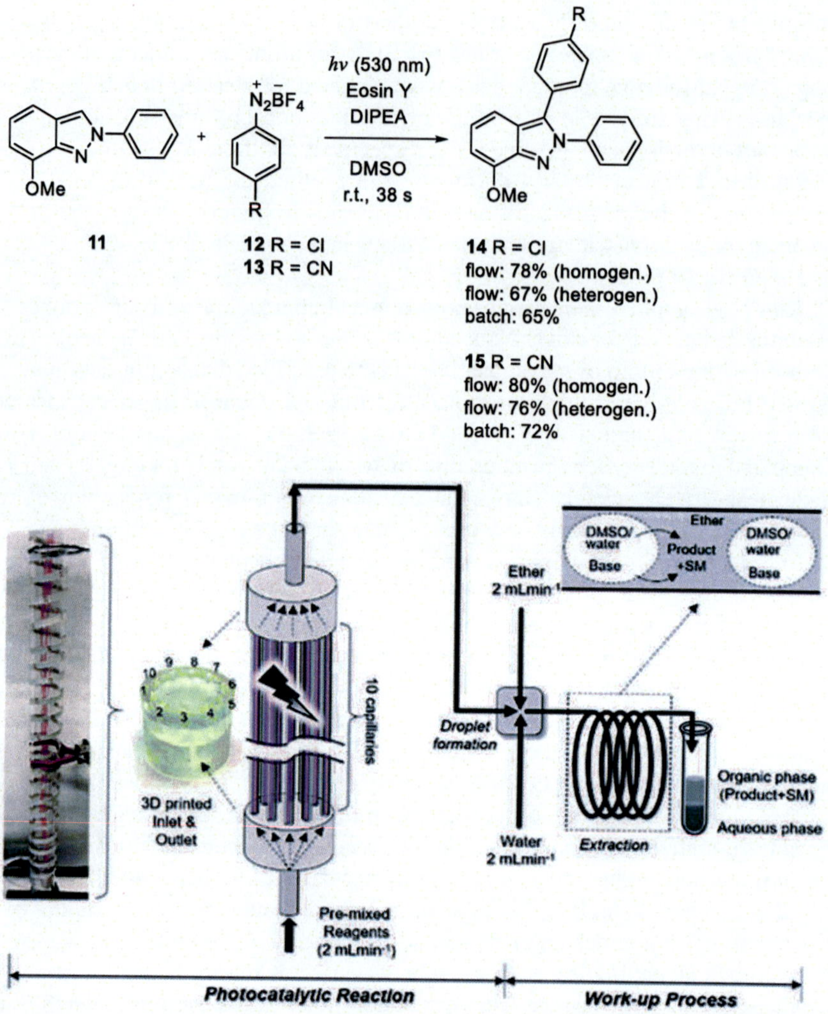

Fig. 3 Continuous flow arylation of 2H-indazoles at C3-position with Eosin Y as immobilized sensitizer on the inner capillary walls. Reproduced with permission from Ref. [81]. Copyright 2019 Royal Society of Chemistry

the amino acid cysteine [83]. The authors already developed a straightforward method for the photocatalysed Ziegler-Stadler reaction for sulphur–sulphur coupling and used this experience for the synthesis of thioether derivatives [84]. Batch reaction screening was performed with N,C-protected cysteine **17** as model substrate in order to evaluate the optimal conditions for the in situ formation of the diazonium salt and its conversion. 4-Fluoroaniline **16** was used as arene source with *t*BuONO and *p*-toluenesulfonic acid for in situ diazotization, acetonitrile as solvent, Eosin Y as metal-free sensitizer. A compact fluorescence lamp was applied as light source.

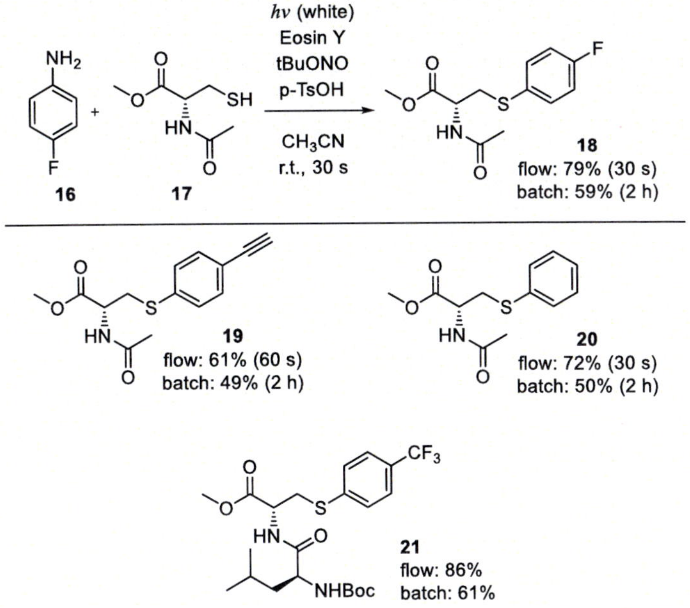

Fig. 4 Arylation of cysteine's thiol group in **17** with photocatalytically activated aryl diazonium salts. Selective arylation of a thiol group in dipeptide *N*-Boc-L-Leu-L-Cys-OMe

Under these conditions, cysteine **17** was arylated in 59% yield. The transfer to continuous flow was performed with a PFA capillary photoreactor (450 μL) and a white light LED stripe (3.12 W) for irradiation. The separate solutions for the diazotization reaction were mixed in a T-piece with *t*BuONO (0.2 mol L^{-1} in acetonitrile) in one stream and 4-fluoroaniline **16** (0.13 mol L^{-1}), Eosin Y (1 mol%), **17** (0.1 mol L^{-1}), and *p*-toluenesulfonic acid (4 mol%) with acetonitrile in the other stream. The reaction solution was irradiated inside the PFA capillary for 30 s, at which the N,C-protected amino acid was converted to the desired thioether **18** in 79% yield (Fig. 4). Based on these optimized process conditions, various aniline derivatives were diazotized in situ for subsequent conversion and coupling with the thiol group of cysteine. In the case of the conversion of 4-ethynylaniline into **19**, a longer residence time of 60 s was necessary to achieve a conversion of 61% in flow against 49% in batch mode. Unsubstituted aniline was converted into **20** in 72% yield compared to 50% in batch mode. Dipeptide Leu-Cys was reacted with 4-trifluoromethylaniline to evaluate the novel method for more complex biological structures. Fortunately, the desired thioether **21** was obtained in 86% in flow mode compared to 61% in batch mode.

In analogy to the charged diazonium salts, diazo compounds are another important class of reactive intermediates in organic chemistry. They are usually stabilized by π-systems or electron-withdrawing groups in close proximity to the diazo group. In the case of non-stabilized diazoalkanes, the in situ preparation is recommended for a safe handling [85]. Ley et al. recently developed a method for the

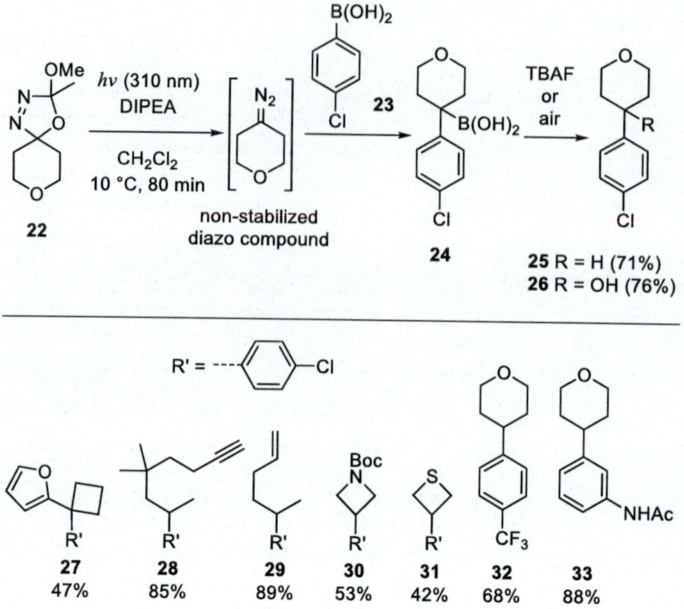

Fig. 5 Top: Non-stabilized diazoalkanes via selective photolysis of oxadiazolines **22** in continuous flow with broad functional-group tolerance. Bottom: Examples of the substrate scope for the photocatalytic C-H arylation via non-stabilized diazoalkanes

photochemical conversion of oxadiazolines into diazoalkanes [86]. This method is based on the foregoing work of Warkentin et al., who irradiated 1,3,4-oxadiazolines selectively with UV light at approx. 300 nm yielding diazoalkanes and alkylacetate as photolysis products [87]. The synthesis of oxadiazolines can be done in a one-pot, two-step procedure with various ketones as starting materials. Hence, Ley et al. prepared several unstabilized diazoalkanes by selective in situ photolysis and converted the resulting reactive intermediates into *p*-substituted arylboronic acids. As one example, 4-chlorophenylboronic acid **23** (0.05 mol L^{-1} in dichloromethane), oxadiazoline **22** (2 eq) and diisopropylethylamine (2 eq) were mixed and transferred into a capillary photoreactor (FEP, 10 mL, 80 min). The reactor was cooled to 10°C and a UV light source was used for selective irradiation at 310 nm (9 W). The reaction was monitored with an inline IR spectrometer analysing the C=O stretch band of methyl acetate at 1,746 cm^{-1} as probe for the photolysis of the oxadiazoline **22**. The resulting tertiary boronic acid **24** was not isolated but stirred either under an air atmosphere or with 3 eq. TBAF (Fig. 5). In the first case, $B(OH)_2$ was substituted with a hydroxy group in 76% overall yield for **26**; in the second case, with a hydrogen atom, in 71% overall yield for **25**. Further derivatizations are possible by trapping the tertiary boronic acid as pinacol ester, which was subsequently converted with 2-lithiofuran for the construction of a quaternary carbon centre in 47% yield for **27**. The mild synthesis method developed by Ley et al. allows the smooth conversion of functionalized diazoalkanes as well, e.g., with C–C triple (**28**) and double bonds

(**29**) as well as diazoalkanes with *O*-, *N*- or *S*-heterocycles and carbocycles in good to excellent yields (**30–33**). The substitution pattern of the aryl boronic acid was varied as well and allowed the integration of, e.g. ,CF_3 or AcNH instead of a chlorine atom (**32,33**). Ley et al. developed a mild and very powerful method for the photochemical in situ formation of unstabilized diazoalkanes under safe and excellently controllable flow conditions. These reactive intermediates have been used in a second step for the metal-free cross-coupling of C(sp^2) and C(sp^3) centres.

2.2 *Photoinduced Metal- and Dye-Catalysed C–C and C–X Bond Formation*

Besides diazonium- and diazo-based arylation reactions induced by light, new variations of well-known transition metal cross-coupling reactions have been developed as well. Dual-catalysis ruthenium or iridium complexes as photosensitizers together were used with another organometallic catalyst for bond formation with the intermediate from the photocatalysed reaction [28–37]. Unfortunately, a sustainable and cost-effective process scale-up is often not possible with rare earth metals. To handle this challenge, Alcázar et al. developed a photosensitizer-free Negishi cross-coupling protocol with earth abundant Zn and Ni complexes [88]. First batch tests were performed with methyl 4-bromobenzoate (**34**) (0.1 mol L^{-1}) and readily available benzylzinc bromide (**35**) (0.2 mol L^{-1}) with fac-Ir(ppy)$_3$ as photosensitizer (1 mol%) and various cross-coupling catalyst on Ni-basis. A commercially available 24 vial photoreactor was used for blue light irradiation resulting in conversions to the desired product **36** from 26% (Ni(cod)) to 65% ($NiCl_2$ glyme, dtbbpy). The pre-screened reaction was then transferred to continuous flow mode with a commercially available capillary photoreactor. The flow rates were set to 250 μL min^{-1} for each stream of starting material solution resulting in a residence time of 20 min. Under these conditions, roughly the same maximum conversion of 64% was obtained in flow mode with Ni(dtbbpy)Cl_2 as cross-coupling catalyst. Interestingly, a higher conversion of 70% was obtained without iridium photosensitizer. Full conversion could be achieved by combining visible light with an increased reaction temperature of 60°C (Fig. 6). Spectral analysis was done for the reaction solution and

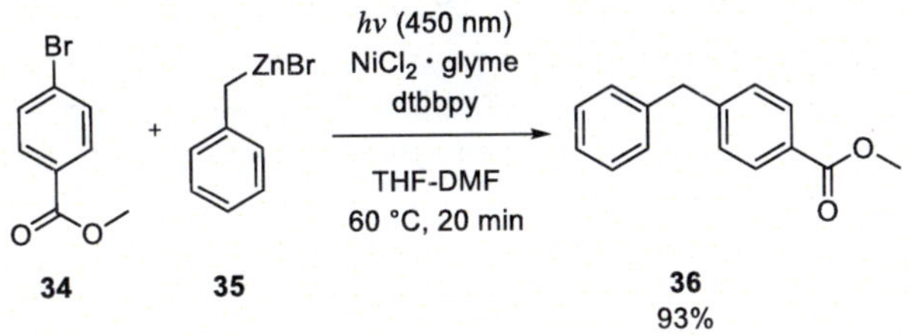

Fig. 6 Acceleration of C(sp^3)-C(sp^2) Negishi cross-coupling reactions by blue light absorption with Ni/Zn complexes

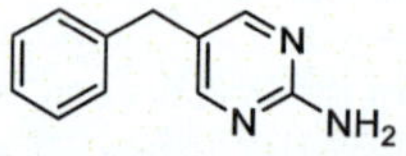

37

94% (~ 80 g d^{-1})

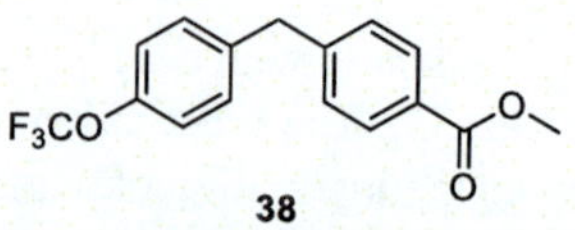

38

93% (~ 130 g d^{-1})

Fig. 7 Examples of scale-up in API synthesis using accelerated Negishi cross-coupling reactions with blue light

for the single compounds. A significant absorption band was revealed in the visible region for the Ni^{2+} complex in combination with the dtbbpy ligand and the organozinc compound. Mechanistic investigations were performed as well, which point in a first step to an accelerated reduction of Ni^{2+} to Ni^0 with the aid of visible light, as well a blue light accelerated reductive elimination within the catalytic cross-coupling cycle.

Besides the above-mentioned Negishi cross-coupling, Alcázar et al. applied a new flow protocol for in situ synthesis of organometallic zinc halides [89]. The novel method was beneficial for the use of 4-bromoanisole, which does not work in standard Ni catalysed reactions of Negishi-type cross-coupling. Hence, the on-demand synthesis of the organometallic zinc reagent was crucial for obtaining the desired cross-coupled product with a conversion of 53%. Contrarily, a zincate reagent from a commercial supplier did not yield any conversion. Chloroarenes as well as bromo- or iodoarenes were converted to the desired bis(hetero)aryl compounds by applying the new continuous flow method. A huge number of the reactions performed clearly benefited from the irradiation with blue light by an increased conversion of up to 90%.

As follow-up, Alcázar et al. then investigated their novel Negishi cross-coupling method on larger scale [90]. A flow photoreactor from a commercial supplier was used in the pilot-scale synthesis of 5-benzylpyrimidin-2-amine **37** and trifluoromethoxylated phenylbenzoate (**38**) (Fig. 7). In this publication, the authors' intention was to highlight a concept for the on-demand production of potential APIs on a scale of multiple 10 g per day for preclinical medical programs. The throughput in synthesis of **37** was increased to 3.4 g h^{-1} (factor 11) corresponding to the synthesis of approx. 80 g per day. A slightly lower factor of 7 was achieved for **38** with a throughput of 5.6 g h^{-1} equivalent to approx. 130 g per day. Alcázar et al. achieved significant progress in continuous flow processing of solid starting materials like zinc, in combination with an accelerated conversion of in situ prepared organometallic intermediates.

The above-mentioned Ni-mediated Negishi cross-coupling methodology is not suitable for the α-arylation of carbonyl compounds due to the deleterious impact of the product's acidity on the catalytic cycle [91]. But unfortunately, this reaction class is a quite important route to several drugs like naproxen, ibuprofen, and tolmetin. Hence, Alcázar et al. adopted palladium catalysis with visible light, originally applied to Heck-type reactions, to facilitate the α-arylation of (2-(*tert*-butoxy)-2-oxoethyl)zinc(II) bromide (**40**) with 4-benzyloxybromobenzene (**39**) [92]. JohnPhos

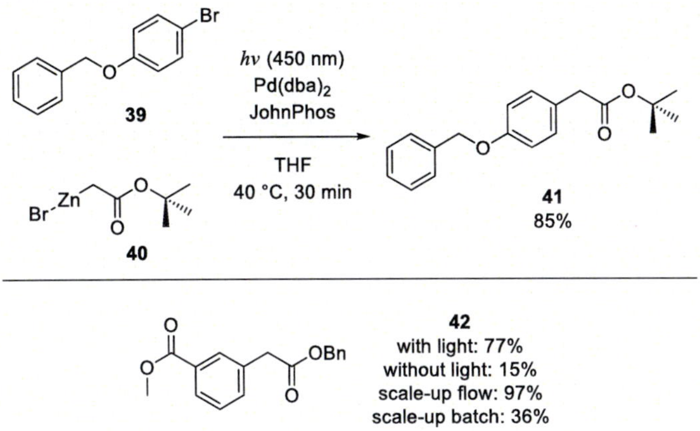

Fig. 8 Photoinduced Pd-catalysed Negishi cross-coupling of organozinc carbonyl compounds with aryl bromides and chlorides

was used as ligand for $Pd(dba)_2$ as catalyst during the first screening in continuous flow mode (Fig. 8 top). Tetrahydrofuran was the solvent and the temperature of the commercially available flow photoreactor (10 mL) was set to 40°C. Irradiation was done with LEDs emitting blue light at 450 nm. Applying a flow rate of 167 μL min^{-1}, full conversion was achieved with 10 mol% of Pd catalyst. With 5 mol% catalyst loading, the same conversion could be achieved, whereas another reduction to 2.5 mol% resulted only in 31% conversion. Alcázar et al. applied this novel methodology to a broad library of aryl bromides and chlorides. For the conversion of the latter ones, XPhos was used as ligand instead of JohnPhos. In nearly all cases, high conversions with very good yields could be obtained, whereas the conversion dropped to very low numbers without blue light irradiation (Fig. 8 bottom). For example, benzyl (3-acetylphenyl)acetate (**42**) could be obtained in 77% yield with 85% conversion in flow mode. Without light, the conversion dropped to 15%. This compound was also used for a two-step scale-up with in situ continuous flow preparation of the organozinc halide prior to the Negishi coupling reaction. The coupled flow process was performed for 5 h in parallel to the batch process under the same chemical conditions. In the first case, an excellent 97% yield was obtained with a corresponding productivity of 520 mg h^{-1}. On the contrary, the batch synthesis yielded only 36% of the product.

A collaboration between the workgroups of Alcázar and Noël resulted in further advancement of the important C–C Kumada coupling reaction. A sustainable continuous flow protocol was developed for the coupling of alkyl Grignard reagents with aryl chlorides. Fe complexes were built in situ with imidazolium ligands for accelerating the reaction (Fig. 9) [93]. Blue light was used for irradiation. Under these conditions, the C–C coupling reaction resulted in considerably higher product formation of, e.g., cyclohexylbenzene (**45**) from cyclohexyl magnesium chloride (**43**) and chlorobenzene (**44**). The yield increased from 39% without light to 99% with blue light irradiation within a short residence time of 5 min. In the case of

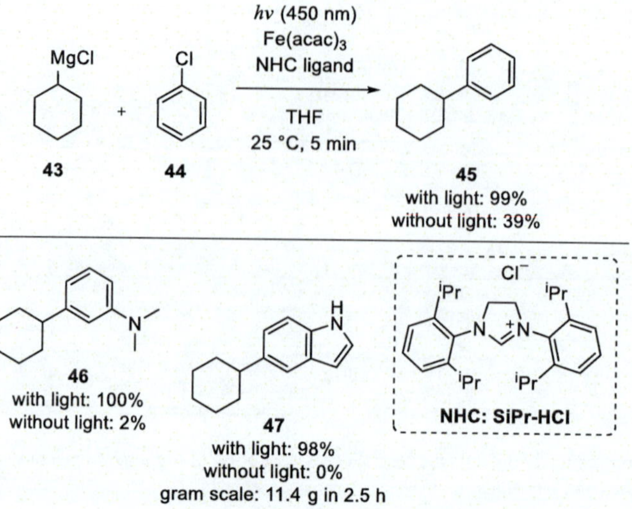

Fig. 9 Kumada coupling reaction products obtained via blue light irradiation

electron-rich aryl chlorides like *N*-methylated anilines or indole, an even stronger tendency to higher product formation could be shown. Usually, these starting materials need high temperatures and long reaction times. *N,N*-Dimethylated chloroaniline was fully converted to **46** (2% without light). Indole **47** was obtained with 98%, but it was not accessible without light. The latter process was also scaled up to approx. 12 g in 2.5 h, corresponding to a space time yield of 454 mg h^{-1} mL^{-1}. These excellent results clearly demonstrate the great benefit for the Kumada C–C coupling by applying light for process acceleration.

Photoredox dual catalysis with iridium and nickel catalysts was used by Boyed et al. for the coupling of alkyl trifluoroborates with aryl bromides in continuous flow mode [94]. Their approach achieved the synthesis of alkyl-substituted quinoxalines in continuous flow with a reduced reaction time of 40 min compared to 16 h in batch mode. Boyed et al. optimized the reaction conditions for the C(sp^2)-C(sp^3) coupling of 5-bromo-2-methylpyrimidine (**48**) with tetrahydropyran-3-trifluoroborate (**49**). As dual catalyst system [Ir{dF(CF_3)ppy_3}$_2$(bpy)]PF_6 (2.5 mol%), $NiCl_2$·dme (5 mol%), and dtbbpy (5 mol%) were used. The exchange of $CsCO_3$ as solid base in dioxane as solvent with 2,6-lutidine in a DMA-dioxane mixture (1/4) resulted in a homogeneous solution that is suitable for flow mode in photocapillary reactors (Fig. 10 top). With slightly increased amounts of catalyst components to 3 mol% for [Ir] and 12 mol% for each [Ni] and dtpbbpy under homogeneous conditions, Boyed et al. were able to increase the yield from 37% (heterogeneous, batch) to 62% (homogeneous, flow) for 2-methyl-5-(tetrahydro-2H-pyran-4-yl)pyrimidine (**50**). With the optimized reaction conditions in hand, the authors performed the coupling of more complex substrates that were not accessible in batch mode yet. A sterically hindered *ortho*-methoxy aryl bromide was coupled with **49** resulting in 90% conversion after 40 min with 46% yield for the desired product **51**. Both heterogeneous

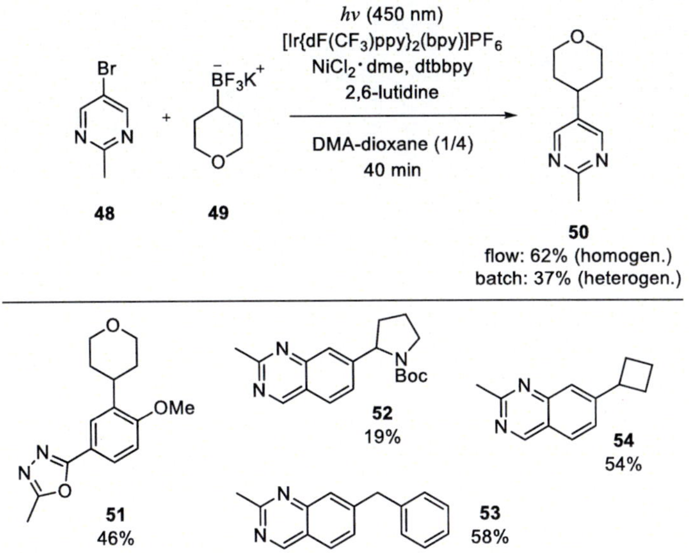

Fig. 10 Top: Reaction conditions for the C(sp^2)-C(sp^3) coupling between aryl bromides and potassium alkyl trifluoroborate salts. Bottom: Quinoxaline derivatives obtained in continuous flow mode via automated library screening

and homogeneous batch tests gave only trace conversion. A small library of nine derivatives was screened with the novel approach yielding 10–35 mg of pure product after inline purification. Yields varied here between 19% (**52**) and 58% (**53**). Boyed et al. could also demonstrate the scale-up capabilities of their methodology. Quinoxaline **54** was synthesized in the continuous flow setup with a productivity of 81 mg h^{-1} or 1.3 g in 16 h, which are necessary for an equal productivity by 48 individual 5 mL batch reactions.

Carbon–boron bond formation should be mentioned as well, since arylboronic acids are important building blocks in synthetic organic chemistry. Such precursors are usually converted with Grignard or Li reagents or applied to (noble) metal catalysis for carbon–carbon or carbon–heteroatom bond formation [95, 96]. Contrary to the known procedures, Li et al. developed a photochemical route for the borylation of aryl halides [97]. Reaction optimization was done in batch mode with a quartz glass tube and with irradiation by a high pressure mercury lamp (365 nm, 300 W). Bis(pinacolato)diboron **55** was selected as borylation reagent (2 eq.) together with 4-iodoanisole (**56**) (0.1 mol L^{-1}) as model substrate. The complex solvent mixture of 0.9 mol L^{-1} acetone in acetonitrile-water (4/1) was used for the photochemical reaction as well as bis(dimethylamino)methane (TMDAM, 50 mol%) was used as additive (Fig. 11). By irradiation with strong UV light for 4 h at room temperature, aryl radicals are formed via homolytic cleavage of the carbon–iodide bond in the excited aryl iodide as proposed by the authors. As another pathway, Li et al. also suggest the decomposition of aryl iodide

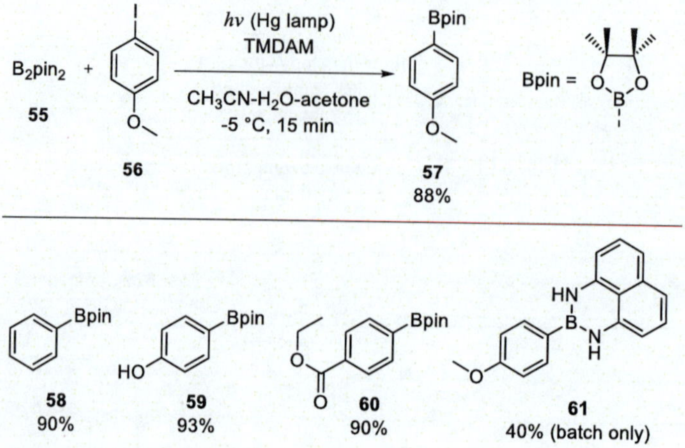

Fig. 11 Top: Photolysis as activation of aryl halides for efficient borylation. Bottom: Photolytic borylation of various arenes with two different borylation reagents

radical anions, which have been formed by a single electron transfer from the amine additive. In the next step the borylation reagent gets cleaved by the aryl radicals, leading to the desired product **57** in 81% yield at 94% conversion. Deiodination takes place as well as another reaction pathway in 10% yield. Li et al. then transferred their new method to continuous flow using a standard FEP capillary (780 μL) wrapped around an immersion well with the same lamp already used for the batch tests. This setup completely outperformed the batch mode system and allowed full conversion after a residence time of only 15 min. Reaction temperature was set to −5°C giving 88% yield for the pinacol-protected boronic acid **57** and 4% yield of anisole by dehalogenation. Under these optimized conditions Li et al. were able to reduce the amount of borylation reagent to 1.5 eq as it was rapidly consumed in the flow reactor, avoiding any decomposition of the reagent by prolonged irradiation time. Borylation on larger gram scale was done with iodobenzene and *p*-iodophenol in a commercially available reactor system with 7.8 mL capillary volume. Excellent yields could be obtained for the unsubstituted compound **58** with 90% and 93% for the phenol derivative **59**. Excellent results were also obtained for the synthesis of **60** from ethyl 4-iodobenzoate in 90% yield. The less reactive ethyl 4-bromobenzoate was converted to **60** in 78% yield. The authors also used various other borylation reagents like bis-boronic acid, bis(neopentanediolato)diboron and unsymmetrical (pinacolate)-(diaminonaphthalene)diboron. Interestingly, in the latter case only the diaminonaphthalene boronate was introduced selectively into **61**. Later on, Li et al. were also able to use their continuous flow photolysis with electron-rich arylchlorides and fluorides, in addition to aryl building blocks with mesylate, triflate, and diethyl phosphate groups used as pseudo halides for radical cleavage [98].

The synthesis of di(hetero)arylmethane derivatives by Alcázar et al. with Zn organometallic precursors applied to flow photochemistry with nickel-based catalysts has already been presented in this section [88]. In general, the di

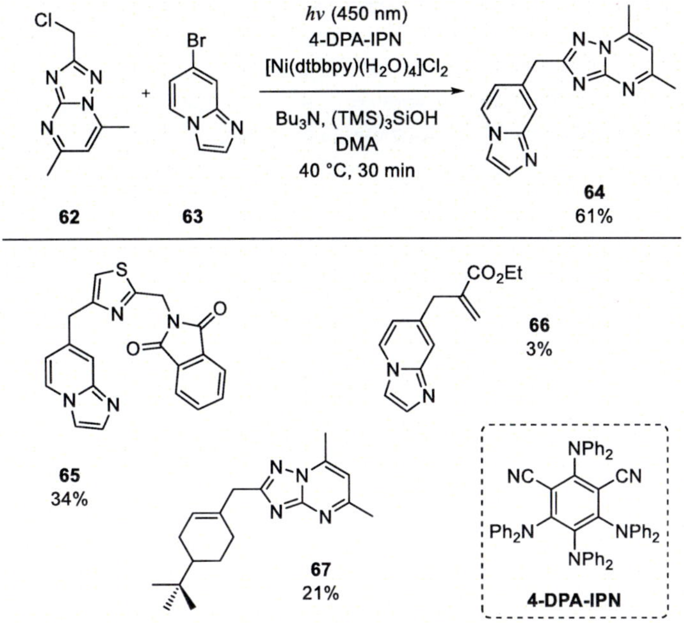

Fig. 12 Top: General reaction conditions for the C(sp^2)-C(sp^3) coupling between aryl bromides and benzylic chlorides. Bottom: Selected examples of the screened di(hetero)arylmethanes obtained in continuous flow mode

(hetero)arylmethane key structure is an important building block for APIs like trimethoprim [99] (antibiotic) or lynparza [100] (cancer treatment). Just recently, Brill et al. applied photoredox dual catalysis with Ni to the C(sp^2)-C(sp^3) coupling of aryl bromides and benzylic chlorides in order to access a broad library of di(hetrero)-arylmethanes for testing in key in vitro pharmacological assays [101]. Based on research from the MacMillan group on photoredox and transition metal catalysed C–C coupling of aryl bromides and benzylic bromides [102], Brill et al. performed intensive reaction optimization in flow for the coupling of pyrimidine **62** (0.025 mol L^{-1}) with pyridine **63** (2.5 eq.). In the case of the photocatalyst, metal-free 4-DPN-IPN was superior to iridium-based catalyst systems. Amine base screening resulted in the use of tributylamine (3 eq.) as optimal additive as well as $(TMS)_3SiOH$ as mild reductant (1.5 eq.) with a lower tendency for hydrodehalogenation of the benzylic chloride substrate compared to $(TMS)_3SiH$. Dimethylacetamide (DMA) was used as solvent, as it was the only one out of eight screened solvents providing a homogeneous solution throughout the complete reaction runtime. An optimal ratio between $[Ni(dtbbpy)(H_2O)_4]Cl_2$ as precatalyst and 4-DPN-IPN as photocatalyst was identified with 5 mol% to 2 mol% (2.5:1). With the optimized reaction conditions in hand, Brill et al. were able to perform the desired coupling reaction within 30 min residence (333 μl min^{-1}) time in a commercially available flow photoreactor setup (10 mL, 450 nm) at 40°C with a yield of 61% for the di(hetero)arylmethane **64** (Fig. 12). The auto sampling modus of the commercially available flow photoreactor setup allowed the library screening of 34 di(hetero)arylmethane derivatives in total with 20 different aryl bromide substrates and 14 different benzylic chlorides (Fig. 12). All reactions were not performed with the intention of achieving the maximum yield, but for obtaining 1–2 mg of pure substance for pharmacological screening. Hence, a yield higher than 51% was not achieved throughout. It should be noted that seven derivatives were not accessible with this methodology, or only in traces detected during LC-MS analysis. The substrate scope was finally extended by exchanging the benzylic chloride with an allylic chloride derivative as well as the aryl bromide with a vinyl bromide. The first reaction could be achieved in low yields of 3% for **66**, whereas the conversion of the vinyl bromide yielded 21% of the desired product **67**. Brill et al. could show with their development that rather fast screening of 34 potential API derivatives can be done with a short reaction time of 30 min for each compound.

In pharmaceutical product synthesis, low residual metal limits are defined for the final product. In the case of iridium, this limit is below 100 μm per day [8, 103, 104]. Despite the costs for, e.g., iridium complexes used as photocatalysts, lengthy and costly purification protocols become necessary to comply with these rules. Hence, it is important as a first step to substitute any metal complexes with less costly organic compounds. One example for this approach was published by Ley et al. for the radical coupling of boronic acid derivatives with methyl vinyl ketone as an example for a radical acceptor [105]. First investigations were done in batch with cyclohexylboronic acid (**68**) as model substrates for the radical coupling with methyl vinyl ketone (**69**) in acetone-methanol (1/1) as solvent mixture. DMAP was used as Lewis base with the iridium complexes, 4CzIPN or Mes-Acr-4 as photocatalysts. Batch screening at 30°C gave after 24 h reaction time excellent results for Mes-Acr-4 as substitute for the iridium catalysts with yields up to 95% (Fig. 13). With these tests, Ley et al. could also prove that no photo bleaching took place for the organic dyes, which is often encountered with other photoreactions. On the contrary, kinetic studies revealed that the iridium catalysts deactivate over time. It is known that radical alkylation of the bpy ligands can occur, as well as ligand substitution by solvent molecules or additives. The authors postulate here that the highly nucleophilic DMAP catalyst can replace the bpy ligand in the iridium complex resulting in the observed deactivation. The transfer to flow mode was done twofold. In the first case a commercially available small-scale flow photoreactor was used (FEP, 10 mL) with blue light irradiation at 420 nm (17 W). Boronic acid **68** (0.1 mol L^{-1}) and methyl vinyl ketone (4 eq.) were mixed with Mes-Acr-4 (2 mol%) and DMAP (20 mol%) in acetone-methanol (1/1) and pumped into the reactor coil at 30°C. With a residence time of 100 min, a maximum yield of approx. 80% was reached (Fig. 13). Based on these good results, Ley et al. used another commercially available reactor with higher irradiation power of 420 W at a blue light emission of 450 nm. Although the PFA coil features only 5 mL inner volume, the massive power increase allowed the residence time to further decrease to 30 min for the same yield of approx. 80%.

To showcase the possibility for application in medicinal chemistry, Ley et al. employed their approach to the synthesis of three precursors for APIs from the γ-aminobutyric acid (GABA) family, namely (±)-baclofen, (±)-phenibut and

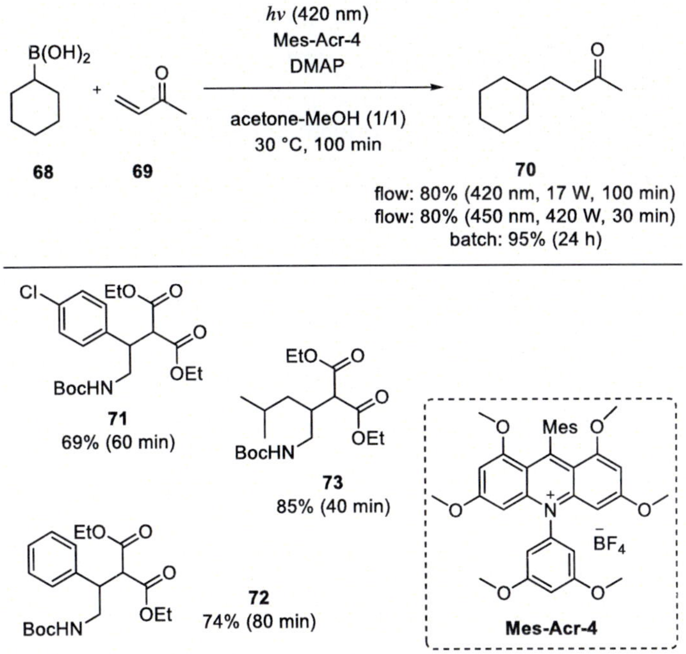

Fig. 13 Top: Model substrates and reaction conditions for the radical coupling of boronic **68** with keton **69**. Bottom: API precursors for baclofen (**71**), phenibut (**72**) and pregabalin (**73**)

(±)-pregabalin [106]. The respective *N*-Boc-aminomethyl boronic acid pinacol esters (0.1 mol L^{-1}) and the appropriate diethyl malonate-derived olefin (0.83 eq.) were subjected under standard conditions to a small-scale low power photoreactor (FEP, 10 mL) for blue light irradiation at 30°C. In all cases the resulting yield was approx. on the same level as for the batch synthesis, but the reaction time was dramatically decreased to 80 min for the phenibut precursor **72**, and 60 and 40 min, respectively, for the baclofen and pregabalin precursors **71** and **73** (Fig. 13 bottom). In summary, besides the substitution of an expensive iridium catalyst by an organic dye, Ley et al. were able to improve the synthesis process of important API building blocks considerably by going to flow and reducing the process time therewith.

The light-mediated amination of (hetero)aryls will be presented here as final example for this sub-chapter. Buchwald et al. applied Ni(II) salt and photoredox dual catalysis for the continuous flow synthesis of tetracaine (**79**), a local anaesthetic, as example for an API [106]. Initial experiments were performed under batch conditions, which led to clogging inside the capillary of the commercially available photoreactor. However, by switching the solvent from DMA to DMSO and further optimization regarding photocatalyst, Ni(II) salt and base additive, a set of optimal reaction conditions was found with 0.2 mol% $Ru(bpy)_3(PF_6)_2$, 5 mol% $NiBr_2$•DME and DABCO (2.0 eq) in DMSO (0.5 mol L^{-1}). Bromobenzene (**74**) (10 mmol) and hexylamine (**75**) (3 eq) were then used as model substrates for the final evaluation of

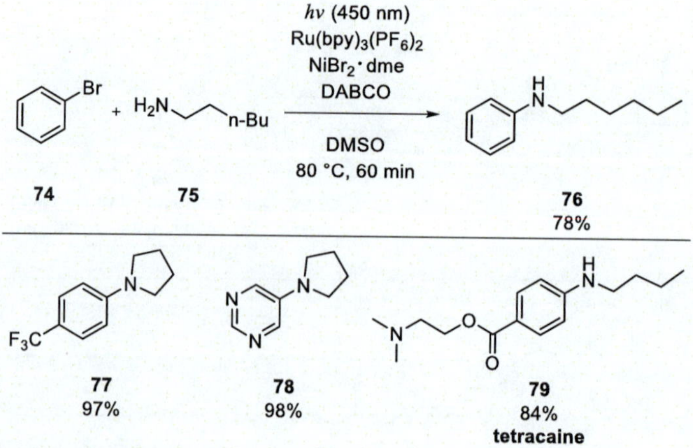

Fig. 14 Top: Dual catalysed amination of (hetero)aryls using Ni^{2+} and $Ru(bpy)_3^{2+}$ as catalysts. Bottom: Selected examples **77** and **78** with pyrrolidine as amine substrate; tetracaine (**79**) as an example for an API

the process conditions under blue light irradiation of 450 nm. Within a residence time of 60 min **74** was aminated at 80°C in 78% isolated yield (Fig. 14). Buchwald et al. screened various bromides (1 mmol scale) with pyrrolidine as amine substrate as it is one of the most common five-membered non-aromatic nitrogen heterocycles in pharmaceutical building blocks [107]. Interestingly, excellent results could be also obtained with a ruthenium catalyst concentration as low as 0.02 mol% for the conversion of, e.g., electron poor 4-(trifluoromethyl)bromobenzene to **77** in 97% yield or for 5-bromopyrimidine to **78** with 98% yield. Both processes were performed with a high flow rate of 1 mL min^{-1} corresponding to a short residence time of only 10 min. Under the same process conditions, the synthesis of tetracaine (**79**) was done in 84% yield. In order to highlight the applicability of the novel flow synthesis, **79** was also synthesized on a tenfold higher scale of 10 mmol. Within the same residence time of 10 min, a similar yield of 84% was reached, which corresponds to a productivity of 19 mmol $h^{-1.}$ The method from Buchwald et al. is an excellent example for the transfer of an already well-recognized synthesis route from batch to flow mode, which gives access to highly relevant aniline derivatives for drug discovery.

2.3 C–C Bond Formation via Photodecarboxylation

Oelgemöller et al. transferred a three-step synthesis route for 3-arylmethylene-substituted isoindolinones from batch to continuous flow mode by coupling a photodecarboxylative reaction step with two thermal dehydration and amination steps (Fig. 15) [108]. In general, such isoindolinone-based building blocks are of

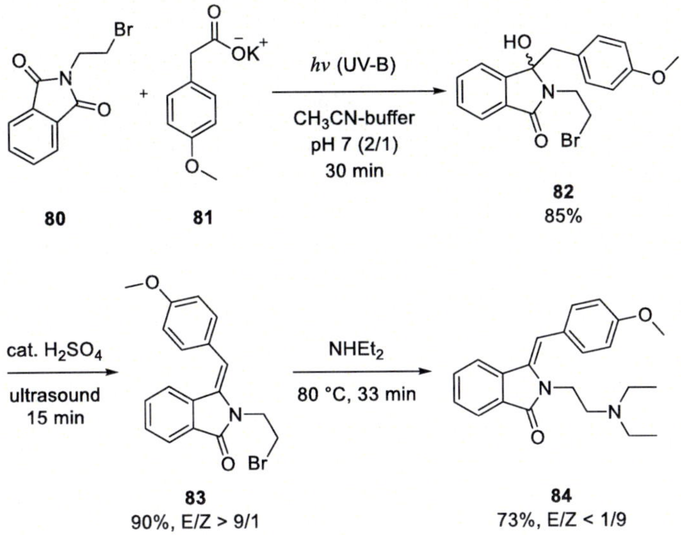

Fig. 15 Three-step reaction sequence for the photochemical and thermal continuous flow synthesis of 3-arylmethylene-2,3-dihydro-1H-isoindolin-1-one derivative **84**

great interest to medicinal chemistry research due to their cardiovascular and anaesthetic activity [109, 110]. As a first step, the transfer to continuous flow necessitates the adaptation of three different solvents to one solvent being most compatible to all reaction conditions. Oelgemöller et al. cross-checked acetone, acetonitrile and DMF in batch for all three reactions with acetonitrile as the final choice for the coupled flow synthesis with yields of 80% for the photodecarboxylation, 83% for the dehydration and 53% for the amination.

The design of the flow setup was established straightforward with three capillary reactors adapted to the physical needs of each reaction step. The first reactor consists of a FEP capillary (I.D. = 0.8 mm, 5 mL) wrapped around a Pyrex glass body with a single UV-B light source (8 W) at its centre. The initial reaction solution of *N*-(bromoethyl)phthalimide (**80**) (0.0165 mol L^{-1}) and potassium (4-methoxyphenyl) acetate (**81**) (2 eq.) with acetonitrile-pH 7 buffer (2:1) was pumped into the reactor with a flow rate of 0.17 mL min^{-1} corresponding to a residence time of 30 min. The desired first intermediate product **82** could be obtained with a yield of 85%. The outlet stream of the photoreactor was contacted in a T-piece with a mixture of 10 mol L^{-1} sulphuric acid-acetonitrile (1/1, 0.17 mL min^{-1}) prior to the inlet of the first thermal flow reactor. This module consisted of another FEP capillary (I.D. = 0.8 mm, 5 mL), which was placed into an ultrasonic bath to prevent any clogging in the small diameter capillary. With a total flow rate of 0.34 mL min^{-1}, the dehydration step was performed within a residence time of 15 min with an overall yield of 90% and very high *E*-selectivity of $E{:}Z \geq 9{:}1$ for the second intermediate product **83**. Prior to the third step, a cartridge with strongly basic Dowex 1-X8 ion exchange resin (20–50 mesh, 50 mL) was integrated to quench residual acid from the

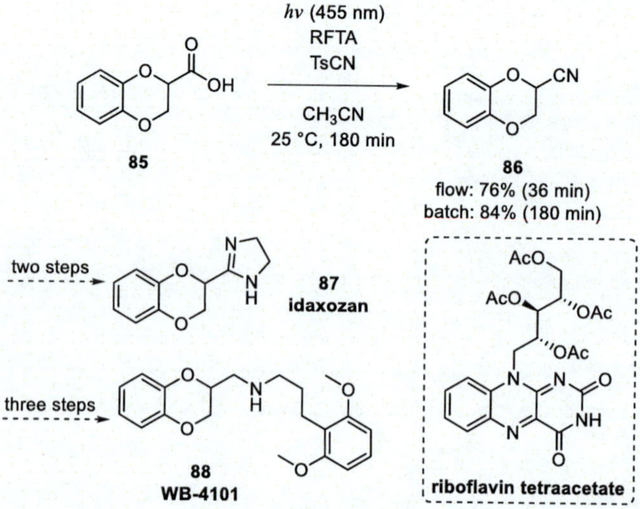

Fig. 16 Riboflavin tetraacetate as sensitizer for the visible light-induced photodecarboxylation of alkyl carboxylic acids with subsequent cyanation

dehydration step. The neutralized reaction solution was then mixed in a T-piece with diethylamine (2 mL in 38 mL acetonitrile, 0.25 mL min^{-1}) and transferred into the last loop reactor consisting of a FEP capillary (I.D. = 1.58 mm, 19.6 mL) placed in a water bath at 80°C. With an overall flow rate of 0.59 mL min^{-1} the final amination step was performed with a residence time of 33 min obtaining the desired product **84** in 73% yield with a nearly complete conversion into the *Z*-isomer with $E{:}Z \leq 1{:}9$. In summary, Oelgemöller et al. could prove with their process that both the transfer to flow mode and process intensification by coupling reaction steps (?) greatly benefits reaction time and yield. Exemplified by the bioactive arylmethylene-substituted isoindolinone derivative AL-12 (**84**), a reduction in reaction time from 9 h to 78 min could be achieved with a good overall yield of 73% compared to max. 45% for the decoupled batch synthesis steps.

The selective cyanation of molecules is a very important method to incorporate a versatile functional group for further derivatization into amines, amides or carboxylic acids [111–113]. Gonzalez-Gomez et al. recently published their work on the photocatalysed decarboxylative cyanation of aliphatic carboxylic acids with riboflavin tetraacetate as purely organic sensitizer [114]. Their approach allowed the synthesis of key intermediate **86** for the subsequent conversion to idazoxan (**87**) or WB-4101 (**88**), both selective antagonists of adrenoreceptors [115, 116]. Decarboxylation reactions with subsequent cyanation have been known for quite a while, but many methods need harsh conditions like high temperature, strong bases, or noble metal catalysts like iridium complexes. Contrary to these approaches, Gonzalez-Gomez et al. developed a mild and versatile method for activating the carboxylic acid with visible light-absorbing riboflavin tetraacetate as photocatalyst and proton acceptor/donor during the catalytic cycle (Fig. 16). α-Oxo-carboxylic acid **85** was used as model substrate for process optimization regarding solvent, additional base, O_2 gas atmosphere, catalyst or substrate loading, and catalyst type.

A 0.1 mol L^{-1} reaction solution of **85** in degassed acetonitrile was irradiated with blue light (455 nm with 5 ± 2 mW cm^{-2}) for 3 h with riboflavin tetraacetate (5 mol %) at room temperature. The desired product could be obtained under these conditions with 1.1 eq. toluenesulfonyl cyanide as CN^- source in 84% yield. Applying deprotected riboflavin as photocatalyst resulted in a low yield due to decreased photostability of the chromophore, lower solubility, and less stable triplet state T_1. Additional organic or inorganic bases led to lower yields, as well as oxygen gas in the reaction solution. A change in substrate or photocatalyst concentration did not yield better results. A switch to other CN^- sources like NaCN or TMSCN did not give better yields as well. No reaction took place in blank tests without light or photocatalysts. With the optimized reaction conditions in hand, Gonzalez-Gomez et al. evaluated the scope of their approach with several α-oxo- and α-thio carboxylic acid, various protected α-amino carboxylic acid and α-carbon carboxylic acid with moderate to good yields [115, 116]. Due to the high molar absorption coefficient of riboflavin tetraacetate with $\varepsilon \approx 13.000$ L mol^{-1} cm^{-1}, it is reasonable to switch to continuous flow mode by applying a capillary photoreactor (PFA, I.D. = 0.5 mm, 1.7 mL) for optimized irradiation. Unfortunately, the yield dropped from 84% in batch to 76% in flow. But the process clearly benefits from a reduced reaction time from 3 h to 36 min in the capillary microreactor. Besides the mild reaction conditions and the absence of additives, the immobilization of the riboflavin chromophore in a continuous flow reactor might give a further impact to this facile approach to medicinal chemistry research due to a reduced effort in product purification.

3 Photochemical Cyclization Reactions

As mentioned in the introduction, molecules with biological activity can often have complex structures like poly(hetero)cycles with various substitution patterns in their scaffold. One possibility to reach such a high degree of structural complexity in a molecule is the cyclization of independent molecules (intermolecular reaction) or the fusion of entities within one molecule (intramolecular reaction) [117–119]. As a first example in this chapter, a photochemical key step in the synthesis route to goniofufurone derivatives will be presented. Ralph, Booker-Milburn et al. used photochemistry in continuous flow mode to illustrate its efficiency for the scale-up of Paternò-Büchi [2 + 2] reactions [120]. (+)-Goniofufurone (**93**) is a natural product used in traditional medicines in the treatment of oedema and rheumatism. Structural analogues of this natural product also have potent antiproliferative effects against a number of cell lines, in particular 7-epi-(+)-goniofufurone-derived oxetane **94** with a greater activity than doxorubicin, the anticancer drug standard. Ralph and Booker-Milburn et al. proposed Paternò-Büchi [2 + 2] photocycloaddition between benzaldehyde (**90**) and bicyclic enol ether **89** (Fig. 17). The authors state that the batch photoreaction on small-scale (310 mg substrate with 0.25 mL benzaldehyde in

Fig. 17 Precursor synthesis for (+)-goniofufurone (**93**) via Paternò-Büchi [2 + 2] photocycloaddition

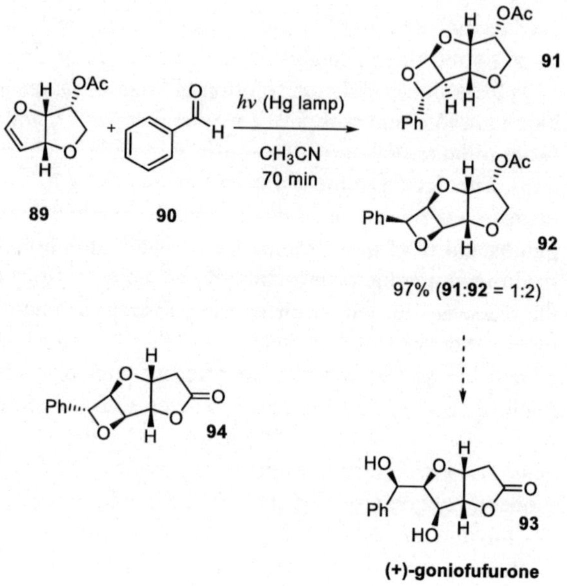

150 mL acetonitrile) gave a good yield of 90%, but it was rather slow with 2 h irradiation time and under high dilution. Consequently, a batch scale-up to 50 g, as desired, was unfavourable.

Based on their foregoing work, the authors transferred the synthesis to flow mode by application of their custom-made three-layer FEP flow photoreactor containing a 400 W medium pressure mercury lamp as source for irradiation. A degassed solution of bicyclic enol ether **89** (0.03 mol L^{-1}) and benzaldehyde (**90**) (1.4 eq.) in acetonitrile was pumped into the reactor with a flow rate of 1 mL min^{-1}, corresponding to a residence time of 70 min. Under these conditions, a yield of 97% could be obtained. In an 83 h long process run, 41 g of product was synthesized as regioisomeric mixture **91** and **92** (1:2) in favour of the desired isomer with the correct stereochemistry at the C-7 position. Four additional steps were finally necessary afterwards for yielding (+)-goniofufurone **93**. The authors state that the rather complex synthesis route development of these steps was only made possible due to the high productivity in the photochemical key step and the availability of large amounts of oxetane starting material.

In analogy to the approach of Ralph and Booker-Milburn, Rutjes et al. also applied a [2 + 2] photocycloaddition as starting point for the identification of lead structures in medicinal chemistry research. Within only three steps, the authors were able to convert 3-amino cyclohexenone derivatives **95** and **98**, respectively, to three tricyclic target scaffolds in very good yields [121]. These precursors were readily available in two steps, firstly via the condensation of cyclohexane-1,3-dione with the free amine in a Dean-Stark apparatus, yielding the corresponding enaminones. The crude products for both amines were then Boc-protected in the second step with 65% overall yield for the allylamine-derived product **95** and 60% for 3-buten-1-amine-

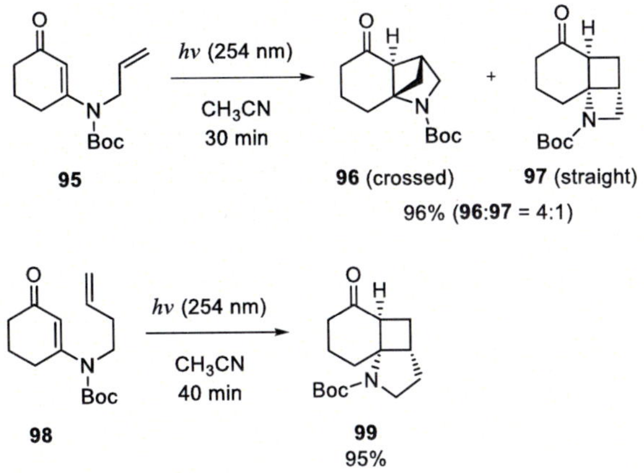

Fig. 18 Intramolecular [2 + 2] photocycloaddition yields three tricyclic aminoketones for further derivatization in medicinal chemistry research

derived product **98**. With the precursor molecules in hand, Rutjes et al. performed first photochemical tests in a custom-made flow photoreactor based on the design from the Booker-Milburn group. A FEP capillary (I.D. = 1.6 mm, 11.6 mL) was wrapped around a condenser which was placed in a Rayonet-type photochemical reactor with two different lamp sets for an irradiation at 254 or 300 nm. Residence time was set according to full conversion as analysed by TLC. Various solvents with different polarity were tested, but in general a mixture of two different tricyclic structures (crossed = **96** and straight = **97**) were obtained for the allylamine-derived precursor **95** (Fig. 18). A ratio of 2.5:1 was obtained with less polar solvents (*n*-hexane, cyclohexane), whereas a ratio of 3.5:1 resulted in more polar solvents like diethyl ether and acetone. Although the temperature was not well regulated in this custom-made reactor, it did not have a strong influence on the product yield. Finally, optimized conditions were found for the photochemical conversion of **95** in a concentration of 0.04 mol L^{-1} in acetonitrile, applying an irradiation at 254 nm for 30 min.

Under these conditions, the desired product mixture of **96** and **97** could be obtained with 96% yield in a ratio of 4:1 (crossed: straight). The second precursor **98** was converted with a slightly longer residence time of 40 min under the same conditions to the straight product **99** in 95% yield. This reactor setup allowed a throughput of approx. 5.6 g per day, which needs to be increased considerably for a larger scale library screening with these building blocks as core structures. As a result, Rutjes et al. fabricated another custom-made reactor system for higher throughput by using a larger commercially available UV-C lamp with 55 W. The capillary volume was increased to 105 mL using an inner diameter of 2.7 mm. The complete reactor system is water-cooled and protected with a metal jacket to block deleterious UV-C radiation from the interior of the reactor. With this new system in hands, the authors were able to convert 10 and 20 g batches of **95** within total running times of 4.75 h and 9.5 h, respectively. The same results were obtained for the precursor **98** within 7.3 and 14.6 h in similar high yields of isolated product. Further studies were conducted by Rutjes et al. regarding the functionalization of the

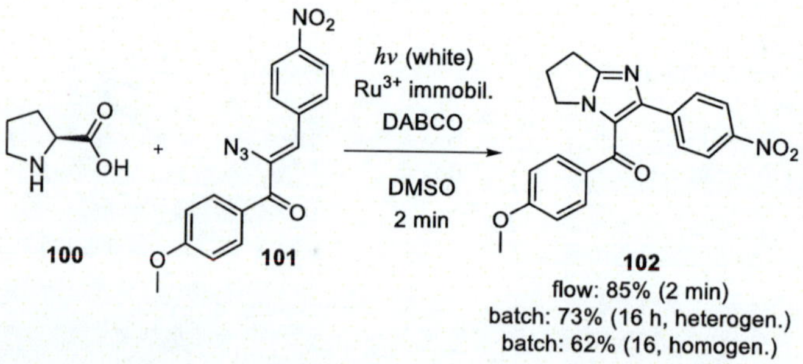

Fig. 19 Visible light-induced denitrogenation of α-azidochalcones with subsequent conversion with L-proline to fused imidazoles

ketone and the amine after Boc-deprotection. In the first case, reductive amination with pyrrolidine was done or a Suzuki coupling to incorporate a phenyl group after converting the ketone to a triflate. The amine was tosylated for future derivatization. In summary, Rutjes et al. could prove the fruitful incorporation of flow photochemistry into the synthesis route of key building blocks for medicinal chemistry development and API screening.

Besides the formation of carbocycles as presented in the first two examples, five-membered nitrogen-containing heterocycles can be made available as well via photochemical approaches. Here, fused imidazole derivatives have attracted great attention as biological building blocks, in commercial drugs and agrochemical agents [122]. Adiyala, Kim et al. designed a new process that allows the photo-induced denitrogenation of α-azidochalcones to 2*H*-azirines, which react with L-proline with subsequent decarboxylation to the desired fused imidazole derivatives [123]. The complete process is photocatalysed with Ru^{3+} ions immobilized in a polyvinyl pyridine matrix on the channel walls of a PDMS-based microreactor. The authors started their investigations with a detailed reaction optimization for the fusion of L-proline (**100**) (0.5 mol L^{-1}) with α-azidochalcone **101** (0.5 mol L^{-1}), the latter one broadly available via Knoevenagel condensation of phenacylazides with aldehydes (Fig. 19). Beside solvent (DMSO, DMF, DCE) and base (K_2CO_3, DABCO, Cs_2CO_3, DBU, 2,6-lutidine) variation, catalyst screening was done as well with Eosin Y, $Ru(bpy)_3(PF_6)_2$ and $Ir(ppy)_3$. Optimized conditions were found for the ruthenium catalyst, DABCO (3 eq.) as base additive and DMSO as solvent (2 mL). Homogeneous photocatalysis with a white light compact fluorescent lamp (23 W) yielded under these conditions 62% of the desired product after 16 h of irradiation in batch. Adiyala, Kim et al. then took the first step for an intensified process and applied a heterogenized ruthenium catalyst for the desired reaction. Here, polyvinylpyridine was used as polymeric matrix for the complexation of Ru^{3+} ions. The immobilization strategy had a positive impact on the reaction process and led to an increased yield of 73%. With these results obtained, Adiyala, Kim et al. designed a PDMS-based microchannel reactor with serpentine structure (length: 1 m, height: 100 μ, width: 500 μm; 50 μL). The walls of the microchannel were coated at first with silicate glass via treatment with allyhydridopolycarbosilane for higher solvent resistance and generation of hydroxy groups on the surface after hydrolysis. Functionalization of the reactive surface with bromo groups was done with 3-bromopropyltrimethoxysilane. Finally, poly(vinylpyridine) was immobilized by quaternization of the external pyridine moieties with the bromo groups on the channel surface. A $RuCl_3$ solution in DMF was flushed through the microchannel and resulted in the desired noble metal complexes in the polymer matrix. Then as prepared flow photoreactor was irradiated either with a CFL (23 W) or with a white light LED array.

With a flow rate of 25 μL min^{-1} and a corresponding residence time of 2 min, the reaction proceeded very well with 85% yield at full conversion. The novel Ru^{3+}-functionalized microreactor clearly outperformed the batch process both in reaction time (2 min vs. 16 h) and yield (85% vs. 73%). Adiyala, Kim et al. then performed a library screening with 20 different α-azidochalcones to react with L-proline in the flow photoreactor. Except for one furan-derived substrate, which did not react at all, all other library members could be converted with very good yields, always being higher in flow mode compared to batch. A wide tolerance to electron-withdrawing and donating functional groups like Cl, Br, NO_2, CN or OMe could be demonstrated as well. Although this novel process allows a great variety for the α-azidochalcones, other amino acids other than L-proline could not be used. Any attempts to convert alanine, glycine, leucine, tryptophan or isoleucine were not successful. The authors assume that the free primary amine group in non-cyclic amino acids leads to the decomposition of the 2H-azirine ring. In summary, this novel photo-induced synthesis process of fused imidazole systems was significantly improved by the switch to flow mode and by a straightforward immobilization strategy for the ruthenium catalyst.

Extending the ring structure of API-relevant compounds to a larger number of interconnected atoms and atom types needs novel synthesis routes and methodologies. Morandi, Bode et al. published a method that allows the conversion of aldehydes with the so-called silicon amine protocol (SLAP) reagents into morpholines, oxazepanes, thiomorpholines and thiazepanes [124]. In their work, iridium-based catalysts have been used alone for the synthesis of piperazines, or with an additional Lewis acid for the synthesis of thiomorpholines. But this catalyst/additive combination could not be applied to the synthesis of morpholines as the required oxidation potential of $ROCH_2SiMe_3$ was still too high, even for the Ir^{IV} species in the catalytic cycle. Interestingly, the change to 2,4,6-triphenylpyrilium tetrafluoroborate (TPP) as inexpensive photoredox catalyst allowed the conversion of the appropriate SLAP reagent. This organic dye has in its excited state a redox potential of $E_{1/2}$ ($PC^{*+}/PC^{\bullet}$) = +2.30 V vs. SCE in acetonitrile, which is high enough to oxidize the $ROCH_2SiMe_3$ moiety with E_p = +1.90 V vs. SCE in acetonitrile. Lewis acid screening proved trimethylsilyl trifluoromethanesulfonate (TMSOTf) as most applicable to the reaction conditions with the organic dye as photocatalyst and the SLAP reagents used. For example, 2-(trimethylsilylmethoxy)ethanamine (**103**) (0.5 mmol) was condensed with *p*-methoxybenzaldehyde (**104**) (1 eq.) to give the appropriate imine **105** that is subsequently transferred with the triphenylpyrilium photocatalyst (5 mol%) and TMSOTf (1.3 eq.) into the glass photoreactor (1.7 mL).

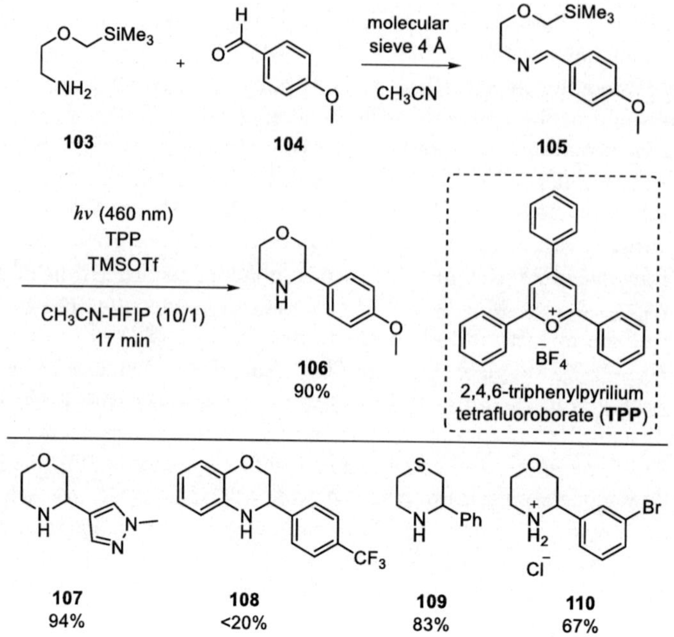

Fig. 20 Top: Imine formation with subsequent photocyclization. Bottom: Selected examples for the photocyclization of SLAP reagents to morpholines and thiomorpholines

A mixture of acetonitrile-hexafluoroisopropanol was used as solvent (10/1). The reactor was irradiated with an air-cooled LED array having a main emission wavelength at 460 nm (Fig. 20).

The best process conditions were evaluated with a flow rate of 0.1 mL min^{-1} corresponding to 17 min residence time. A yield of 90% at full conversion was obtained for **106** and proved that the system is running properly. Various morpholine derivatives were synthesized with yields between 55% and 94%. In the case of the oxazepane derivatives, a higher photocatalyst loading of 10 mol% was necessary as well as a lower flow rate of 0.06 mL min^{-1} (28 min). In analogy to these structures, Morandi, Bode et al. synthesized substituted oxazepanes, benzomorpholines, thiomorpholines and thiazepanes as well. The yields range from <20% for the CF_3-substituted benzmorpholine **108** to 83% for an arylated thiomorpholine **109**. The broad substrate scope of this methodology was finally demonstrated by the scale-up for the synthesis of 3-(3-bromophenyl)morpholine as chloride salt **110**. For this example 4.4 g of 2-(trimethylsilylmethoxy)ethanamine was condensed with 5.6 g of 3-bromobenzaldehyde and photocyclized to 5.5 g (67%) of the desired morpholine after hydrochloride salt formation in diethylether. This example clearly demonstrates the versatility of the process both in substrate variations and process scale.

In the last example, the flow synthesis of 6(5*H*)-phenanthridinone derivatives will be presented, which has been developed by Tranmer et al. for a reliable access to

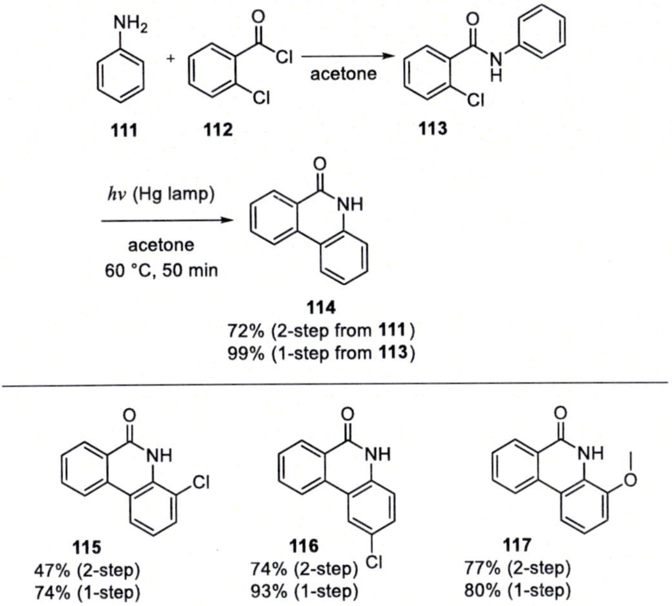

Fig. 21 Top: 2-Step synthesis of 6(5*H*)phenanthridinones **114** via 2-chloroamide building block **113**. Bottom: The effective yield of the 2-step reaction is determined by the substitution patterns of the starting material. "1-step" = starting with aryl amide, "2-step" = starting with aryl acid chloride

these building blocks as potential poly(ADP-ribose) polymerase inhibitors [125]. Known methodologies based on the Schmidt reaction use explosive HN_3. In the case of the Ullmann and Suzuki coupling reaction only moderate yields can be achieved. Hence, Tranmer et al. applied a photochemical cyclization route which proved to be superior to the mentioned protocols. 2-Chlorobenzamides (e.g. **113**, synthesized from aniline **112** and benzoyl chloride **111**) were used to optimize the reaction conditions by investigating the impact of the solvent, substrate concentration, flow rate and type of UV light filter equipment. In the end, the authors used acetone as solvent and a substrate concentration of 5 mM with a flow rate of 0.2 mL min^{-1} in a capillary photoreactor (FEP, 10 mL, 50 min). A quartz glass filter was applied with a medium pressure Hg lamp as light source. These process conditions were used for the substrate screening applying a reactor temperature of 60°C (Fig. 21, second step). Fully unsubstituted 2-chlorobenzamide **113** was converted in **114** with a high yield of 99%. Moderate (74%) to very good yields (93%) were achieved for aryl groups with an electron-withdrawing chlorine atom (**115**, **116**) or with an electron-donating methoxy group (**117**).

The process was then improved by combining the photocyclization reaction with the precursory amidation reaction between aniline (**111**) and 2-chlorobenzoyl chloride (**112**). The starting materials were dissolved in acetone and pumped via a T-piece into a thermal coil reactor (10 mL, 50 min), which was set to 60°C. No additional base like NEt_3 was necessary to trap the hydrogen chloride resulting from the amidation in the first step. The raw product solution with the amide precursor for the second step was directly transferred into the capillary photoreactor, which was converted into the phenanthridinone derivatives, with a slightly lower yield of 77% for the methoxy derivative **117** (80%) or with a clearly lower yield of 72% for the unsubstituted phenanthridinone **114** (99%). Although the coupled process results in lower yields, but the developed multi-step flow synthesis is a fast and important tool for rapid drug discovery under sustainable conditions, without purification of the intermediates.

4 Photochemical Rearrangement Reactions

Rearrangements are often part of complex synthesis routes for natural products [126–128]. The first example in this sub-chapter is about the photochemical transformation of pyridinium salts into bicyclic aziridines published by Siopa, Afonso and co-workers [129]. Bicyclic aziridines are key intermediates in the synthesis route of aminocyclopentitols. The photolysis of an alkylated pyridinium salt **118** results in a *cis*-fused cyclopentenoaziridine allylic cation **119**, which can react with poor nucleophiles, e.g. water or alcohols from the solvent. This step leads stereochemically controlled to the *cis*-fused cyclopentenoaziridine **120** with the nucleophile in *trans* position to the nitrogen atom of the aziridine (Fig. 22). Subsequent reaction with another nucleophile completes the reaction path to the desired trisubstituted cyclopentene isomers **121** and **122**.

Unfortunately, the photolysis as the first step in this reaction sequence is the limiting one, making such bicyclic aziridines not accessible in large quantities via batch synthesis. For example, *n*-propylated pyridinium perchlorate gives a yield of only 33% in methanol as solvent, with a productivity of 0.22 g L^{-1} h^{-1} in a batch photoreactor with a 450 W medium pressure mercury lamp with vycor glass filter. In order to overcome this bottleneck in the synthesis of bicyclic aziridines, Siopa,

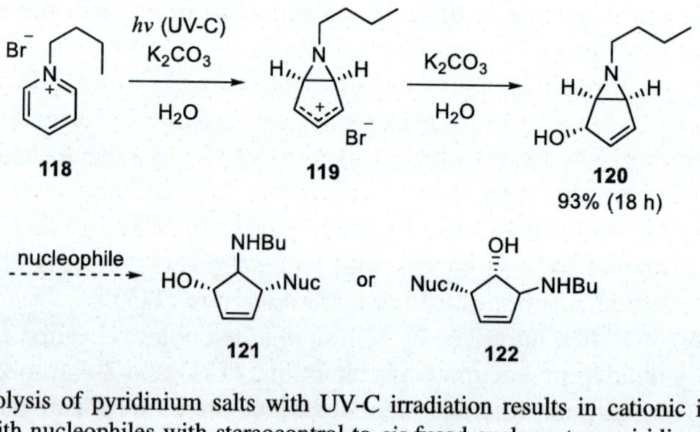

Fig. 22 Photolysis of pyridinium salts with UV-C irradiation results in cationic intermediates, which react with nucleophiles with stereocontrol to *cis*-fused cyclopentenoaziridines. Subsequent nucleophilic attack gives trisubstituted cyclopentene scaffolds

Afonso et al. design three different custom-made flow photoreactors either with a FEP capillary (length: 22.96 m, I.D. = 4 mm) or with quartz glass tubes (24 tubes with 22.5 cm in length and I.D. = 4 mm, or 32 tubes with 23 cm and I.D. = 2 mm) interconnected with polymer tubings. The flow devices were adapted to a Rayonet RPR-100 reactor with 16 UV-C light sources with 8 W for each lamp. Batch experiments were performed with single quartz glass tubes and FEP capillary of equivalent length and diameter. Substrate solutions of pyridinium bromide **118** in water (0.06 mol L^{-1}) with 1.2 eq. potassium carbonate were filled into the tubes and capillaries and irradiated for 8 h inside the UV-C light source. For all three batches, approx. the same conversions between 88% and 92% were obtained, corresponding to nearly the same productivities between 0.96 and 1.09 g L^{-1} h^{-1}. The photolysis of a highly concentrated substrate solution with 0.3 mol L^{-1} resulted in different conversions due to the physical restriction of light absorption depending on the pathlength through the reaction solution. As a result, both the tube and the FEP capillary with 4 mm diameter gave significantly lower yields of 17% and 13%, compared to 33% for the 2 mm thick quartz glass tube. Hence, the latter one has a high productivity of 1.81 g L^{-1} h^{-1}. Recirculating continuous flow experiments were then performed as well with all three flow devices. A given volume of reaction solution was circulated through the flow devices until the conversions of approx. 90% were reached for all reactors. The large diameter glass tube reactor gave the best productivity with 3.7 g L^{-1} h^{-1} after 9 h, the small diameter glass reactor achieved 3.3 g L^{-1} h^{-1} after 10 h. The FEP reactor had the lowest productivity with 0.5 g L^{-1} h^{-1} after 19.5 h. Finally, single continuous flow through experiments were done solely with the large diameter quartz glass tube reactor feeding a substrate solution of 0.062 mol L^{-1}. Although it was estimated to have a conversion greater than 90% after a residence time of 9 h, the results here were more or less disappointing with a low conversion of 56%. The flow rate was reduced from 0.56 mL min^{-1} to 0.28 mL min^{-1} resulting in an increased conversion of 93% after a residence time of approx. 18 h. Throughout the complete process run time of 3.4 days under these conditions, 1.4 L substrate solution was irradiated with an overall conversion greater than 60%. In summary, Siopa, Afonso et al. were able to improve the synthesis of bicyclic aziridines from photolyzed pyridinium salts by applying flow photochemistry to this process. High productivities of up to 3.7 g L^{-1} h^{-1} clearly outperform the literature-known batch syntheses of equivalent substrates.

Lattes and Aubé studied a two-step synthesis route as an interesting variant to the well-known Beckmann rearrangement. They started from a ketone and converted it into an oxaziridine with subsequent photolytic rearrangement into the desired lactam [130–132]. Based on this work, Cochran et al. investigated the use of continuous flow technology for the optimization of the photolytic rearrangement of chiral oxaziridines [133]. For process development and optimization, batch tests were done with prochiral 4-*t*-Butylcyclohexanone. Its conversion into the corresponding imine was done with *S*-α-methylbenzylamine. After heating both compounds in toluene, the diastereomeric product was then selectively oxidized with *m*-chloroperbenzoic acid (*m*-CPBA) to the oxaziridine **123**. A three-layer capillary

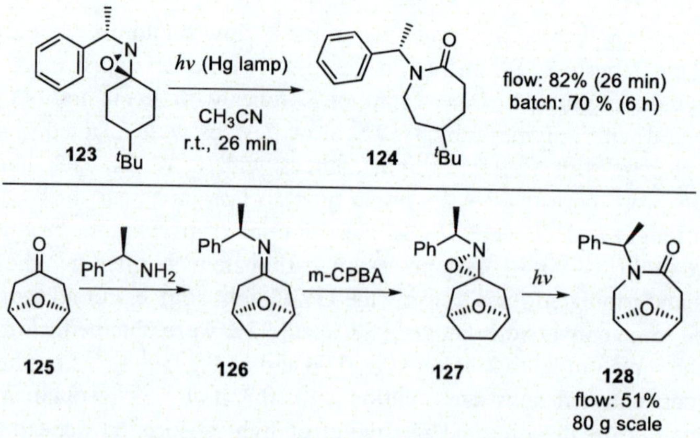

Fig. 23 Top: Synthesis route to monocyclic lactams via photolytic oxaziridine rearrangement. Bottom: Conversion of bicyclic ketone **125** into the chiral bicyclic lactam **128**

photoreactor (FEP, 130 mL) was finally used for the photochemical rearrangement with a medium-pressure Hg lamp as light source (450 W). A cold stream of nitrogen gas was flushed around the capillary photoreactor to maintain the reactor temperature at room temperature placed inside a Dewar flask. The reaction solution with **123** (0.1 mol L^{-1} in acetonitrile) was pumped through the capillary with a flow rate of 5 mL min^{-1} (26 min). Under these conditions, the process yielded 82% of the isolated *S*-isomer **124** on one-gram scale, whereas in 6 h batch reaction time only 70% yield was possible (Fig. 23). Based on these optimized reaction conditions, a scale-up of the same synthesis route was done as well, obtaining 20 g of the desired lactam with a yield of >80%.

The authors also applied their method to the synthesis of bicyclic lactams. Oxabicyclooctanone **125** was condensed with *R*-α-methyl-benzylamine to imine intermediate **126**, which was then converted with *m*-CPBA to **127** as single oxaziridine diastereomer with a yield of 51%. As last step the photolysis was performed on larger scale with 80 g starting material. Surprisingly, the photolysis of **127** was rather slow compared to foregoing experiments and gave approx. 40 g of product **128** (51% yield). Chromatographic separation of the product allowed the recovery of the starting material and therefore the effective yield of the chiral lactam was increased to 99%.

Not only lactams, but also amide groups in non-cyclic compounds can be accessed by the rearrangement of oxaziridines induced by light. Jamison et al. designed a synthesis route for the conversion of nitrones into amide groups via oxaziridine. The nitrones were obtained by the condensation of amino acid-based hydroxyl amines with aldehydes. After photochemical conversion of nitrone **129** into **130**, homolytic photolysis of the nitrogen–oxygen bond and subsequent hydrogen-atom migration results in a final rearrangement into the amide group in **131** (Fig. 24) [134]. Jamison et al. used this synthesis strategy as coupling reagent-free amide bond formation for (oligo)peptide synthesis without usually necessary

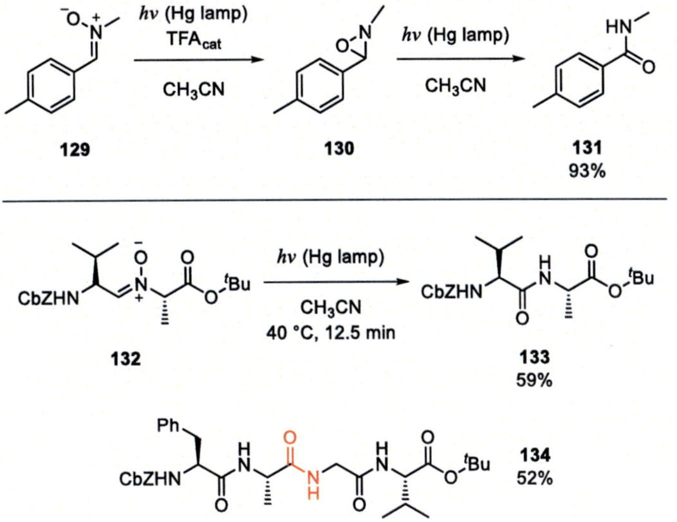

Fig. 24 Top: Light-induced rearrangement of nitrone and oxaziridine as key steps in the amide group formation with 10 min residence time. Bottom: Coupling of two or four protected amino acid fragments to L-Ala-L-Val (**133**) and L-Val-Gly-L-Ala-L-Phe (**134**), respectively, as example for protection group tolerant photochemical synthesis

separation steps (**133**, **134**). The flow experiments were done in a customized quartz glass tubing (I.D. = 0.762 mm, 0.625 mL) instead of a UV-transparent polymer capillary. A medium pressure Hg lamp was applied (450 W) as light source without filter equipment. Nitrone **129** was used for flow process development, readily prepared from *p*-tolualdehyde and methyl hydroxylamine hydrochloride in CH_2Cl_2 and NEt_3 as additional base. The starting material was dissolved in acetonitrile (0.05 mol L^{-1}) with additional 0.25 eq TFA as catalytically active acid. The desired amide could be obtained in 93% yield at full conversion with a residence time of 10 min at a rather high temperature of 92°C.

Amide coupling was then performed under these optimized reaction conditions for L-alanine and L-valine. The desired nitrone **132** was obtained by the condensation L-alanine hydroxylamine and an L-valine-derived aldehyde. The nitrone (0.1 mol L^{-1} in acetonitrile) was pumped into the photoreactor without any additive. With a residence time of only 12.5 min, the nitrone was fully converted to dipeptide **133** with a yield of 59%. Larger peptides can be also accessed, e.g. Val-Gly-Ala-Phe (**134**) with a yield of 52%. Finally, Jamison et al. proposed to apply their novel methodology to the engineering of proteins. They suggested to use a bis-aldehyde as linker molecule, which connects two proteins with hydroxylamine moieties at the protein surface for nitrone formation.

5 Incorporation of Fluorine and Fluorine-Containing Groups

Late-stage fluorination reactions are essential for the direct structural modification of high-value molecules regarding, e.g. lipophilicity, stability and acid-base behaviour [135, 136]. With a more detailed knowledge about the benefit of fluorinated molecules in drug applications, the research on selective and mild fluorination methods increased as well [137, 138]. Various electrophilic and nucleophilic reagents for monofluorination were developed as well as strategies for the installation of, e.g., CF_2H groups, (S, O)CF_3, or perfluorinated alkyl chains [139–143]. Photochemistry became a method of choice as it offers mild reaction conditions with a broad repertory of photocatalysts with fine-tuneable redox properties. As a first example, Britton et al. combined tetrabutylammonium decatungstate (TBADT), a well-known hydrogen-atom transfer (HAT) photocatalyst, with *N*-fluorobenzenesulfonimide (NFSI) for the selective monofluorination of benzylic C-H positions [144]. Intensive reaction optimization showed that a mild base like sodium hydrogen carbonate is necessary to prevent by-product formation of acetamides via Brønsted acid-catalysed fluorine substitution with an acetonitrile solvent molecule. Britton et al. applied their novel methodology to a library of 15 varying substituted arenes like 4-ethylphenyl acetate (**135**) (0.6 mol L^{-1}) with NFSI (3 eq.), TBADT (2 mol%) and $NaHCO_3$ (1 eq.) suspended in acetonitrile. The degassed reaction solution was irradiated for a minimum of 16 h in batch mode with UV light at 365 nm until the reaction progress stopped as indicated by NMR analysis (Fig. 25 top). Under these conditions, 4-ethylphenyl acetate (**135**) was monofluorinated to **136** in 75% yield at 92% conversion. Britton et al. then tested their synthesis route in flow with the monofluorination of ibuprofen methyl ester **137** as an example for an API precursor (Fig. 25 bottom). Although the reaction mixture is a slurry, the flow procedure worked out well with a custom-made photoreactor (FEP, I.D. = 0.7 mm, wrapped around a black light bulb). With a reduced substrate concentration of 0.2 mol L^{-1}

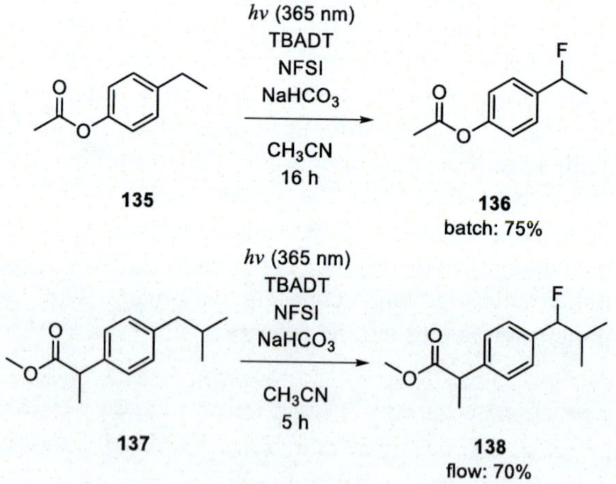

Fig. 25 Decatungstate-based C-H activation of benzylic positions for the selective monofluorination of **135** and ibuprofen methyl ester **137**

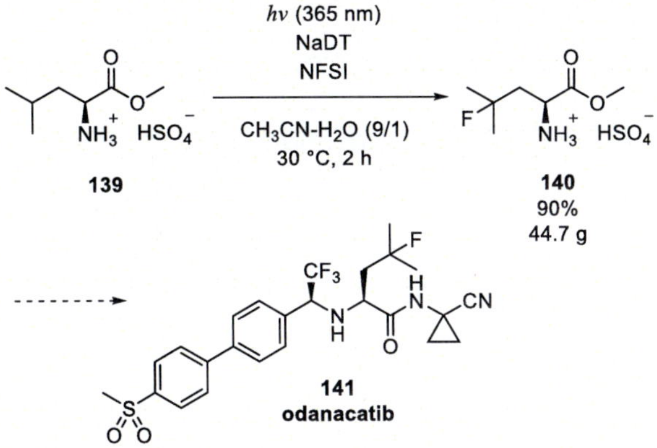

Fig. 26 Photocatalysed monofluorination of L-leucine in flow on multigram scale as one key step in the synthesis route of drug candidate odanacatib

and only 2 eq. NFSI, the monofluorination was achieved in 70% yield within a shorter residence time of 5 h compared to the examples in the batch synthesis with 16–48 h irradiation time.

In a similar work, Britton, DiRocco et al. used the same combination of photocatalyst and fluorination reagent for the selective fluorination of leucine methyl ester (**139**) on large gram scale within the synthesis route of odanacatib, a drug candidate in clinical trials for the treatment of osteoporosis [145]. Based on this work, Britton, DiRocco et al. changed the solvent from pure acetonitrile to a water-acetonitrile mixture (1/2). Instead of 3 eq. NFSI, only 1.2 eq. was necessary in addition to 2 mol% of TBADT. The hydrochloride salt of L-leucine was converted to the desired γ-fluoro leucine methyl ester in 83% yield. High throughput experimentation was then applied to further optimize the reaction conditions regarding the amino acid and the decatungstate counter ion, concentration, catalyst loading and water content. Interestingly, the experiments showed that an exchange of both counterions to sodium for decatungstate and bisulfate for the amino acid is beneficial for the process. This new composition also allowed a lower amount of water in the solvent mixture of 1:9 instead of 1:2 (Fig. 26). These new reaction conditions were then applied to the gram scale synthesis of γ-fluoro leucine methyl ester (**140**) in continuous flow mode. Britton, DiRocco et al. design a photoreactor with a FEP capillary (I.D. = 1.6 mm, 60 mL) and an integrated UV light source for 365 nm emission wavelength (32 W).

The flow synthesis was performed on the 190 mmol scale with 46.5 g of L-leucine methyl ester bisulfate salt **140** dissolved in aqueous acetonitrile (0.2 mol L^{-1}), with a flow rate set to 0.5 mL min^{-1} (2 h). The reaction solution was collected, concentrated under reduced pressure prior to an azeotropic distillation with 2-MeTHF as added solvent. With this purification step, 44.7 g of **140** (90% yield) could be prepared and isolated by direct precipitation. Once again, this example clearly demonstrates the capability of flow photochemistry for mild chemical processing on a larger scale. The method of Britton, DiRocco et al. outperforms the known multi-step synthesis routes by far in simplicity and yield.

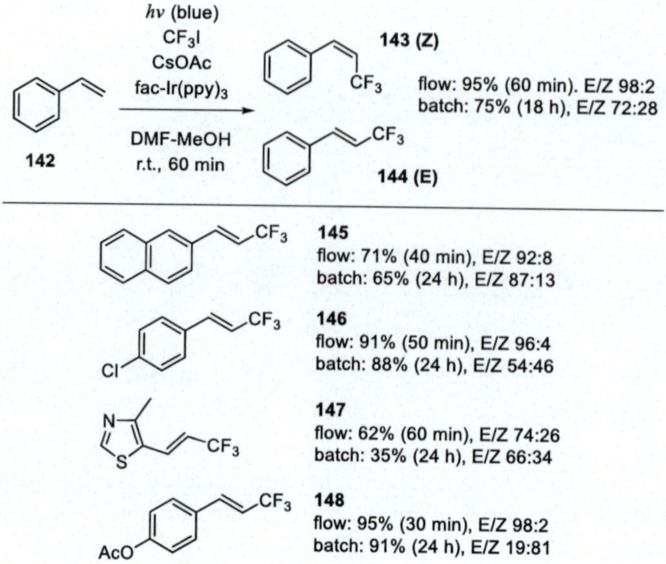

Fig. 27 Top: CF_3I as source for selective trifluoromethylation of styrene (**142**) with high selectivity for the *E* isomer within short irradiation time. Bottom: Examples of styrene building blocks and analogues applied to the Ir-catalysed trifluoromethylation

Besides the trifluoromethylation of aromatic thiols [146], and the perfluoroalkylation of heteroaryls [147], the workgroup of Noël also developed a flow synthesis for allylic alcohols from Grignard reagents with the difluoroalkylation of the vinyl double bond in the second step [148]. Noël et al. also used their expertise in photocatalysis for light-induced activation of CF_3I as an inexpensive, gaseous reagent for the trifluoromethylation or hydrotrifluoromethylation of the vinyl double bonds in styrene derivatives [149]. Screening tests were done with styrene (**142**) as substrate (0.5 mmol) with various solvents, catalysts and bases. Finally, fac-Ir(ppy)$_3$ was applied as photocatalyst (1 mol%), with DMF as solvent (0.1 mol L^{-1}) and caesium acetate as mild base (2 eq). Styrene (**142**) was converted in batch at room temperature and atmospheric pressure within 18 h into the trifluoromethylated target compounds **143** (Z) and **144** (E) in 75% yield and an *E*/*Z* ratio of 72:28 (Fig. 27). As intended, the following transfer to continuous flow mode had also a strong impact on yield and reaction time. Noël et al. used a capillary photoreactor with a standard PFA tubing (1.25 mL) and blue LEDs (6.24 W). Gas and liquid phases were mixed with a T-piece as static mixing element ($f_l = 1.25$ mL min^{-1}, $f_g = 1.20$ mL min^{-1}). Within a residence time of 60 min, styrene was converted with an increased yield of 95% to the desired products (Fig. 27). More complex starting materials were converted as well with equivalent results to the desired products **145–148**. Besides the increased productivity of the flow photoreactor, the *E*/*Z* ratio also tends towards more *E* isomer than *Z* isomer. For the conversion of styrene (**142**), the *E*/*Z* ratio raised from 72:28 to 98:2. The *E*/*Z* product ratio changed for the conversion of 4-acetoxysytrene to **148**

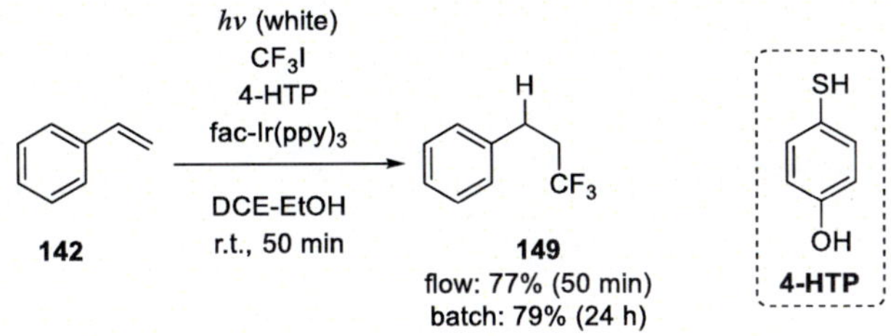

Fig. 28 Selective hydrotrifluoromethylation of **142** accessed via 4-hydroxythiophenol as additive

even more drastically from 19:81 in batch mode to 98:2 in the flow photoreactor. This notable increase in favour of the *E* isomer is a result of the shorter residence time in the flow photoreactor and shortened irradiation time. Hence, under flow conditions, the photocatalysed isomerization of the *E* isomer to the *Z* isomer is not dominant and results in a higher amount of *E* isomer [150].

In order to use the newly developed method for the selective hydrotrifluoromethylation of styrenes, Noël et al. used a mixture of DCE-ethanol (9/1) instead of DMF and added 1.2 eq of 4-hydroxythiophenol (Fig. 28). This hydrogen-atom donor circumvents the formation of by-products resulting from overoxidation and dimerization of the styrene substrates. 3,3,3-Trifluoropropylbenzene (**149**) was obtained with a yield of 77%, slightly lower than 79% in batch mode. Again, a significant decrease in reaction time from 24 h to 50 min was possible in flow mode. In summary, both trifluoromethylation and hydrotrifluoromethylation are possible with a broad variety of styrene derivatives or vinyl group containing substrates under these very mild reaction conditions.

Highly functionalized (hetero)arenes are of great interest for medicinal chemistry research. Indeed, complex and rather challenging compounds are often ignored substrates during the development of novel functionalization methods like fluorination. Low substituted arenes with inactive groups are often used as benchmark substrates here without clear perspective for further derivatization. Especially in the case of API-relevant building blocks, it is interesting to have reactive substituents like iodine or bromine at the arene core for further derivatization via classical noble metal catalysed C–C or C–X coupling. In order to achieve this goal for the trifluoromethylation of highly substituted heteroarenes, Alcazar, Noël et al. refined their experience in photocatalysed trifluoromethylation with iridium photocatalysts and sodium trifluorosulfinate as shelf-stable CF_3 source (Fig. 29) [151]. Caffeine (**150**) (0.1 mol L^{-1} in DMSO) was used as model substrate for the reaction optimization in flow applying a commercially available photoreactor (FEP, 10 mL) and 450 nm irradiation (24 W). With 1.5 eq of CF_3SO_2Na and [Ir{dF(CF_3)ppy}$_2$(dtbbpy)]PF_6 (1 mol%) as photocatalyst, only 12% yield was achieved as analysed with LC-MS (Fig. 29). The addition of 1 eq. of $(NH_4)_2S_2O_8$ as oxidant gave an improved yield of 48%. The authors assume that this reagent is beneficial for the re-aromatization of the radical intermediate after addition of the CF_3 group to the heterocycle. Finally, with 3 eq. of CF_3SO_2Na a maximum yield of 54% could be obtained for the trifluoromethylated benchmark substrate **151**.

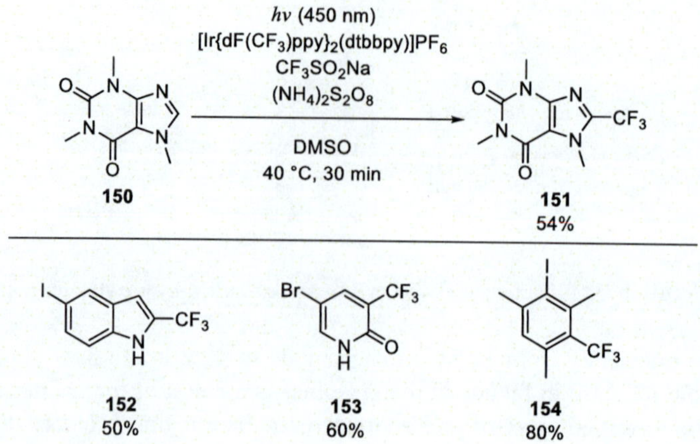

Fig. 29 Top: Trifluoromethylation of caffeine (**150**) as benchmark substrate. Bottom: Potential API-relevant (hetero)arenes

With these conditions in hand, Alcazar, Noël et al. screened a large variety of indole, pyridine and pyrimidine derivatives with and without halogen substituents. For example, the presence of an iodo substituent on the indole was tolerated and gave a C2/C3 regioisomeric mixture of the CF_3-substituted 5-iodoindole (**152**). 5-Bromopyridone was selectively trifluoromethylated at the C3-position to form **153** in 60% yield. An unactivated arene like iodomesitylene was converted into the CF_3 derivative **154** within 30 min with a high yield of 80%. This compound class usually need long batch reaction times of multiple hours. But in flow mode, the improved irradiation on the microscale accelerates such reactions considerably.

As last example in this chapter, the difluoromethylation of heteroarenes will be discussed with special focus on the use of ^{18}F isotopes in radiopharmaceuticals for PET studies. Luxen, Genicot et al. applied photoredox catalysis for the radical addition of a radiolabelled $CHF^{18}F$ group into *N*-heteroaromatic building blocks [152]. Based on the work of Hu on the use of benzothiazole sulfone reagents for the introduction of CHF_2 groups into alkenes [153, 154], the authors first performed a two-step thermal synthesis to produce the ^{18}F-radiolabelled benzothiazol sulfone **156** via halogen exchange (HALEX) reaction and subsequent oxidation of the sulphur atom. The HALEX reaction was done with $[^{18}F]KF$, K_{222}, K_2CO_3 in acetonitrile at 120°C. For the oxidation of the sulphur atom, $RuCl_3$ with $NaIO_4$ was used in water at room temperature. Luxen, Genicot et al. report a radiochemical yield (RCY) of $15.2 \pm 0.3\%$ for the first step, and $13.4 \pm 0.4\%$ for the second step. With the radiolabelled reagent in hands with good overall yields, the second step was optimized in flow mode with the anti-herpetic drug acyclovir (**157**) as substrate (Fig. 30). Variation of the solvent (DMSO, DMF, acetonitrile, DCE, with or without aliquots of water), the photocatalyst ($Ir(ppy)_3$, $Ru(bpy)_3$, benzophenone) and the temperature or flow rate gave finally the best results for DMSO as most compatible solvent and $Ir(ppy)_3$ as most productive catalysts (0.01 μmol) [155].

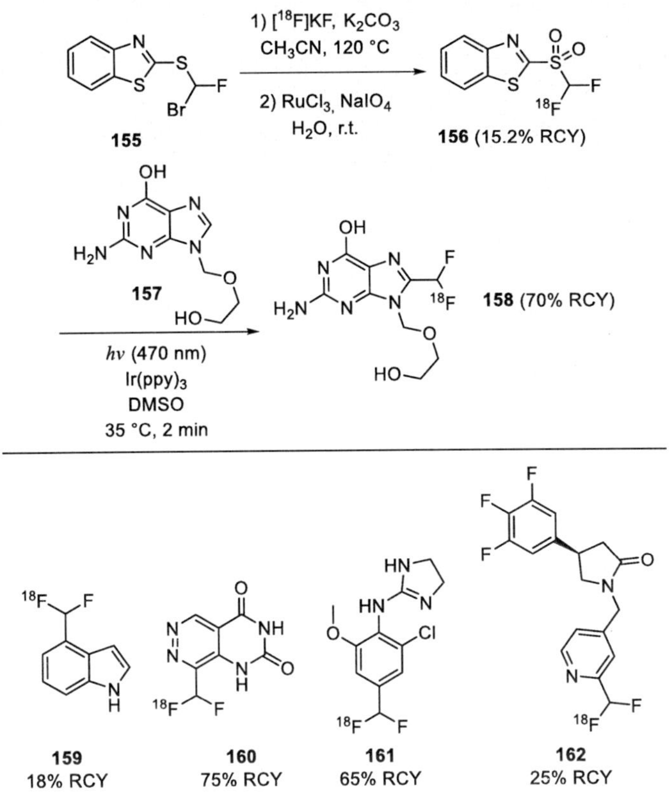

Fig. 30 Top: Radical addition of $CHF^{18}F$ into acyclovir (**157**) as important example for *N*-heteroarenes to be radiolabelled for medicinal chemistry applications in PET studies. Bottom: Selected examples synthesized via sulfone **156** incl. Moxonidine analogue **161** and SV_2A PET tracer **162**

With a short residence time of 2 min and a temperature of 35°C, acyclovir **157** (20 μmol) was converted with a radiochemical yield of 70 ± 7% into its $CHF^{18}F$-substituted derivative **158** applying blue light at 470 nm for exciting the iridium catalyst. Based on these very positive results, Luxen, Genicot et al. screened a broad variety of 24 different indoles, benzimidazoles, azaindoles, pyridines and pyrimidines. With the novel methodology low to good yields could be obtained for, e.g., 1*H*-indole **159** (18%) or pyrimido[4,5-*d*]pyridazine-2,4(1*H*,3*H*)-dione **160** (75%). Besides acyclovir, two other drugs were functionalized with $CHF^{18}F$ as well: the moxonidine analogue **161** for the treatment of hypertension, and the SV_2A PET tracer **162** with SV_2A as target. Both compounds could be converted with RCY of 65 ± 4% and 25 ± 3%, respectively. In the latter case, it was necessary to separate the isomeric mixture by HPLC, which yielded the desired isomer with the $CHF^{18}F$ group in the 3-position of the pyridine ring in 1.5 ± 0.1% RCY. Finally, Luxen, Genicot et al. recently published their development of an automated continuous flow photoredox ^{18}F-difluoromethylation procedure of *N*-heteroarenes. With the integration of their novel methodology into a fully automated system, it is now possible to produce the desired ^{18}F-radiolabelled benzothizol sulfone in one module, directly coupled to a second module for the photoredox difluoromethylation of heteroarenes [155]. In this case, the complete synthesis sequence was done for acyclovir **157** as model substrate in 95 min. This automation for the onsite production of radiolabelled drugs is a great progress for preclinical development and human PET studies with new radiotracers.

6 Trend to Photochemical-Assisted Biocatalysis

In this last short chapter, a new trend in photochemistry should be mentioned as well, which features a very interesting progress in the modality to use and apply photochemistry and photocatalysis, respectively [156]. In medicinal chemistry research, many disciplines have joint forces to design and produce libraries of chemical compounds which are essential for the development of drug candidates with very specific mode of action against various diseases [157]. Often, special functional groups in the molecule like the above-mentioned fluorine-containing moieties or stereochemical centres define such mode of action and lead to, e.g., an improved pharmacokinetics or transport into or inside cells. Especially the stereochemical conformation of a drug molecule can have massive impact on the mode of action. In this case, either complex synthesis routes with expensive chiral catalysts are necessary and/or a cost-intensive purification must be done afterwards. Hence, the request for more straightforward routes to access such chiral molecules resulted in the adaptation of biological catalysts, enzymes, for the stereocontrolled synthesis of complex molecules [158, 159]. In conjunction with the development of visible light photocatalysis and the plethora of available metal-based photocatalysts or purely organic dyes, photochemical-assisted biocatalysis is becoming a hot spot of current research in synthetic (bio)organic chemistry [160, 161]. Visible light as low energy carrier does not interrupt the enzyme activity due to light absorption by aromatic amino acid moieties of the enzyme as it is the case for UV light [162]. Hence, photocatalysis and biocatalysis are energy-independent from each other and allow a very powerful synergetic use of both modes of catalysis. Recent examples illustrate this combination in various modes. Ward, Wenger et al. combine a novel water-soluble iridium photocatalyst containing sulfonate groups on the ligand with a monoamine oxidase enzyme for the cyclic enrichment of a cyclic chiral amine (Fig. 31) [163]. 5-Cyclohexyl-3,4-dihydro-2*H*-pyrrole (**163**) (0.01 mol L^{-1}) was used as model substrate with $Na_3Ir(sppy)_3$ as photocatalyst (1 mol%) and ascorbic acid (0.2 mol L^{-1}) as hydrogen-atom source in phosphate buffer at pH 8. *E. coli* cells with recombinantly expressed monoamine oxidase (MAO-N-9) were used as whole cell biocatalyst. In the photocatalytic cycle, the imine is converted into the amine as both *R*- and *S*-isomer via radical imine reduction and subsequent hydrogen-atom transfer from the ascorbic acid to the α-amino alkyl radical. In the enzymatic catalysis, only the *S*-isomer is converted to imine, which again enters the

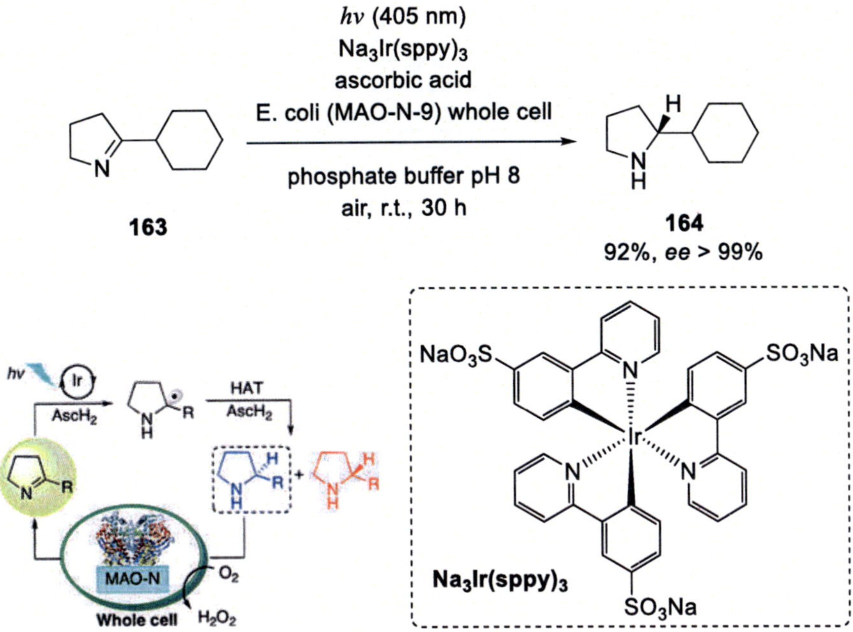

Fig. 31 Enantioselective synthesis of amine **164** by the combination of photoredox catalysis with whole cell biocatalysis. Reproduced with permission from Ref. [163]. Copyright 2018 Royal Society of Chemistry

photocatalytic cycle. After 10 h of irradiation with blue light at 405 nm under air, the imine was converted with 92% into the desired chiral amine **164** with 91% *ee* for the *R*-isomer. Prolonged irradiation by another 20 h led to an enantiomeric excess greater than 99%.

A second example features the photocatalysed thio-Michael addition in the first step with a biocatalytic ketoreduction to enantiomerically pure 1,3-mercaptoalkanoles in the second step. Castagnolo et al. convert various α,-β-unsaturated carbonyl compounds and thiols into the respective thioethers with $Ru(bpy)_3Cl_2$ as photocatalyst (0.3 mol%) [164]. Highly selective ketoreductases (KRED) were used to convert the obtained ketones into enantiomerically pure secondary alcohols (Fig. 32). As part of the biocatalytic cycle, 2-propanol was used as hydrogen source for the in situ reduction of the necessary KRED co-factor $NADP^+$ to NADPH. In the case of methylvinylketone **69** and thiophenol **165** as substrates, a conversion of 99% was possible at 73% yield with an *ee* > 99% for the *R* isomer of the desired 1,3-mercaptoalkanol **166**.

Hartwig et al. designed experiments for the cooperative asymmetric reduction of the vinyl double bond in 2-phenylsuccinates, their related diesters or cyanoacrylate substrates [165]. In this example, an iridium-based photocatalyst (1 mol%) was selected for the efficient *Z*- to *E*-photoisomerization of dimethyl 2-phenylsuccinate (**167**) as alkene model substrate (0.005 mol L^{-1}), since only the *E* isomer was

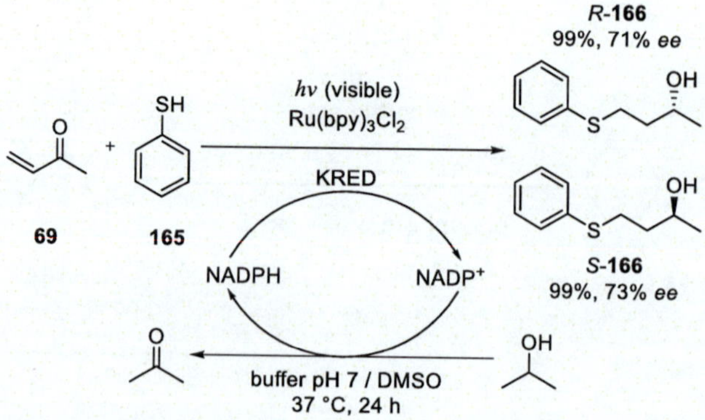

Fig. 32 One-pot photo-biocatalysed cascade reaction for the synthesis of 1,3-mercaptoalkanol **166**

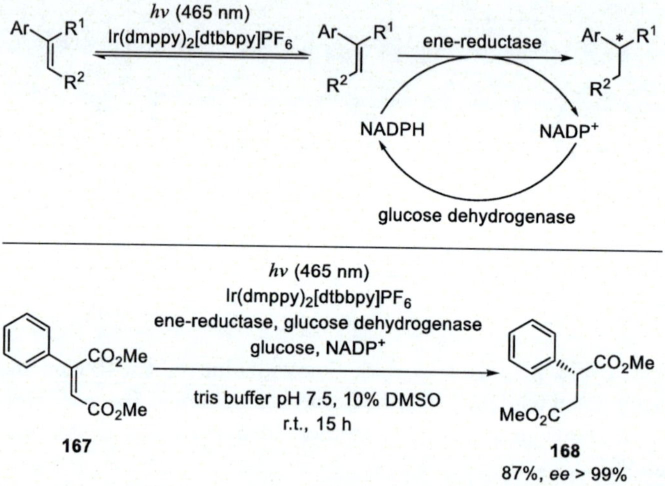

Fig. 33 Combining photocatalysis and enzymatic catalysis for the cooperative isomerization and reduction of dimethyl 2-phenylsuccinate (**167**) as alkene model substrate

accessible for the ene-reductase (0.5 mol%) applied in the following biocatalytic step (Fig. 33). The necessary co-factor $NADP^+$/NADPH was continuously regenerated by a second enzymatic system using glucose as substrate for the glucose dehydrogenase. With this strategy Hartwig et al. could prove a very good compatibility between photocatalysis and biocatalysis with a high *ee* > 99% for, e.g., dimethyl (2*R*)-2-phenylbutanedioate (**168**) with 78% isomerization yield and 87% all-over yield.

The last example is based on the direct use of hydrogen peroxide as mild and stable oxidant in biocatalytic processes. Hydrogen peroxide is a real green reagent, since water and oxygen are the only by-products upon its consumption during a reaction. Hence, hydrogen peroxide driven biocatalysis has become an important research area in recent years [166]. There are several photocatalytic systems known,

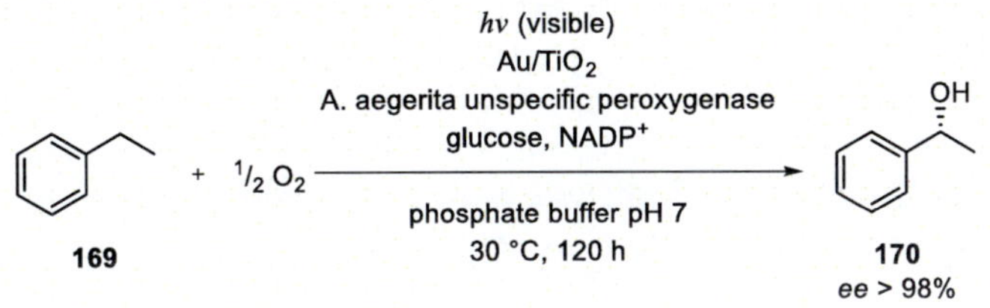

Fig. 34 Hydrogen peroxide formation via TiO_2 photocatalysis combined with unspecific peroxygenase biocatalysis allow the stereoselective hydroxylation of molecular building blocks

which are applied for the in situ production of hydrogen peroxide, e.g. TiO_2, flavin mononucleotide/riboflavin tetraacetate or carbon nitride [167–169]. Hollmann et al. combined the aerobic water oxidation on the surface of Au-loaded TiO_2 for hydrogen peroxide production with the conversion thereof by unspecific peroxygenases (Fig. 34) [170]. The authors screened dissolved and immobilized enzymes (rAaeUPO) in combination with anatase or rutile Au-TiO_2 and optimized their combined photo- and biocatalytic system on the stereoselective hydroxylation of ethyl benzene (**169**) as model substrate (0.015 mol L^{-1}).

The best conditions were found with free rAaeUPO (0.150×10^{-6} mol^{-1}) and rutile Au-TiO_2 (5 g L^{-1}) dissolved in phosphate buffer (pH 7.0). The reaction suspension was irradiated with visible light >400 nm for up to 120 h yielding (*R*)-1-phenyl ethanol (**170**) with an *ee* > 98%. Besides the screening of various other substrates, the authors also performed cascade reactions, which involved, e.g., the conversion of the intermediate ketone into a chiral amine by the application of a transaminase in a second biocatalytic step.

Already with these few examples, one can assume the plethora of possible reactions applied to photochemical-assisted biocatalysis like, e.g., halogenations [171] or cyanations [172]. The mentioned reactions open the door for mild and highly selective reactions as necessary for the synthesis of drug candidates with complex structure. Photocatalysis, especially with visible light, has found its way into many syntheses of academic interest and inspires industry and catalyst developers for either real-life industry applications or novel and even more specific photocatalytically active materials [3, 11]. On the other hand, biochemistry applied to synthetic organic synthesis is a fully accepted and highly elaborated research field, which reaches out for new directions as it is the case for the synergetic collaboration with photocatalysis [173, 174]. For both research fields, the transfer of the synthesis methodology to a technological application has taken place already. Flow chemistry with adjunct micro reaction technology has proven many advantages for photochemistry and photocatalysis [175–177] as well as for biocatalysis [178–180]. Hence, it is obvious to start thinking about photochemical-assisted biocatalysis in continuous flow mode. One can assume here the transfer of a biological cell with its chloroplasts and other biochemical compartments into a macroscopic reactor technology for the synthesis of chemical compounds. An improved physical contacting of reactive phases in a continuous flow microreactor with an advanced process control of the physical parameters of the flow process (t, T, p, ...) will realize a far better conversion and selectivity with less waste compared to batch processing. Immobilization methodologies are available for photocatalysts and enzymes. This will lead to an easier product separation from the reaction solution and, in consequence, to less cost-intensive processes for real-life industry applications. Enzymes substitute costly (noble) metal catalysts, which often have a complex ligand sphere and need a tedious synthesis upfront as well. From a technological point of view, the use of reactor systems with immobilized catalyst systems is excellently suitable for the scale-up of a production process. And, in addition, such a technology platform exhibits a very good flexibility regarding the catalyst combinations used for dedicated synthesis processes. Hence, photochemical-assisted biocatalysis has the best prospect to become a highly relevant methodology for complex molecular synthesis for library screening and follow-up syntheses in continuous flow mode inspired by nature's "technology".

7 Summary

As mentioned in the introductory chapter, medicinal chemistry and the on-going search for new drug candidates and APIs are inherently connected to great efforts in synthetic organic chemistry. Over the last two decades, novel catalysis methodologies and enabling technologies have paved the way for very mild and selective reactions in complex molecular synthesis. This chapter gives an overview about continuing efforts exemplified for photochemistry and photocatalysis in continuous flow mode. The combination of both research fields resulted in a most useful tool in organic synthesis with access to improved safety and sustainability as well as specific functionalization of molecules, which are not possible via a typical thermal route. Flow photochemistry has proved its potential as key step in a multi-step synthesis route due to its compatibility of physical process parameters. It is noteworthy as well that with photochemistry and photocatalysis it is possible to generate large compound libraries that eventually lead to the selection of drug candidates, which are later on produced for pharmacokinetic and toxicology studies in larger quantities by applying continuous flow photoreactors. Several examples of important reactions classes have been presented in this chapter to detail the benefits of flow photochemistry. Carbon–carbon bond formation between variously hybridized carbon atoms is essential for the synthesis of biologically active molecules as well as the coupling between carbon and heteroatoms. Reactive intermediates like diazo compounds or organometallic species are provided in situ and can be easily converted in continuous flow mode. Cyclization and rearrangement reactions induced by light allow the construction of complex molecules under mild conditions. Fluorination reactions and the incorporation of fluorinated groups are highly relevant for drug discovery due to their strong impact on the physicochemical characteristics and pharmacokinetics of the molecules. These reactions have become available as well with photochemistry, and especially important with late-stage fluorinations in continuous flow mode. In general, the mild and selective reaction conditions enable the conversion of molecules with a broad tolerance of other functional groups. Finally, some examples were mentioned for the photochemical-assisted biocatalysis as another highly interesting combination of two well-established catalysis methodologies. Although the synergy by this concept has been proven for batch mode, a real transfer to continuous flow mode has not yet been achieved. But as described in the short discussion of this trend, the synergy between flow technology, photocatalysis and biocatalysis is easily conceivable and might play an important role for the synthesis of future drug candidates. Photochemistry with visible light is still a rapidly growing research field and offers in close conjunction with flow chemistry and micro reaction technology a great potential for medicinal chemistry and drug discovery to solve today's challenges in growing mankind and nature's response.

Compliance with Ethical Standards

Ethical Approval: This manuscript is a review of previously published accounts, as such, no animal or human studies were performed.

Informed Consent: No patients were studied in this chapter.

References

1. Bogdan A, Dombrowski A (2019) J Med Chem 62:6422–6468
2. König B (2013) Chemical photocatalysis. DeGruyter, Berlin
3. Stephenson C, Yoon T, MacMillan D (2018) Visible light photocatalysis in organic chemistry. Wiley-VCH, Weinheim
4. Kisch H (2015) Semicondurctor photocatalysis – principles and applications. Wiley-VCH, Weinheim
5. Razeghifard R (2013) Natural and artificial photosynthesis – solar power as an energy source. Wiley, Weinheim
6. Prier C, Rankic D, MacMillan D (2013) Chem Rev 113:5322–5363
7. Hari D, König B (2014) Chem Commun 50:6688–6699
8. Romero N, Nicewicz D (2016) Chem Rev 116:10075–10166
9. Kärkäs M, Porco J, Stephenson C (2016) Chem Rev 116:9683–9747
10. Ramamurthy V, Sivaguru J (2016) Chem Rev 116:9914–9993
11. Marzo L, Pagire S, Reiser O, König B (2018) Angew Chem Int Ed 57:10034–10072
12. Ravelli D, Fagnoni M, Albini A (2013) Chem Soc Rev 42:97–113
13. Brimioulle R, Lenhart D, Maturi M, Bach T (2015) Angew Chem Int Ed 54:3872–3890
14. Beatty J, Stephenson C (2015) Acc Chem Res 48:1474–1484
15. Staveness D, Bosque I, Stephenson C (2016) Acc Chem Res 49:2295–2306
16. Chen J-R, Hu X-Q, Lu L-Q, Xiao W-J (2016) Acc Chem Res 49:1911–1923
17. Reiser O (2016) Acc Chem Res 49:1990–1996
18. Zou Y-Q, Hörmann F, Bach T (2018) Chem Soc Rev 47:278–290
19. Yoon T, Ischay M, Du J (2010) Nat Chem 2:527–532
20. Wenger O (2019) Chem Eur J 24:6043–6052
21. Otto S, Nauth A, Ermilov E, Scholz N, Friedrich A, Resch-Geger U, Lochbrunner S, Opatz T, Heinze K (2017) ChemPhotoChem 1:344–349
22. Treiling S, Wang C, Förster C, Reichenauer F, Kalmbach J, Boden P, Harris J, Carrella L, Rentschler E, Resch-Genger U, Reber C, Seitz M, Gerhards M, Heinze K (2019) Angew Chem Int Ed 58:18075–18085

23. Ravelli D, Fagnoni M (2012) ChemCatChem 4:169–171
24. Barona-Castaño J, Carmona-Vargas C, Brockson T, de Oliveira K (2016) Molecules 21:310–337
25. de Gonzalo G, Fraaije M (2013) ChemCatChem 5:403–415
26. Riente P, Noël T (2019) Cat Sci Technol 9:5186–5232
27. Tan H, Abdi F, Ng Y (2019) Chem Soc Rev 48:1255–1271
28. Tellis J, Kelly C, Primer D, Jouffroy M, Patel N, Molander G (2016) Acc Chem Res 49:1429–1439
29. Lang X, Zhao J, Chen X (2016) Chem Soc Rev 45:3026–3038
30. Skubi K, Blum T, Yoon T (2016) Chem Rev 116:10035–10074
31. Milligan J, Phelan J, Badir S, Molander G (2019) Angew Chem Int Ed 58:61526163
32. Perry I, Brewer T, Sarver P, Schultze D, DiRocco D, MacMillan D (2018) Nature 560:70–75
33. Meng Q-Y, Wang S, König B (2017) Angew Chem Int Ed 56:13426–13430
34. Hopkinson M, Sahoo B, Li J-L, Glorius F (2014) Chem Eur J 20:3874–3886
35. Fabry D, Rueping M (2016) Acc Chem Res 49:1969–1979
36. Silvi M, Melchiorre P (2018) Nature 554:41–49
37. Hering T, Mühldorf B, Wolf R, König B (2016) Angew Chem Int Ed 56:5342–5345
38. Areco E, Jurberg I, Álvarez-Fernández A, Melchiorre P (2013) Nat Chem 5:750–756
39. Pirnot M, Rankic D, Martin D, MacMillan D (2013) Science 39:1593–1596
40. Hoa H, Shen X, Wang C, Zhang L, Röse P, Chen L-A, Harms K, Marsch M, Hilt G, Meggers E (2014) Science 515:100–103
41. Metternich J, Gilmour R (2016) J Am Chem Soc 138:1040–1045
42. McManus J, Nicewicz D (2017) J Am Chem Soc 139:2880–2883
43. Wozniak L, Magagnano G, Melchiorre P (2017) Angew Chem Int Ed 56:1068–1072
44. Ghosh I, Shaikh R, König B (2017) Angew Chem Int Ed 56:8544–8549
45. Goti G, Bieszczad B, Vega-Peñaloza A, Melchiorre P (2018) Angew Chem Int Ed 57:1213–1217
46. Chisholm T, Clayton D, Dowman L, Sayers J, Payne R (2018) J Am Chem Soc 140:9020–9024
47. Kerzig C, Guo X, Wenger O (2019) J Am Chem Soc 141:2122–2127
48. Ichiishi N, Caldwell J, Lin M, Zhong W, Zhu X, Streckfuss E, Kim H-Y, Parish C, Krska S (2018) Chem Sci 9:4168–4175
49. Li J, Zhu D, Lv L, Li C-J (2018) Chem Sci 9:5781–5786
50. Singh K, Staig S, Weaver J (2014) J Am Chem Soc 136:5275–5278
51. Liu B, Lim C-H, Miyake G (2017) J Am Chem Soc 139:13616–13619
52. Rombach D, Wagenknecht H-A (2018) ChemCatChem 10:2955–2961
53. Jurberg I, Davies H (2018) Chem Sci 9:5112–5118
54. Ghosh I, Ghosh T, Bardagi J, König B (2014) Science 346:725–728
55. Ghosh I, König B (2016) Angew Chem Int Ed 55:7676–7679
56. Haimerl J, Ghosh I, König B, Vogelsang J, Lupton J (2019) Chem Sci 10:681–687
57. Jähnisch K, Hessel V, Löwe H, Baerns M (2004) Angew Chem Int Ed 43:406–446
58. Darvas F, Hessel V, Dorman G (2014) Flow chemistry – fundamentals (vol. 1) & applications (vol. 2). DeGruyter, Berlin
59. Plutschak M, Pieber B, Gilmore K, Seeberger P (2017) Chem Rev 117:11796–11893
60. Rossetti I, Compagnoni M (2016) Chem Eng J 296:56–70
61. Hessel V, Renken A, Schouten J, Yoshida J-I (2013) Micro process engineering: a comprehensive textbook. Wiley-VCH, Weinheim
62. Hessel V, Kralisch D, Kockmann N (2015) Novel process windows: innovative gates to intensified and sustainable chemical process. Wiley-VCH, Weinheim
63. Vaccaro L (2017) Sustainable flow chemistry. Wiley-VCH, Weinheim
64. Rehm TH (2014) Darvas F, Hessel V, Dorman G (eds) Flow chemistry – applications. DeGruyter, Berlin, pp 63–98

65. Mizuno K, Nishiyama Y, Ogaki T, Terao K, Ikeda H, Kakiuchi K (2016) J Photochem Photobiol C Photochem Rev 29:107–147
66. Ziegenbalg D, Kreisel G, Weis D, Kralisch D (2014) Photochem Photobiol Sci 13:1005–1015
67. Mo F, Dong G, Zhang Y, Wang J (2013) Org Biomol Chem 11:1582–1593
68. Felpin F-X, Sengupta S (2019) Chem Soc Rev 48:1150–1193
69. Hari D, König B (2013) Angew Chem Int Ed 52:4734–4743
70. Liang Y-F, Steinbock R, Yang L, Ackermann L (2018) Angew Chem Int Ed 33:10625–10629
71. https://roempp.thieme.de/lexicon/RD-04-00097. Accessed 9 Sept 2020
72. Cerecetto H, Gerpe A, González M, Arán V, De Ocáriz C (2005) Mini-Rev Med Chem 5:869–878
73. Schmidt A, Beutler A, Snovydovych B (2008) Eur J Org Chem:4073–4095
74. Li X, Chu S, Feher VA, Khalili M, Nie Z, Margosiak S, Nikulin V, Levin J, Sprankle KG, Tedder ME, Almassy R, Appelt K, Yager MK (2003) J Med Chem 46:5663–5673
75. Tanitame A, Oyamada Y, Ofuji K, Kyoya Y, Suzuki K, Ito H, Kawasaki M, Nagai K, Wachi M, Yamagishi J-I (2004) Bioorg Med Chem Lett 14:2857–2862
76. Picciola G, Ravenna F, Carenini G, Gentili P, Riva M (1981) Farmaco Ed Sci 36:1037–1056
77. Keisner S, Shah S (2011) Drugs 71:443–454
78. Wu F, Shu C (2014) Xuanzhu Pharma Co., Shandong. WO2014040373A1
79. Vidyacharan S, Ramanjaneyulu B, Jang S, Kim D-P (2019) ChemSusChem 12:2581–2586
80. The process intensification is given by the ratio between batch reaction time (24 h) to residence time (30 s) in the capillary photoreactor: $(24 \times 60 \times 60)$ s/30 s = 2880
81. Jang S, Vidyacharan S, Ramanjaneyulu B, Gyak K-W, Kim D-P (2019) React Chem Eng 4:1466–1471
82. Chen H, Jiang W, Zeng Q (2020) Chem Rec 20:1269–1296
83. Bottecchia C, Rubens M, Gunnoo S, Hessel V, Madder A, Noël T (2017) Angew Chem Int Ed 56:12702–12707
84. Wang X, Cuny G, Noël T (2013) Angew Chem Int Ed 52:7860–7864
85. Yang H, Martin B, Schenkel B (2018) Org Process Res Dev 22:446–456
86. Greb A, Poh J-S, Greed S, Battilocchio C, Pasau P, Blakemore D, Ley S (2017) Angew Chem Int Ed 56:16602–16605
87. Majchrazk M, Bekhazi M, Tse-Sheepy I, Warkentin J (1989) J Org Chem 54:1842–1845
88. Abdiaj I, Fontana A, Gomez M, de la Hoz A, Alcázar J (2018) Angew Chem Int Ed 57:8473–8477
89. Berton M, Huck L, Alcázar J (2018) Nat Protoc 13:324–334
90. Abdiaj I, Horn C, Alcázar J, Org J (2019) Chem 23:4748–4753
91. Wu Y, Fu WC, Chiang C-W, Choy PY, Kwong FY, Lei A (2017) Chem Commun 53:952–955
92. Abdiaj I, Huck L, Mateo J, de la Hoz A, Gomez V, Díaz-Ortiz A, Alcázar J (2018) Angew Chem Int Ed 57:13231–13236
93. Wie X-J, Abdiaj I, Sambiagio C, Li C, Zysman-Colman E, Alcázar J, Noël T (2019) Angew Chem Int Ed 58:13030–13034
94. DeLano T, Banderage U, Palaychuck N, Green J, Boyd M (2016) J Org Chem 81:12525–12531
95. Xu L, Zhang S, Li P (2015) Chem Soc Rev 44:8848–8858
96. Fyfe J, Watson A (2017) Chem 3:31–55
97. Chen K, Zhang S, He P, Li P (2016) Chem Sci 7:3676–3680
98. Chen K, Cheung M, Lin Z, Li P (2016) Org Chem Front 3:875–879
99. https://roempp.thieme.de/lexicon/RD-20-02945. Accessed 15 Sept 2020
100. https://roempp.thieme.de/lexicon/RD-15-01743. Accessed 15 Sept 2020
101. Brill Z, Ritts C, Mansoor U, Sciammetta N (2020) Org Lett 22:410–416
102. Zhang P, Le C, MacMillan D (2016) J Am Chem Soc 138:8084–8087
103. Majek M, von Wangelin J (2015) Angew Chem Int Ed 54:2270–2274
104. Anderson NG (2012) Practical process research and development: a guide for organic chemists. Academic Press, Oxford

105. Lima F, Grunenberg L, Rahman H, Labes R, Sedelmeier J, Ley S (2018) Chem Commun 54:5606–5609
106. Park B, Pirnot M, Buchwald S (2020) J Org Chem 85:3234–3244
107. Vitaku E, Smith D, Njardarson J (2014) J Med Chem 57:10257–10274
108. Mumtaz S, Robertson M, Oelgemöller M (2019) Molecules 24:4527–4536
109. Hatoum F, Engler J, Zelmer C, Wißen J, Motti C, Lex J, Oelgemöller M (2012) Tetrahedron Lett 53:5573–5577
110. Anamimoghadam O, Mumtaz S, Nietsch A, Saya G, Motti C, Junk P, Qureshi A, Oelgemöller M (2017) Beilstein J Org Chem 13:2833–2841
111. Fleming F (1999) Nat Prod Rep 16:597–606
112. Fleming F, Yao L, Ravikumar P, Funk L, Shook B (2010) J Med Chem 53:7902–7917
113. Otto N, Opatz T (2014) Chem Eur J 41:13064–13077
114. Ramirez N, König B, Gonzalez-Gomez J (2019) Org Lett 21:1368–1373
115. Mammoli V, Bonifazi A, Del Bello F, Diamanti E, Giannella M, Hudson A, Mattioli L, Perfumi M, Piergentili A, Quaglia W, Titomanlio F, Pigini M (2012) Bioorg Med Chem 20:2259–2265
116. Fumagalli L, Pallavicini M, Budriesi R, Bolchi C, Canovi M, Chiarini A, Chiodini G, Gobbi M, Laurino P, Micucci M, Straniero V, Valoti E (2013) J Med Chem 56:6402–6412
117. Lempenauer L, Lemière G, Duñach E (2019) Adv Synth Catal 361:5284–5304
118. Thebtaranonth C, Thebtaranonth Y (1990) Tetrahedron 46:1385–1489
119. Zhang B, Studer A (2015) Chem Soc Rev 44:3505–3521
120. Ralph M, Ng S, Booker-Milburn K (2016) Org Lett 18:968–971
121. Blanco-Ania D, Gawade S, Zwinkls L, Maartense L, Bolster M, Benningshof J, Rutjes F (2016) Org Process Res Dev 20:409–413
122. Zhang L, Peng X-M, Damu G, Geng R-X, Zhou C-H (2014) Med Res Rev 34:340–437
123. Adiyala P, Jang S, Vishwakarma N, Hwang Y-H, Kim D-P (2020) Green Chem 22:1565–1571
124. Jackl M, Legnani L, Morandi B, Bode J (2017) Org Lett 19:4696–4699
125. Fang Y, Tranmer G (2016) Med Chem Commun 7:720–724
126. Bach T, Hehn J (2011) Angew Chem Int Ed 50:1000–1045
127. Zhang X-M, Tu Y-Q, Zhang F-M, Chen Z-H, Wang S-H (2017) Chem Soc Rev 46:2272–2305
128. Chen Z-M, Zhang X-M, Tu Y-Q (2015) Chem Soc Rev 44:5220–5245
129. Siopa F, António J, Afonso C (2018) Org Process Res Dev 22:551–556
130. Lattes A, Oliveros E, Rivière M, Belzecki C, Mostowicz D, Abramskj W, Piccinni-Leopardi C, Germain G, van Meerssche M (1982) J Am Chem Soc 104:3929–3934
131. Wolfe M, Dutta D, Aubé J (1997) J Org Chem 62:654–663
132. Zeng Y, Smith B, Hershberger J, Aubé J (2003) J Org Chem 68:8065–8067
133. Cochran J, Waal N (2016) Org Process Res Dev 20:1533–1539
134. Zhang Y, Blackman M, Leduc A, Jamison T (2013) Angew Chem Int Ed 52:4251–4255
135. Campbell M, Ritter T (2014) Org Process Res Dev 18:474–480
136. Neumann C, Ritter T (2015) Angew Chem Int Ed 54:3216–3221
137. Purser S, Moore P, Swallow S, Gouverneur V (2008) Chem Soc Rev 37:320–330
138. Jeschke P (2004) ChemBioChem 5:570–589
139. Kirk K (2008) Org Process Res Dev 12:305–321
140. Champagne P, Desroches J, Hamel J-D, Vandamme M, Paquin J-F (2015) Chem Rev 115:9073–9174
141. Szpera R, Moseley D, Smith L, Sterling A, Gouverneur V (2019) Angew Chem Int Ed 58:14824–14848
142. Tomashenko O, Grushin V (2011) Chem Rev 111:4475–4521
143. Ma J-A, Cahard D (2008) Chem Rev 108:PR1–PR43
144. Nodwell M, Bagai A, Halperin S, Martin R, Knust H, Britton R (2015) Chem Commun 51:11783–11786
145. Halperin S, Kwon D, Holms M, Regalado E, Campeau L-C, DiRocco D, Britton R (2015) Org Lett 17:5200–5203

146. Straathof N, Tegelbeckers B, Hessel V, Wang X, Noël T (2014) Chem Sci 5:4768–4773
147. Straathof N, Gemoets H, Wang X, Schouten J, Hessel V, Noël T (2014) ChemSusChem 7:1612–1617
148. Wei X-J, Noël T (2018) J Org Chem 83:11377–11384
149. Straathof N, Cramer S, Hessel V, Noël T (2016) Angew Chem Int Ed 55:15549–15553
150. Singh A, Fennell C, Weaver J (2016) Chem Sci 7:6796–6802
151. Abdiaj I, Bottecchia C, Alcázar J, Noël T (2017) Synthesis 49:4978–4985
152. Trump L, Lemos A, Lallemand B, Pasau P, Mercier J, Lemaire C, Luxen A, Genicot C (2019) Angew Chem Int Ed 58:13149–13154
153. Rong J, Deng L, Tan P, Ni C, Gu Y, Hu J (2016) Angew Chem Int Ed 55:2743–2747
154. Fu W, Han X, Zhu M, Xu C, Wang Z, Ji B, Hao X, Song M (2016) Chem Commun 52:13413–13416
155. Trump L, Lemos A, Jacq J, Pasu P, Lallemand B, Mercier J, Genicot C, Luxen A, Lemaire C (2020) Org Process Res Dev 24:734–744
156. Hong B-C (2020) Org Biomol Chem 18:4298–4353
157. Blakemore D, Castro L, Churcher I, Rees D, Thomas A, Wilson D, Wood A (2018) Nat Chem 10:383–394
158. Adams J, Brown M, Diaz-Rodriguez A, Lloyd R, Roiban G-D (2019) Adv Synth Catal 361:2421–2432
159. Abdelraheem E, Busch H, Hanefeld U, Tonin F (2019) React Chem Eng 4:1878
160. Seel C, Gulder T (2019) ChemBioChem 20:1871–1897
161. Schmermund L, Jurkaš V, Özgen F, Barone G, Büchsenschütz H, Winkler C, Schmidt S, Kourist R, Kroutil W (2019) ACS Catal 9:4115–4144
162. Douglas J, Sevrin M, Stephenson C (2016) Org Process Res Dev 20:1134–1147
163. Guo X, Okamoto Y, Schreier M, Ward T, Wenger O (2018) Chem Sci 9:5052–5056
164. Lauder K, Toscani A, Qi Y, Lim J, Charnock S, Korah K, Castagnolo D (2018) Angew Chem Int Ed 57:5803–5807
165. Litman Z, Ang Y, Zhao H, Hartwig J (2018) Nature 560:355–359
166. Burek B, Bormann S, Hollmann F, Bloh J, Holtmann D (2019) Green Chem 21:3232–3249
167. Hou H, Zeng X, Zhang X (2020) Angew Chem Int Ed 59:17356–17376
168. Schmaderer H, Hilger P, Lechner R, König B (2009) Adv Synth Catal 351:163–174
169. Girhard M, Kunigk E, Tihovsky S, Shumyantseva V, Urlacher V (2013) Biotechnol Appl Biochem 60:111–118
170. Zhang W, Fernández-Fueyo E, Ni Y, van Schie M, Gacs J, Renirie R, Wever R, Mutti F, Rother D, Alcalde M, Hollmann F (2018) Nat Catal 1:55–62
171. Seel C, Králík A, Hacker M, Frank A, König B, Gulder T (2018) ChemCatChem 10:3960–3963
172. Zhang W, Fernandez Fueyo E, Hollmann F, Martin L, Pesic M, Wardenga R, Hohne M, Schmidt S (2019) Eur J Org Chem:80–84
173. Reetz M (2013) J Am Chem Soc 135:12480–12496
174. de Souza R, Miranda L, Bornscheuer U (2017) Chem Eur J 23:12040–12063
175. Cambié D, Bottecchia C, Straathof N, Hessel V, Noël T (2016) Chem Rev 116:10276–10341
176. Rehm TH (2020) ChemPhotoChem 4:235–254
177. Rehm TH (2020) Chem Eur J 26:16952–16974
178. Britton J, Majumdar S, Weiss G (2018) Chem Soc Rev 47:5891–5918
179. Zhu Y, Chen Q, Shao L, Jia Y, Zhang X (2020) React Chem Eng 5:9–32
180. Guajardo N, Domínguez de María P (2019) ChemCatChem 11:3128–3137

Top Med Chem (2021) 38: 121–172
https://doi.org/10.1007/7355_2021_115

Published online: 17 April 2021

Electrochemistry in Flow for Drug Discovery

Bethan Winterson and Thomas Wirth

Contents

Abstract Electrochemistry, first proposed over two centuries ago, is one of the oldest forms of reaction set-ups explored in a laboratory. Electrochemical methods possess many benefits in comparison with traditional reagent-based transformations, such as (1) innate scalability and sustainability, (2) high functional group tolerance, (3) mild reaction conditions, (4) high versatility and the ability to carry out reactions that were otherwise inaccessible. While preparative organic electrosynthesis has been an active area of research over the past century, the adoption of electrochemical methods has been remarkably under-resourced, particularly in drug discovery. This phenomenon seems to arise from the lack of expertise in this area and the seemingly complex reaction set-ups.

Current movements in modern society demand new synthetic processes, namely in terms of clean alternative processes, where organic electrochemistry shows clear

B. Winterson and T. Wirth (✉)
School of Chemistry, Cardiff University, Cardiff, UK
e-mail: WintersonB@cardiff.ac.uk; wirth@cf.ac.uk

potential. Here, toxic oxidants or reductants are replaced by electrons, a characteristically sustainable and cost effective 'reagent'. Electrons facilitate reactions that were conventionally inaccessible because the narrow redox potential of common oxidants/reductants can be circumvented by finely tuning the electrolysis parameters. The growing availability of electrochemical equipment on the market, the emergence of new suppliers, and the search by developing countries for new clean technologies make organic electrosynthesis a competitive technology. The ability to achieve high product yields with considerably no emissions, avoiding environmentally harmful by-products and reactants as well as re-useable materials and equipment are important pre-conditions for the realisation of contemporary processes in industry.

Flow electrochemistry, which combines the intrinsic scalability of flow chemistry with the environmentally benevolent properties of electrochemistry, has received unprecedented development in the last few years. Automated flow electrolysis platforms have already been established meaning reactions can be performed with minimal human intervention, which greatly accelerates process optimisation and work-up procedures.

Automated flow electrochemical platforms which integrate machine-learning algorithms can be foreseen as the next big development in this field. Intelligent self-optimising systems using novel algorithms with real-time reaction analysis and their combination with automation bring exciting new opportunities. Early drug discovery is typically practiced through the synthesis of a large library of compounds for a greater chance of success. Here, the ability to quickly synthesise large and structurally diverse libraries through automated platforms is easily recognisable, which clearly lends itself to efficient drug development.

Although flow electrochemistry has yet to be fully resourced, the commercial availability of simple and easy to use electrochemical platforms is sure to facilitate its widespread adoption.

Keywords Automation, Drug development, Electrochemistry, Flow

1 Introduction

1.1 General Introduction

The efficient design and synthesis of novel bioactive compounds that are potential drug molecules is a key encompassment in medicinal chemistry. Drug research requires an interdisciplinary effort from synthetic organic chemistry, biology and medicine. Development of a large library of structurally diverse molecules is crucial for the success of drug discovery efforts. Modern drug discovery programmes typically rely on screening compound collections derived from a known set of chemical reactions. As such, the role of synthetic organic chemistry in subsequent drug discovery and optimisation has been severely under-resourced. Synthetic organic chemists are employed for three main steps in early drug discovery, including (1) identification of screening hits from up to 10,000 drug candidates towards a biochemical mechanism involved in a disease condition; (2) large library synthesis, focused on the synthesis of similar compounds like the initial 'hit' drug and (3) optimisation of those molecules to increase affinity, selectivity, potency and metabolic stability, etc. by structural modification. A key caveat to any potential drug molecule is the un-expected metabolic generation of toxic and highly reactive metabolites inside a patient which compels structural alteration of the compound whilst maintaining the desired activity. Whilst early drug discovery is typically practiced through the synthesis of a huge number of compounds for a greater chance of success, medicinal chemistry approaches have so far been defined by established reaction types. The known chemical space represented by the number of substances listed in the Chemical Abstracts Service (CAS) is growing faster than the human population. In 2007, roughly three million chemists published more than 800,000 papers [1]. From a medicinal chemistry perspective there remains serious under-exploration of chemical space for drug discovery. In 1995, half of the compounds used in drug development could be described by a measly 32 frequently occurring frameworks [2]. By 2010, there was virtually no improvement in these characteristics; the top 50 frameworks still covered 48–52% of approved and experimental drugs [3]. Schneider et al. confirmed that a few reactions dominate contemporary practice, and the top frameworks and ring systems in drugs have remained fairly unchanged with time [4]. Analysis of a comprehensive set of 1.15 million unique reactions in pharmaceutical patent literature (published after the year 2000) concluded that the medicinal chemistry toolkit is biased towards a manageable set of standard reaction types. While oxidations and heterocycle formations are rare (2.2 and 1.6%, respectively), heteroatom alkylations/arylations and acetylations represent nearly 50% of all synthetic strategies employed. Carbon–carbon bond formations (10.1%), deprotections (14.5%) and functional group interconversions (7.8%) are other common transformations used in drug development processes. Consequently, even modern drug discovery efforts are largely driven by attempts to 'capture' expertise from the past, which hinders its advancement.

The use of a restrained set of reactions in drug development approaches, derived from commercially available building blocks, compromises the structural diversification of potential drug compounds. Accordingly, the possibility for the discovery of novel and much more active molecules is narrow. Late-stage functionalisation of elaborate and complex drug molecules is an on-going struggle in drug development since the functional group tolerance of the conventional reactions is limited. The limitation of reaction parameters in synthetic organic chemistry undoubtedly leads to structural constraint. The impact could be substantial, restricting the implementation of therapeutically beneficial functional groups which could intensify the therapeutic activity of a drug. In this respect, the advancement in synthetic organic chemistry is key for the progression of drug discovery efforts.

Modern advances in synthetic technologies facilitate more efficient exploration and modification of scaffolds for medicinal chemistry [5]. The discovery of new chemistries is evolving, enabling the synthesis of novel compounds by expanding the constrained range of conditions available. Most importantly, the emphasis of synthetic efficiency in modern synthetic chemistry research hugely benefits drug discovery, leading to quicker synthetic routes and faster drug synthesis for screening. Analysis of synthetic strategies employed in more than 6.5 million organic reactions shows that key criteria must be met for a methodology to be widely accepted [1]. These include reaction time, temperature, pressure, cost and anthropogenic influence. Half of the reactions analysed were complete within <3 h and 90% of reactions were run at atmospheric pressure, typically between −80°C and + 200°C [1].

Drug developmentalists have more recently been seeking to take out new chemical real estate with more complex hybrid structures. The use of electricity to induce reactions is an intriguing alternative to traditional approaches and in the last decade, organic electrosynthesis has seen a renaissance. Electrochemical synthesis is a versatile synthetic tool and offers a sustainable alternative to the use of stoichiometric quantities of chemical oxidants. The significance of this development to the course of drug discovery is evident upon examination of its advancements. The method can substitute procedures when traditional methods fail, greatly expanding the scope of transformations with its unique reactivity. Electrochemical protocols remain, however, remarkably scarce in early synthetic drug discovery. Although, the technique is gaining broader utility in organic chemistry and hence its incorporation into medicinal synthetic strategies is more prominent. The increased adoption of this methodology is mainly owing to the vast improvements into practical electrochemical technology in recent years. As part of the process, there have been enormous developments into ease of use electrochemical reactors that are now commercially available [6–8]. Practical organic electrochemical synthesis has also seen large recent improvements. Previously, installation of an electroauxillary functional group, e.g. α-silyl, arylthiol or organostannane, to control electron transfer events was a strong disincentive to its adoption in medicinal chemistry. They are abnormal reagents in drugs and require substantial optimisation of each target compound. In this context, the synthesis of large libraries of potential drug targets would be hugely inefficient. Very recent developments in synthetic organic electrochemistry means that common building blocks can be utilised with no electroauxillary, substantially expanding the scope of synthetic reactions. Along with the ability to access unique transformations, electrochemistry brings a wealth of other advantages to medicinal chemistry. Some key benefits from a pharmaceutical development perspective are:

- immense functional group tolerance for late-stage functionalisation;
- facilitation of single-site modification of elaborate drug type molecules through careful parameter choice;
- flow synthesis for kilogram scales of key fragments;
- lowered impurity and environmental waste product burdens by resorting to metal-free green chemistry.

Access to fast purification techniques, especially for large library synthesis often guides synthetic choices in drug discovery. The replacement of oxidising/reducing reagents and the reduction of by-product formation in electrochemistry fulfil such criteria. Moreover, these reactions are highly efficient and given electrolysis can be performed in the absence of reagents, purification is often possible by simple removal of the reaction solvent.

In the last decade, electrochemistry has been integrated with flow chemistry, giving rise to a highly superior and exciting synthetic methodology. Flow electrochemistry itself can be integrated into a fully automated platform meaning a range of conditions can be efficiently examined to produce large libraries of target compounds [9]. Enabling the synthesis of novel compounds by expanding the range of reactivities facilitates much quicker drug discovery, with vastly improved exploration of chemical space. Electrosynthetic flow reactors and the more widespread adoption of electrochemistry by the synthetic community will undoubtedly accelerate the development of advanced drug and drug-like molecules.

1.2 Introduction to Electrochemistry

Electroorganic synthesis is synthetic organic chemistry enabled by the direct use of electricity to transform or generate organic molecules. Over the last two decades, the method has experienced a renaissance in the field and preparative organic electrosynthesis is nowadays becoming more conventional. Electrochemical methods possess many benefits over traditional reagent-based transformations, such as innate scalability and sustainability, inherent selectivity, reduced toxicity, and increased cost-effectiveness. Electrons, the basis of electrochemical synthesis, are a clean reagent and among the alternative reagents used today have the lowest cost per unit charge, approximately £ 0.17/mol. Electrons facilitate reactions that were conventionally inaccessible because the narrow redox potential of common oxidants/reductants can be circumvented by finely tuning the electrolysis parameters [10]. Moreover, these reactions often occur more rapidly and with comparable or better yields. Electrochemistry is selective and tolerant. The significance of functional group tolerance is realised in the late-stage functionalisation of drug molecules. The potential for highly selective single site modifications by finely tuned electrolysis is unparalleled by other synthetic strategies.

In times of increasing environmental awareness, organic electrochemistry has become a central tool aimed at developing sustainable methods. Hence, the most substantial attraction to this field are the environmentally benevolent principles of electrochemistry. The use of electrical current avoids the generation of reagent waste (by removing the demand for oxidants/reductants) and by-product formation which increases atom economy. In general, the reaction conditions are mild, and reactions are performed under ambient temperature and atmospheric pressure. In industrial terms, the substitution of heavily resourced reagents by electrons, particularly in molecular synthesis, provides a much cheaper and sustainable alternative.

Unfortunately, electrochemistry has some drawbacks. Mainly, procuring standardised instrumentation is costly and far outweighs the cost of standard glassware and some reagents. Non-standardised equipment, although cheaper, is highly undesirable due to problems with reproducibility from laboratory to laboratory. The limited progression of electrochemical synthesis stems from the intrinsic restrictions with conventional batch cells. Scale-up in regular batch type cells can be hugely problematic because issues such as overheating due to increased electrical demand and the requirement of large concentrations of supporting electrolytes can make the whole process cumbersome. The best electrode materials often involve precious metals because of their inertness and durability. Since electrochemical reactions are initiated by heterogeneous electron transfer at the electrode surface, the electrode surface area is a limiting factor. Preparative scale batch electrolysis requires the utilisation of large electrodes which are expensive.

Some limitations of electrosynthesis are overcome using electrochemical flow reactors. Continuous flow electrochemical synthesis is inherently scalable. Low concentrations of supporting electrolytes are permitted because the distance between the electrodes is small, which reduces efforts in work-up procedures. Column chromatography, a fundamental purification technique in organic synthesis which utilises an extensive amount of organic solvents, can often be avoided. Not only is this technique wasteful, but the environmental and biological impacts of some of these solvents are concerning. A continuous flow process avoids the demand for high currents/voltages meaning the process is safer. Flow electrochemical synthesis comes into its own for industrial applications, particularly the pharmaceutical industry. Resorting to metal-free flow synthesis will vastly reduce the environmental waste burden, and more substantially, it can greatly reduce the levels of toxic impurities in subsequent drug molecules.

1.3 Fundamentals of Organic Electrochemistry

In the following discussion, electrolysis with respect to organic electrochemical synthesis will be prioritised. Electrochemistry is a field of chemistry that facilitates the interchange between electrical and chemical energy. Electrochemical synthesis is the passing of an electrical current through a conductive solution to perform chemical transformations. Typically, an electrochemical cell has a power source (i.e. battery, potentiostat) connected to three electrodes: the anode, cathode and reference electrode. Electrodes are critical to organic electrosynthesis as all substrate molecules undergo direct electron transfers with the electrode surface. Such reactions are heterogeneous and can impose a high kinetic barrier [11]. The electrode where the desired reaction takes place is known as the 'working' electrode, while the other is referred to as the 'counter' electrode, which performs the balancing redox reaction. A working electrode will be chosen within the potential window of the desired electrochemical transformation. An important aspect of the working electrode is that it is composed of a redox inert material in the potential range of interest. The concept of current (measured in Ampere [A]) refers to the rate of electron movement. The potential is measured between two electrodes and is measured in

Reduction

Oxidant + e^- → product

Oxidation

Reductant → product + e^-

Fig. 1 The two components of a redox reaction

Fig. 2 Molecular-orbital diagram for oxidation/reduction electron transfer

units of volts [V], whereby 1 V is the potential difference necessary to generate 1 Coulomb [C] from 1 Joule [J] of energy. Both are inextricably linked by Ohm's law (Eq. ch.1):

$$V = I \bullet R \tag{1}$$

where $V =$ voltage, $I =$ current and $R =$ resistance.

The power source pushes electrons into the cathode from the anode, resulting in an oxidative environment at the anode and a reductive environment at the cathode. Only the surfaces of the electrodes facing each other are electrochemically active and at the backside of the electrode no transformation takes place. When a molecule loses an electron, its oxidation state is increased and this constitutes oxidation (Fig. 1). When a substance gains an electron, its oxidation state decreases, thus it is reduced. In terms of molecular orbitals, anodic oxidation processes are the transfer of electrons from the highest occupied molecular orbital (HOMO) of the substrate to the anode (Fig. 2). Alternatively, a cathodic reduction is the transfer of electrons from the cathode to the lowest unoccupied molecular orbital (LUMO) of a substrate. In simple terms, for a reduction reaction, heterogeneous electron transfer from the electrode is favourable when the 'electrons' from the electrode are at a higher energy than the LUMO of the substrate. The energy of the electrons in the electrode is controlled by the potentiostat; their energy can be increased until electron transfer becomes favourable. The thermodynamic driving force of reduction therefore is the population of a lower energy state, which is more stable (Fig. 3).

A circuit is formed when the electrodes are immersed in a conductive solution. Each electrochemical reaction is the sum of two half-reactions: oxidation and reduction. To ensure electroneutrality, both half-reactions must occur. Typically, only one of these reactions involves formation of a product, whilst the other entails a redox transformation of the solvent, electrolyte or sacrificial species. Thus, in an anodic oxidation, oxidation at the working electrode is balanced by reduction at the

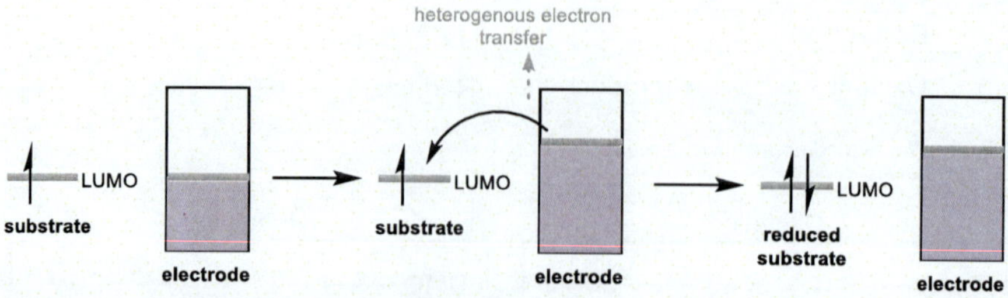

Fig. 3 Heterogeneous electron transfer from an electrode to a substrate

cathode; electrons donated by the substrate at the anode move around the circuit to reduce solvent molecules, protons, etc.

Electrochemical reactions are driven by single-electron transfer processes which are initiated at the surface of an electrode. Electrically charged species arising from one electron transfers are known as radicals. Radical cations are formed by single-electron anodic oxidation and have both a positive formal charge and an unpaired electron. Analogously, radical anions are formed by single-electron transfer from the cathode and have a formal negative charge and an unpaired electron. Reactions occurring in the space near the electrode surface (known as a double layer or diffusion layer) lead to the accumulation of high energy radical cations or anions at the surface. Some of these reactive species can diffuse back into the bulk electrolysis solution for downstream transformations, whilst others may decompose. Unlike conventional organic synthesis, the formation of a double layer allows greater control of the reactive radical species (e.g. intramolecular reactions over intermolecular) leading to exquisite selectivity. Radical intermediates can also undergo further oxidation or reduction (i.e. one electron transfer, followed by another) to form cations or anions. The formation of cations promotes electrophilic/nucleophilic substitutions, addition, elimination and cyclisation reactions (Fig. 4).

Like traditional organic synthesis, the choice of solvent is crucial for electrochemical reactions. The potential windows of stability of common organic solvents vary dramatically, and the solvent of choice must be stable towards oxidation/reduction in the potential range of the experiment. Acetonitrile has one of the largest potential windows in electrochemistry. A high dielectric constant is desirable because it can prevent ohmic resistance of the solution. Polar protic solvents can facilitate H^+ reduction as an ideal cathodic reaction, and some solvents can even stabilise radical intermediates [12].

The low conductivity of most organic solvents means that conductivity must be assisted by the addition of salts. They are referred to as electrolytes. If the electrolyte is not involved in the reaction, i.e. it just facilitates conductivity, it is referred to as a supporting electrolyte. In conventional batch electrolysis, large amounts of supporting electrolytes are necessary to minimise ohmic drop. Ohmic drop is observed due to the additional resistance encountered by the electrical current

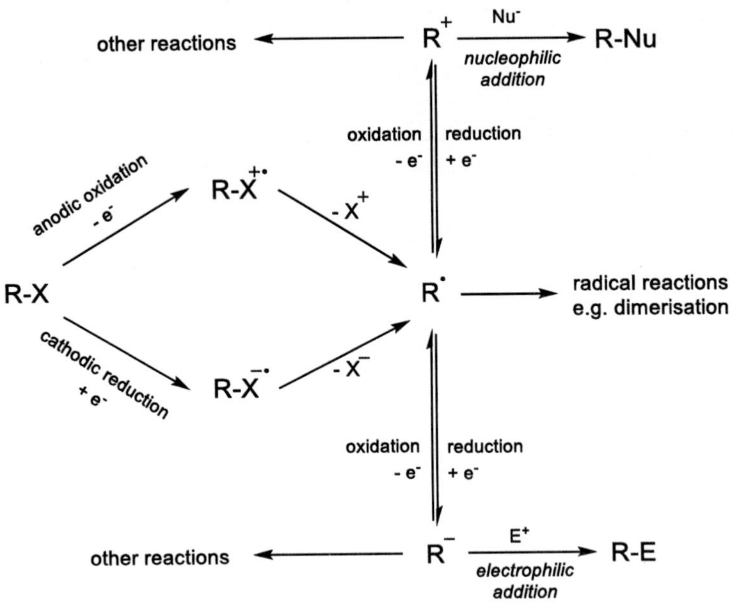

Fig. 4 Main reactions in organic electrosynthesis

when travelling through a liquid phase between two electrodes. The contributing factors to ohmic resistance can be recognised by (Eq. 2):

$$\Delta U = IR_{\text{drop}} = I\frac{1}{A_e}\frac{d}{\kappa} \tag{2}$$

where ΔU is the potential, I is the current, A_e is the electrode surface, R_{drop} is the ohmic drop, κ is the solution conductivity and d is the interelectrode gap. A good supporting electrolyte is chemically and electrochemically inert, has good solubility in the desired solvent and is easily removed for purification. Given the low solubility of most inorganic salts in organic solvents, organic salts are typically employed. For example, Bu_4NClO_4, Bu_4NBF_4, Et_4NPF_6 are common, although some inorganic salts are also used, e.g. $LiClO_4$. For polar solvents such as dichloromethane and acetonitrile, tetrabutylammonium salts are used. Whilst for less polar solvents like benzene, tetrahexylammonium salts are more soluble. The selection of the anion is relevant in anodic oxidations, and it needs to be unreactive in the potential range of the desired transformation. Perchlorates, hexafluorophosphates, tetrafluoroborates or nitrates have high discharge potentials, hence are anodically stable.

Electrochemical reactions depend not only on how reactive the generated species are, but also on how those intermediates interact with the chemical environment around the electrode. How the molecules diffuse away from the electrode and interact with the gradient of the chemical environments encountered depends on the nature of the electrochemical field inside the cell. This, in turn, depends on the electrode configuration. Varying the experimental conditions (i.e. electrolyte concentration, temperature, applied potential, and/or current) and reaction characteristics (electrochemical construction materials, electrode materials, and their size and shape) optimises production of the desired product.

Passivation of the electrode surface caused by polymerisation or decomposition is a major limiting event. Given the surface availability of the electrode for the heterogeneous electron transfer is diminished, further reactivity is impeded. To circumvent these problems, a mediator (or redox catalyst) can be used.

An important cost consideration for electrochemical processes is the cell energy consumption, mainly owing to cell resistance and operating current density. The cell potential (E_c) can be calculated as follows (Eq. 3):

$$E_c = -\sum E_e - \sum \eta - \sum IR_e \tag{3}$$

where E_e is the thermodynamically required potential for a transformation, η is the over-potential and R_e is the cell internal resistance. The overpotential is the potential difference between the thermodynamically required potential of the reaction (E_e) and the experimental potential at which the actual reaction occurs. The total overpotential comprises of several factors, including an electron transfer overpotential, a concentration overpotential and a resistance overpotential. The presence of overpotentials results in an increased energy consumption of the cell, which increases costs. Various methods to decrease energy consumption will be discussed throughout this chapter.

1.4 Methods for Organic Electrosynthesis

1.4.1 Controlled Potential Electrolysis (Potentiostatic)

Two major electrochemical approaches with reference to electroorganic synthesis are used: controlled potential (CP) and constant current electrolysis (CCE). Controlled potential electrolysis is more complicated since the electrochemical potential has no natural fix point. These reactions require a three-electrode set-up consisting of a working electrode, where the desired reaction occurs, a counter electrode, which preserves the cell circuit with a balancing electrochemical reaction, and a reference electrode. The reference electrode provides a constant reference potential as a fix point. During controlled potential experiments, the power source maintains the potential between the working and reference electrode and the current may vary. A wide variety of reference electrodes are used, but the most commonly encountered for organic synthesis are based on an Ag^+/Ag couple. In order to perform potentiostatic electrolysis, the redox potential (E) of the starting material in a solvent must be known. Developments in analytical processes such as cyclic voltammetry allow the electrochemical potentials (E) for individual functionalities to be accurately determined. Thus, selective manipulation of functional groups is feasible. For instance, the cell potential can be held at the desired voltage for the duration of the reaction. Under potentiostatic conditions, the applied potential is set to the desired value and is realised at the surface of the working electrode. Importantly, this value should not be misinterpreted with the terminal potential (U) in galvanostatic conditions, where the potential here drops through internal resistance of the electrochemical cell. Since the potential is the driving force behind a redox process, controlled potential methods result in enhanced selectivity and allow single site modification. A major disincentive of this method is that the cell current can drop to compensate for the increased cell resistance which dramatically extends reaction times. For this reason, controlled potential experiments often poorly translate to preparative scale synthesis and thus the appeal of this technique for medicinal chemistry is limited. Although, the technique still has some applications with respect to drug development as exquisite selectivity can be achieved which is perfect for late-stage functionalisation of elaborate drug molecules. However, with respect to early drug discovery, the inefficiency caused by long reaction times for building large libraries of compounds is extremely undesirable.

1.4.2 Constant Current Electrolysis (Galvanostatic)

Notwithstanding, the use of constant current or galvanostatic electrolysis (CCE) is the best currently available option for drug discovery, owing to the operational simplicity and reduced reaction times. In this respect, the combinatorial technique of automated flow electrosynthesis (discussed later) using CCE is an extremely efficient method for synthesising large compound collections. Constant current electrolysis is easy to set up as only two electrodes, an anode and cathode, are required. The power supply applies a constant current and the cell potential is allowed to vary in order to maintain the desired current. Since the cell potential will fluctuate as cell resistance changes during the reaction, lower selectivities are observed in comparison with controlled potential experiments. This in itself can be viewed as an advantage because structural diversification is possible. Product selectivity is controlled by regulating the current density (A cm^{-2}) and the amount of electricity (charge [F]) passed through the cell. Since the electrolysis potential of the transformation is not controlled, at low current densities the material with the lowest redox potential is transformed first. At a certain point, the mass transport through the electrochemical double layer by stirring and diffusion becomes insufficient. When the current is maintained beyond this point, the galvanostat will increase the applied voltage to ensure maintenance of the current. Consequently, the material with the next lowest redox potential will be transformed next. To ensure reproducibility, when CCE is performed in organic synthesis it is essential to report the current applied over the course of the reaction, electrode surface area, amount of applied charge (F) and the distance between the electrodes. In theory, 1 mol of product needs 96,485 C (Faraday constant $F =$ charge of 1 mol of electrons), multiplied by the stoichiometric number of electrons required in the reaction. The quantity of electricity needed in a reaction can be calculated by Faraday's law (Eq. 4):

$$Q_{\text{theoretical}} = z \cdot N \cdot F \tag{4}$$

where $Q_{\text{theoretical}}$ is (C, 1 C = 1 As), z is the number of electrons for a reaction, N is the number of moles of the compound being transformed and F is the Faraday constant (96,485 C mol^{-1}). When this calculated electricity ($Q_{\text{theorectical}}$) is divided by the applied current (I [A]), the time of reaction can be calculated (Eq. 5).

$$t = \frac{Q}{I} \tag{5}$$

For an analogous flow cell, the current required can be calculated by (Eq. 6):

$$I_{(A)} = n \bullet F \bullet Q_V \bullet C \tag{6}$$

where n is the number of electrons required in the electrochemical transformation; F is the Faraday constant (A s mol^{-1}); Q_v is the volumetric flow rate of the reaction solution (mL s^{-1}) and C is the concentration of the reaction solution (mol cm^{-3}). For the theoretical current, the theoretical number of Faraday should be inserted, i.e. for 2 e^- transfer, 2 F (2 × 96,485 A s mol^{-1} should be inserted).

Often, an excess of charge is needed to complete a reaction. Current efficiency is a parameter to measure the degree of excess required, a highly important consideration in industrial processes. This is calculated from the equation below (Eq. 7):

$$\text{Current efficiency} = \frac{X \cdot Q_{\text{theoretical}}}{Q} \tag{7}$$

where X is the yield of the transformation, $Q_{\text{theoretical}}$ is the calculated charge and Q is the real charge passed through the cell.

1.5 Cyclic Voltammetry

Cyclic voltammetry (CV) is a versatile electroanalytical technique used to guide electrolysis parameters. In electrochemistry, CV is invaluable to study electron transfer-initiated reactions. Three electrodes are required in a cyclic voltammetry study: a working electrode, counter electrode and reference electrode. Here, a good quality reference electrode is a necessity to allow accurate electro-potentials to be determined.

To measure and control the potential difference applied, the potential of the working electrode is varied while the potential of the reference electrode remains fixed. Reference electrodes are generally separated from the reaction solution by a porous fit and have a well-defined, stable equilibrium potential. For organic synthesis, i.e. mostly non-aqueous solvents, commonly employed reference electrodes are based on an Ag^+/Ag couple. These consist of a silver wire in a solution containing an

Ag^+ salt such as $AgNO_3$ or AgCl. Precipitation and clogging of these salts can be problematic and oscillation of the potential results in uncontrolled electrolysis. Since the reference electrode is compartmentalised in a separate glass tube, junction potentials should be minimised by matching the solvent and electrolyte in the reference compartment to the one used in the experiment [13].

For the CV experiment, a potentiostat is used to linearly sweep the potential between the working and reference electrodes until a preset limit. It is then swept back in the opposite direction, switching potentials, and the changing current between the working and counter electrodes is measured in real time. Hence the x-axis is the applied potential (E [V]), whilst the y-axis is the current response (I [mA]) or current density (mA cm^{-2}). When the scan is applied increasingly with a negative potential, this is termed the negative scan, and this is when reduction occurs. Alternatively, the scan in which the potential becomes increasingly positive is termed the positive scan, and this is when oxidation occurs. One forward and reverse sweep in CV is known as a scan, and multiple scans are often performed to improve the accuracy of data [13].

The current response depends on the concentration of the redox species at the working electrode; it is described by a combination of Fick's first law of diffusion and Faraday's Law (Eq. 8): [14]

$$i_d = nFAD_0\left(\frac{\delta C_0}{\delta x}\right)_0 \tag{8}$$

where i_d is the diffusion limited current, A is the electrode surface area, D_0 is the diffusion coefficient (of the substrate) and $(\partial C_0/\partial x)_0$ is the concentration gradient at the electrode surface.

The trace in Fig. 5 is a cyclic voltammogram. The shape of the voltammogram is dependent on how the ions and substrates move between the surface of the electrode and the bulk solution, contributing to the 'duck' shaped voltammogram. Initially, the applied potential is not enough to oxidise the analyte, and a straight line is observed (point **a**). Once the onset of oxidation (E_{onset}) is reached, the current rises exponentially as the analyte is oxidised at the working electrode surface. At point **b**, the current is linearly increasing with increasing voltage with a constant concentration gradient of the analyte within the diffusion layer. The volume of solution at the surface of the electrode containing the oxidised species is called the diffuse double layer and it continues to grow through the scan. The diffuse double is composed of both orientated electric dipoles and ions that counteract the charge on the electrode. The current response decreases from linearity as the analyte is depleted and a maximum peak potential is observed (point **c**). An applied negative scan results in a cathodic peak potential (E_{pc}) of the oxidant to occur at −0.06 V. An anodic peak potential (E_{pa}) of the reductant occurs at +0.06 V. The growing double layer slows down mass transport of the analyte to the electrode. Thus, scanning to more positive potentials (beyond E_{pa}) results in a decreasing current, as the rate of diffusion of the analyte from the bulk solution to the anode is slowed. When the switching potential

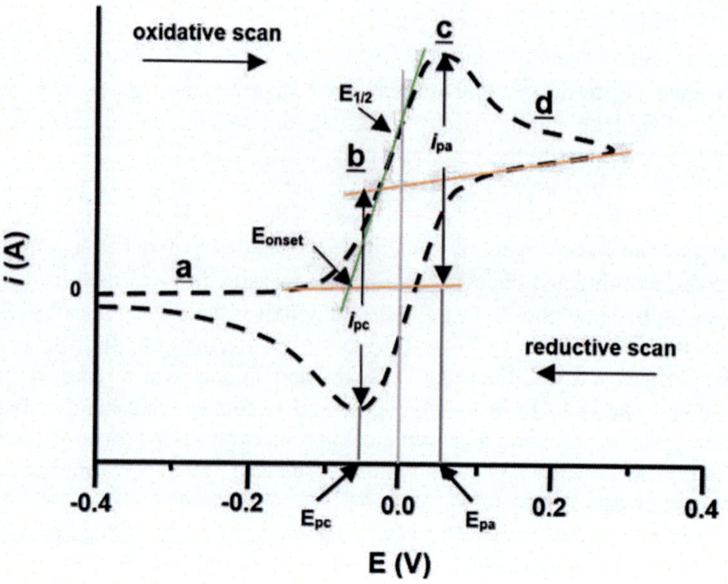

Fig. 5 A generic cyclic voltammogram

is met the scan direction is reversed, and the potential is scanned in the negative (cathodic) direction. The reverse scan almost mirrors in forward scan.

Notably, supporting electrolytes are added to the electrochemical solution in molar excess to the substrate, such that the ionic strength is sufficient to obey the Nernst equation. Excess electrolyte ensures the potential drops to a negligible level within nanometres of the working electrode, resulting in a precise current response at the electrode surface. Moreover, a significant parameter is the scan rate (υ [mV s^{-1}]), which controls how fast the applied potential is scanned. Faster scan rates lead to a decrease in the size of the diffusion layer and as a consequence, higher currents are observed [15, 16].

The Nernst equation (Eq. 9) relates the potential of an electrochemical cell (E) to the standard potential of a species (E^0). In the equation, R (8.314 J K^{-1} mol^{-1}) is the universal gas constant, T (K) is the temperature, n (unitless dimension) is the number of electrons transferred in the cell reaction, F (96,485.333 C mol^{-1}) is the Faraday constant and Q_r is the reaction quotient of the cell reaction.

$$E = E^0 - \frac{RT}{nF} \ln Q_r \tag{9}$$

Peak to peak separation (ΔE_p) defines the difference between the anodic and cathodic peak potentials, and it is used to infer the barrier to electron transfer (electrochemical reversibility). Electrochemical irreversibility arises when the electron transfer at the electrode surface is slow compared to mass transport. Since the electron transfer reactions are sluggish, there is a high barrier to electron transfer. As a result, a ΔE_p larger than 57 mV (at 25°C, 2.22 RT/F) for a one electron redox couple indicates an irreversible process [13]. Electrochemically reversible processes, where ΔE_p is small and the electron transfers are fast, follow the Nernst equation. They are referred to as 'Nernstian'. The anodic and cathodic peak current (i_{pa} and i_{pc}) should be of equal magnitude with opposing signs if the process is truly reversible. This occurs when the electron transfer rates are sufficiently fast such that the concentration of the oxidised and reduced species is in equilibrium. The Nernstian equilibrium is established immediately upon any change in applied potential.

1.6 Direct vs Indirect Electrolysis

An electrochemical reaction can be classified as direct or indirect depending on the type of electron transfer (Fig. 6). Direct reactions mean that the electron transfer occurs directly between the surface of the electrode and the organic substrate and, thus, it is an intrinsically heterogeneous process. In this context, the electrochemical parameters are optimised to transform the starting material directly. For indirect electrolysis, first, a mediator (or redox catalyst) undergoes heterogeneous electron transfer with the electrode surface to form a stabilised intermediate. A following homogenous reaction of the mediator and substrate facilitates the desired transformation. Therefore, we can think of mediators as electron shuffles. Often, simultaneous regeneration of the mediator is observed upon reaction with the desired substrate, meaning they can be used in catalytic quantities. For a cathodic reduction, the electron transfer between a mediator and substrate is thermodynamically favourable if the LUMO of the substrate is at a lower energy than the HOMO of the mediator. The transfer of electrons in solution is driven by the difference in

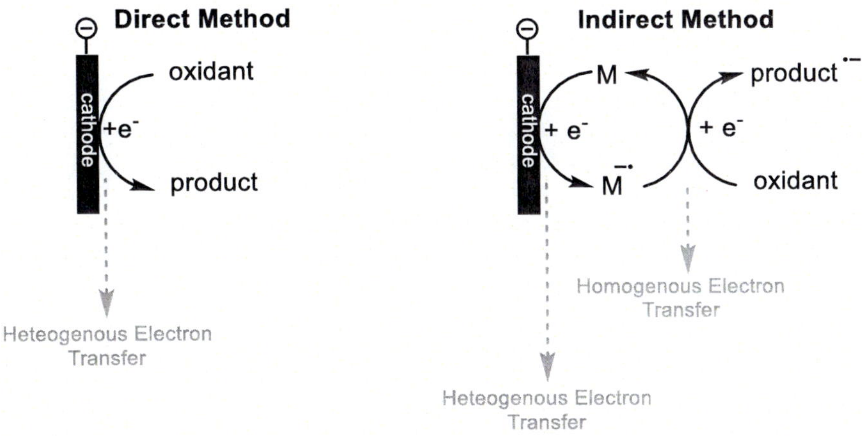

Fig. 6 Schematic representation of direct vs indirect electrolysis

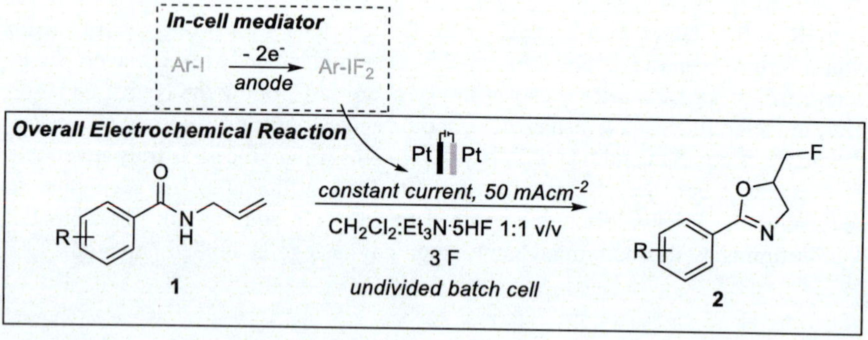

Fig. 7 Hypervalent iodine mediated fluorocyclisation [21]

energy levels, i.e. the electrons move to a lower energy orbital and this is more stable.

The electron transfer to and from the electrode is heterogeneous and can be physically or kinetically hindered. In such cases, homogenous electron transfer to the substrate via a mediator can be useful. Mediators must fulfil three key pre-requisites [17]:

- The mediator must be stable in both oxidation states over the course of the reaction;
- the homogenous electron transfer to/from the substrate must not have a significant kinetic barrier;
- the heterogeneous electron from the electrode to/from the mediator must not be associated with high over-potentials.

If the criteria are fulfilled, the electrochemical reaction can be performed at much lower redox potentials in comparison with the direct transformation of the substrate. Moreover, if decomposition or polymerisation leads to passivation in direct electrolysis, the utilisation of a mediator can negate this. Mediators are commonly employed to avoid over-oxidation typically associated with decomposition and polymerisation.

While indirect electrolysis using inorganic mediators found applications as early as 1900 using chromium salts to facilitate the anodic synthesis of quinones [18], powerful organic redox mediators were not recognised until much later. Organic redox mediators such as triaryl amines [19] and nitroxyl radicals [20] were popularised during the 1970s and 1980s. Nowadays, organic compounds such as polycyclic aromatic compounds, triarylamines, 2,2,6,6-tetramethylpiperidine-*N*-oxide (TEMPO), $I^{(I/III)}$, $Fe^{(II/III)}$ compounds are recognised as 'greener' alternatives. Transition metal complexes and ionic halides represent two other types of common mediators.

Applications of mediated processes span a whole range of organic electrochemistry. For example, Waldvogel et al. developed a sustainable synthesis of 5-fluoromethyl-2-oxazolines **2**, by electrochemical fluorocyclisation of *N*-allylcarboxamides using a hypervalent iodine (III) mediator (Fig. 7) [21]. The iodoarene mediator was applied in an in-cell manner, meaning both the mediator and substrate were present during electrolysis. Theoretically, the anodic oxidation of iodine (I) to iodine (III) requires a charge amount of 2 F. In the absence of the iodoarene (ArI) compound, no oxazoline formation was observed, indicating that anodic oxidation of the mediator was crucial for the conversion of the substrate.

Indirect electrolysis can be performed using two discrete cells: in-cell or ex-cell. Electrolysis carried out only for activation of the mediator is called an ex-cell methodology. This technique may be used when the redox potential of the substrate is lower than that of the mediator. In indirect in-cell electrolysis, a homogenous redox reaction occurs in solution in the presence of the mediator and substrate. This is distinct from a heterogeneous electron transfer between an electrode and the substrate in a direct approach. Critically, the redox potential of the mediator must be lower than that of the substrate, ensuring the mediator is activated before the substrate. Lower reduction/oxidation potentials permit milder reaction conditions (i.e. lower current densities) avoiding the generation of by-products. Electrolysis in this manner is particularly useful in the presence of functional groups that need to be preserved as it commonly avoids overoxidation.

For example, Brown et al. developed a general flow procedure for the TEMPO-mediated electrooxidation of primary and secondary alcohols under microflow conditions at ambient temperature in an environmentally acceptable reaction medium (Fig. 8) [22]. The active species in the TEMPO-mediated oxidation of alcohols is the oxoammonium ion **4**, generated from TEMPO through a single-electron transfer process at the anode. The following mediated oxidation of the alcohol gives rise to the hydroxylamine by-product **5** of TEMPO. This compound is rapidly oxidised at the anode to regenerate the TEMPO radical **3**, hence catalytic use of the mediator was possible.

Electrolysis was performed in the Flux electrolytical cell supplied by Syrris LTD. [23] Various functional groups were tolerable, and several primary and secondary alcohols were selectively transformed to the aldehyde or ketone in yields of up to 87%. Overoxidation to the corresponding carboxylic acid, a common problem with traditional chemical reagents, was avoided by development of this electrochemical protocol.

1.7 Types of Cells

1.7.1 Batch Cells

Although electrolysis in batch cells is not the focus of this chapter, it is important to understand their construction to appreciate their limitations. As such, a brief discussion will be included. Batch electrochemical cells are easily constructed from components readily available in most organic and medicinal laboratories, therefore they are the most widely adopted approach to date. A batch cell is an umbrella term which encompasses two further types of cells, these are termed undivided and

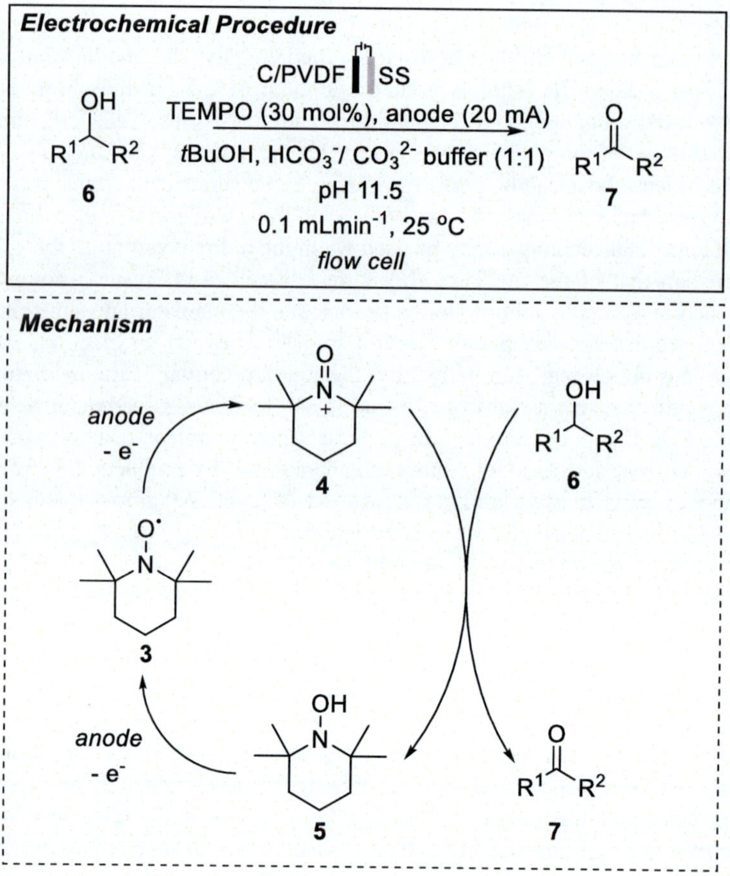

Fig. 8 Electrochemical TEMPO-mediated oxidation of alcohols [22]

divided cells, each with their own strengths in terms of operational simplicity (undivided) and enhanced control (divided) (Fig. 9). In the simplest cell, working and counter electrodes reside in the same chamber and this is called an undivided cell. However, if, for example, high energy intermediates generated at the working electrode are prematurely reduced at the cathode before purification, this cell is not applicable. This may be overcome with divided cell set-ups. In divided cells, the anodic and cathodic solutions are in different vessels, separated by a small porous frit or salt bridge that allows charge transfer. Therefore, the two half reactions occur separately. An early example in 1889 by Maigrot and Stabate describes an example of membrane electrosynthesis [24]. For a reduction, a sacrificial anode consisting of oxidisable materials (e.g., Mg, Zn or Fe) forestalls the undesired oxidation of desired intermediates by preferential oxidative dissolution of the anode.

Batch cells bear several disadvantages to analogous flow reactors: they only allow slow conversion of reactant to product, require longer reaction times, require large

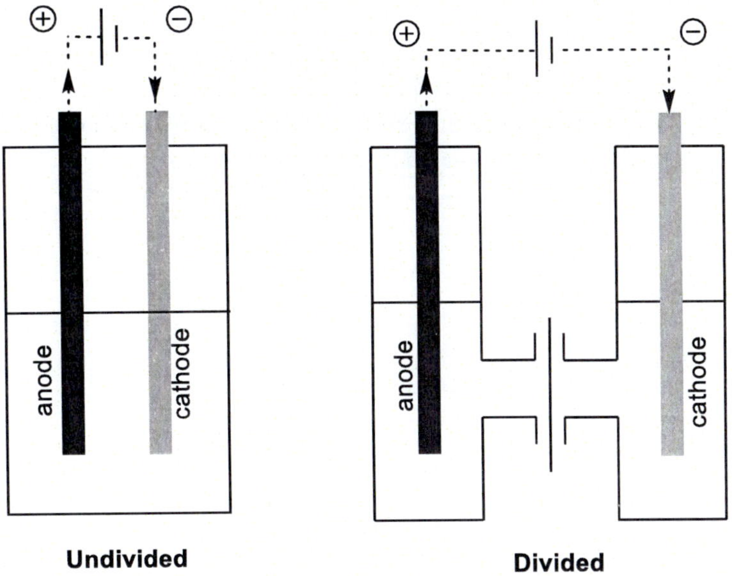

Fig. 9 Schematic representation of a divided and an undivided batch electrolysis cell

electrodes, and demand high concentrations of electrolytes. This impacts the scalability of such cells.

Batch electrolysis does, however, have some advantages over flow cells. The simplicity of the cell and availability of materials means that they are convenient for small scale synthesis. Another important advantage of batch cells over flow-through cells is that solvent and gas parameters can be controlled more easily. In the case of gas evolution, pressure can be easily released through the small headspace above the reaction solution.

Often precise descriptions of these cells in the literature lack precision. Information including the distance between electrodes, electrode geometry or electrode position are not reported, which leads to irreproducibility of results from laboratory to laboratory.

1.7.2 Flow Electrochemical Reactors

Parallel Plate Flow Cells

Electrochemical flow reactors are a more sophisticated approach to overcome the issues surrounding batch electrochemistry. Nowadays, laboratory flow reactors are commonly constructed using a 'parallel plate' design based on two electrodes separated by a spacer, such as a polymer sheet, whereby its centre is cut away to form a flow channel (Fig. 10). The cell is sealed between two metal conductive plates slightly larger than the electrodes, and compression of the sandwich allows leak-free

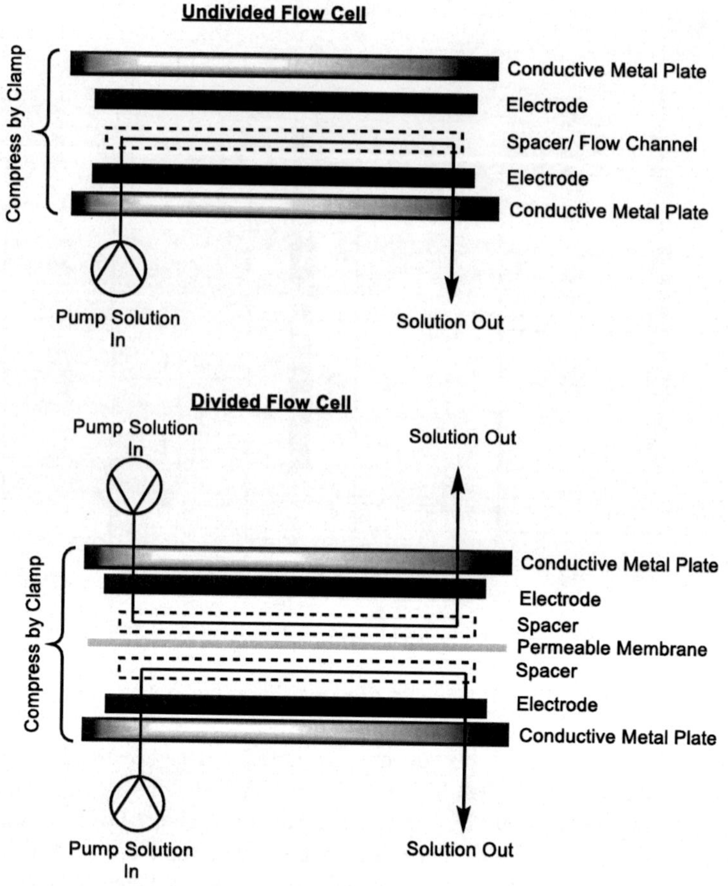

Fig. 10 Schematic representation of a parallel plate designed divided and undivided flow electrochemical cell

operation. The reactor normally includes entry and exit holes at opposing sides of the reactor, ensuring a uniform flow over the whole surface of the electrodes. Here, parameters such as residence time and flow rate can be manipulated which give a greater degree of control over electrochemical reactions. The distance between the electrodes can be controlled by the thickness of the separating spacer.

In flow electrolysis, the aim is to obtain full conversion of the reactant to the desired product in the minimum amount of time. Thus, high flow rates are desirable because they facilitate high productivities (i.e. $>1\ g\ h^{-1}$). Flow electrolysis is nearly always performed in a galvanostatic manner due to the complexity of incorporating a reference electrode into flow reactor designs.

Flow electrochemical cells have convenient advantages over batch cells, including:

- A very high electrode surface to reactor volume ratio;
- fast conversions and hence short reaction/residence times – intrinsically an observation linked to the enhanced electrode surface area;
- no or low concentrations of supporting electrolyte – permitted by the small interelectrode gap which facilitates easy purification;
- uniform current distribution for greater control of electrochemical reactions;
- easier scale-up – a continuous flow protocol is ideal for integration into an automated flow synthesis platform.

A commonly observed problem in batch cells is ohmic drop caused by uncompensated resistance between the working and counter electrodes that is, in turn, caused by the low solution conductivity and a large distance between the electrodes. The electrode gap in flow electrochemical reactors is very small, normally below 500 μm. Given that the ohmic drop is influenced by the distance between the electrodes, the small electrode distance results in a significant reduction in the ohmic resistance avoiding the presence of large current gradients, which leads to a uniform current distribution [25]. Moreover, flow electrochemical reactions can be carried out in the absence of or in very low concentrations of supporting electrolytes since the two diffusion layers of anode and cathode overlap. Large amounts of supporting electrolytes make electrochemical processes more complicated, as the supporting electrolyte needs to be separated from the product during purification, and preferentially should be recycled to minimise the cost of the process. A substantial contribution to reduced reaction times in these cells can be attributed to the better temperature control. Temperature control is aided by smaller channels which promote efficient heat transfer. These factors combined mean that electrochemical reactions in flow can occur much faster than the analogous batch process.

Some disadvantages are still apparent, but these issues are quickly being addressed. The limited controlled over gas in flow reactors can be problematic. Hydrogen generation at the cathode is a common balancing electrochemical reaction for organic reactions, however, the generation of gas in flow dramatically effects the flow rate. The observed flow rate in this context is higher than the applied flow rate which can affect the reproducibility of the reaction. Back-pressure regulators applied after the electrochemical reactor can help, but they require reactors with good pressure tolerance to avoid leakage. These drawbacks limit the suitability of electrochemical flow reactors for reactions which use or generate large volumes of gas.

Like batch cells, electrochemical flow cells can be operated in a divided or undivided manner. Due to operational simplicity, most flow electrochemical cells to date are operated in an undivided manner whereby a single reactant solution is pumped through the narrow gap between the electrodes and collected from the solution outlet. Divided cells contain an additional solution compartment separated by an ion permeable membrane or porous polymer separator, e.g. Nafion. The porous separator enables the transfer of ions or small molecules but blocks any significant mixing of solutions between the working and counter electrode compartments.

In cases where high flow rates are used, and the residence time is low, undivided flow reactors can be more tolerable than undivided batch cells. The continuous flow of fresh substrate limits over-oxidation at the electrode surface, although this cannot be prevented completely. Moreover, in the case of an anodically generated species that could be itself reduced, since the residence time is short, the species may exit the reactor before the undesirable reduction has a chance to occur.

Thin-Layer Flow Cells

The following discussion on thin-layer and porous flow cells is important for metabolite inspired drug development (Fig. 11). A major limitation of thin-layer cells for synthesis purposes is their small surface area and cell volume. In synthetic terms, these reactors are difficult to scale up and are rarely used for preparative scale reactions. Instead, these electrochemical flow cells have been proposed for continuous, online monitoring and applications in flow injection analysis and chromatography [26]. Modern electrochemistry offers a wide range of analytical methods that can be used for continuous measurements in flowing liquids [27]. Thin-layer flow cells have vast applications in amperometric measurements, but for purpose of this chapter they have also been used to study drug metabolism [28–31]. Thin-layer flow cells exhibit working electrodes situated at the top of a flat flow chamber of cylindrical [32, 33], channel like [34] or other shape and the planar working and counter electrodes are separated by a thin spacer (typically 10–100 μm). The working electrodes can be removed and a range of electrodes including Pt, Au, Ag, Cu, glassy carbon and BDD, which are relevant for drug oxidation, can be exchanged. In comparison with parallel plate designs, the working electrode surface area is in the mm^2 range, which limits scalability. Moreover, the small surface area

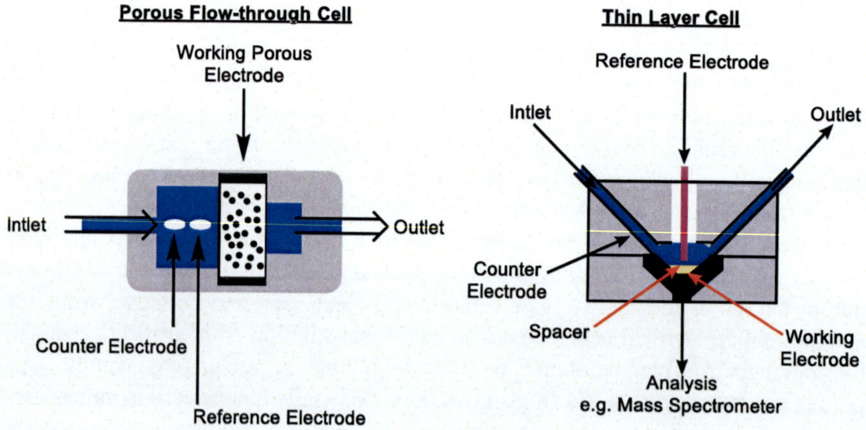

Fig. 11 Schematic representation of a porous flow-through cell and thin-layer cell

and cell volume demands low flow rates (<100 μL min^{-1}) in order to obtain high conversions [35].

Porous Flow Cells

Porous flow-through cells display many advantages compared to thin-layer cells (Fig. 11). For instance, the large surface area owing to the porosity of the working electrode (estimated to be in the region of the cm^2 range) allows extremely high conversion rates. For some compounds, flow rates of up to 0.5 mL min^{-1} accompanied up to 100% conversion [36, 37]. A relatively high flow rate through the electrode facilitates enhanced mass transfer rates. Electrochemical synthesis of oxidative drug metabolites has thus far mostly been performed using commercially available coulometric flow cells with a porous glassy carbon electrode [38]. A wide range of porous electrode structures and materials are available, for example, reticulated carbons and metals [39], packed beds [40], cloths and felts [41] and sintered metals [42]. In practice, the effective electrode bed depth can be limited by the potential drop in the fluid, which can restrict the conversion per pass. Two basic electrode configurations are possible: a flow-through porous electrode configuration is one where the fluid flows parallel to the current, whereas 'flow by' arrangements are where the current and electrolyte flows are perpendicular. Flow porous cells suffer with limited control over gas parameters which can only be adjusted prior to infusion into the cell. These drawbacks limit the suitability of flow-through cells for large-scale metabolite synthesis [26].

Automated Flow Electrolysis Platforms

Automatisation refers to a 'hands off' synthesis approach, whereby synthesis and even real-time analysis are driven by a computer or robotic equipment which allows unattended operation. The capacity for automation to improve the scalability, safety, efficiency and reproducibility of synthetic processes is obvious. A recent analysis of synthetic facets of medicinal chemistry demonstrates that medicinal chemistry is inevitably biased towards synthetic routes where constraints in resources, time and budget are limited. Indeed, rapid cycle times in the pharmaceutical industry, linked to performance metrics and deadlines strongly influence the adoption of any technology. Technologies that enable greater automation of compound synthesis, endorsing a wide range of reaction conditions are extremely desirable as they vastly reduce constraints in resources and are inevitably cost effective.

Flow electrochemistry has already been merged into many automated commercially available platforms [8, 43]. Most simply, these machines are built with a computer which controls the machinery and parameters (i.e. flow rate via control of the pump, pressure, current, potential), various pumps (HPLC or peristaltic – depending on needs), flow tubing reactors, pressure sensors and regulators, heating and cooling systems – all conveniently connected together by flow tubing. Such automated platforms can be integrated with online spectroscopy, such as HPLC, mass spectroscopy, IR, UV-vis, etc., which allows real-time analysis. These platforms can often be controlled by remote computers, meaning the experiments can be easily monitored and manipulated. For dangerous reactions, the benefit of the 'hands off' approach is significant. Such reactions can be performed remotely with no risk to the experimentalist.

Despite these advantages, the main drawback of automated systems is the large amount of data that is created in a short amount of time. For this reason, 'design of experiment' approaches are increasingly employed in automated platforms coupled with online analysis, allowing the rapid optimisation of reaction parameters [44].

Increasing the productivity from laboratory scale to production scale is a formidable challenge in the pharmaceutical industry. Automated flow electrochemical synthesis in this respect could be recognised as a foundation for developing a continuous integrated small molecule optimisation platform that would greatly enhance drug optimisation programs. The most ambitious vision for drug discovery would integrate machine-learning algorithms coupled to an automated synthesis module. The progress of machine learning refers to the idea that a system can 'learn' from data sets, identify patterns and then make decisions with minimal human intervention. In this regard, a continuous loop of newly designed and highly optimised drug molecules could be recognised.

There are still, however, many poorly manageable practical hurdles such as work-up, product separation and purification that have not been fully addressed. Since electrochemistry can be performed in the absence of reagents and by-product formation is typically avoided, these reactions can often by-pass problematic purification. In this context, automated flow electrochemical synthesis where possible, as opposed to automated flow syntheses (in the presence of reagents) is more conceivable for a truly fully automated process.

2 Flow Electrochemistry for Drug Discovery

Electrosynthesis in flow cells has been used to synthesise many natural products, drug molecules and metabolites. The width of electrochemical synthesis for general organic chemistry is enormous. For this context, we will highlight some broad synthetic electrochemical strategies and platforms that have been developed over the last decade that are valuable tools for medicinal chemists and have considerable further potential in drug discovery. The breadth of these strategies is far too wide to discuss them all, but the most important or interesting reactions with respect to this circumstance will be conferred by providing at least one example of each method.

2.1 *Flow Electrosynthesis of Pharmaceutically Relevant Scaffolds/Fragments*

Structural modification of known drug molecules and drug related scaffolds is a key approach in drug development. In this section, drug molecules that are similar to the reaction products/scaffolds that are discussed herein are displayed in a dashed line box next to each figure with the heading 'context'.

2.1.1 Electrochemical Synthesis of Nitrile-Containing Scaffolds

Over 30 nitrile-containing pharmaceuticals are available for a diverse array of medicinal conditions, with more than 20 additional nitrile-containing leads in clinical development [45]. As the number of nitrile-containing pharmaceuticals has increased, trends in identifying the role of the nitrile functionality has emerged. The nitrile group is quite robust and in most cases is highly metabolically stable. Hence, the nitrile group in most nitrile-containing drugs is unchanged when passing through the body.

The prevalence of nitrile-containing pharmaceuticals and the continued stream of potential drugs in the clinic has influenced the development of more efficient synthetic protocols. In the past, the formation of nitriles from oximes (of the most convenient starting materials) required strongly corrosive dehydration reagents such as phosphorous pentoxide, thionyl chloride, benzenesulfonyl chloride, acetic anhydride, etc. These reagents were often used in excess, which both represents an unattractive waste implication and promotes difficulties with purification. The flow electrochemical synthesis represents an intriguing alternative. Flow electrochemistry both reduces the waste, hence cost burden of the process, and diminishes the safety concerns for the experimentalist.

Waldvogel et al. demonstrated a halogen-free procedure to generate nitriles **10** from oximes **8** by the application of inexpensive and easily available electrode materials (Fig. 12) [46]. The electrolysis was carried out in an undivided flow electrolysis cell bearing a graphite anode and lead cathode. Mechanistically, the authors suggest the reaction proceeds in a domino oxidation-reduction sequence whereby the oxime **8** is first oxidised at a graphite anode to the corresponding nitrile-*N*-oxide **9**, which is directly reduced at the cathode to the desired nitrile **10**. The intermediate, **9**, was observed and identified via GC-MS. The key for the desired reaction sequence seemed to be the deoxygenation of the nitrile oxide **9** formed. Whereas lead turned out to be an excellent cathode material, the deoxygenation was less favoured at the other electrode materials such as Pt, Ni and BBD. Aldoximes are first oxidised to form iminoxyl radicals, which can dimerise to form the undesired aldazine bis-*N*-oxides. These aldazine bis-*N*-oxides can react further to form the corresponding aldehydes by the loss of nitrogen. Stabilisation of the reactive intermediate could be achieved by adding bulky groups *ortho* to the aldoxime moiety. Here, the formation of the nitrile oxides **9** by further oxidation of the iminoxyl

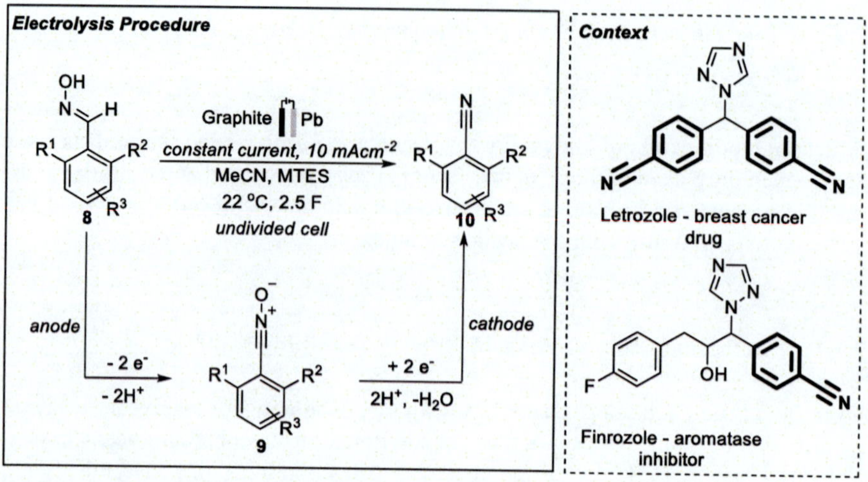

Fig. 12 Synthesis of benzylic nitriles by anodic oxidation by Waldvogel et al. [46]

radical was favoured, which underwent deoxygenation at the cathode leading to the desired nitrile products **10**. Whilst mesitylaldoxime (81%), 2,4,6-trimethoxybenzaldoxime (75%) and 2,6-dimethylbenzaldoxime (73%) could be transformed in good yields, 2,6-dichlorobenzaldoxime (41%) and starting materials bearing bromine substituents were more problematic. All of these substrates vary in the stability of the formed nitrile oxides, as well as their preference for dimerisation. The unexpectedly low yield of the stabilised 2,6-dichlorobenzaldoxime was attributed to the competing dehalogenation reaction. In general, stabilised substrates resulted in higher yields of the products **10** than non-stabilised congeners. However, the corresponding aldehyde, formed by the undesired dimerisation process, was detected in all reactions. Nonetheless, the developed protocol was applicable for a wide range of substrates under constant current and ambient conditions, to give the nitrile products **10** in moderate to excellent yields.

2.1.2 Electrochemical Synthesis of Benzoxazoles and Benzothiazoles

A large number of marketed drugs are available bearing a benzoxazole as a core active moiety. For example, drugs incorporating this important scaffold include flunoxaprofen (anti-inflammatory), benoxaprofen (anti-inflammatory), calcimycin (antibiotic), boxazomycin B (antibacterial) and chlorzoxazone (muscle relaxant). Several electrosynthetic procedures for the synthesis of benzoxazoles have been reported in the literature using conventional batch techniques. The Waldvogel and Moeller groups exploited the electrochemical generation of amidyl radicals from anilides, enabling the efficient synthesis of benzoxazoles [47]. Methanol and other solvents commonly applied in the electrochemical generation of amidyl radical lead to degradation of the anilides. The authors suggest this observation is indicative with

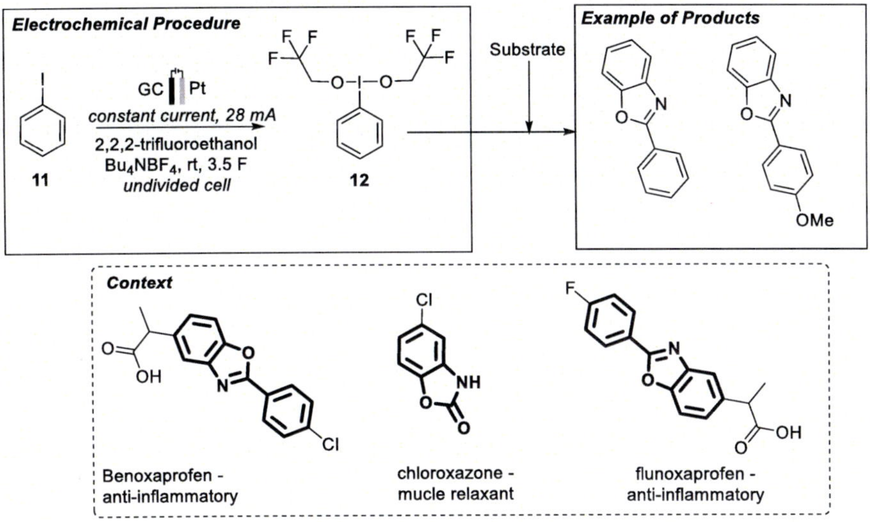

Fig. 13 Hypervalent iodine mediated synthesis of benzoxazoles in flow [48]

a slow intramolecular cyclisation process, and thus sufficient stabilisation by a solvent was required. Interestingly, 1,1,1,3,3,3-hexafluoro-2-propanol (HFIP) was found to efficiently stabilise the electrochemically generated radical intermediates [12]. The almost quantitative recovery of this particular solvent (b.p. 56°C) allowed minimisation of the fluorine footprint. Several functionalities were tolerated by the method leading to the synthesis of a variety of benzoxazoles in good to excellent yields. Notably, typical leaving group functionalities, i.e. chloro and triflate moieties, were compatible. Such reactions expose the opportunity to further functionalise and diversify the scaffolds using an assortment of subsequent reactions, i.e. cross-coupling reactions, which is a key incentive to medicinal chemistry.

Very recently, the Wirth group disclosed an electrochemical generator of hypervalent iodine reagents and their subsequent utilisation in a diverse array of reactions. The flow synthesis of benzoxazoles using an electrochemically generated hypervalent iodine mediator **12** was demonstrated (Fig. 13) [48]. The electrolysis was performed in a commercially available flow reactor using a glassy carbon anode and platinum cathode [6]. The anodically generated, unstable hypervalent iodine reagent derived from iodobenzene could be generated quantitatively in HFIP in flow. However, passivation of the electrode with the use of HFIP hindered the production of the subsequent I (III) reagent over time. Using 2,2,2-trifluoroethanol resolved this issue, and the production of **12** was stable for over 4 h. Notably, fluorinated alcohols were again targeted for the electrochemical experiments because they display several advantages compared to common organic solvents, i.e. they are stable to anodic oxidation and have a high conductivity. Their relatively low pKa means H^+ reduction at the cathode is an ideal balancing electrochemical reaction, since H_2 gas could in theory be captured and resourced. Furthermore, fluorinated alcohols are known to

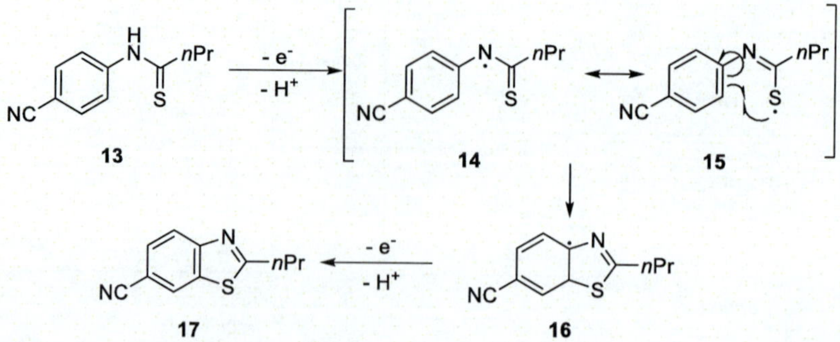

Fig. 14 Proposed mechanism for the anodic generation of benzothiazoles from *N*-arylthioamides

play an active role in stabilising I (III) species [12]. The generation and immediate use of these reagents **12** in flow was highly advantageous. These reagents were not bench stable and decomposed immediately upon removal of the solvent. Also, by coupling a second step in flow, the developed '*ex-cell*' procedure means electrochemical functional group tolerance of the substrates was not a concern. Hence, the demonstrated synthesis was highly versatile and gave access to a diverse array of products.

Instinctively, Wirth et al. reported a catalyst and supporting electrolyte-free electrosynthesis of benzothiazoles and thiazolopyridines in continuous flow (Fig. 14) [49]. Benzothiazole is a versatile heterocycle scaffold that has aroused much interest in drug discovery. It is the backbone of many anticonvulsants, neuroprotective drugs, analgesics, anti-inflammatory drugs, antimicrobials and anticancer drugs. Two benzothiazolamines, riluzole and lubeluzole, are known voltage-gated sodium channel blockers. A broad range of *N*-arylthioamides were converted to the corresponding benzothiazoles in good to excellent yields with high current efficiencies. Mechanistically, the authors propose the thioamide undergoes a single-electron anodic oxidation to form the thioamidyl radical **14** (Fig. 14). Next, the radical intermediate undergoes intramolecular cyclisation and further anodic oxidation to form the corresponding benzothiazole **17**.

Interestingly, there is no requirement for an inert atmosphere in flow and laboratory grade solvents could be utilised without prior de-gassing. To demonstrate the scale-up potential of the developed method, a large scale (13 mmol) reaction was performed, giving the product **19** in an 87% yield (2.4 g) (Fig. 15). This work highlights three key advantages of electrochemical flow systems, including (1) a supporting electrolyte-free reaction which aids purification, (2) easy scale-up without the requirement of a larger reactor and (3) enhanced reaction rates facilitated by increased mixing in flow compared to batch. The method was demonstrated for the synthesis of a large library of versatile molecules (28 substrates) giving benzothiazoles and thiazolopyridines in high average yields.

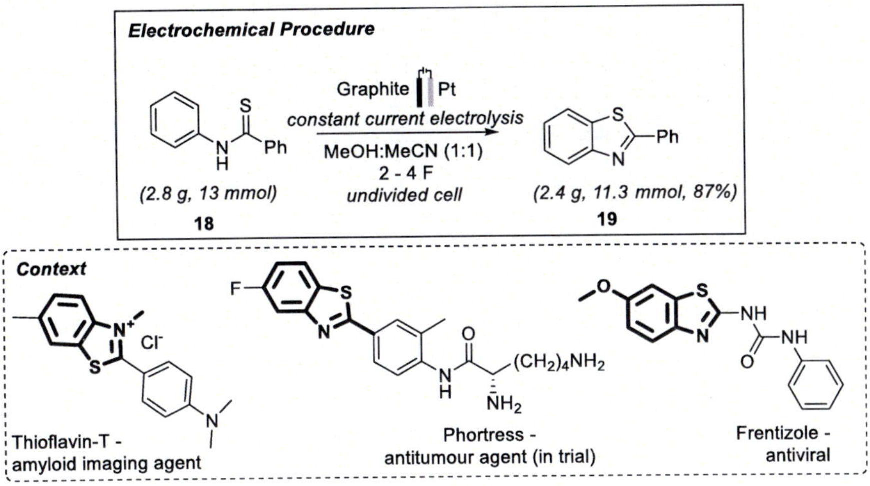

Fig. 15 Flow electrochemical synthesis of benzothiazoles [49]

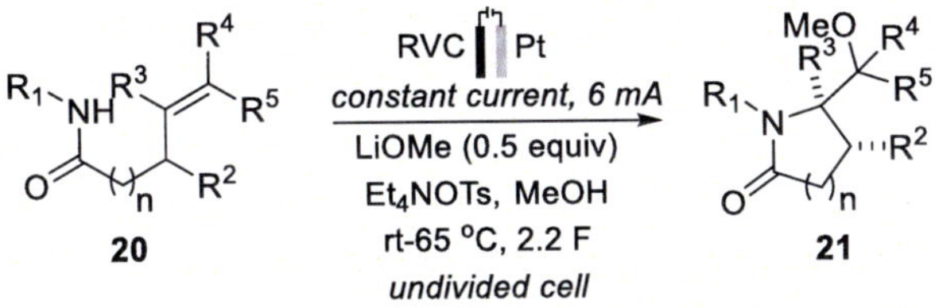

Fig. 16 Electrochemical generation of amidyl radicals for synthesis of γ- and δ-lactams by Moeller [50]

2.1.3 Electrosynthesis of N-Containing Heterocycles by Nitrogen-Centred Radicals

Unlike the previously discussed examples, nitrogen-centred radicals have historically been underutilised in synthetic chemistry. The term nitrogen-centred refers to a species where the unpaired electron is localised on a nitrogen atom. Nitrogen-centred radicals are versatile intermediates, for example, they have been demonstrated to add alkenes, alkynes, dienes and engage in C–H functionalisation. Their applicability in academic and industrial settings is however limited, owing to the lack of mild and reliable methods for their generation.

Electrochemistry has shown to be an efficient and mild methodology to create nitrogen-centred radicals. The addition of electrochemically generated nitrogen-radicals to double bonds has been known for the last century and is now reasonably established in batch synthesis. For example, the Moeller group disclosed a protocol for the synthesis of γ- and δ-lactams using *O*-benzyl hydroxamates or *N*-phenyl amides in batch (Fig. 16) [50]. Here, anodically generated nitrogen-centred amidyl radicals undergo cyclisation reactions with electron-rich olefins to form carbon-centred radicals. The radicals undergo further single-electron transfer with the

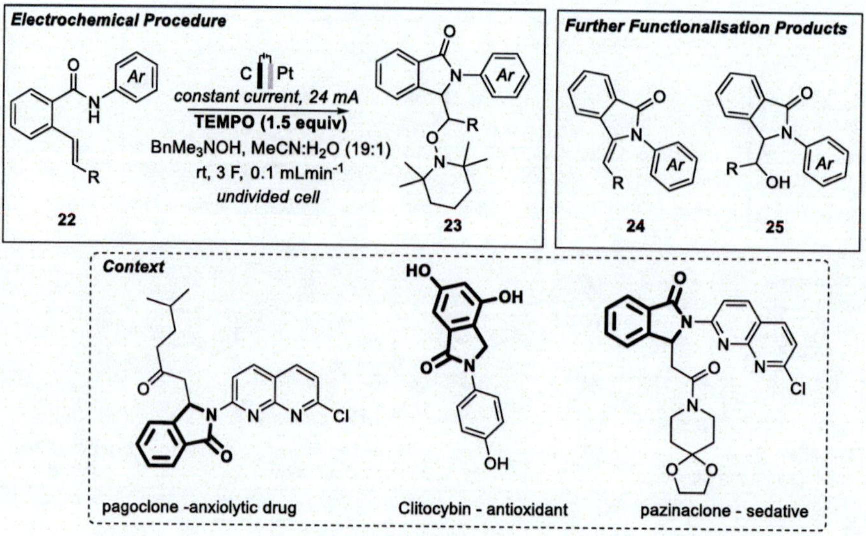

Fig. 17 Synthesis of isoindolinones using an electrochemical flow microreactor [51]

anode, to furnish the carbocation, which is trapped by nucleophiles to provide the lactam products **21**.

Instead, the generated carbon-centred radicals can be trapped with TEMPO, to give the oxyamination products. Wirth et al. demonstrated a facile flow electrochemical synthesis of amidyl radicals which were used in intramolecular hydroaminations to produce isoindolinones **23** [51]. Isoindolinone motifs are found in many pharmaceuticals and biologically active molecules (Fig. 17). In this report, TEMPO has a dual function, being the mediator and oxygen source. An electrochemical microflow reactor was designed and manufactured; a large electrode area (25 cm^2) and easily exchangeable electrode materials made the reactor both flexible and efficient. Connecting the outlet of the flow reactor to an inline mass spectrometer via an automatic sampling valve facilitated rapid electrolysis optimisation. It was determined that at least 3 F was necessary to achieve full conversion of the starting material. Aromatic substituents on the amidic nitrogen were crucial to stabilise the initially formed amidoyl radical and achieve the desired cyclisation. Here, aliphatic substituents on the amidic nitrogen (e.g. nBu) led to no product formation. Interestingly, when electron deficient alkenes were used (e.g. R = *p*-NO_2-C_6H_4), TEMPO addition was not observed attributed to the higher stability of the generated carbon radicals. Instead, subsequent reduction at the cathode occurred. For further product diversification, the isoindolinone products were subjected to subsequent functionalisation, mainly elimination of TEMPO and the reduction of the N–O bond by integration of a second step in flow. The reduction was performed under flow conditions by using AcOH and a heated zinc cartridge. Since the concept of a further steps in flow was established, the developed procedure was useful to develop a diverse array of products **23, 24** and **25**.

2.1.4 Electrochemical Anodic Aryl-Aryl Cross-Coupling in Flow Cells

Aromatic C–C cross-coupling reactions are central tools for the synthesis of ligands, polymers and natural products [52, 53]. Although a number of synthetic approaches for aromatic C–C cross-couplings have been realised, so far, transition metal catalysed reactions predominate [54]. Transition metal catalysts are often toxic and expensive, hence a more practical synthesis is highly desirable.

On the other hand, the electrochemical activation of C–H bonds is an environmentally and economically attractive alternative. Since electrons are the sole reagent, toxic transition metal catalysts are avoided. The direct electrochemical cross-coupling of two non-symmetrical aromatic compounds suffers from two key factors: (1) homocoupled reactions occur by non-selective oxidation of starting materials, which limits the selectivity and hence yield and (2) the cross-coupled products generally have a lower oxidation potential than the starting materials, hence over-oxidation of the product is often un-avoidable. Flow electrolysis excels over conventional batch synthesis.

The Atobe group, in collaboration with the Waldvogel group, demonstrated a flow electrochemical anodic phenolic arene C–C cross-coupling procedure [55]. The reaction was achieved in inexpensive and sustainable media such as methanol, acetic acid or formic acid. Depending on the solvent and additive, the undesired homocoupled product or desired phenolic product could be obtained. However, the selectivity for the production of the cross-coupled product could be easily controlled by fine-tuning the electrochemical conditions. The electrolysis was carried out in a flow electrochemical microreactor with a boron-doped diamond anode (BDD) and nickel cathode and allowed a good range of products to be synthesised.

In a similar reaction, Atobe and colleagues disclosed a parallel laminar flow mode in a two-inlet flow microreactor for efficient aromatic C–C cross-coupling (Fig. 18) [56]. When two solutions are introduced through the two inlets, a stable liquid–liquid interface could be formed and mass transfer between input streams occurred only via diffusion. Here, the flow microreactor allowed selective anodic oxidation as the undesired oxidation of the aromatic nucleophile **26** was effectively prevented, whilst the aromatic substrate **27** was selectively oxidised at the anode. In addition, the

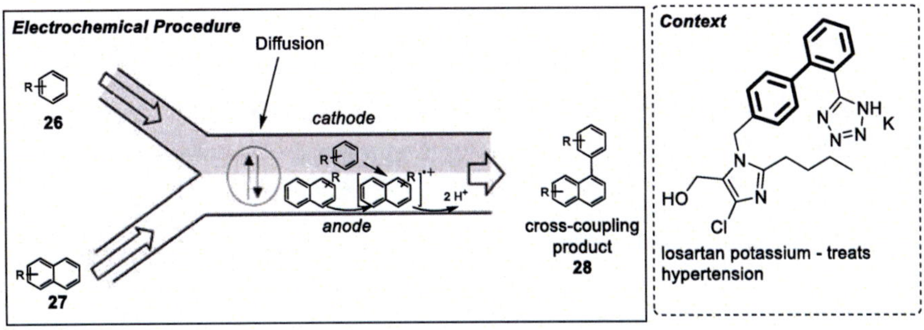

Fig. 18 Anodic aryl-aryl electrochemical cross-coupling procedure by Atobe et al. [56]

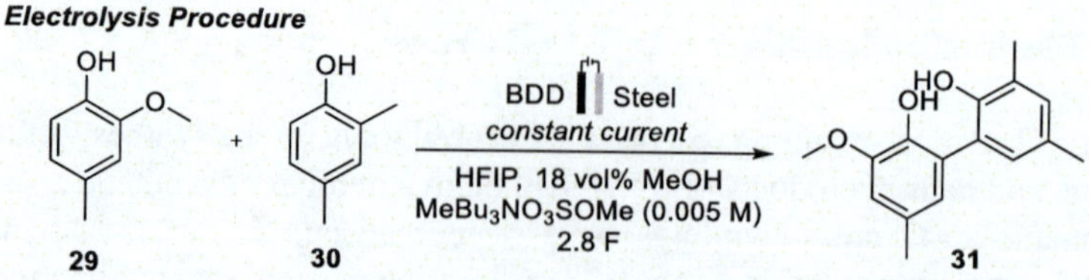

Fig. 19 Electrochemical anodic aryl-phenol cross-coupling in a flow cell [57]

overoxidation of the cross-coupling product could be avoided due to the flow operation. Several cross-coupled products **28** were obtained in moderate to excellent yields. The yields were a significant improvement in comparison with conventional batch synthesis.

More recently, Waldvogel and co-workers developed a modular parallel plate electrochemical flow cell, whereby the temperature of the electrodes could be controlled from the backside via an external cooling circuit [57]. The practicability of the novel cell was validated by three different anodic phenol-phenol cross-coupling reactions (Fig. 19). The electrochemical cross-coupling between phenols and arenes was performed using a boron-doped diamond (BDD) anode and stainless-steel cathode in fluorinated media. In this reaction, employing water or methanol as a mediator represents the key improvement for achieving non-symmetrical biaryls. Interestingly, using no supporting electrolyte was unsuccessful and gave only 5% of the product **31**. The addition of trace amounts (0.005 M) of tributylammonium methylsulfate ($MeBu_3NO_3SOMe$) increased the yield to 44%. Cooling the system to 10°C increased the yield to 48%. Instead of a single pass, pumping the electrolyte through the cell multiple times was more beneficial, however, a maximum yield of 59% was achieved. Notwithstanding, the flow system was able to increase the productivity of the cross-coupling reaction from 0.12 g h^{-1} (25 mL beaker-type cell) to 2.53 g h^{-1} (4 cm × 12 cm flow cell). Analogously, the group later demonstrated a straightforward approach for performing oxidative cross-coupling reactions of electron-rich benzene derivatives [58]. Using molybdenum pentachloride ($MoCl_5$), biphenyls could be obtained in 21–91% isolated yield.

2.1.5 Electrosynthesis of *N*-Containing Heterocycles by Shono Oxidations

Nitrogen-containing heterocycles are essential structural components in drugs and are commonly found in bioactive natural products, and they have broad application in the pharmaceutical and chemical industry [59, 60]. Shono's seminal report in 1975 has spawned continual interest in the anodic oxidation of carbamates to *N*-carbamoyl iminium ions (Fig. 20) [61, 62]. The electrolysis was performed using 2 carbon electrodes as the anode and cathode, tetraethylammonium *p*-toluenesulfonate (Et_4NOTs) as electrolyte and methanol as a solvent. Electrochemical anodic oxidation proceeds via initial formation of a nitrogen-centred radical **33**

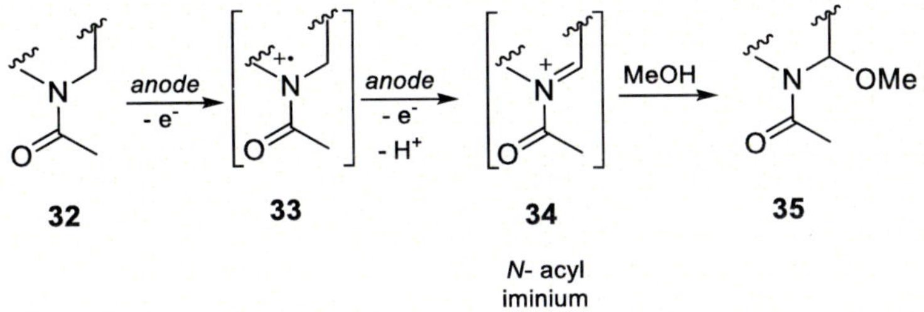

Fig. 20 Shono oxidation mechanism [61, 62]

that is subsequently oxidised to an *N*-acyliminium intermediate **34**, which is further trapped by an alcohol-based solvent. Hence, the anodic oxidation requires a theoretical charge of 2 F. The method facilitates functionalisation of the α-position adjacent to nitrogen atoms in heterocycles.

Whilst carrying out the electrolysis in methanol gives access to the α-methoxylated product **35**, trapping the *N*-acyl iminium ion with cyanide [63–65], fluoride [66], isocyanides [67], furans [68], trialkylphosphites [69], silyl enol ethers [70], etc. has also been well established (discussed later). Noteworthily, a variety of attempts have focused on developing asymmetric type Shono oxidations using various chiral auxiliaries including oxazolines [71], chiral cyclic dipeptides [72], or chiral phosphorus-based structures [73] to give reasonably high stereoselectivities. Extensive batch electrolysis studies have been developed, whereby the Shono oxidations have even been used in metabolic studies of drug molecules. Royer and co-workers showed that the methoxylated analogues of two anticancer drugs, ifosfamide and cyclophosphamide, could be obtained through anodic Shono-type oxidation in high yields [74]. Under galvanostatic conditions, the oxidation took place in a chemo-selective manner at the α-position of the tertiary nitrogen. Alternatively, functionalisation can be carried out to yield a diversity of scaffolds [75, 76].

Limitations with early Shono oxidations hindered the scope of the transformations possible. Selective oxidation to the iminium ion when electrolysis is performed in the presence of the amine and nucleophile demands that the amine has a lower oxidation potential than the nucleophile to avoid the undesired oxidation of the nucleophile. In the early 2000s, the scope of the Shono oxidation was considerably expanded with the development of the 'cation pool' method. Here, the anodic oxidation reactions are carried out by low temperature electrolysis, which allows for the accumulation of the iminium species in a so-called cation pool. As no nucleophile is present during the initial electrolysis, competing nucleophile oxidation is avoided. Yoshida et al. have reported the application of a cation pool and microflow system for flow electrochemical combinatorial organic syntheses [77, 78]. The cell consisted of two compartments: one with a carbon felt anode and the other with a platinum cathode. The two-compartment cell was divided by a polytetrafluoroethylene (PTFE) membrane, and the whole cell could be dropped into dry ice to achieve the desired low temperatures. A schematic diagram of the

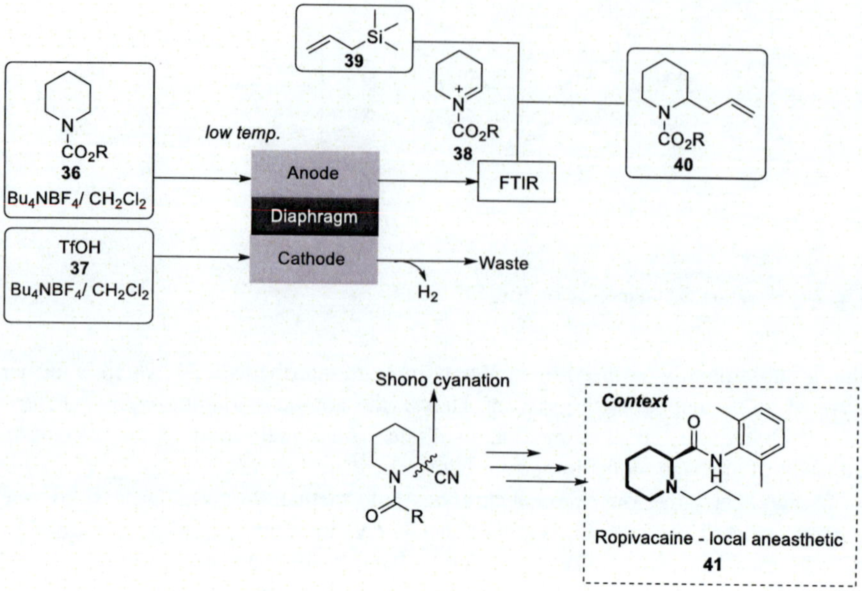

Fig. 21 Shono oxidation via the cation pool method in a microflow reactor [77, 78]

semi-flow process is shown in Fig. 21. Trifluoromethanesulfonic acid **37** is used in the cathodic chamber as a sacrificial compound to provide the balancing redox reaction. Following the typical Shono mechanism, the method demonstrated a one-step step α-allylation of amides using various allyltrimethylsilanes. The anodically generated acyliminium cation **38** could be analysed by inline FTIR spectrometry and then transferred to a reservoir with nucleophile **39**, where the coupling reaction takes place to give the final product **40**. The established flow synthesis allowed rapid generation of a large library. Various allylic silanes and carbamates were tolerable giving a diverse range of products. Interestingly, the addition of cyanide onto the electrogenerated iminium cation followed by reductive amination allows asymmetric syntheses of ropivacaine **41**, mepivacaine and levobupivacaine, which are commonly used local anaesthetics [79].

Inspired by these early efforts, Atobe and co-workers developed a novel electrosynthetic system for the anodic reaction of cyclic carbamates by using parallel laminar flow in a microflow reactor (Fig. 22) [80]. As opposed to Yoshida's efforts, the authors used a platinum (Pt) pair for the anode and cathode. The flow system enabled nucleophilic reactions to overcome restraint, such that the oxidation potential of the nucleophile and stability of the cationic intermediate were not a concern. For example, the anodic substitution reaction of *N*-(methoxycarbonyl)pyrrolidine (E_{pa} = 1.91 V vs Ag/AgCl) with allyltrimethylsilane (E_{pa} = 1.75 V vs Ag/AgCl) could be conducted in good yields. Since the substrate **42** could be less easily oxidised than the nucleophile **39**, the use of a parallel laminar flow mode was key to their success. When the carbamate and the nucleophile are introduced in different

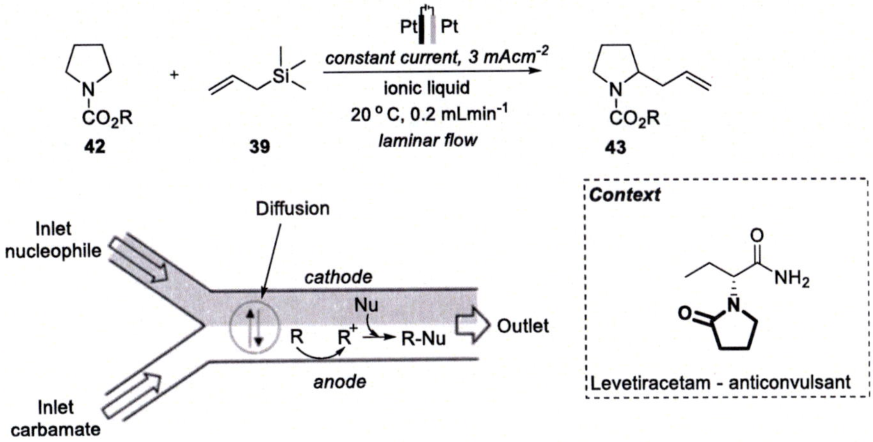

Fig. 22 Anodic Shono oxidation by Atobe et al. using laminar flow [80]

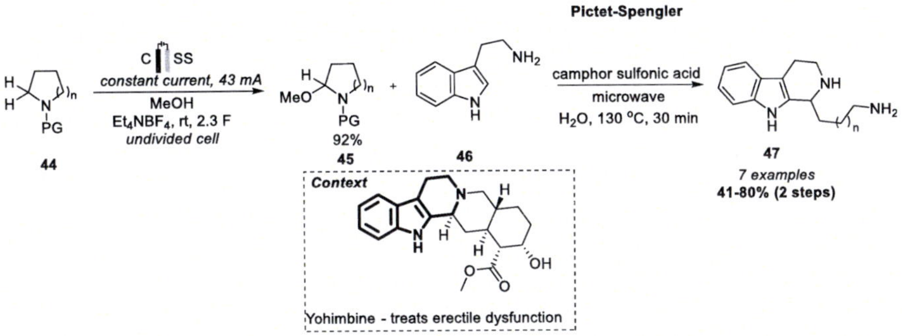

Fig. 23 Shono oxidation of various *N*-protected cyclic amines by Ley et al. [81]

streams, mass transfer between these separate streams can only occur via diffusion between the liquid–liquid contact area. Due to the small size of the flow channel, the area remains stable and laminar. As the nucleophile **39** is spatially removed from the anode, the undesired anodic oxidation is minimised. The anodic substitution reactions of other carbamates were also possible in low to good yields, given the high tolerance of the developed electrolysis system. From a drug development perspective, the highly tolerant and scalable system for α-functionalisations of pyrrolidine motifs could be recognised as an opportunity for the structural modification, hence diversification of many known drug molecules.

More recently, in 2014, Ley et al. identified flow electrolysis as an enabling technology for the Shono oxidation of various *N*-protected cyclic amines (Fig. 23) [81]. The electrolysis was carried out in the commercially available Syrris flux electrosynthesis module [8], using a carbon anode (C) and stainless-steel cathode (SS). Employing a flow electrochemical cell permitted low loadings of electrolyte (20 mol%) which aided purification. Moreover, the system was robust for overnight

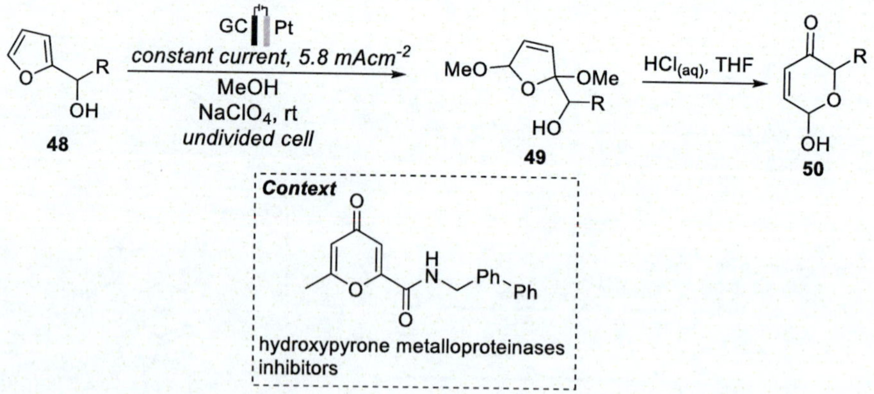

Fig. 24 An electrochemical flow cell for the convenient oxidation of furfuryl alcohols [85]

operation and large-scale preparation is feasible. The electrochemical flow technique permitted the rapid generation of a compound library through the preparation of protected cyclic α-methoxyamines **44**. Here, pyrrolidine, piperidine, azepane and morpholine rings could all be successfully methoxylated using the optimised conditions in high yields (90–98%) and purity. Applying a subsequent Pictet-Spengler reaction to the electrochemically derived products **45** provided the biologically active natural product nazlinine **47** and several other related unnatural congeners (41–80% over two steps). Here, microwave irradiation as an enabling technology reduced the typical 15 h reaction time to just 30 min. The reported scaffolds could lead to the preparation of further unnatural relatives of the indole alkaloids tryptargine, indoloquinolizidine, komaroidine, isokomarvine and schobercine.

In a similar fashion, other alkaloids have been prepared, including (−)-crispine A [82] and (±)-pumiliotoxin C [83]. Pumiliotoxin C was prepared in 15 steps from a commercially available 4-piperidone ethylene ketal in an overall 5% yield. Here, an anodic Shono cyanation furnished the desired α-amino nitriles. The electrolysis was carried out in a flow electrochemical cell fitted with a graphite-felt anode to yield the Shono oxidation product in a 78% yield. Further manipulations furnished the desired alkaloid pumiliotoxin C, a potent frog toxin.

Electrochemical flow methods applicable for industrial scale reactions have also been developed. For example, Pletcher and co-workers developed a large microflow electrolysis cell for anodic Shono-type oxidations on a multigram scale [84]. The electrochemical cell was based on two circular electrodes with a diameter of 149 mm and a spiral electrolyte flow channel 2,000 mm long, 5 mm wide and 0.5 mm interelectrode gap. Using the methoxylation of *N*-formylpyrrolidine as a model reaction, the cell approached 100% conversion in a single pass and an astonishing product formation rate of >20 g h^{-1}.

Interestingly, when furfuryl alcohols are used as the anodic substrates, the resulting de-aromatised products when subjected to acid hydrolysis furnished various hydroxypyrones. In 2018, Robertson et al. developed an electrochemical flow cell for the 'Shono-type' oxidation of furfuryl alcohols (Fig. 24) [85]. The electrolysis was performed using a glassy carbon anode (GC) and platinum cathode (Pt) in methanol. In general, separate oxidations could be conducted back-to-back over the course of a day without the need to disassemble the cell and clean the electrodes. Notably, only low flow rates, i.e. 165 μL min^{-1}, were tolerable. In contrast to the typical flow procedures, experiments in the absence of supporting electrolytes were inconsistent because the solution conductivity was too low for useful current densities to be reached. It was found that the addition of trace amounts of $NaClO_4$ as an electrolyte considerably improved the results. For the anodic oxidation, alkanes, alkenes and alkynes were all well tolerated. Whilst phenyl substituents were compatible, these substrates gave lower yields. This was attributed to a competing decomposition pathway resulting in the extrusion of benzaldehyde. The resulting products were converted into the hydroxypyrones by acid hydrolysis and in most cases, the yield over the two steps paralleled that for the oxidation step.

2.1.6 Flow Electrochemical Synthesis for Chiral Selectivity

The important trend towards single-enantiomer drugs has brought asymmetric synthesis to the forefront as a theme in drug discovery and development [86]. Optimisation of enantioselective transformations for the production of a single enantiomer can be challenging. As many other synthetic choices become important, i.e. solvent choice, catalyst choice and temperature, optimisation of these reactions in comparison with production of a racemic compound is much more time-consuming.

Self-optimising automated platforms would clearly excel here [87–89]. In this context, Wirth et al. developed an efficient method to quickly optimise asymmetric flow electrochemical transformations via a DoE approach using an online multidimensional HPLC [44]. The enantioselective electrochemical oxidation of *N*-aryl carbonylated l-proline to various enantiomerically enriched methoxylated amides was disclosed (Fig. 25). Electrochemical synthesis was performed in a commercially available flow reactor [6] using a glassy carbon anode and platinum cathode. Whilst carbon-based anodes gave better yields than platinum or boron-doped diamond, the glassy carbon anode stood out for its impressive impact on the stereoselective nucleophilic attack. Using the optimal conditions developed, compound **53** could be obtained with 58% *ee*. The memory of chirality was not constrained to cyclic amino acids, albeit the effect was only moderate (up to 14% *ee*) for the acyclic derivatives. A variety of enantioenriched products could be obtained with moderate to good yields. Importantly, given the stereoselective reactions could benefit from a hugely accelerated optimisation with online HPLC and DoE, the method clearly leads itself for efficient library synthesis. In the pursuit for drug discovery, the method is ideal for quick structural modification of many proline containing drugs (Fig. 25).

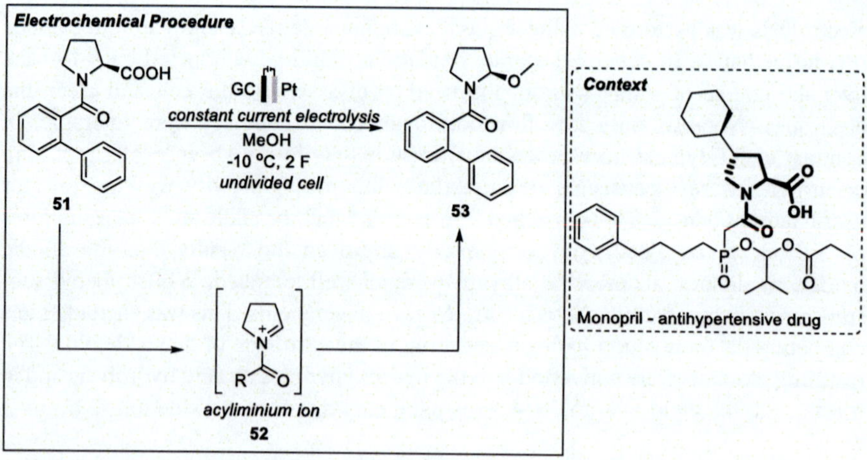

Fig. 25 'Memory of chirality' electrochemical transformation by Wirth et al. [44]

2.2 *Electrochemistry 'On-Route' to Small Molecule Drugs*

Flow electrochemistry is a fast emerging technique. However, only in the past 3–4 years have commercially available platforms been developed. As such, the application of flow reactors for the direct synthesis of small drug molecules is rather sparse, owing to newness of the technique. Previously, utilisation of flow electrochemistry required substantial specialist knowledge of engineering for the building of 'home made reactors' and electrochemistry. As such, the development of commercially available reactors is sure to encourage its widespread adoption. Here, we will cover just three interesting synthetic strategies that give a broad idea of the potential of the technique.

Licarin A and (−)-licarin A are promising compounds that could be used in the development of many drugs. For example, licarin A is a candidate compound for the treatment of immediate hypersensitivity [90], prevents against Leishmania major associated with immunomodulation [91] and is a potentially active anti-tuberculosis agent [92]. Nishiyama and co-workers were able to transform isoeugenol into licarin A in just a single-step reaction in a flow cell (Fig. 26) [93]. The reactions were conducted under galvanostatic conditions and using a BDD anode and platinum cathode, although the highest yield obtained was just 8% in flow. Glassy carbon and platinum anodes performed significantly worse. The very low yield of the reaction mainly results from the reaction between MeOH and the formed radical. In this case, the side product was likely formed by the reaction of radical intermediate **55** through sequential electron transfer and deprotonation followed by reaction with methoxy radicals formed in high concentrations at the surface of the BDD electrode. Interestingly the authors suggest the size of the diffusion layer in the flow cell is responsible. Since the diffusion layer in a flow cell is smaller than in batch, there could be a higher probability for the reaction of the radical **55** and methoxy radicals,

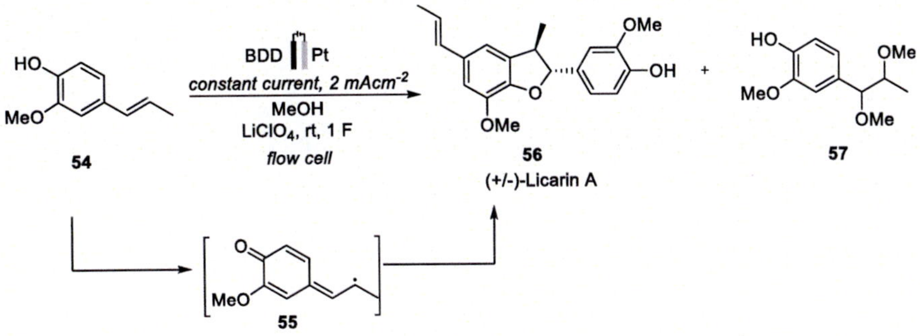

Fig. 26 Flow electrosynthesis of licarin A from isoeugenol by Nishiyama et al. [93]

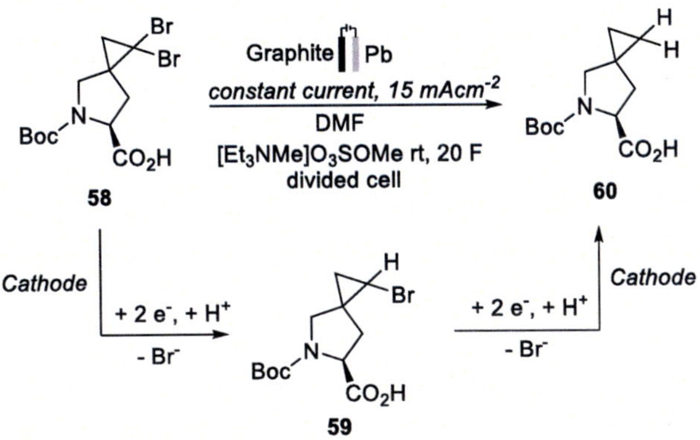

Fig. 27 Electrochemical double dehalogenation of a cyclopropane derivative in a flow cell [95]

which accounts for the lower yield in flow (8%), as opposed to batch (40%). Nevertheless, the results indicate that the electrooxidative reaction of isoeugenol **54** using the BDD electrode could produce licarin A **56** in higher yields than previously reported. Instead, the coupling product **57** was formed in high yields, facilitated by the high concentrations of methoxy radicals at the anode.

Ledipasvir is an orally available inhibitor of the hepatitis C virus. Upon oral administration and after intracellular uptake, ledipasvir binds to and blocks the activity of the NS5A protein. The NS5A inhibitor compounds such as Ledipasvir are highly important for the complete cure of hepatitis C [94]. Waldvogel and co-workers disclosed an electroreductive flow process for double dehalogenation which is important for the synthesis of a key intermediate for NS5A inhibitors (Fig. 27) [95]. The electrolysis was performed in a divided flow cell whereby the anodic and cathodic were separated by a Nafion membrane. Leaded bronze (15% Pb) was used as the cathode, and graphite-felt was used as the anode. Using this gap cell, the anolyte was pumped in a cycle whereas the catholyte was collected after passage through the flow cell, and the oxidation of methanol served as a balancing anodic reaction. The isolated yield for the dehalogenated product **60** in the continuous

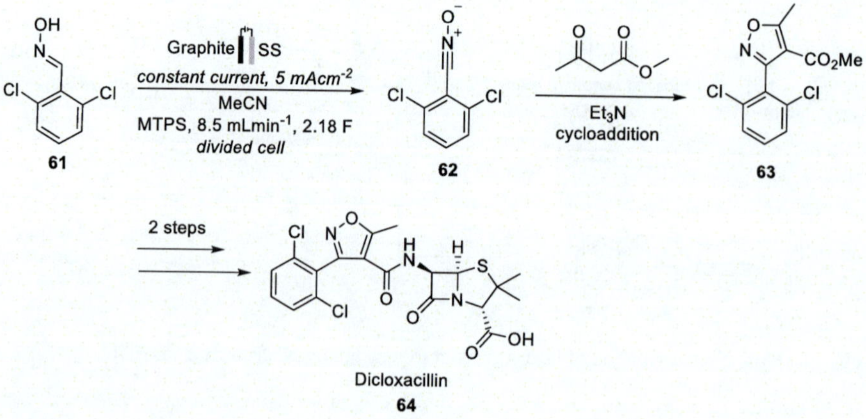

Fig. 28 Electrochemical oxidation in a divided cell followed by a 1,3-dipolar cycloaddition and further formal transformation to dicloxacillin [96]

approach was 70% with 99% *ee*, which means that no racemisation took place during the electrolysis. For technical application in the pharmaceutical industry, both the synthesis and the work-up for a reaction must be scalable. For recycling of the electrolyte and isolation of pure product the authors developed a simple work-up protocol. First, removal of the reaction solvent DMF was performed under reduced pressure and the solvent could be distilled for re-use. Next, the product was isolated from the solid residue by extraction with toluene or ethyl acetate, which is further removed under reduced pressure. Following filtration through a short silica gel plug using ethyl acetate the product exhibited a trace metal content lower than 10 ppm, thus meeting the tight residual impurity requirements for the production of pharmaceuticals.

Dicloxacillin is a narrow-spectrum β-lactam antibiotic of the penicillin class. The antibiotic is used to treat many different types of infections caused by bacteria such as bronchitis, pneumonia or staphylococcal infections, etc. Recently, in 2017, the Waldvogel group developed an efficient electrochemical flow protocol for an important pharmaceutical building block towards dicloxacillin **64** (Fig. 28) [96]. Using the previously discussed electrosynthesis for benzylic nitriles (Sect. 2.1.1), but instead using a divided cell set-up including a separator, the reduction of the nitrile-*N*-oxide **62** to the nitrile was prevented. Consequently, the nitrile-*N*-oxide **62** was allowed to accumulate in the anolyte. Compound **62** is an excellent precursor for 1,3-dipolar cycloadditions and upon treatment with methyl acetoacetate gave the corresponding product **63**. The test reactions revealed that the cycloaddition was highly selective in a mixture of acetonitrile/water (4:1) and triethylamine, giving a 100% yield. For the electrolysis of **61** to **62**, a stainless-steel cathode and graphite anode were employed. The catholyte solution consisted of methyltriethyl ammonium methylsulfate (MTPS) in acetonitrile and water, and for the anolyte, the substrate 2,6-dichlorobenzaldoxime and the electrolyte MTPS were dissolved in acetonitrile. Stacking two electrochemical cells together allowed the flow rate to be doubled and the productivity increased from 176 to 353 mg h^{-1}. After electrolysis and cycloaddition, the product **63** was obtained in an isolated yield of 60%. In comparison with conventional synthesis, the yield was moderately increased from 53 to 60%, however, the scalability and facile purification (without column chromatography) means this new synthesis is far more appealing for industry.

2.3 *Flow Electrochemistry for Late-Stage Functionalisation*

The late-stage functionalisation (LSF) of drug molecules is often the most tedious and time-consuming synthetic step in organic synthesis. A technique that enables selective transformations at a late-stage is highly desirable for the transformation of elaborate drug molecules, and in this sense, the astonishing tolerance of electrochemical synthesis could greatly speed up the drug development process. LSF approaches offer the promise of rapid exploration of chemical space and greatly decrease the experimental efforts of the researcher. Often, the limited functional group tolerance of conventional reactions means that an enormous amount of optimisation of synthetic steps in the late stages of a 'decorated' drug molecule's synthesis is required. A technique with greater functional group tolerance would massively reduce the time taken for optimisation and the number of synthetic steps required which in turn decreases the cost burden. The scalability and huge functional group tolerance of electrochemistry in this respect is invaluable. For this section, we will discuss two very recent examples of where electrochemical approaches are used for the transformations of large compounds as opposed to conventional reagents. Here, electrochemistry allows single-site modification whilst leaving the rest of the highly functionalised scaffold unchanged.

In 2017, Su and co-workers developed an electrochemical bromination of late-stage intermediates and drug molecules in flow (Fig. 29) [97]. Aromatic C–H bromination is one of the applications of late-stage functionalisation that provides precursors for generation of radio-labelled compounds and supports drug metabolism and pharmacokinetic studies. Conventionally, brominated arenes are prepared via electrophilic aromatic substitution but several limitations exist, including (1) side reactions such as benzylic bromination, (2) over bromination, i.e. double bromination, (3) non-selective bromination and the generation of multiple regioisomers. Instead, electrochemistry offers a facile and selective alternative process since the reactivity can be finely tuned by varying the potential. Here, the electrochemical mono-bromination was conducted on late-stage intermediates and drug molecules such as cytidine, uridine, tenofovir, MK-4618, Sch48973 and MK-8457 in moderate to excellent yields. The reactions were conducted under galvanostatic electrolysis conditions in aqueous sodium bromide using acetonitrile as a co-solvent. Electrolysis was performed in an electrochemical microflow cell with centrifugal pumps to recycle the reaction mixture in both cells thereby increasing contact time with the electrode surfaces. Tenofovir derivatives **65** are medications used to treat chronic hepatitis B. The mono-brominated product **66** could be detected in a 35% yield using

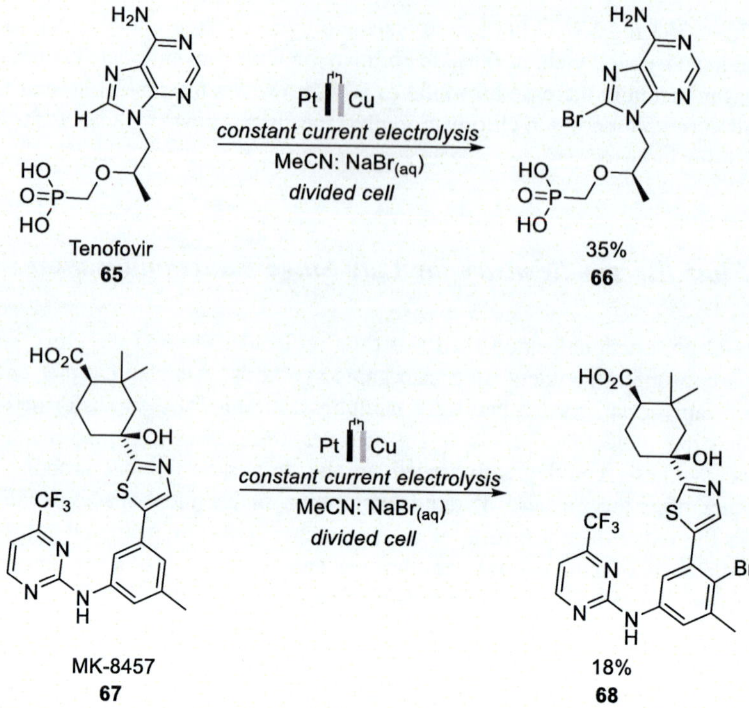

Fig. 29 Electrochemical late-stage mono-bromination of 'decorated' drug molecules [97]

the optimal conditions. Other more complex drug molecules were also tolerable, and electrochemical bromination of MK-461818 gave the corresponding aryl bromide as the major mono-brominated product. Interestingly, the mono-bromination of cytidine was disclosed in a 78% yield. Despite the somewhat low to moderate yields, this methodology can be applied to a diverse set of substrates to provide facile access to brominated drug molecules. This, in turn, presents opportunities for further functionalisation or use as precursors to produce tritium labelled material.

In 2019, Opatz et al. demonstrated the total synthesis of (−)-oxycodone via anodic aryl−aryl coupling (Fig. 30) [98]. Oxycodone, sold under the brand name OxyContin, is an opioid medication used for treatment of moderate to severe pain. An extensive screening of reaction parameters comprising of temperature, current density, reactant concentration and stoichiometry of an acidic additive was disclosed. Their electrolysis was performed under constant current conditions in an undivided flow cell, using a BDD anode in combination with a platinum cathode in acetonitrile. Small amounts of aqueous HBF_4 as an acidic electrolyte were necessary to prevent amine oxidation. BDD is often a superior choice for dehydrogenative couplings and has emerged as powerful electrode material for numerous transformations [99]. A molybdenum anode, which was previously shown to be a suitable material for the dehydrogenative coupling of oxygenated arenes, caused rapid

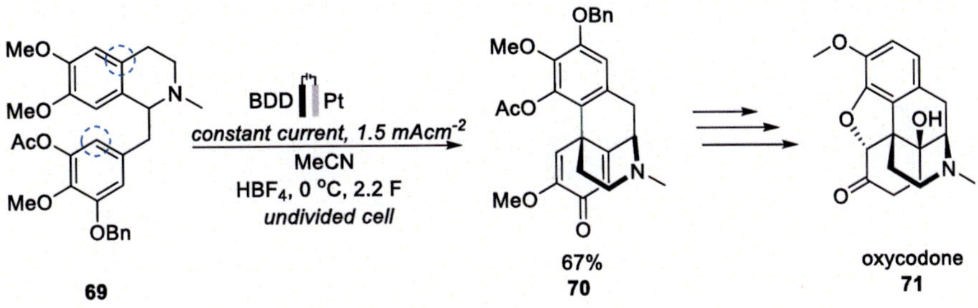

Fig. 30 Anodic aryl–aryl coupling by Opatz et al. [98]

passivation of the electrode surface. Exchanging the solvent from acetonitrile to HFIP resulted in complex product formation. With the optimal electrolysis conditions, diastereoselective anodic oxidation afforded the ring closed product **70** in a respectable 67% yield. The remarkably selective anodic coupling was operationally simple and almost reagent-free, which facilitated purification. Further functionalisation afforded the desired drug oxycodone **71**.

2.4 *Flow Electrochemistry for Metabolic Studies*

Electrochemistry has emerged as a powerful analytical technique for chemical analysis of biologically active molecules and metabolites. Metabolism comprises of all the redox and chemical reactions involved in the biotransformation of a drug within the body [100]. Often, the toxicity of a new drug or of its metabolites is the main reason most candidate drugs are removed from the development process, hence a quick and facile analysis procedure would greatly speed up drug discovery. The intrinsic redox properties of electrochemistry lend the technique as an ideal candidate to study drug metabolism, but it is much less developed compared to biological methods. Predominantly, the first step towards elimination of compounds in vivo occurs through first pass hepatic oxidation and this constitutes phase I metabolism. For clarity, phase I metabolism consists of oxidative, reductive or hydrolytic reactions that add a polar group or expose a previously masked polar group within a drug [100]. The use of electrochemistry in this respect has yet to be fully realised. Key advantages include (1) the study of drugs in a non-cellular environment which simplifies purification and reproducibility, (2) the potential to synthesise metabolites on a preparative scale and (3) the ability to gain mechanistic insight through well-established analytical techniques, i.e. cyclic voltammetry or spectroelectrochemistry. Flow electrochemistry is highly desirable in this field since many analytical techniques can be easily employed in an online manner which negates the need for purification of such unstable intermediates.

On the whole, preparative electrosynthesis is not usually the main purpose of these studies, however, a few studies exist. Herein, we discuss some interesting contributions to this area that exploit flow electrochemistry as a key synthetic tool to present at least a flavour of the capability. A representative example was described by Stalder and Roth who discuss the continuous flow electrosynthesis of the phase I metabolites of several commercially available drugs (Fig. 31) [101]. The electrolysis was performed in an electrochemical microreactor using a platinum cathode and a carbon or platinum anode. Interestingly, phase II glutathione (GSH) adducts are also shown to be synthetically accessible and isolable. Five drugs were used to investigate the oxidative electrochemistry in flow: diclofenac, tolbutamide, primidone, albendazole and chlorpromazine, whereby each drug was selected based on first pass hepatic oxidation at unique metabolic sites. For any compound, the product selectivity of electrochemical oxidation is governed by the most redox-active sites on the molecule. The anthelminthic drug albendazole **80**, depicted in Fig. 31, has two phase I metabolites reported in vivo, which result from *S*-oxidation of the thioether: the sulfoxide and the sulfone. Applying 1.5 equivalents of electrons afforded the sulfoxide metabolite **81** in 38% yield after purification, with a reaction output of 65 mg/h. The product was confirmed by ^{1}H NMR spectroscopy; the methylene protons *alpha* to the S atom observed at 2.80 ppm spectrum of the product became diastereotopic, whereby the newly formed S – O bond induced chirality. At higher electron equivalents, a distribution of five by-products was observed by LC – MS. This included a compound with a molecular ion mass of 298 (MH+), which presumably corresponded to the sulfone metabolite. However, the high number of by-products hindered isolation of any single compound. Also, chlorpromazine **78**, an antipsychotic medication, underwent *S*-oxidation upon electrochemical exposure. The oxidation product was identified as the sulfoxide metabolite **79** which was isolated in 83% yield with a reaction output of 33 mg/h. Diclofenac (DCF) **72**, an anti-inflammatory drug, is metabolised in the liver by aromatic hydroxylation of either of its two phenyl rings *para* to the nitrogen atom. When treated to the developed electrochemical conditions, depending on the concentration of sodium bisulfite used, the product obtained is DCF-5-OH **73** or the quinone imine DCF-5-QI. As a mild reducing agent, sodium bisulfite may have played an important role in preventing overoxidation in the cell. Quinone imines are known to be first hepatic metabolites, but they are considered toxic because of their electrophilicity [102]. Since DCF-5-QI was stable enough to be isolated, its reaction with glutathione was attempted targeting known conjugation metabolites of DCF. The continuous flow technology excelled here because the output of the reactor could be directly mixed with a flow of glutathione solution with minimal time for degradation. In contrast, primidone **76**, an anticonvulsant, underwent alkyl oxidation in the electrochemical flow cell. The metabolite phenobarbital **77**, formed by oxidation of the methylene linking the two amide nitrogens, could be isolated in a 24% yield after electrolysis. A limited reaction output of 7 mg h^{-1} was achieved due to the poor solubility of the drug. Overall, the developed electrochemical flow cell was applicable for a diverse array of substrates achieving moderate to excellent yields in all cases. It is demonstrated that such metabolites could be synthesised by flow electrolysis at the 10–100 mg scale and the purified products could be all fully characterised.

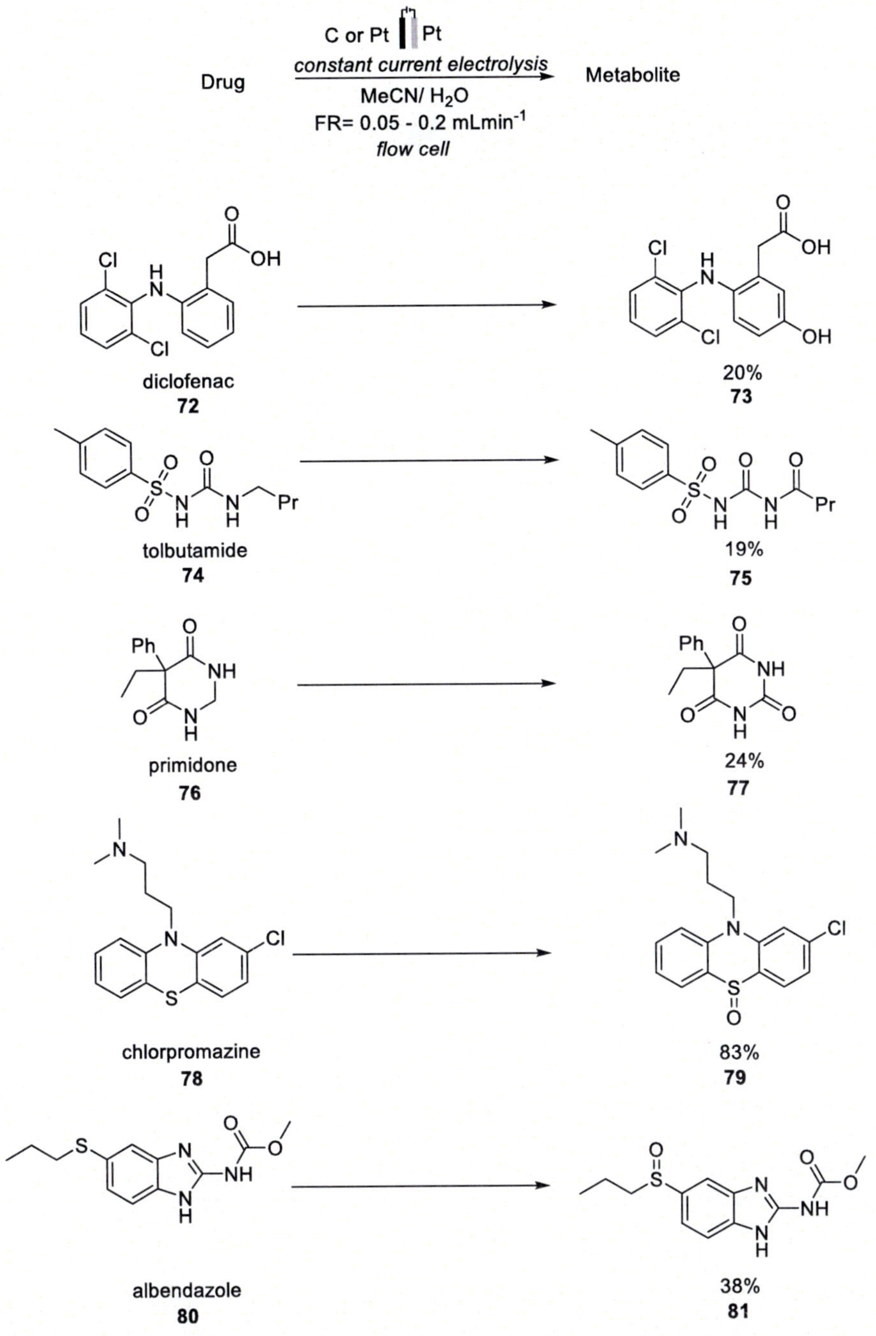

Fig. 31 Electrosynthetic generation of metabolites from different commercial drugs in flow cells [101]

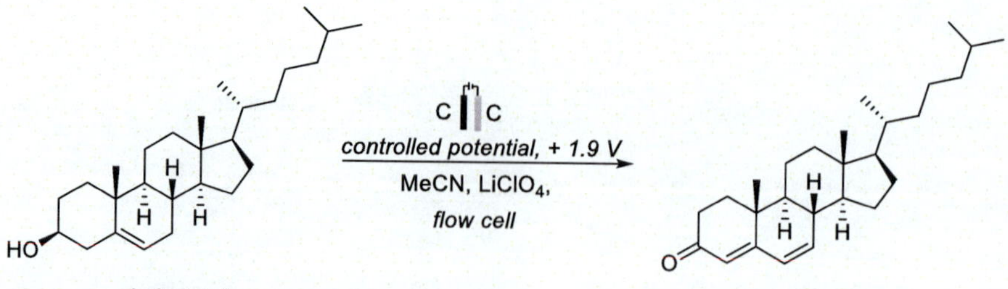

Fig. 32 Electrosynthesis of cholesta-4,6-dien-3-one from cholesterol by Kusu and co-workers [103]

Kusu et al. disclosed the electrosynthesis of cholesta-4,6-dien-3-one from cholesterol on a laboratory scale (Fig. 32). Here, the authors disclose one of the few approaches encompassing a reference electrode into a flow cell for potentiostatic experiments [103]. Using a flow-through column electrolysis system, the anodic oxidation of cholesterol **82** was performed in a four-electron, four-proton process at the applied potential of 1.9 V vs Ag/AgCl with a flow rate of 2.5 mL min^{-1}, providing a green tool for the synthesis of cholesta-4,6-dien-3-one **83** in a 43% yield. Although the product was obtained only on a mg scale, this was enough for structural elucidation of the product and for in vitro biological experiments.

2.5 *Automated Flow Electrosynthesis*

Optimising chemical reactions and exploring the most efficient ways to produce compounds for clinic testing and drug commercialisation can be tedious, time-consuming and expensive. In this sense, technologies that enable greater automation and more effective exploration of chemical space are highly desirable in drug discovery efforts. One clear benefit here is the simplicity and automation of a developed flow protocol as it could also be easier used by untrained personnel. These approaches can be advanced by recent building block initiatives and 'design of experiment' approaches could be used to rapidly define the parameter range of the reactions. Here, integration of an online analytical technique, such as HPLC, would allow real-time analysis. To make this process even broader and more efficient, computational tools that harness the power of artificial intelligence and machine learning could play a key role in facilitating the synthesis of previously inaccessible compounds. Purification in automated synthesis is still a limiting factor that hinders the potential for a fully automated platform. In this respect, flow electrochemistry performed in the absence of reagents is ideal. Facile purification procedures are easily recognisable and since these reactions are often high yielding, evaporation of the solvent or simple recrystallisation for complete purification is feasible. Although

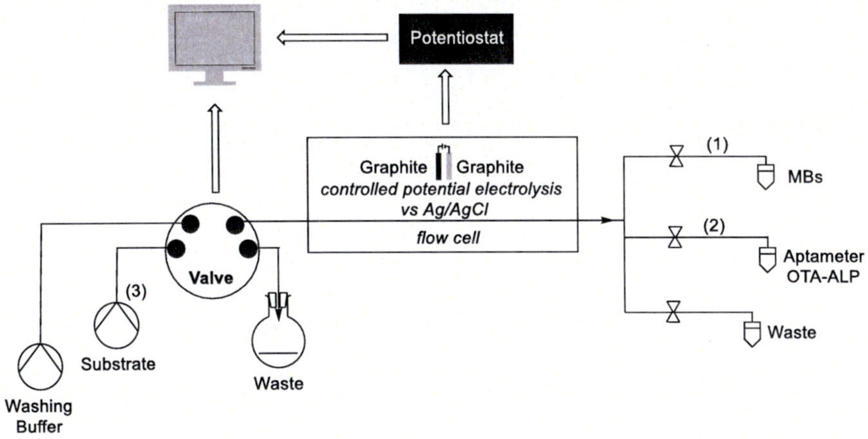

Fig. 33 Automated flow-based electrochemical of OPA; (1) Injection of the modified MBs through valve, (2) injection of the ALP labelled OTA and (3) injection of substrate to generate current signal [104]

these methods support the automatisation of possible synthetic pathways, their use has been limited so far.

Few examples of automated flow electrochemical chemical synthesis as a means to drug discovery have been published. In this respect, this short section will cover two fully automated platforms that although not directly applicable, demonstrate the potential of the technique. Marty and co-workers reported the development of a novel automated flow-based electrochemical aptasensor based on magnetic beads (MBs) for the online detection of ochratoxin A (OTA) in beer samples [104]. The schematic representation of the automated synthesis is shown in Fig. 33. Ochratoxins are the most dangerous mycotoxins found in food and beverages. The system was designed by injecting functionalised MBs onto the surface of a screen-printed carbon electrode (SPCE) integrated into a central flow cell and the device was connected with a flow injection system. Here, MBs have been employed as a solid surface for immobilisation because of their high surface area and ability to accommodate high numbers of ochratoxins. Amperometric detection based on competitive assays was performed for the sensitive and online detection of OTA. The electrolysis was performed with a graphite working and counter screen-printed electrode and an Ag/AgCl reference electrode employing direct and indirect competitive strategies. The incorporation of the aptamer into flow device increased the sensitivity of the system to determine OTA at low concentration in comparison with batch protocols.

Selenium compounds have received substantial attention in medicinal chemistry, especially in experimental chemotherapy, both as cytotoxic agents and adjuvants in chemotherapy [105]. In 2020, Wirth et al. demonstrated an automated electrochemical synthesis of organoselenium compounds (Fig. 34) [9]. The automation facilitated rapid optimisation of the electrolysis conditions followed by efficient synthesis

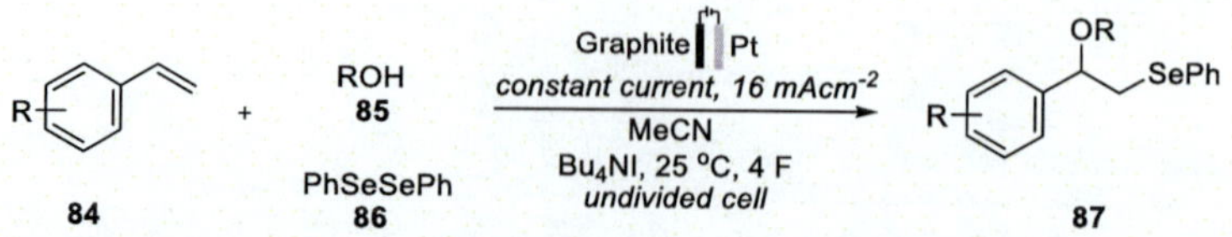

Fig. 34 Automated electrochemical selenenylations by Wirth and co-workers [9]

of a large library of drugs in a fully autonomous fashion. The electrochemistry was performed in a Vapourtec automated flow system with an integrated ion electrochemical microflow reactor under galvanostatic conditions [6]. Graphite was employed as the anode and platinum was used as the cathode. Here, Bu_4NI served as a mediator and could be used in sub-stoichiometric quantities. The authors predict iodide is anodically oxidised sequentially to the iodine radical, then to the iodine cation, i.e. $I^- \rightarrow I^\bullet \rightarrow I^+$. Then, the iodine cation is thought to activate diphenyl diselenide **86** by generating PhSeI or $PhSe^+$ that reacts with the substrate **84**, and finally the nucleophile **85** to form the desired products **87**. Other diselenides such as dibenzyl diselenide could also be used. The optimised electrochemical conditions were tolerable to many different substituted styrene derivatives giving the desired products in good to excellent yields. For disubstituted alkenes, however, no product formation was observed. Large product diversification was possible because primary, secondary or tertiary alcohols, water, formic acid, acetic acid and benzotriazole could all participate as the nucleophile. This methodology was also expanded to intramolecular cyclisations and was found to be efficient to obtain a variety of cyclised *O*-heterocycles in moderate to excellent yields (30–73% yield). A large library of 54 molecules from electrochemical alkoxy- and aza-selenenylations was obtained, ranging from intermolecular reactions to intramolecular reactions for the synthesis of heterocyclic compounds in good to excellent yields. Finally, to demonstrate the potential of the automated synthesis a gram-scale reaction was also performed. 1.52 g of (2-methoxy-2-phenylethyl)(phenyl)selane could be obtained in 100 min, corresponding to an 87% yield. The convenience and efficiency of the fully automated is undoubtedly appealing for industrial applications.

2.6 Conclusions

Electrochemistry is one of the oldest forms of reaction set-ups explored in a laboratory and constitutes an extremely intimate way of interacting with molecules. The history of electroorganic synthesis can be traced all the way back to the 1830s, when Faraday used current to drive nonspontaneous organic reactions [106]. During this period, fundamental electrolysis nomenclature was devised including anode, cathode and even electrolyte. Even today, modern electrochemical cells bear the same fundamental principles, whereby a power source is connected to a reaction mixture through an electrode, where heterogeneous electron transfer events take place. The benefits of 'reagent free' synthesis was clear: the method was versatile, no reagents meant the cost implications of the experiments could be reduced, and easy tuning of electrolysis parameters meant exquisite selectivity could be imagined.

These early forays paved the way to modern electrochemistry. Nowadays, contemporary synthesis is focused on sustainable chemistry, more commonly termed 'green' chemistry, where the benefits of electrochemistry can clearly excel. The replacement of heavily resourced toxic oxidants and reductants with just electrons can drastically reduce the waste burden of organic synthesis, particularly in industrial terms. Flow electrochemistry has received unprecedented attention in the last few years and the development of commercially available platforms is on a growing trend. Here, the intrinsic scalability of flow chemistry and the environmentally benevolent properties of electrochemistry are combined, producing a highly superior methodology. The extraordinary versatility and adaptability of the technique has been demonstrated throughout this chapter, and the ever-growing interest in this area is sure to advance the technique even further. Automated flow electrochemical synthesis can be foreseen as the next big development in this field. The most ambitious vision would integrate machine-learning algorithms, whereby new synthetic strategies could be discretely 'calculated' and optimised in a fully autonomous manner. These properties clearly lend themselves as highly desirable from a pharmaceutical development and industrial perspective.

Although electrochemistry has yet to be fully resourced in industry, the advancements in this technique are expanding fast. For instance, the commercial availability of simple and easy to use electrochemical platforms is sure to facilitate its widespread adoption.

Compliance with Ethical standards

Funding: No funding was received for this chapter.

Informed Consent: All procedures in this manuscript were not performed with human participants.

Ethical Approval: All procedures in this manuscript were not performed with human participants, Nor any other animals.

References

1. Keserü GM, Soós T, Kappe CO (2014) Chem Soc Rev 43:5387–5399
2. Bemis GW, Murcko MA, Med J (1996) Chem 39:2887–2893
3. Wang J, Hou T (2010) J Chem Inf Model 50:55–67
4. Schneider N, Lowe DM, Sayle RA, Tarselli MA, Landrum GA (2016) J Med Chem 59:4385–4402
5. Cernak T, Dykstra KD, Tyagarajan S, Vachal P, Krska SW (2016) Chem Soc Rev 45:546–576
6. https://www.vapourtec.com/products/flow-reactors/ion-electrochemical-reactor-features/. Accessed Apr 2020
7. https://www.ika.com/en/Products-Lab-Eq/Electrochemistry-Kit-csp-516/ElectraSyn-20-pro-Package-Videos-cpvd-40003261/. Accessed Apr 2020
8. https://syrris.com/product/asia-electrochemistry-flow-chemistry-system/. Accessed Apr 2020

9. Amri N, Wirth T Synthesis. https://doi.org/10.1055/s-0039-1690868
10. Wiebe A, Gieshoff T, Mçhle S, Rodrigo E, Zirbes M, Waldvogel SR (2018) Angew Chem Int Ed 57:5594–5619
11. Yan M, Kawamata Y, Baran PS (2017) Chem Rev 117:13230–13319
12. Dohi T, Yamaoka N, Kita Y (2010) Tetrahedron 66:5775–5785
13. Elgrishi N, Rountree KJ, McCarthy BD, Rountree ES, Eisenhart TT, Dempsey JL (2018) J Chem Educ 95:197–206
14. Compton RG, Banks CE (2007) Understanding voltammetry. World Scientific, Singapore
15. Bard AJ, Faulkner LR (2001) Electrochemical methods: fundamental and applications.2nd edn. Wiley, Hoboken
16. Savéant J-M (2006) Elements of molecular and biomolecular electrochemistry. Wiley, Hoboken
17. Hilt G (2020) ChemElectroChem 7:395–405
18. Le Blanc MZ (1900) Elektrochem Angew Phys Chem 7:287–290
19. Seo ET, Nelson RF, Fritsch JM, Marcoux LS, Leedy DW, Adams RN (1966) J Am Chem Soc 88:3498–3503
20. Semmelhack MF, Chou CS, Cortes DA (1983) J Am Chem Soc 105:4492–4494
21. Haupt JD, Berger M, Waldvogel SR (2019) Org Lett 21:242–245
22. Hill-Cousins JT, Kuleshova J, Green RA, Birkin PR, Pletcher D, Underwood TJ, Leach SG, Brown RCD (2012) ChemSusChem 5:326–331
23. https://syrris.com/modules/asia-flux-electrochemistry-module/. Accessed Apr 2020
24. Paidar M, Fateev V, Bouzek K (2016) Electrochim Acta 209:737–756
25. Paddon CA, Atobe M, Fuchigami T, He P, Watts P, Haswell SJ, Pritchard GJ, Bull SD, Marken F (2006) J Appl Electrochem 36:617–634
26. Gul T, Bischoff R, Permentier HP (2015) Trends Anal Chem 70:58–66
27. Haghighi B, Aghajari H, Bozorgzadeh S, Gorton L (2011) Anal Lett 44:258–270
28. Baumann A, Lohmann W, Schubert B, Oberacher H, Karst U (2009) J Chromatogr A 1216:3192–3198
29. Mielczarek P, Raoof H, Kotlinska JH, Stefanowicz P, Szewczuk Z, Suder P, Silberring J (2014) Eur J Mass Spectrom 20:279–285
30. Baumann A, Lohmann W, Rose T, Ahn KC, Hammock BD, Karst U, Schebb NH (2010) Drug Metab Dispos 38:2130–2138
31. Faber H, Melles D, Brauckmann C, Wehe CA, Wentker K, Karst U (2012) Anal Bioanal Chem 403:345–354
32. Jusys Z, Massong H, Baltruschat H (1999) J Electrochem Soc 146:1093–1098
33. Baltruschat H (2004) J Am Soc Mass Spectrom 15:1693–1706
34. Fuhrmann J, Zhao H, Holzbecher E, Langmach H, Chojak M, Halseid R, Jusys Z, Behm J (2008) Phys Chem Chem Phys 10:3784–3795
35. Modestov AD, Gun J, Savotine L, Lev O (2004) J Electroanal Chem 565:7–19
36. Johansson T, Jurva U, Grönberg G, Weidolf L, Masimirembwa C (2009) Drug Metab Dispos 37:571–579
37. Kertesz V, Van Berkel GJ, Granger MC (2005) Anal Chem 77:4366–4373
38. Bussy U, Ferchaud-Roucher V, Tea I, Krempf M, Silvestre V, Boujtita M (2012) Electrochim Acta 69:351–357
39. Pletcher D, Whyte I, Walsh FC, Millington JP (1991) J Appl Electrochem 21:667–671
40. Bennion DN, Newman J (1972) J Appl Electrochem 2:113–122
41. Roberts EPL, Yu H (2002) J Appl Electrochem 32:1091–1099
42. Vilar EO, Coeuret F (1995) Electrochim Acta 40:585–590
43. https://www.vapourtec.com/products/flow-reactors/ion-integrated-electrochemical-reactor/. Accessed Apr 2020
44. Santi M, Seitz J, Cicala R, Hardwick T, Ahmed N, Wirth T (2019) Chem A Eur J 25:16230–16235
45. Fleming FF, Yao L, Ravikumar PC, Funk L, Shook BC (2010) J Med Chem 53:7902–7917

46. Hartmer MF, Waldvogel SR (2015) Chem Commun 51:16346–16348
47. Gieshoff T, Kehl A, Schollmeyer D, Moeller KD, Waldvogel SR (2017) Chem Commun 53:2974–2977
48. Elsherbini M, Winterson B, Alharbi H, Folgueiras-Amador AA, Génot C, Wirth T (2019) Angew Chem Int Ed 58:9811–9815
49. Folgueiras-Amador AA, Qian X-Y, Xu H-C, Wirth T (2018) Chem A Eur J 24:487–491
50. Xu H-C, Campbell JM, Moeller KD (2014) J Org Chem 79:379–391
51. Folgueiras-Amador AA, Philipps K, Guilbaud S, Poelakker J, Wirth T (2017) Angew Chem Int Ed 56:15446–15450
52. Kotha S, Lahiri K, Kashinath D (2002) Tetrahedron 58:9633–9695
53. Corbet JP, Mignani G (2006) Chem Rev 106:2651–2710
54. Negishi E (1982) Acc Chem Res 15:340–348
55. Kashiwagi T, Elsler B, Waldvogel SR, Fuchigami T, Atobe M (2013) J Electrochem Soc 160: G3058–G3061
56. Arai T, Tateno H, Nakabayashi K, Kashiwagi T, Atobe M (2015) Chem Commun 51:4891–4894
57. Gleede B, Selt M, Gütz C, Stenglein A, Waldvogel SR Org Process Res Dev. https://doi.org/10.1021/acs.oprd.9b00451
58. Beil SB, Uecker I, Franzmann P, Müller T, Waldvogel SR (2018) Org Lett 20:4107–4110
59. Asif M (2017) Int J Bioorganic Chem 2:146–152
60. Joule JA, Mills K (2010) Heterocyclic chemistry.3rd edn. Wiley-Blackwell, Hoboken
61. Shono T, Hamaguchi H, Matsumura Y (1975) J Am Chem Soc 97:4264–4268
62. Shono T, Matsumura Y, Tsubata K, Sugihara Y, Yamane S-I, Kanazawa T, Aoki T (1982) J Am Chem Soc 104:6697–6703
63. Libendi SS, Demizu Y, Matsumura Y, Onomura O (2008) Tetrahedron 64:3935–3942
64. Libendi SS, Demizu Y, Onomura O (2009) Org Biomol Chem 7:351–356
65. Tereshchenko AD, Myronchuk JS, Leitchenko LD, Knysh IV, Tokmakova GO, Litsis OO, Tolmachev A, Liubchak K, Mykhailiuk P (2017) Tetrahedron 73:750–757
66. Cao Y, Suzuki K, Tajima T, Fuchigami T (2005) Tetrahedron 61:6854–6859
67. Shono T, Matsumura Y, Tsubata K (1981) Tetrahedron Lett 22:2411–2412
68. Shono T, Matsumura Y, Tsubata K, Takata J (1981) Chem Lett:1121–1124
69. Shono T, Matsumura Y, Tsubata K (1981) Tetrahedron Lett 22:3249–3252
70. Shono T, Matsumura Y, Tsubata K (1981) J Am Chem Soc 103:1172–1176
71. Kanda Y, Onomura O, Maki T, Matsumura Y (2003) Chirality 15:89–94
72. Papadopoulos A, Heyer J, Ginzel K-D, Steckhan E (1989) Chem Ber 122:2159–2164
73. Sierecki E, Turcaud S, Martens T, Royer J (2006) Synthesis:3199–3208
74. Paci A, Martens T, Royer J (2001) Bioorg Med Chem Lett 11:1347–1349
75. For synthetic applications of *N*-acyl iminium ions, see: Maryanoff BE, Zhang H-C, Ohen JH, Turchi IJ, Maryanoff CA (2004) Chem Rev 104:1431–1628
76. Wu P, Nielsen TE (2017) Chem Rev 117:7811–7856
77. Suga S, Okajima M, Fujiwara K, Yoshida J (2005) QSAR Comb Sci 24:728–741
78. Suga S, Okajima M, Fujiwara K, Yoshida J, Am J (2001) Chem Soc 123:7941–7942
79. Shankaraiah N, Pilli RA, Santos LS (2008) Tetrahedron Lett 49:5098–5100
80. Horii D, Fuchigami T, Atobe M (2007) J Am Chem Soc 129:11692–11693
81. Kabeshov MA, Musio B, Murray PRD, Browne DL, Ley SV (2014) Org Lett 16:4618–4621
82. Louafi F, Moreau J, Shahane S, Golhen S, Roisnel T, Sinbandhit S, Hurvois J-P (2011) J Org Chem 76:9720–9732
83. Girard N, Hurvois J-P, Moinet C, Toupet L (2005) Eur J Org Chem:2269–2280
84. Green RA, Brown RCD, Pletcher D (2015) Org Process Res Dev 19:1424–1427
85. Syntrivanis L-D, del Campo FJ, Robertson J (2018) J Flow Chem 8:123–128
86. Farina V, Reeves JT, Senanayake CH, Song JJ (2006) Chem Rev 106:2734–2793
87. Echtermeyer A, Amar Y, Zakrzewski J, Lapkin A (2017) Beilstein J Org Chem 13:150–163
88. Fabry DC, Sugiono E, Rueping M (2016) React Chem Eng 1:129–133

89. Mateos C, Nieves-Remacha MJ, Rincón JA (2019) React Chem Eng 4:1536–1544
90. Matsui T, Ito C, Masubuchi S, Itoigawa M (2015) J Pharm Pharmacol 67:1723–1732
91. Néris PLN, Caldas JPA, Rodrigues YKS, Amorim FM, Leite JA, Rodrigues-Mascarenhas S, Barbosa-Filho JM, Rodrigues LC, Oliveir MR (2013) Exp Parasitol 135:307–313
92. León-Díaz R, Meckes M, Said-Fernández S, Molina-Salinas GM, Vargas-Villarreal J, Torres J, Luna-Herrera J, Jiménez-Arellanes A (2010) Mem Inst Oswaldo Cruz 105:45–51
93. Sumi T, Saitoh T, Natsui K, Yamamoto T, Atobe M, Einaga Y, Nishiyama S (2012) Angew Chem Int Ed 51:5443–5446
94. Link JO, Taylor JG, Xu L, Mitchell M, Guo H, Liu H, Kato D, Kirschberg T, Sun J, Squires N, Parrish J, Keller T, Yang Z-Y, Yang C, Matles M, Wang Y, Wang K, Cheng G, Tian Y, Mogalian E, Mondou E, Cornpropst M, Perry J, Desai MC (2014) J Med Chem 57:2033–2046
95. Gütz C, Bänziger M, Bucher C, Galvão TR, Waldvogel SR (2015) Org Process Res Dev 19:1428–1433
96. Gütz C, Stenglein A, Waldvogel SR (2017) Org Process Res Dev 21:771–778
97. Tan Z, Liu Y, Helmy R, Rivera NR, Hesk D, Tyagarajan S, Yang L, Su J (2017) Tetrahedron Lett 58:3014–3018
98. Lipp A, Selt M, Ferenc D, Schollmeyer D, Waldvogel SR, Opatz T (2019) Org Lett 21:1828–1831
99. Waldvogel SR, Elsler B (2012) Electrochim Acta 82:434–443
100. Rahman MH, Bal MK, Jones AM (2019) ChemElectroChem 6:4093–4104
101. Stalder R, Roth GP (2013) Med Chem Lett 4:1119–1123
102. Speck K, Magauer T (2013) Beilstein J Org Chem 9:2048–2078
103. Hosokawa Y-Y, Hakamata H, Murakami T, Kusu F (2010) Tetrahedron Lett 51:129–132
104. Rhouati A, Hayat A, Hernandez DB, Meraihi Z, Munoz R, Marty J-L (2013) Sens Actuators B 176:1160–1166
105. Spengler G, Gajdács M, Marć MA, Domínguez-Álvarez E, Sanmartín C (2019) Molecules 24:336
106. Faraday M (1834) Ann Phys Leipzig 47:438

Top Med Chem (2021) 38: 173–222
https://doi.org/10.1007/7355_2021_128

Published online: 23 July 2021

Recent Advances of Asymmetric Catalysis in Flow for Drug Discovery

Yuki Saito and Shu Kobayashi

Contents

Abstract Enantioselective continuous-flow catalysis enables highly efficient synthesis of optically active compounds. The early examples of this field were mainly limited to homogeneous catalysis, and enantioselective heterogeneous flow catalysis remained challenging in terms of activity, selectivity, and lifetime. However, there have been continuous developments in recent years toward highly active and selective chiral heterogeneous catalysts. Besides, technology development enabled in-line workup and analysis of flow reactions to make the process more efficient. This review summarizes the recent achievements of enantioselective flow catalysis mainly focusing on chiral heterogeneous catalysts. Successful examples in recent 5 years are categorized and discussed based on the types of reactions, including transition metal catalysis, organocatalysis, and enzymatic reactions. Multistep-flow

Y. Saito and S. Kobayashi (✉)
Department of Chemistry, School of Science, The University of Tokyo, Tokyo, Japan
e-mail: y-saito@chem.s.u-tokyo.ac.jp; shu_kobayashi@chem.s.u-tokyo.ac.jp

synthesis of optically active compounds using enantioselective catalysis is also discussed in the last chapter.

Keywords Chiral heterogeneous catalyst, Continuous flow, Enantioselective catalyst, Flow reaction, Multistep-continuous flow catalysis, Sequential flow

1 Introduction

Continuous-flow synthesis offers significant advantages over conventional batch synthesis in terms of efficiency, safety, and environmental compatibility. In general, reactions in a confined space can achieve rapid mixing of reagents and heat transfer, which provide ideal chemical reaction environments. Microflow photochemistry has an additional benefit of high light intensity. The productivity can be easily controlled by tuning flow parameters such as flow rate, concentration, and amount of reactor, and sometimes one reactor can be adapted to operate from milligram to multi-hundred gram-scale reactions. It also enables "on-demand" synthesis and overproduction can be suppressed. Thanks to relatively small size of reactors, hazardous chemicals can be employed, and the risk of accident can be minimized. A recent development in flow technology enables various kinds of in-line operations including quenching, separation, and monitoring of reactions. When combined with AI technology, reaction automation and self-optimization can be achieved for simple reactions and parameters.

Due to rapid developments in both chemistry and chemical engineering, flow synthesis has been growing rapidly in recent years. In the field of enantioselective catalysis, in particular, there have been significant achievements in both heterogeneous and homogeneous catalyses. One of the main aims of flow synthesis research is to construct high productivity systems that can also be used for scaled-up synthesis. However, some reports also emphasize that flow synthesis can be employed for rapid construction of compound libraries by flashing different substrates consecutively. This would be especially helpful in the field of medicinal chemistry and drug discovery.

A lot of useful review papers have been published in recent years; therefore, this review covers and summarizes reports on catalytic enantioselective flow reactions published from 2016 to 2020, and the aim of this review is to provide a state-of-the-art summary of enantioselective flow catalysis. The following sections are divided into three parts: homogeneous catalysis, heterogeneous catalysis, and multistep-flow synthesis. Each chapter is further categorized by the types of reactions.

2 Homogeneous Catalysis

Homogeneous enantioselective flow reactions are a simple application of chiral homogeneous catalysts in flow reactions. Although catalysts have to be introduced continuously, recent technology enables in-line separation of catalysts and products, and in-line recycling and reuse become possible. Moreover, conditions of batch reactions can be directly transferred to flow reactions, which simplify investigations of flow reactions. On the other hand, residence time is usually limited to less than 1 h in most cases due to limitations in the size of the reactor, and only highly efficient catalysts can be successfully applied for homogeneous catalysis in flow.

2.1 *Hydrogenation and Hydroacylation*

Transition-metal-catalyzed hydrogenation is one of the most frequently used asymmetric reactions for the synthesis of optically active molecules. Thanks to the extensive research efforts for the development of efficient metal catalysts, various types of unsaturated compounds such as ketones, imines, and functionalized olefins can be reduced to the corresponding alcohols, amines, and alkanes with high enantioselectivity and excellent S/C (mol substrate/mol catalyst) ratio. The reaction can be directly applied for continuous-flow reactions using homogeneous catalysts because it does not require stoichiometric amounts of additives and produces no side products in most cases. Use of compact flow reactors can improve safety issues in large-scale production compared with batch reactors with respect to the use of flammable hydrogen gas.

In 2015, the Jensen group developed a system for the continuous-flow asymmetric hydrogenation of tetralone equipped with in-line catalyst recycling apparatus based on nanofiltration (Scheme 1) [1]. They employed poly(ether ether ketone) (PEEK) nanofiltration membranes' filtration, which is suitable for the separation of large molecules (200–1,000 amu) and is compatible with a wide range of solvents and strong bases. The flow reaction was performed by feeding ruthenium-diamine/diphosphine precatalyst, PPh_3, KO^tBu, and substrate solution with H_2 gas into a packed-bed reactor. The resulting solution containing the product and the catalyst was passed through an in-line gas/liquid separator and a membrane separator. As a result, the active catalyst with large amu was separated from the product and reused continuously. In this system, the H_2 pressure was maintained over 2.76 MPa and prevented decomposition of the active catalyst. Although the catalyst deactivation could not be suppressed completely, TON of 4,750 with 93–97% enantioselectivity was achieved throughout the 24 h experiments by feeding in a small amount of fresh precatalyst. Contamination of ruthenium in the product was less than 200 ppb and 99.6% of catalyst was reused, which indicates the high efficiency of the in-line nanofiltration. This in-line separation/reuse of the unstable catalyst highlights the advantage of continuous-flow reaction.

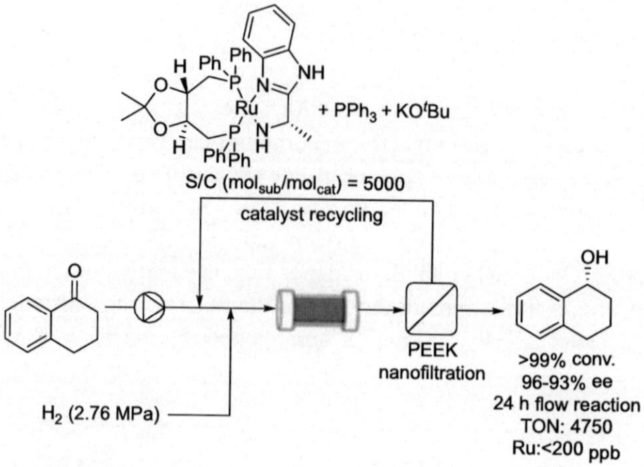

Scheme 1 Asymmetric hydrogenation of tetralone using a system equipped with in-line catalyst recycling apparatus

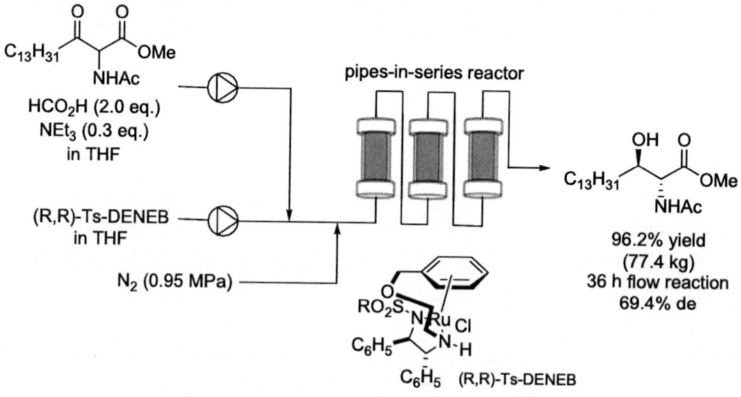

Scheme 2 Asymmetric transfer hydrogenation by dynamic kinetic resolution

In 2019, Touge et al. at Takasago International Corporation developed a continuous-flow asymmetric transfer hydrogenation by dynamic kinetic resolution (Scheme 2) [2]. The flow reaction was performed by feeding methyl 2-acetamido-3-oxooctadecanoate, HCOOH, NEt_3, and the catalyst in THF into a pipes-in-series reactor. Under the optimized reaction conditions, the reaction was performed in a 100 L reactor with 0.1 mol% of the ruthenium-diamine catalyst to produce 77.4 kg of the product with 97% ee and 69% de. It should be emphasized that higher conversion was observed than under batch conditions under the same reaction/residence time. The authors concluded that efficient removal of CO_2, which was generated as a by-product of the reaction, can improve the conversion. Several-step transformation under the batch conditions gave their target bioactive molecule (D-erythro-CER [NDS]) with perfect diastereoselectivity and enantioselectivity.

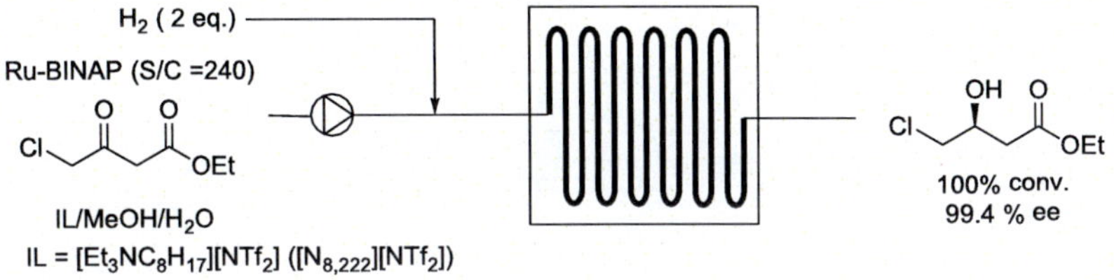

Scheme 3 Asymmetric hydrogenation using a microfluidic chip reactor

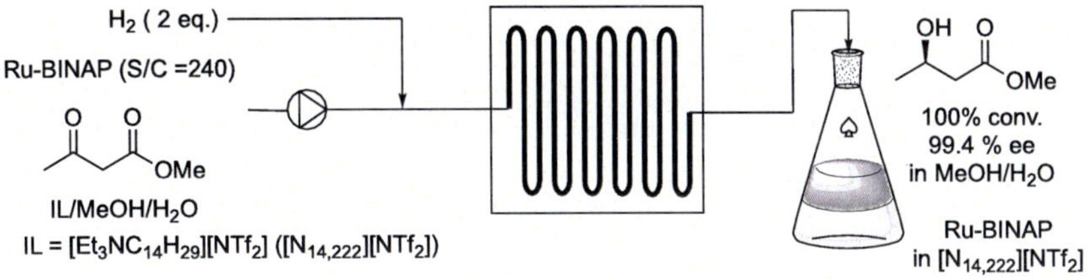

Scheme 4 Asymmetric hydrogenation of methyl acetoacetate

In the same year, the Kluson group developed a continuous-flow asymmetric hydrogenation system using a microfluidic chip reactor (Scheme 3) [3]. The flow reaction was performed by feeding 4-chloro-acetoacetate and the ruthenium-BINAP catalyst in $[N_{8,\ 222}][Tf_2N]$/MeOH/water cosolvent with H_2 gas at 20 bar into 5–19.5 μL chip reactors. The authors identified that the use of ionic liquid as a third cosolvent was crucial to improving enantioselectivity. To calculate the activation energy of the reaction, an Arrhenius plot was constructed using the microfluidic chip reactor. The residence time was precisely controlled by adjusting the flow rate of the solution. The microflow reactor enabled the collection of the product using microscale amounts of materials.

More recently, the same group further expanded this chemistry for the continuous-flow asymmetric hydrogenation of methyl acetoacetate (Scheme 4) [4]. In this report, they examined the structure of the ionic liquid and found that the reaction mixture becomes biphasic after cooling the temperature using an ionic liquid with long alkyl chain without loss of enantioselectivity. Moreover, the authors found that the reaction mixture became biphasic and the product dissolved in the MeOH/water phase while >90% of Ru-BINAP catalyst remained in the ionic liquid phase. Unfortunately, reuse of the recovered catalyst resulted in decreased conversion, although enantioselectivity was maintained.

Along with the development of flow asymmetric hydrogenation, flow asymmetric hydroformylation was reported by the Landis group in 2016 (Scheme 5) [5]. One of the challenges of hydroformylation under flow conditions is achieving the efficient mixing of gaseous reagent with a solution. This aspect is crucial for this transformation because the liner/branch selectivity and enantioselectivity are reported to be sensitive to CO pressure. To address this issue, the authors employed a pipes-in-series reactor, which is known to be suitable for gas–liquid phase reactions with long

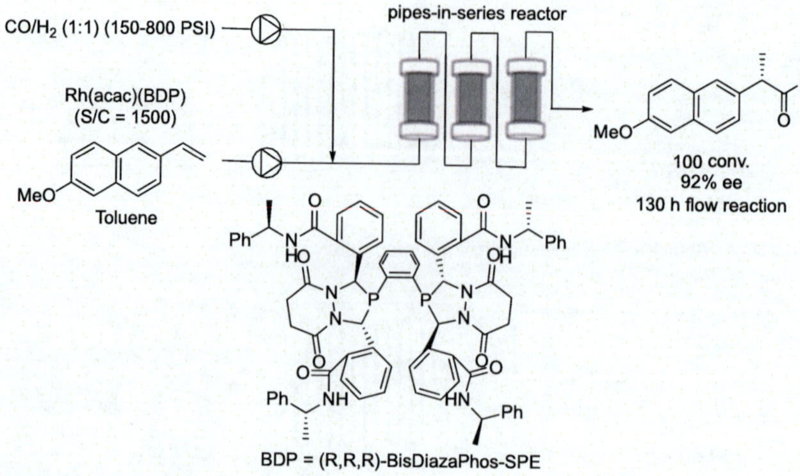

Scheme 5 Asymmetric hydroformylation

residence times. The flow reaction was performed by feeding 2-vinyl-6-methoxynaphthalene and Rh-diphosphine catalyst in toluene with H_2 gas into a pipes-in-series reactor. Under the optimized reaction conditions with 1,500 S/C ratio, the reaction was continued for a total 130 h and the target compound was obtained in excellent yield and selectivity. It should be noted that almost the same levels of conversion, liner/branch selectivity, and enantioselectivity were obtained as under batch conditions, which suggests efficient mixing under flow conditions.

2.2 *Oxidation Reactions*

In 2016, the Gao group developed a flow asymmetric sulfoxidation system with chiral Mn(II) catalyst (Scheme 6) [6]. The flow reaction was performed by feeding sulfide, H_2O_2, adamantane carboxylic acid, and catalyst in MeCN and ^{i}PrOH into a microflow reactor. The authors found that the reaction/residence time could be significantly decreased compared with batch conditions due to the efficient mixing and heat transfer while maintaining yield and enantioselectivity. Under the optimized conditions, the reaction was completed within 2 min using only 0.35 mol% of the catalyst. The reaction showed a broad substrate scope and reactions completed within 1 min residence time in some substrates. Interestingly, the authors developed a microreactor system with four microreactors working in parallel. This system ensures efficient mixing and heat transfer with high productivity, and 5 g of product was obtained in 20 min continuous-flow reaction. This parallel micro reactor also improved the safety of the procedure.

Scheme 6 Asymmetric sulfoxidation

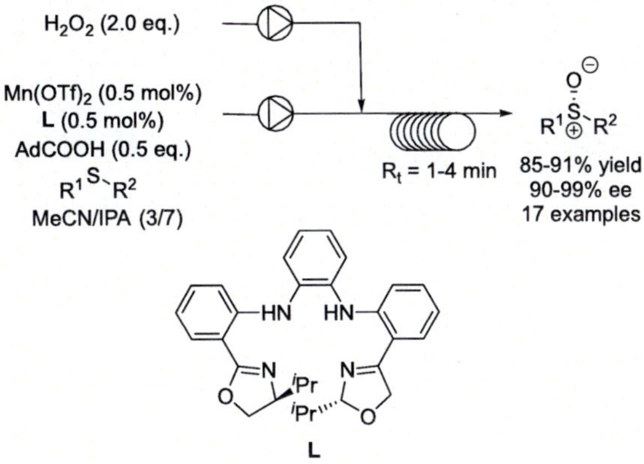

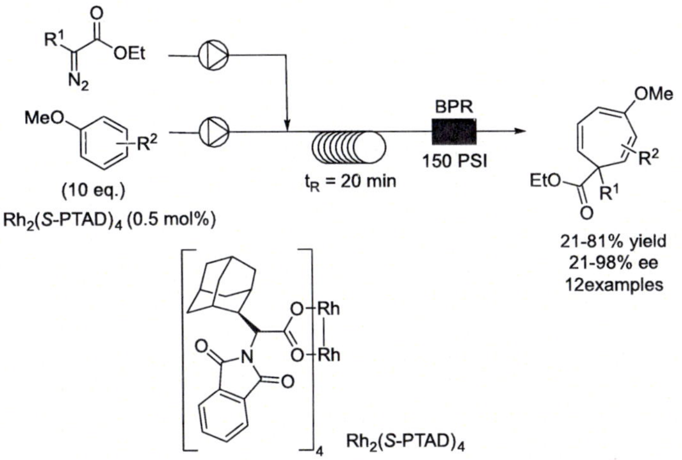

Scheme 7 Asymmetric intermolecular Buchner ring expansion

2.3 *Other Types of Reactions with Transition-Metal Catalysts*

In 2017, the Beeler group developed Rh(II) dimer catalyzed flow asymmetric intermolecular Buchner ring expansions (Scheme 7) [7]. The flow reaction was performed by feeding ethyl diazoacetate, arene, and Rh catalyst in DCE into a tube reactor. The yield and enantioselectivity was highly dependent on the structure of both substrates, but moderate yield with excellent enantioselectivity was observed for most of the substrates with 0.5 mol% of catalyst and 20 min of residence time. Interestingly, the authors found that the reaction under batch conditions gave decreased regioselectivity compared with that under flow conditions. The authors concluded that heat generated from the exothermic reaction increased the reaction temperature under batch conditions at the initial stage of the reaction to decrease the selectivity, and efficient heat transfer under flow conditions kept the reaction

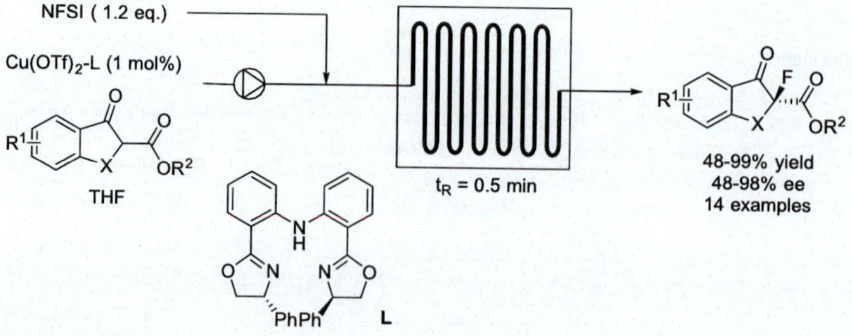

Scheme 8 Asymmetric fluorination of β-keto ester

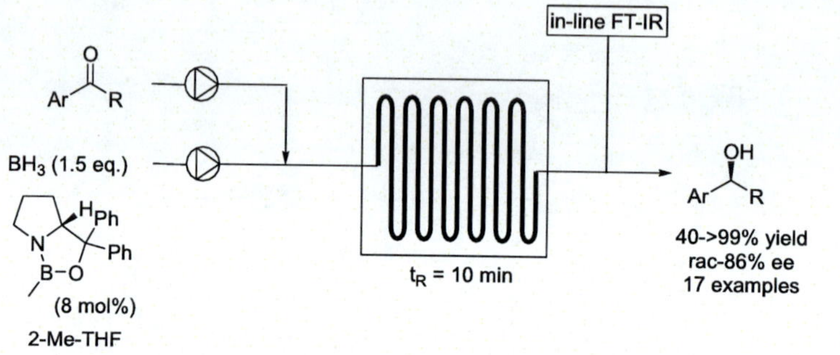

Scheme 9 CBS reduction in a microreactor-flow system

temperature constant throughout the reaction. This observation highlights one of the advantages of flow reactions.

In 2020, the Xu group developed flow asymmetric fluorination of β-keto esters (Scheme 8) [8]. The flow reaction was performed by feeding β-keto ester, NFSI, and Cu(II)-BOX catalyst in THF into a microflow reactor. Under the optimized reaction conditions, the reaction completed within 30 min with 1 mol% catalyst. The reaction demonstrated a broad substrate scope, with excellent yield and excellent enantioselectivity. The authors also succeeded in a gram-scale synthesis using the same reactor without loss of either activity or selectivity.

2.4 *Other Types of Reactions with Organocatalysts*

In 2016, Luisi et al. developed Corey–Bakshi–Shibata (CBS) reductions in microreactor-flow systems (Scheme 9) [9]. The safety of the process regarding the use of hazardous BH_3 reagent could be improved by using a microflow reactor. In-line monitoring of the reaction by IR measurement allowed automation of the reaction, and it was found that the reaction was completed in 10 min with 8 mol%

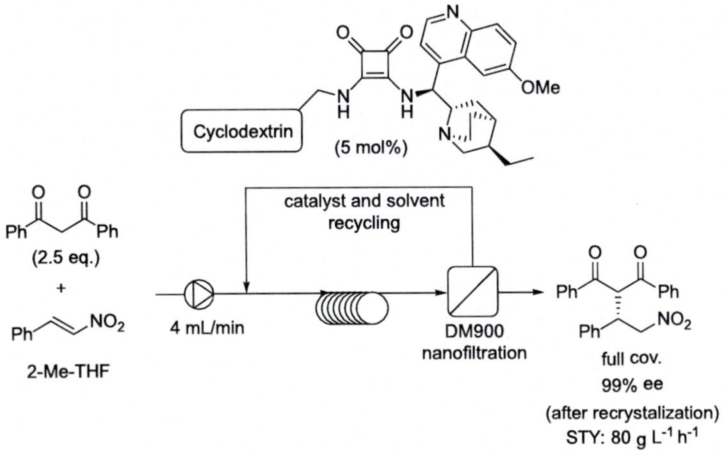

Scheme 10 Enantioselective 1,4-addition of 1,3-diketone to nitro olefin

catalyst. Under the optimized reaction conditions, various kinds of aryl ketones were reduced to the corresponding alcohols in good yields and enantioselectivity. Finally, the authors successfully integrated the microflow reactor and in-line workup and separation units. The resulting solution was quenched by the in-line addition of ethyl acetate and water, and over 90% of the desired product was recovered after in-line liquid–liquid separation.

In 2019, Kupai and Szekely et al. developed enantioselective 1,4-addition of 1,3-diketones to nitro olefins with recyclable H-bonding catalysts under continuous-flow conditions (Scheme 10) [10]. The active squaramide catalysts were covalently attached on the cyclodextrin. The prepared catalysts were introduced into a coiled tube plug reactor together with substrates, and the outlet was connected to an in-line membrane nano-filter system. The product was allowed to pass through the membrane and recrystallized in the permeance. On the other hand, the catalyst stayed in the solution, which allows in-line catalyst and solvent recycling. As a result, the product was obtained in a stable, high yield for over 24 h continuously without external addition of catalyst during the flow reaction. Under the optimized reaction conditions, productivity reached 80 g L^{-1} h^{-1}, with excellent enantioselectivity.

In 2019, Groger et al. developed the first asymmetric organocatalytic reaction with hydrophobic substrates in aqueous medium under flow conditions (Scheme 11) [11]. The biggest challenge facing the development of flow reactions in aqueous medium is the solubility of substrates and catalyst. To solve this problem, the authors investigated extensively the solvent system and found that the use of a buffer/2-propanol system dissolved both organocatalyst and aromatic aldehyde sufficiently. The authors also investigated the effect of reactor size and flow rate and found that a higher flow rate and longer residence time improved the conversion of the reaction. They rationalized that higher flow rate led to more efficient mixing, which leads to a higher TOF of the catalyst. Under the optimized reaction conditions, the flow

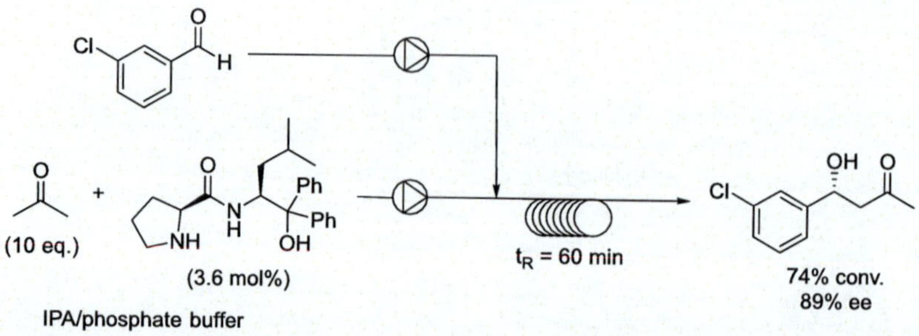

Scheme 11 Asymmetric organocatalytic reaction with hydrophobic substrate in aqueous medium

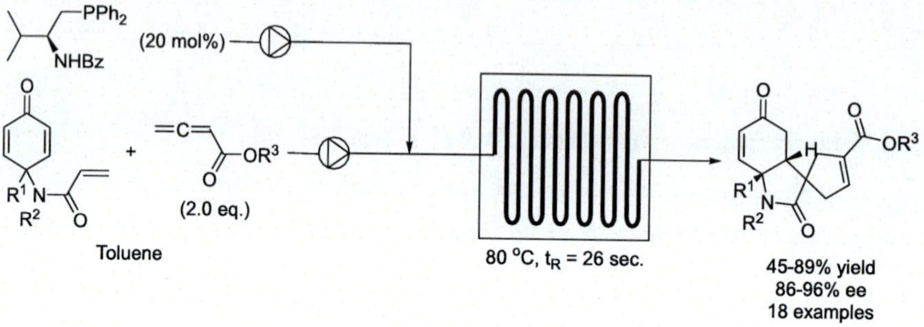

Scheme 12 Combination of new technologies of flow reaction and machine-learning

reactions could achieve almost the same level of conversion and enantioselectivity as in the batch reaction.

In 2019, Washio, Takizawa, and Sasai et al. combined the new technologies of flow reaction and machine-learning to identify the best reaction conditions with a minimum number of reactions for multi-parameter screening (Scheme 12) [12]. As flow reactions have more individual parameters such as flow rate and reactor size than batch reactions, the conventional optimization process, which is to vary the reaction parameters individually while keeping the others constant, requires significant effort. To overcome this problem, they focused on machine-learning as a robust and reliable tool to achieve efficient optimization. Particularly, the authors employed Gaussian process regression (GPR), which is a kernel-based statistical learning algorithm. Their target reaction was the organocatalyzed Rauhut–Currier reaction and [3 + 2] annulation sequence. Due to the high complexity of this domino reaction and various possible side reactions, the optimization under batch conditions resulted in only 20% yield, albeit with high selectivity. To further improve the yield, the authors employed machine-learning in flow reaction. As a result of machine-learning, by changing the flow rate, reaction temperature, and stoichiometry of reagents, the yield could be improved to 76% yield with 94% ee. The optimized conditions could be applied for various substrates, and this demonstrated the power of machine-learning.

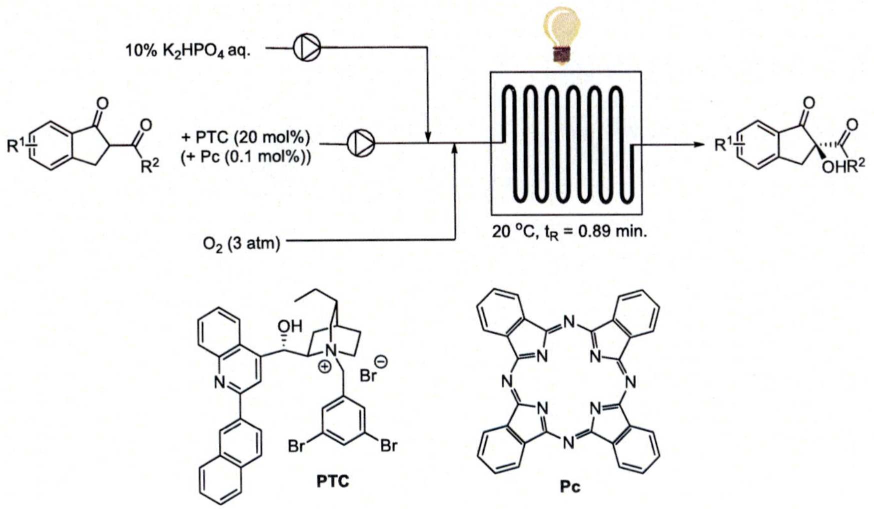

Scheme 13 Enantioselective photooxygenation of 1,3-dicarbonyl compound

In 2019, Meng et al. developed enantioselective photooxygenation of 1,3-dicarbonyl compounds in a flow photoreactor (Scheme 13) [13]. There are several challenges for scaling-up photoreactions under batch conditions due to the attenuation effect of photon transport according to the Bouguer–Lambert–Beer law. Microflow photoreactor can solve this problem; however, most of the reported examples are limited to the synthesis of achiral compounds. In this report, the authors took on the challenge of a more complex reaction environment, which is the combination of photocatalyst and phase-transfer catalyst involving gas–liquid–liquid tri-phasic catalysis. After optimization under batch conditions, a flow reaction was investigated using a flow photoreactor. It was shown that the microflow reactor could realize this complex reaction, and residence time could be reduced from 8 h in batch to 0.89 min in flow reaction without significant loss of either yield or enantioselectivity. In the same study, they revealed that their cinchona-derived phase-transfer catalyst could also work as a photocatalyst and performed a flow reaction without the external addition of photocatalyst [14].

3 Heterogeneous Catalysis

Enantioselective flow reaction with heterogeneous catalysts is the most efficient method to prepare optically active compounds. In an ideal system, the target products can be obtained continuously in pure form by simply flowing substrates in a catalyst column. In reality, however, there are many challenges facing the realization of such an ideal system. The major problems in this field are catalyst activity, selectivity, and durability. Even highly active and selective catalysts often show inferior catalytic activity when they are heterogenized. This is mainly because solid supports usually have a detrimental influence on original catalysts, and chemical modification of an original catalyst for immobilization typically also decreases performance. The durability of homogeneous catalysts is often ignored; however, it becomes an important parameter for flow heterogeneous catalysts. To overcome these difficulties, chemists have devoted much effort to the development of novel and efficient heterogeneous catalysts. For example, covalent-bond immobilization onto solid supports such as polystyrene or silica has been a main strategy, but new immobilization methods that do not rely on such a covalent bond have been developed. For some catalysts, binding to support actually has a positive effect; indeed, sometimes catalytic performance can exceed original homogeneous catalysts, which indicates that supports are no longer just a solid phase.

3.1 *Hydrogenation and Hydroformylation*

In 2019, Suarez and Godard et al. developed heterogeneous Rh catalysts immobilized on carbon materials by π-π stacking for enantioselective hydroformylation of norbornene (Scheme 14) [15]. First, chiral diphosphate ligands derived from furanose were examined under homogeneous batch conditions, and the best ligand gave good activity with 144 TOF and 71% enantioselectivity. For immobilization onto carbon material, the pyrene group was introduced to the ether group of the furanose, and this new ligand showed slightly decreased activity and selectivity compared with pyrene-free ligand. After the complex formation with cationic Rh precursor, the complex was immobilized on carbon materials simply by mixing it in an organic solvent. The choice of organic solvent was crucial for achieving efficient immobilization, and a polar solvent such as ethyl acetate gave good loading efficiency. Three kinds of carbon materials such as MWCNT, reduced-

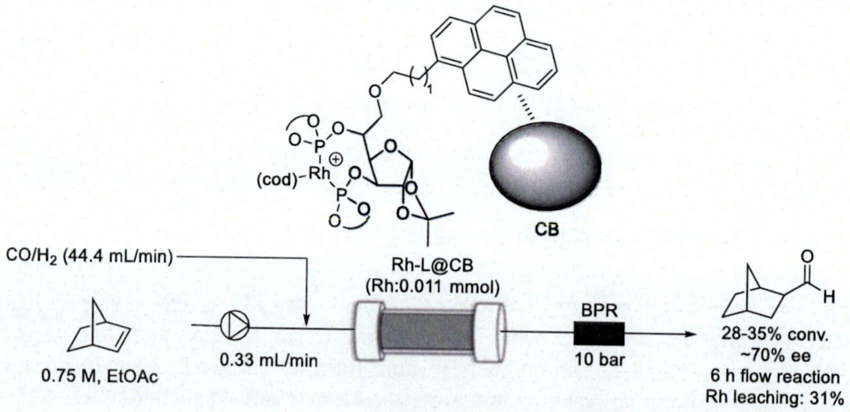

Scheme 14 Heterogeneous Rh catalysts immobilized on carbon materials by π-π stacking

graphene oxide, and carbon black were employed and evaluated under batch and continuous-flow conditions. Comparison between batch and flow conditions revealed that much higher enantioselectivity was observed under the flow conditions. It is known that enantioselectivity of hydroformylation is highly dependent on the gas pressure, and efficient mass transfer under flow conditions may contribute to the improved enantioselectivity. However, deactivation of the catalyst was observed after 60 min, while enantioselectivity was maintained. It was found that higher pressure and CB support was effective in prolonging the lifetime of the catalyst, and, finally, 360 min of flow reaction was performed using the best catalyst to give 28–35% of target product in 70% ee. However, 31% of Rh was leached after 360 min of flow reaction.

In 2015, Francio and Leitner et al. developed a heterogeneous Rh catalyst with supported ionic liquid for the enantioselective hydrogenation of enol esters (Scheme 15a) [16]. Supported ionic liquid phase (SILP) can be easily prepared by simply mixing IL with a porous material such as silica and can be used as an effective support for cationic metal complexes due to the strong electrostatic interaction. The authors especially focus on the use of supercritical CO_2 ($scCO_2$) as a mobile phase because the metal complex stays on the SILP. First, the best chiral ligand and the ionic liquid were determined by investigations with homogeneous catalysts. It was revealed that the use of NTf_2 anion was crucial to achieving high enantioselectivity and activity. The catalyst was also employed in flow conditions and demonstrated excellent catalyst activity. The catalyst maintained its activity for more than 211 h with 90% conversion and 84% ee. As a result, total TON reached up to 70,400 and the leaching of Rh was under the detection limit. This immobilization method was further studied for the synthesis of a key intermediate of API by the same group. In this report, they faced the challenge of the asymmetric hydrogenation of enamide, which is insoluble in $scCO_2$. To improve the solubility, modified CO_2 ($modCO_2$), which is a mixture of $scCO_2$ and organic solvent, was employed and demonstrated the good solubility and compatibility with SILP (Scheme 15b) [17]. After the organic solvent screening, toluene was found to be the best solvent to maintain the enantioselectivity and prevent Rh leaching, while significant leaching and decreased catalytic performance were observed using methanol or DCM as organic solvent. The catalyst retained activity and selectivity for 90 h of continuous-flow reaction, and TON was over 10,000.

The same group employed another immobilization method of Rh complex and developed enantioselective hydrogenation for the synthesis of the key intermediate of an API (Scheme 16a) [18]. The immobilization method, called Augustine's method, employs Al_2O_3 immobilized heteropoly acid as a heterogeneous counter anion. The cationic Rh complex was simply immobilized by mixing the Rh complex with support material. The catalyst was evaluated in flow conditions and showed excellent activity and 99% ee. The catalyst retained activity for 18 h and total TON was 7,700. The heterogeneous Rh catalyst immobilized by Augustine's method was also studied for the asymmetric hydrogenation of itaconate by Bakos et al. in 2018 (Scheme 16b) [19]. The immobilization was also effective for the Rh complex with monophosphate ligands. The catalyst showed a TOF of 80 under flow conditions to

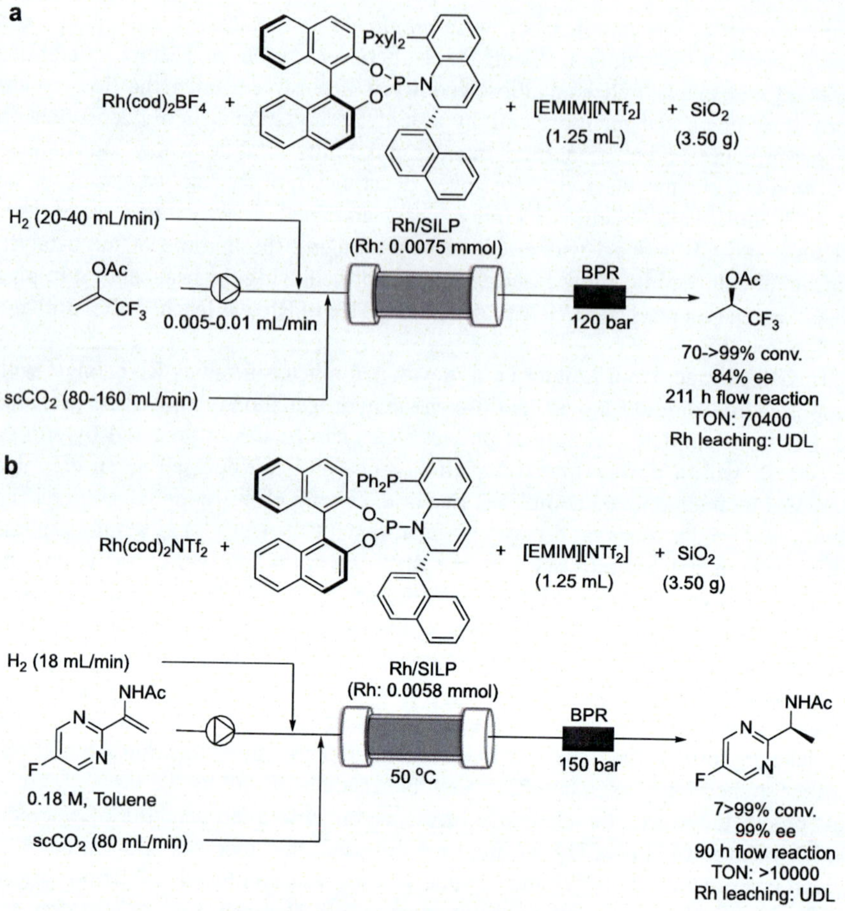

Scheme 15 (**a**) Heterogeneous Rh catalyst with supported ionic liquid. (**b**) Heterogeneous Rh catalyst with supported ionic liquid with mixed $scCO_2$

give >99% conversion with 92–99% ee. However, catalyst deactivation was observed after 380 min and total TON was 509 after 680 min flow reaction.

In 2020, Kobayashi et al. developed a new immobilization method for the cationic Rh complex using noncovalent interactions, for the enantioselective hydrogenation of enamides (Scheme 17) [20]. The authors employed amine-functionalized mesoporous silica as a support for heteropoly acids using acid–base interactions, and Rh complexes were immobilized through electrostatic interactions by simply mixing them with heteropoly acid anchored support. The authors demonstrated that the new support has a much higher catalyst lifetime than those of catalysts prepared by conventional Augustine's method. The use of KIT-6, which is a mesoporous silica, was also important to achieve high catalytic activity and robustness. Under the best conditions, various kinds of enamides could be converted into chiral amides in flow

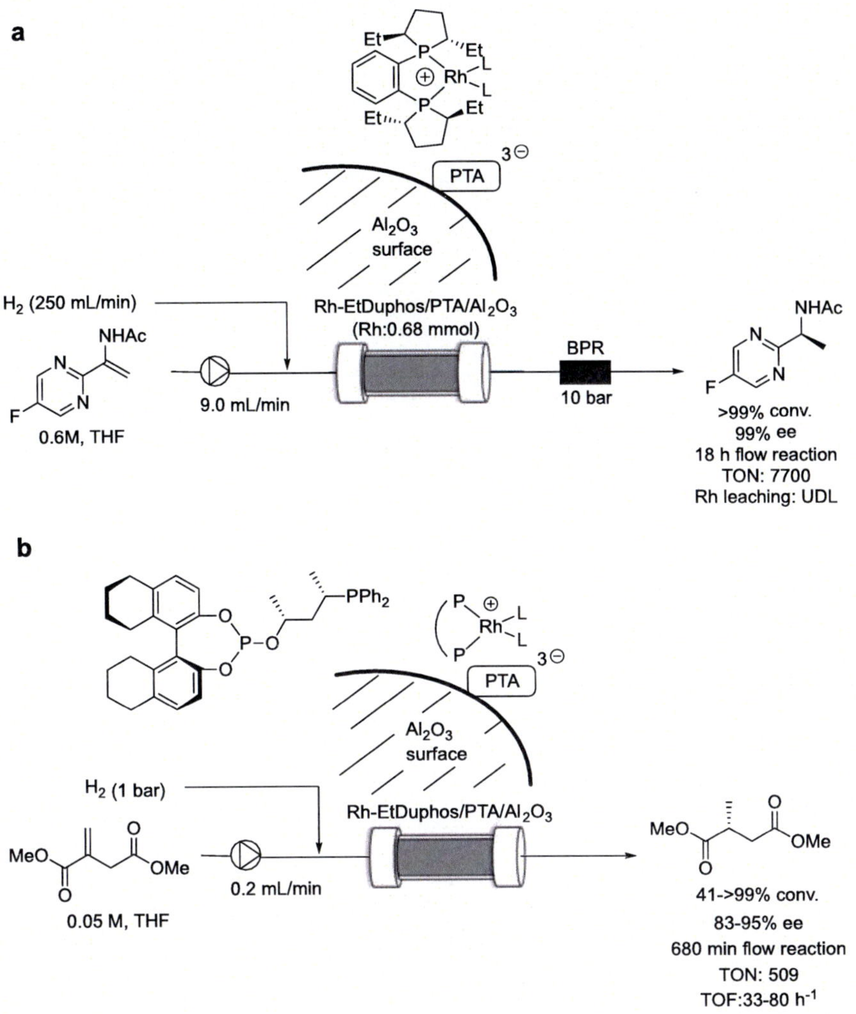

Scheme 16 **(a)** Enantioselective hydrogenation for the synthesis of the key intermediate of an API. **(b)** Asymmetric hydrogenation of itaconate

conditions with TOF of 120 and TON of 2,640. The authors also demonstrated the tunability of chiral ligands to achieve selective hydrogenation with challenging substrates. Finally, lifetime experiments showed that the catalyst remained active for 90 h with a TON of 10,000 in total.

In 2019, Kobayashi et al. developed polystyrene immobilized Ir catalyst combined with homogeneous chiral phosphoric acid (CPA) catalyst for the enantioselective asymmetric hydrogenation of imines and reductive amination (Scheme 18) [21]. The chiral diamine ligand was immobilized by the copolymerization of styrene tagged diamine, polystyrene, and DVB. The heterogeneous catalyst

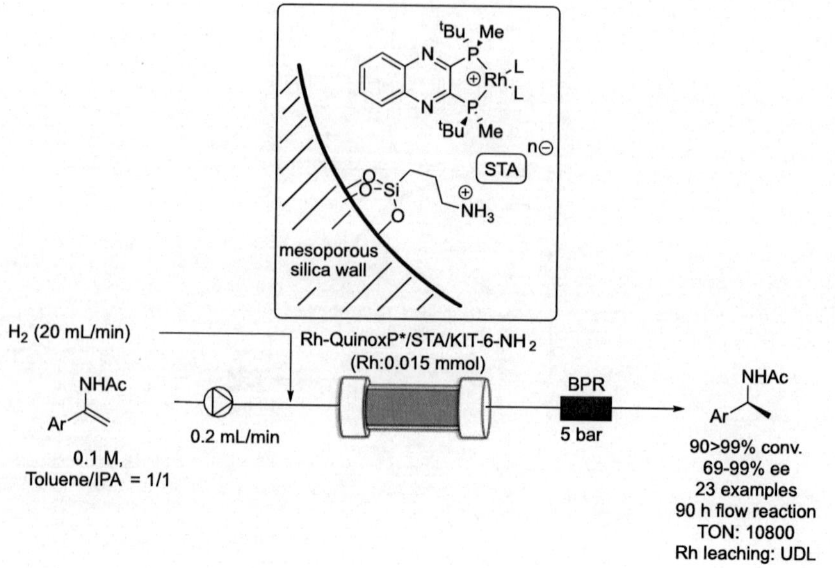

Scheme 17 Enantioselective hydrogenation of enamides

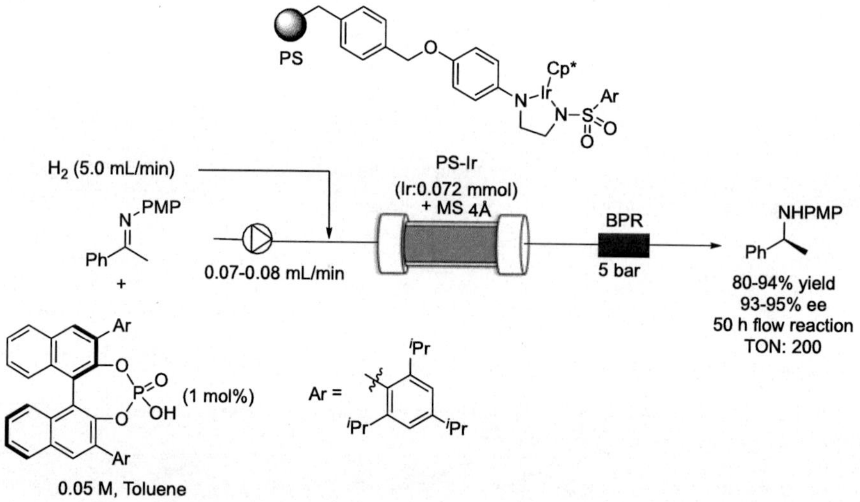

Scheme 18 Enantioselective asymmetric hydrogenation of imine and reductive amination

was evaluated under batch conditions in the presence of a homogeneous CPA catalyst. It was found that the distance between styrene tag and diamine unit has an effect on both activity and selectivity, and the introduction of a spacer improved the catalyst performance. The catalyst showed high activity under flow conditions for 50 h without significant loss of activity and selectivity with a TON of 200, and

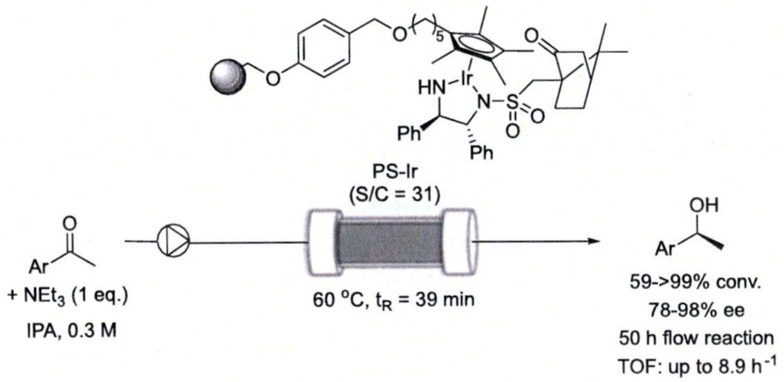

Scheme 19 Asymmetric transfer hydrogenation of ketone

hydrogen pressure could be decreased compared with batch reaction, emphasizing the effectiveness of mass transfer in flow conditions. It should be noted that a homogeneous CPA catalyst could also be recovered by insertion of a basic resin column at the end, and it could be reused without loss of activity. The catalyst could also be employed for enantioselective reductive amination.

In the same year, heterogeneous Ir catalyst covalently immobilized on a diamine ligand was also studied by Blacker et al. for the asymmetric transfer hydrogenation of ketones (Scheme 19) [22]. The diamine ligand tagged with alkyl triflate was immobilized on the Wang resin by the S_N2 reaction and evaluated under batch and flow conditions. It was found that ligands with a shorter spacer had a longer lifetime than catalysts with longer spacers, and the catalyst maintained the activity for 5 days by the use of 5 mol% KO^{t}Bu, while deactivation was observed after 60 h by the use of 3 equivalents of NEt_3. On the other hand, stable and higher ee were accomplished by the use of NEt_3.

3.2 *Organocatalysis*

Various kinds of efficient chiral organocatalysts have been developed such as the MacMillan catalyst, Hayashi–Jorgensen catalyst, Akiyama–Terada catalyst, and thiourea catalyst [23]. Now, asymmetric organocatalysis holds a central position in asymmetric catalysis and has become one of the most powerful tools to synthesize chiral compounds. Thanks to their stability and non-metallic easily modifiable nature, various covalent immobilization methods of active species onto solid supports have been developed. Especially, post modification of functionalized polystyrene resins has become a common technique. In the last 5 years, much effort has been devoted to expanding the scope of catalysts and types of catalysis.

3.2.1 Enamine-Iminium Catalysis

In 2015, Fulop et al. developed continuous-flow enantioselective α-amination of aldehydes with solid immobilized dipeptide catalysts (Scheme 20) [24]. First, polystyrene immobilized dipeptides having proline and acidic amino acids units were prepared by a well-established solid-phase peptide synthesis method. The prepared catalysts were evaluated under batch conditions, and Pro-Asp dipeptide was found to be the best catalyst in terms of enantioselectivity. The authors further investigated the flow conditions and found that both equivalence of aldehyde and flow rate affected both conversion and enantioselectivity significantly, which was rationalized by the epimerization induced by the catalyst. It should be noted that conventional batch reactions do not allow precise control of reaction time and, indeed, resulted in lower enantioselectivity than achieved under flow conditions. It was also demonstrated that backpressure affected the conversion. It was established that flow reactions are under diffusion control, and that swellable supported catalysts and high pressure improve the mass transfer into the polymer matrix. Under the optimized reaction conditions, TOF reached 5.38–6.25 h^{-1} and activity was maintained for over 20 h of flow reaction giving 3.48 g of target product in 81% yield with 88–92% ee.

In 2015, Pericas et al. developed polystyrene-supported amino quinine derivatives for asymmetric 1,4-addition of ethyl nicotinate to unsaturated ketones under continuous-flow conditions (Scheme 21) [25]. The active amino cinchona alkaloids were immobilized onto polystyrene by copper-catalyzed alkyne-azide cycloaddition. Prepared catalysts were first examined under batch conditions and a quinine derived catalyst was found to give the best enantioselectivity. The choice of solvent was also crucial due to the swelling of polystyrene, and chloroform was selected as the optimal solvent. With the best catalyst in hand, the continuous-flow reaction was examined. The catalyst demonstrated excellent enantioselectivity under flow conditions within 40 min of residence time, resulting in the formation of 12.9 mmol of the desired product with TON of 32 for 21 h reaction. The authors also succeeded in the preparation of a library of chiral compounds aimed at the lead discovery. Under the

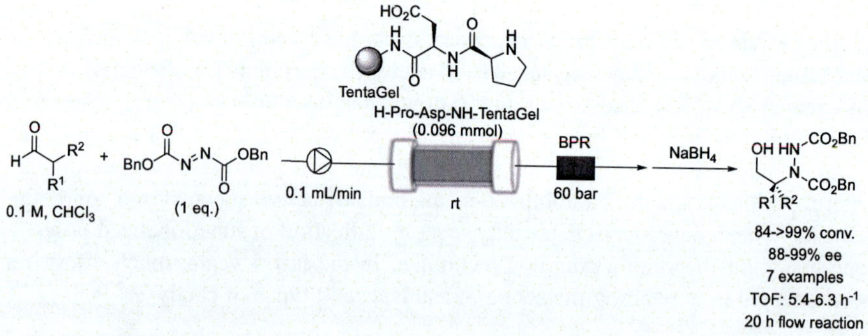

Scheme 20 Enantioselective α-amination of aldehyde with solid catalyst

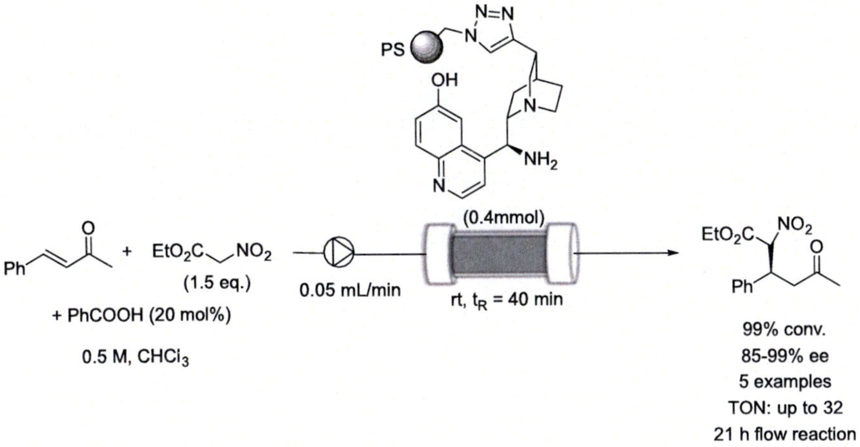

Scheme 21 Asymmetric 1,4-addition of ethyl nicotinate to unsaturated ketone

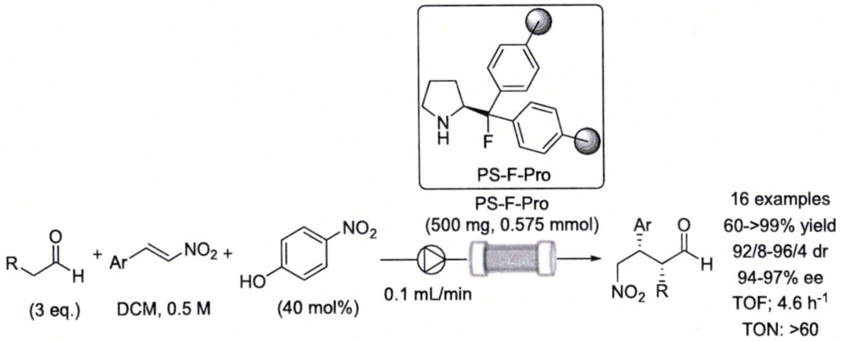

Scheme 22 Robust enamine catalyst for continuous-flow reaction

optimized flow conditions, five kinds of products were prepared by combination of different enones and nucleophiles. Each solution was circulated through the system for 2 h with 50 μL/min flow rate. As a result, products were obtained in 85–99% ee with TOF of 1.4–3.2.

The lifetime of heterogeneous catalysts remains as a big issue facing the realization of sustainable flow synthesis. The same group addressed this point and developed a robust enamine catalyst for continuous-flow reaction in 2015 (Scheme 22) [26]. It is known that the immobilized Hayashi–Jorgensen catalyst is gradually deactivated during recycling experiments due to the desilylation of TMS ether. To overcome this point, a robust homogeneous catalyst was developed by Gilmour et al. by the replacement of TMS ether with F substitution. It was found that such a catalyst was much more robust than the original catalyst without significant loss of selectivity. Therefore, the authors investigated the immobilization of the fluorinated catalyst for 1,4-addition of aldehydes to nitro styrenes. As a result, the polymer-

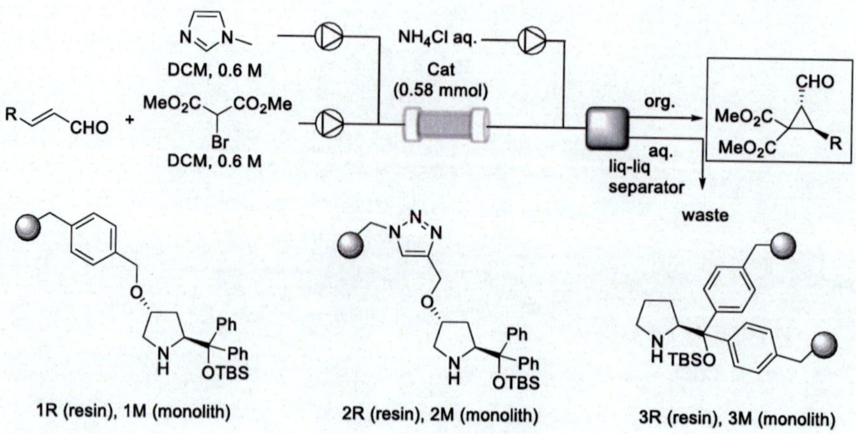

Catalyst	Conv. (%)	Ee (%)
1R	71	93
1M	34	96
2R	94	87
2M	79	90
3R	50	94
3M	20	97

Scheme 23 Anchor effects in asymmetric cyclopropanation

supported catalyst demonstrated excellent activity and selectivity under batch conditions. Recycling experiments reveal that the catalyst remained active until the eighth run, although gradual deactivation was still observed. Control experiments revealed that deactivation was caused by the oxidation of enamine intermediate by oxygen from the air. Therefore, the authors decided to employ continuous-flow conditions, which can minimize the contact with air. As expected, the catalyst maintained its activity for more than 13 h and TON reached over 60 without loss of enantioselectivity. Library construction was also performed using different substrate solutions and the same catalyst column.

The same group also investigated the effect of anchoring strategy and the type of polymeric support for the catalyst activity. Asymmetric cyclopropanation was selected as a model reaction and several types of immobilized Hayashi–Jorgensen catalysts were evaluated under continuous-flow conditions [27]. Notably, significant activity and selectivity differences were observed for each catalyst. As a general tendency, the monolith catalyst showed lower activity but slightly higher selectivity. Immobilization through the diaryl part resulted in lower activity, probably due to the close distance between support and active site (Scheme 23).

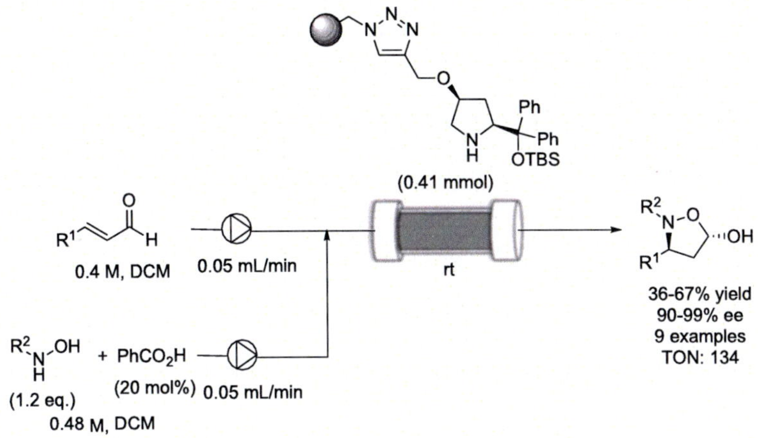

Scheme 24 Tandem 1,4-addition and cyclization reaction

Substrate scope was investigated using **1R** as the heterogeneous catalyst for the construction of the product library. Different substrate solution was flowed into the reactor every 6 h using the same catalyst column. The resulting product solution could be directly connected to flow HWE reaction without any purification.

In 2018, Escrich and Pericas et al. developed immobilized *cis*-4-hydroxy diphenylprolinol silyl ethers for tandem 1,4-addition and cyclization reactions (Scheme 24) [28]. *trans*-Hydroxy proline was conventionally used for the immobilization of proline and its derivatives due to its availability. However, the author's group discovered that *cis*-Hydroxy proline derivatives demonstrated higher activity and selectivity compared with the corresponding *trans*-proline derivatives. Accordingly, *cis*-4-hydroxy diphenylprolinol silyl ethers were immobilized onto polystyrene polymer. The activity and selectivity were evaluated for tandem aza-Michael and cyclization reaction and showed that a *cis*-configured catalyst demonstrated higher activity and selectivity than the *trans*-configured catalyst. The catalyst was also employed for the rapid preparation of chiral building blocks under flow conditions. Ten types of optically active products were prepared by passing solutions of different aldehydes and hydroxylamines through the catalyst column. Total TON reached 134 and, remarkably, the catalyst maintained its activity for a period of 2 months without apparent deactivation.

The similar *cis*-prolinol silyl ether catalyst was further examined by Ochiai and Nishiyama et al. in 2020 [29]. 1,4-Addition of aldehyde to nitro alkene was studied. The catalyst was prepared by the copolymerization of styrene substituted 4-hydroxy prolinol silyl ether monomer, styrene, and DVB. The prepared catalysts were evaluated under batch conditions and it was found that a *cis*-configured catalyst showed higher activity and selectivity than the corresponding *trans*-catalyst. The catalyst was even more selective than a homogeneous catalyst, although decreased activity was observed. The catalyst was also employed under flow conditions,

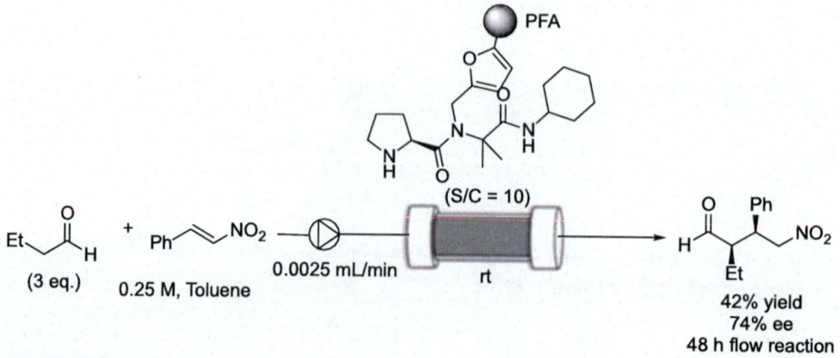

Scheme 25 Polyfurfuryl alcohol immobilized peptide catalysts for asymmetric 1,4-addition of aldehyde with nitro olefin

showing the high stereoselectivity. However, catalyst deactivation started just after the flow reaction started.

In 2019, Torre, Rivera, and Paixao et al. reported the polyfurfuryl alcohol immobilized peptide catalysts for asymmetric 1,4-addition of aldehydes with nitro olefins (Scheme 25) [30]. Polyfurfuryl alcohol is derived from renewable resources such as biomass and, therefore, is compatible with the principles of green chemistry. The catalytically active proline was coupled with furfuryl amine or isocyanide via Ugi four-component reactions. The obtained peptide with furyl functionality was further polymerized with furfural with TFA catalyst. The prepared material was thermally stable up to 100°C, but started to degrade at 100–300°C. The prepared catalysts were evaluated under batch conditions. It was revealed that the catalysts were compatible with both hydrophilic and hydrophobic solvents, and it also showed activity in water solvent. The catalyst also worked under flow conditions. The catalyst maintained its activity during a 24 h reaction, giving 74% ee with up to 50% conversion. However, catalyst deactivation was observed after 24 h, while stereoselectivity was maintained.

A novel type of chiral primary amine catalyst was developed by the same group in 2017 for asymmetric Robinson annulation (Scheme 26) [31]. A *t*-leucine derived 1,2-diamine was chosen as a target catalyst for immobilization, based on reported homogeneous catalysis. A novel immobilized catalyst was synthesized in six steps starting from *t*-leucine and evaluated for Robinson annulation. During the optimization of reaction conditions, the authors revealed that the activity of a heterogeneous catalyst depended significantly on the reaction temperature and that the reaction completed within 1 h at 55°C. Interestingly, such temperature dependence was not observed using a homogeneous catalyst, and lower activity was observed. It was suggested that the nonpolar environment of the polymer had a positive effect on catalyst activity and that higher temperature enhanced the mass transfer in the polymer matrix. Continuous-flow catalysis was performed using meso triketones formed in situ because vinyl ketones were found to be a catalyst poison due to

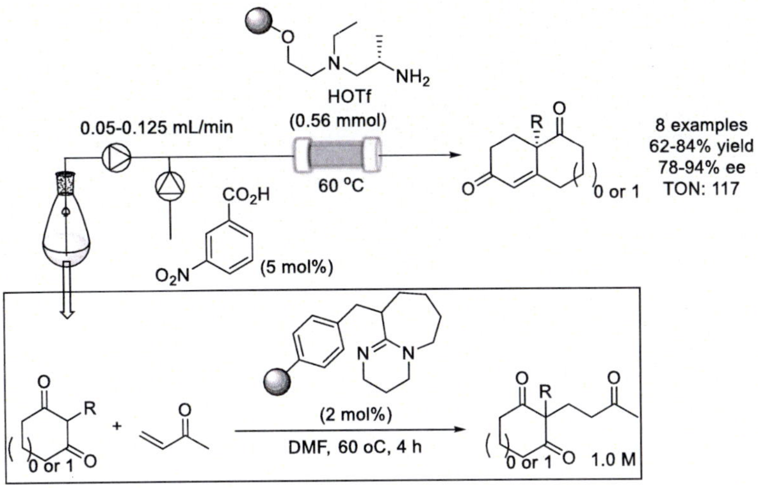

Scheme 26 Asymmetric Robinson annulation

aza-Michael addition reaction. Eight types of chiral diketones could be synthesized in moderate to good yield with good enantioselectivity.

3.2.2 Chiral Phosphoric Acid

BINOL derived chiral phosphoric acid (CPA) catalysts, which are commonly known as Akiyama–Terada catalysts, developed independently in 2004, have been revealed to be highly active, highly selective, and robust chiral Brönsted acid catalysts. They can activate imines, aldehydes, and α,β-unsaturated carbonyl compounds by protonation and, at the same time, create an ideal chiral environment for enantioselective catalysis. Over the last 5 years, there have been significant advances in the field of flow CPA catalysis, although there are few reported examples.

One example of heterogeneous CPA catalysis is asymmetric allylboration of aldehydes reported in 2016 from Pericas group (Scheme 27) [32]. In this report, the preparation method of heterogeneous catalysts was improved by changing the anchor structure. Using the same intermediate for the preparation of homogeneous catalysts, bromination at the 6,6′-position followed by Suzuki–Miyaura coupling afforded the styrene introduced BINOL. It was then copolymerized with styrene and DVB to obtain chiral BINOL immobilized polymer. Finally, postfunctionalization of PS-BINOL with $POCl_3$ gave the heterogeneous CPA catalyst (PS-TRIP). The heterogeneous catalyst was found to be a highly active and selective catalyst under both batch and continuous-flow conditions. Especially, no catalyst deactivation was observed during 28 h, and the target product was obtained in excellent yield and selectivity, with a TON of 282.

In 2017, Liu and Cui et al. developed metal-organic framework incorporated chiral phosphoric acid catalysts for enantioselective alkylation of indoles (Scheme

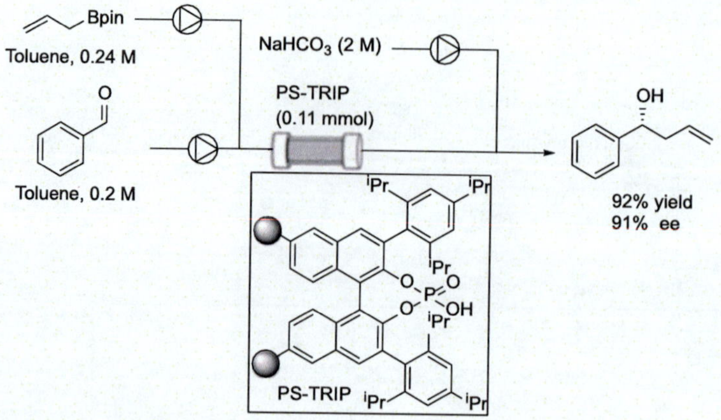

Scheme 27 Asymmetric allylboration of aldehyde

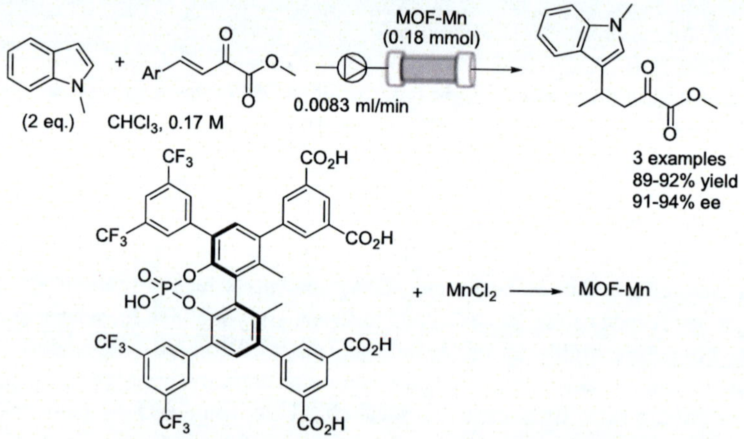

Scheme 28 Metal-organic framework incorporated chiral phosphoric acid catalyst for enantioselective alkylation of indole

28) [33]. MOFs have some attractive features such as well-defined and controllable nanostructure and high surface area. Therefore, it is a promising material, especially for heterogeneous chiral catalysts. For the preparation of MOF material, organic linkers with biphenyl phosphoric acid as an active site were designed and synthesized. Prepared linkers were reacted with $MnCl_2$ to form MOFs with $Mn_2(CO_2)_4(PO_4)(H_2O)_2$ as an edge structure. The prepared materials have high thermal stability as well as large surface area. On the other hand, it was revealed that substituents on the linker had a significant impact on chemical stability. MOFs with a larger linker maintained the structure even upon treatment under basic conditions for 24 h, while other MOFs lost the crystallinity under aqueous conditions. The prepared heterogeneous catalysts were evaluated for enantioselective Friedel–Crafts alkylation of

indoles with unsaturated keto ester. The catalyst with the larger linker had the highest activity and enantioselectivity, and even showed higher performance than with a homogeneous linker, which indicates the important role of the MOF structure. The MOF catalyst also showed excellent activity under continuous-flow conditions. The catalyst showed slightly higher activity but decreased selectivity compared with batch conditions. It was rationalized that substrate channeling may improve the diffusion, while silica, used for packing, may be unfavorable for the active site. The system could be reused seven times of each 24 h run without loss of activity or enantioselectivity.

3.2.3 Other Types of Organocatalysis

In recent years, several other types of organocatalysts have been adapted to heterogeneous systems and applied for continuous-flow catalysis. Such examples are still limited compared with enamine catalysis or CPA catalysis, but provide a new and unique methodology to synthesize chiral molecules.

In 2015, the Pericas group were the first to develop a polymer immobilized benzotetramisole catalyst as a Lewis base catalyst for flow reaction (Scheme 29) [34]. The catalyst could be prepared from enantiopure alkynyl epoxy ether in four steps and immobilized on Merrifield resin by Huisgen cyclization. The prepared catalyst was evaluated for domino Michael addition/cyclization reaction between acid anhydride and α,β-unsaturated imines. It was revealed that stoichiometry of PivCl was crucial to prevent the catalyst deactivation, and, interestingly, demonstrated higher diastereoselectivity compared with a homogeneous catalyst. Continuous-flow catalysis was performed using a solution of in-line formed mixed anhydride. In-line quenching and the inclusion of a liquid–liquid separator allowed continuous production and collection of the crude organic phase. The target product was obtained in 70% yield with >99% ee with a TON of 22.5.

In 2017, the Puglisi group developed heterogeneous *N*-picolylimidazolidinines as a new class of chiral heterogeneous Lewis base catalysts (Scheme 30) [35]. The same group previously discovered that *N*-picolylimidazolidinines could act as asymmetric

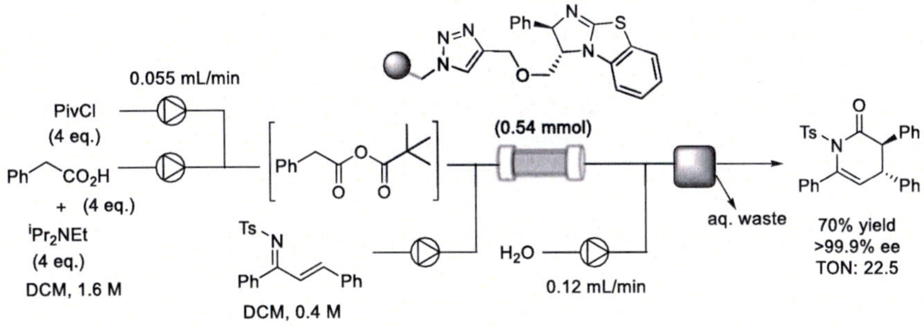

Scheme 29 Polymer-immobilized benzotetramisole catalyst

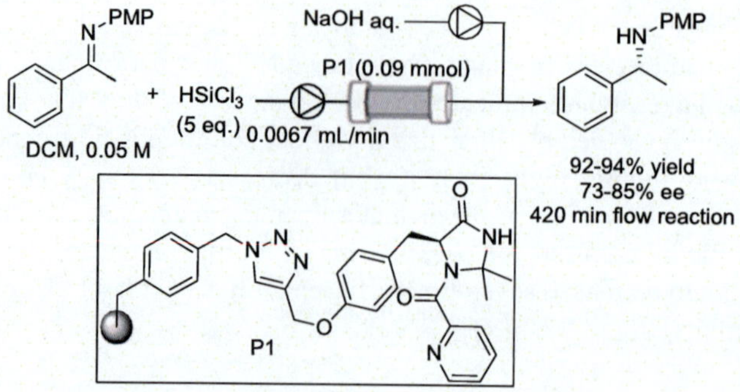

Scheme 30 Heterogeneous *N*-picolylimidazolidinines as a new class of chiral heterogeneous Lewis base catalyst

hydrosilylation catalysts for imines, and they succeeded in heterogenizing their catalysts for continuous-flow reaction in this report. Three heterogeneous catalysts could be prepared from tyrosine and immobilized on different supports and different anchors. The catalysts were evaluated for the asymmetric hydrosilylation of a ketimine under batch conditions. It was found that polystyrene-supported catalyst showed higher activity and selectivity than silica-supported catalysts. Especially, the triazole-anchored catalyst showed significantly lower selectivity. The best catalyst was also applied for a continuous-flow reaction. Although activity remained excellent and constant, a decrease in enantioselectivity was observed over time. The cause of deactivation was suggested to be partial degradation of the imidazolidinone ring.

Hydrogen-bonding organocatalysts have gained attention as robust acid/base catalysts. Unlike enamine catalysis, they do not make a covalent bond with the substrate; therefore, the undesired catalyst decomposition pathway can be minimized. Such a feature is also attractive for those who investigate heterogeneous catalysts because stability is a key factor in the development of efficient and recyclable catalysts. The Pericas group developed the first heterogeneous thiourea catalyst as a representative H-bonding catalyst and employed it for the asymmetric amination reaction of 1,3-dicarbonyl compounds (Scheme 31) [36]. Their group often uses Huisgen cyclization as an anchoring reaction to immobilize the active site and polystyrene. However, the initial investigation revealed that the heterogeneous thiourea catalyst prepared by such protocol delivered inconsistent results. The authors reasoned that Cu contamination during Huisgen cyclization was an issue and prepared the triazole-free thiourea catalyst. As expected, the newly prepared catalyst showed excellent activity and selectivity for α-amination of 1,3-dicarbonyl compounds. Although deactivation was observed during recycling experiments, the authors identified that the deactivation was caused by partial protonation of tertiary amine rather than degradation of thiourea part, and the catalyst could be reactivated by amine treatment. Therefore, continuous-flow catalysis was performed in the presence of a catalytic amount of triethylamine to minimize catalyst deactivation and to enhance reactivation. As a result, the catalyst remained active for 7.5 h operation with >90% enantioselectivity.

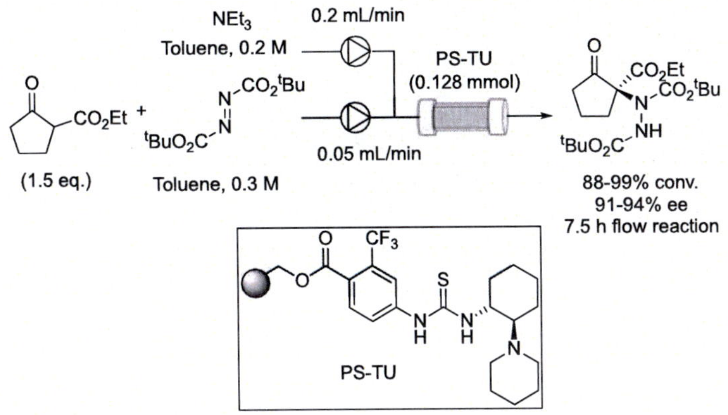

Scheme 31 Heterogeneous thiourea catalysts as a representative H-bonding catalyst

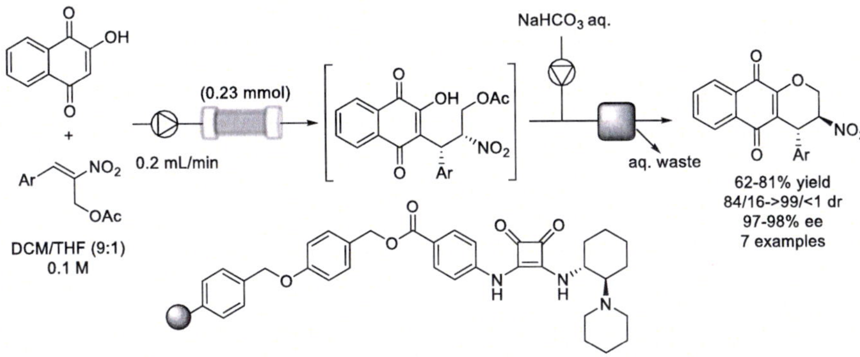

Scheme 32 1,4-Addition to nitro styrene

The same group further improved the immobilization protocol in 2016 (Scheme 32) [37]. In this report, they tried to avoid the use of expensive CF_3 substituted carboxylic acid linker as well as the use of Huisgen cyclization. Chiral squaramide was chosen as a target H-bonding catalyst, and immobilization was performed in a similar manner to that of a previous thiourea catalyst. Prepared catalysts were evaluated for the 1,4-addition to nitro styrenes. Interestingly, it was found that the CF_3 group on the linker has little impact on catalyst performance, and simple 4-carboxylate worked as a sufficient linker. Further increase in catalyst performance was achieved by introducing a longer linker between the active site and the polymer backbone. Continuous-flow catalysis was examined using the best catalyst, and excellent activity and selectivity were observed without any deactivation. Furthermore, the resulting product solution could be connected to base-mediated oxy-cyclization reactions to synthesize chiral pyranonaphthoquinones in a stereoselective manner.

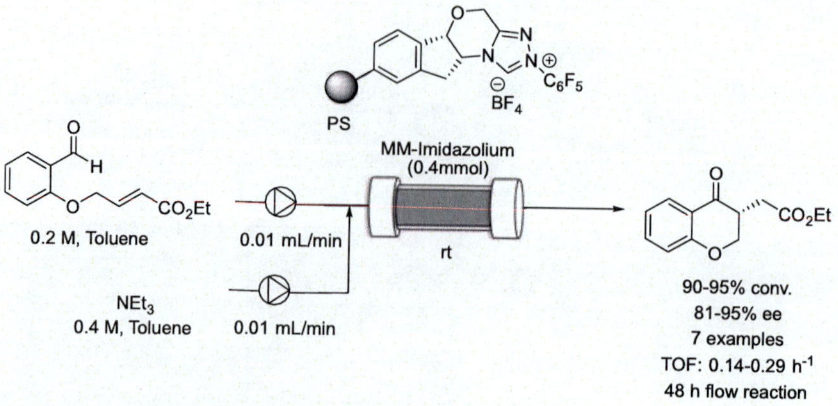

Scheme 33 Asymmetric heterogeneous NHC catalysis

Asymmetric NHC catalysis provides a new and non-conventional synthetic strategy to produce chiral molecules due to its umpolung nature. The Massi group reported the first asymmetric heterogeneous NHC catalysis under continuous-flow conditions in 2017 (Scheme 33) [38]. The heterogeneous catalyst was prepared from indanol and immobilized by copolymerization with styrene as a BF_4 salt. The prepared heterogeneous catalysts were employed for intramolecular Stetter reaction under batch conditions. Reactions were performed in the presence of a catalytic amount of base to form NHC in situ, and KHMDS was found to be the best initiator. Under the optimized reaction conditions, the target cyclized compound could be obtained in good yield with excellent enantioselectivity. For the continuous-flow reaction, a monolith catalyst was prepared by copolymerization in the catalyst column. The flow catalysis was initially investigated in the presence of 50 mol% KHMDS in the substrate solution. However, an undesired side product generated by the base-promoted vinylogous aldol condensation was obtained. On the other hand, decreasing the amount of KHMDS resulted in poor conversion of the starting material. Finally, it was found that NEt_3 is a suitable base to form active NHC from imidazolium salt, and at the same time, to suppress an undesired base-promoted side reaction. Under optimized reaction conditions, seven types of chiral chromanones could be obtained in excellent yield and selectivity.

In 2017, Wang and Lu et al. developed porous carbon nanosheet supported chiral squaramide for an enantioselective Friedel–Crafts reaction (Scheme 34) [39]. Porous carbon nanosheets possess open and interconnected macropores that can offer free space for facilitating the mass transfer while keeping the original chiral environment of the homogeneous catalysts. Unlike conventional inorganic supports such as silica, the surface of PCN is hydrophobic and chemically inert, which may be preferable for some catalysts such as hydrogen bonding catalysis and also would provide concentrated substrate near the surface due to the good affinity with organic molecules. Furthermore, it has high chemical and thermal stability, and various reaction conditions can be applicable. The surface modification was performed by reaction with

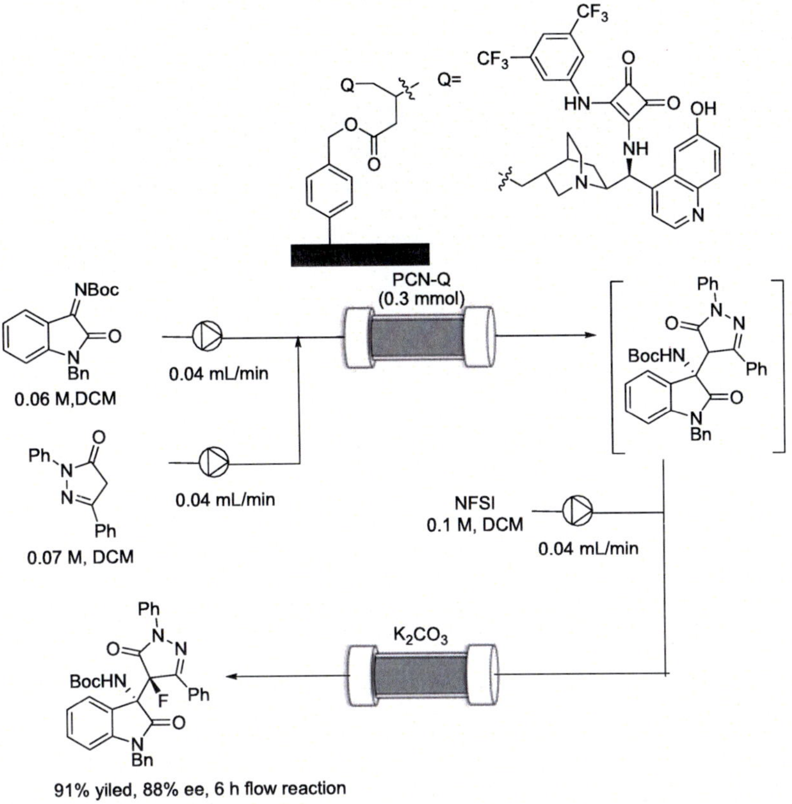

Scheme 34 Porous carbon nanosheet supported chiral squaramide for enantioselective Friedel–Crafts reaction

diazonium salts formed in situ. After the introduction of the acrylate functional group on the surface, radical polymerization was performed with quinine derived squaramide to afford heterogeneous H-bonding catalysts. The prepared catalysts were evaluated for the asymmetric Friedel–Crafts reaction under batch conditions. It was revealed that copolymerization with divinylbenzene during surface immobilization, which prevents the self-oligomerization of catalysts, was effective for achieving higher activity. The effect of the PCN support was clarified by comparison with SiO_2 supported catalyst and activated carbon-supported catalyst; this showed that PCN supported catalyst exhibited the best activity and selectivity, and even slightly higher activity was observed than the homogeneous catalyst. The catalyst could also work in flow conditions, and 478 mg of product was obtained with 0.7 $mmol^{-1}$ g^{-1} of productivity for 6 h. However, enantioselectivity was slightly decreased due to the background reaction without the catalyst.

In 2018, Pericas, Hahner, Smith et al. developed polymer-supported isothiourea catalysts for acylative kinetic resolution of alcohols (Scheme 35) [40, 41]. The homogeneous isothiourea catalyst called HyperBTM, which was developed by the

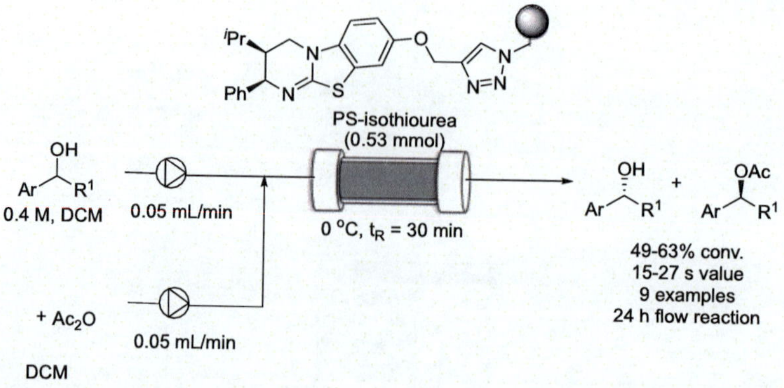

Scheme 35 Polymer-supported isothiourea catalysts for acylative kinetic resolution of alcohol

author's group previously, was immobilized onto a Merrifield resin by the copper-catalyzed azide-alkyne cycloaddition reaction to afford a heterogeneous catalyst. The prepared catalyst demonstrated high activity and enantioselectivity in the acylative kinetic resolution of alcohol with selectivity values (s-values) of 100, which is comparable with original homogeneous catalysts. The catalyst was also employed in a continuous-flow reaction. The high conversion and good enantioselectivity with s-values of 27–30 were achieved with 30 min of residence time. The catalyst remained active for a 24 h reaction to give target compounds in 28.8 mmol. The catalyst was also employed for the preparation of a library of chiral compounds. Each kinetic resolution with nine kinds of alcohols was performed on 4 mmol scales using the same catalyst and by simply washing between reactions. As a result, high conversion with s-values of 11–200 was obtained without loss of catalyst activity.

In 2019, Ciogli et al. developed bifunctional catalysts of quinine derived chiral amine and benzoic acid immobilized on a silica surface. In most of the enamine catalyses the addition of Brønsted acid as a co-catalyst is required to improve the activity (Scheme 36) [42]. Such Brønsted acid was added as a homogeneous catalyst in conventional immobilized chiral amine catalysts, which requires an additional purification step to remove acid co-catalysts. In this report, both the active amine catalyst and acid co-catalyst were immobilized onto the same support, and the products were obtained by simple filtration of a heterogeneous catalyst. Bifunctional catalysts were prepared by the post functionalization of surface thiol functionalized silica. The post functionalization was performed by the thiyl radical addition to vinyl anchor in the presence of AIBN. The prepared catalyst showed good stereoselectivity but low TOF of 0.02 h^{-1}. The catalyst was employed for the synthesis of the API Warfarin in flow conditions. The product was obtained in 97% yield with 78% ee with 16 h residence time.

In 2020, Andres, Pedrosa et al. developed polystyrene-supported cinchona alkaloid derived thiourea catalysts for Friedel–Crafts reaction with naphthols (Scheme 37) [43]. The immobilization was performed by the post functionalization of

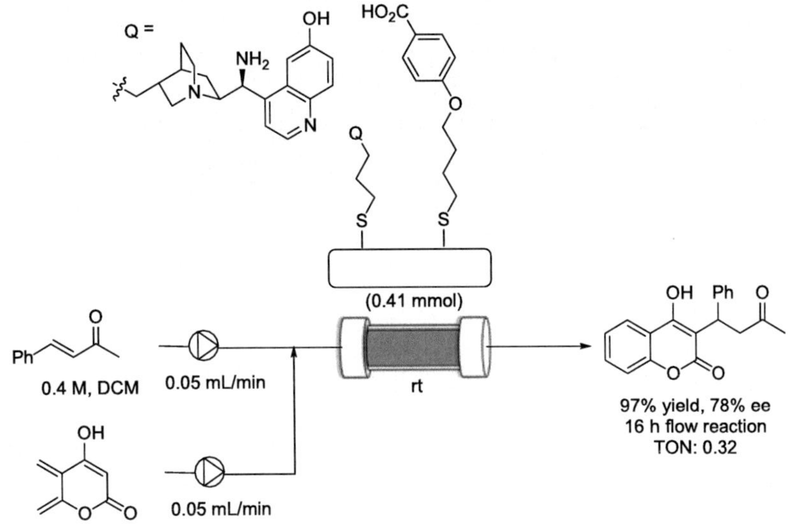

Scheme 36 Bifunctional catalysts of quinine derived chiral amine and benzoic acid immobilized on a silica surface

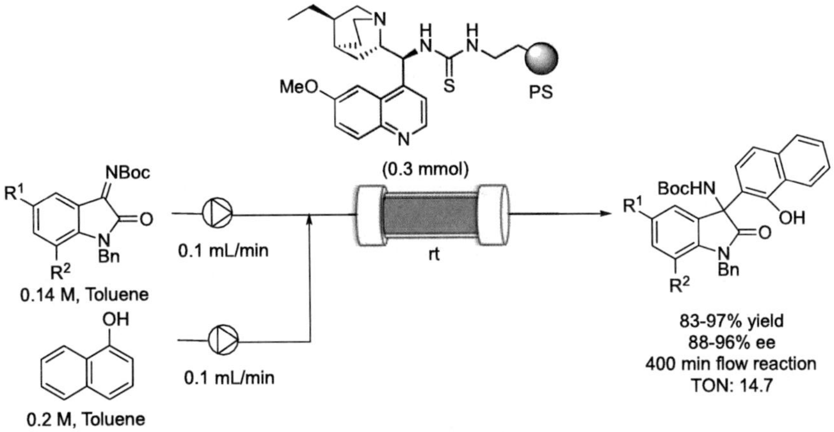

Scheme 37 Polystyrene-supported cinchona alkaloid derived thiourea catalyst

isocyanate functionalized polystyrene and primary amine via thiourea formation. The prepared catalysts were evaluated under batch conditions for the enantioselective Friedel–Crafts reaction. Among various kinds of catalysts derived from different cinchona alkaloids, the catalyst derived from hydro quinine showed the best enantioselectivity, with 100% conversion. Continuous-flow reactions were performed with two different kinds of substrates using the best catalyst. The catalyst maintained high activity and selectivity for 200 and 400 min, respectively, without

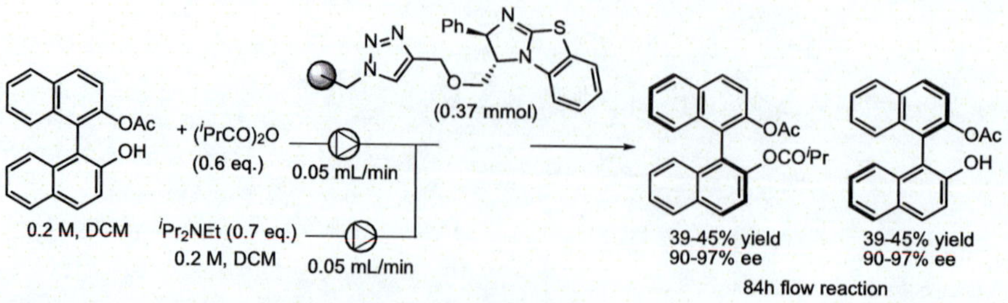

Scheme 38 Preparation of enantiomerically pure BINOL

significant loss of either yield or enantioselectivity. As a result, TON reached up to 7.4 and 14.7 for each substrate with a TOF of 2.2.

In 2020, Pericas et al. developed a continuous-flow preparation of enantiomerically pure BINOLs by acylative kinetic resolution using heterogeneous isothiourea catalysts (Scheme 38) [44]. The isothiourea unit reacted with azide functionalized polystyrene via copper-catalyzed azide-alkyne cyclization to form heterogeneous catalysts. The catalyst was employed for the acylative kinetic resolution of mono-acetylated BINOL using isobutyric anhydride under batch conditions. The catalyst showed good conversion with a good s value of 29 in $CHCl_3$ solvent. The catalyst showed high activity under flow conditions with a slightly decreased s value of 20. It was highly rewarding to find that the catalyst did not show any deactivation over 84 h of flow reaction to give 50 mmol of each target compound.

3.3 *Transition-Metal Catalysis*

In 2015, Zhang, Miller, Su et al. developed self-assembled Mn-salen heterogeneous gel catalysts for the oxidative kinetic resolution of phenethyl alcohols (Scheme 39) [45]. The Mn(III) Salen complex with two aldehydes units was condensed with tetrakis-(4-aminophenyl)-methane in the presence of a catalytic amount of acetic acid to form a self-assembled gel through the formation of a polyimine structure. For the flow reaction, the gel was coated on the amine-functionalized silica capillary by the formation of imine. The material had around 2 μm thickness of gel on the surface of the capillary based on the SEM analysis. The capillary catalyst was employed for the microflow oxidative kinetic resolution of 1-phenethylalcohol. The reaction completed with 15 min residence time to give ketone and enantioenriched alcohol in 65% conversion with 91% ee. The capillary was recycled by simply washing with DCM and was reused for various substrates. Under the optimized reaction conditions, conversion was 63–66% with 93–99% ee over 120 min of reaction time without the leaching of Mn species.

In 2018, Uozumi et al. developed an amphiphilic resin-supported chiral diene Rh complex for the enantioselective 1,4-addition of aryl boronic acids to enone in

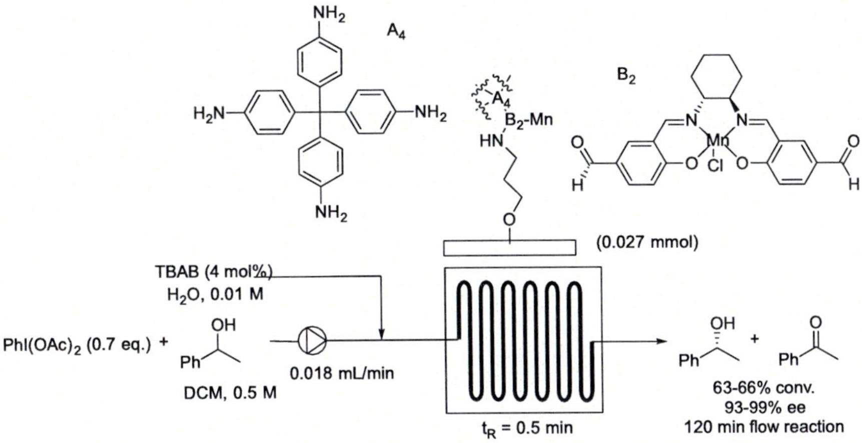

Scheme 39 Mn-salen heterogeneous gel catalyst for oxidative kinetic resolution

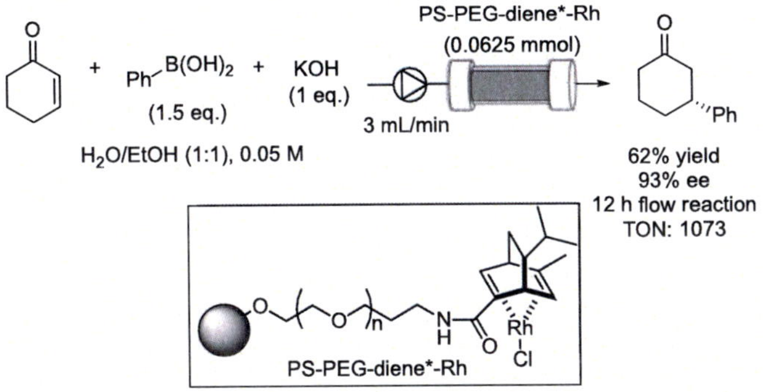

Scheme 40 Amphiphilic resin-supported chiral diene Rh complex

aqueous medium under continuous-flow conditions (Scheme 40) [46]. The chiral carboxylic acid with diene unit was condensed with amine-substituted polyethylene glycol (PEG) tethered polystyrene (PS) to form chiral amide. The catalyst was prepared by the complexation of Rh with the heterogeneous ligand. The amphiphilic PS-PEG is expected to provide a suitable reaction environment for the organic reaction in an aqueous solvent. The catalyst showed high catalytic activity in pure H_2O under batch conditions with 95% enantioselectivity. However, the addition of organic solvent significantly decreased the yield while maintaining selectivity. The flow reaction was performed in water–ethanol mixed solvent system to improve the solubility of substrates. It was found that the addition of 1 equivalent of KOH significantly improved the reactivity, and the target product was obtained in 79% yield within 10 s residence time. Under the optimized flow conditions, the reaction was continued for 12 h to give the target product in 62% yield with 93% ee and 1,073 of TON.

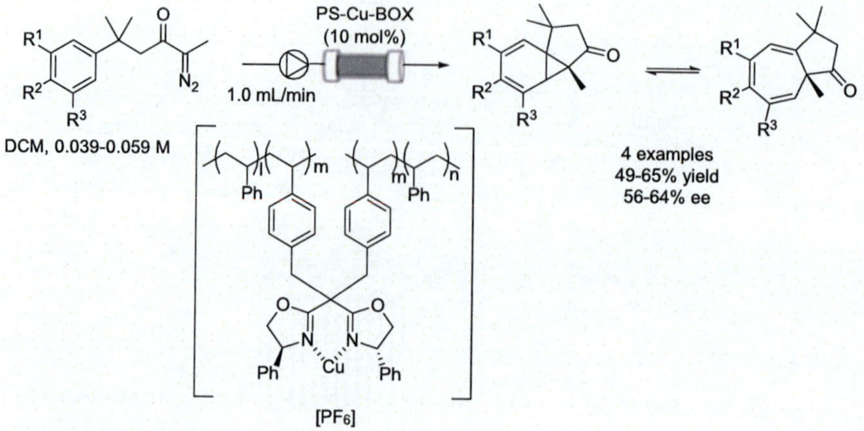

Scheme 41 Intramolecular asymmetric C–H insertion reaction

In 2018, the Maguire group developed an intramolecular asymmetric C–H insertion reaction with heterogeneous Cu catalyst (Scheme 41) [47]. Chiral Cu-BOX complex was immobilized onto polystyrene by copolymerization of styrene-tethered BOX and styrene. The prepared catalyst was first examined under batch conditions. The activity was comparable to that of the homogeneous catalyst, while decreased selectivity and partial leaching of Cu were observed. Detailed comparison of homogeneous batch results, heterogeneous batch results, and heterogeneous flow results were investigated, and, as a general tendency, homogeneous catalysts showed the highest enantioselectivity, while few differences were observed between batch conditions and flow conditions using a heterogeneous catalyst.

Another important example for carbene insertion catalysis was reported by Davis and Jones et al. in 2018 (Scheme 42) [48]. Recently, chiral modification by the introduction of chiral carboxylate has enabled highly enantioselective transformations with various kinds of inert substrates. In this context, silica-supported chiral carboxylate was developed and applied for flow asymmetric catalysis. Chiral carboxylate was immobilized by silane coupling reaction with silica, and three kinds of asymmetric catalysis were examined. In these experiments, diazo compounds formed in line by the oxidation of hydrazones were employed, and the catalyst showed excellent performance not only for simple cyclopropanation reactions, but also for more challenging inert C–H insertion reactions.

3.4 *Lewis Acid Catalysis*

In 2018, Harada et al. developed a silica-immobilized H_8-BINOL ligand for the enantioselective 1,2-addition of organo-Ti species to aldehydes (Scheme 43) [49]. C2 functionalized-H_8 BINOL was immobilized onto silica by dehydrogenation

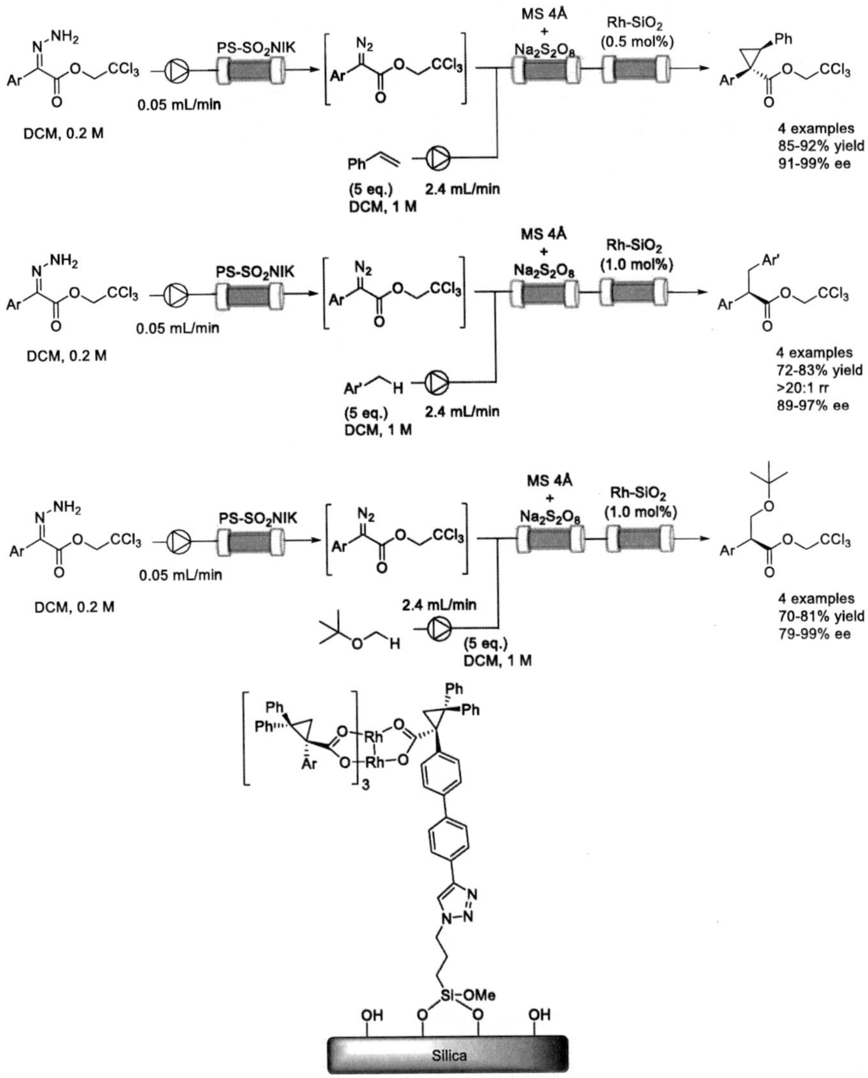

Scheme 42 Silica-supported chiral Rh(II) carboxylate catalyst

between silane and surface silanol. Remaining surface silanol was further capped by a subsequent dehydrogenation with a small silane reagent to improve the compatibility. The prepared catalysts were used for the enantioselective 1,2-addition of phenyl titanium to aromatic aldehydes under continuous-flow conditions. An organotitanium reagent was introduced into the catalyst column to form immobilized Ti complex, and excess organometallic reagent and aldehyde was flowed into the column. The target alcohol was obtained with good conversion, high

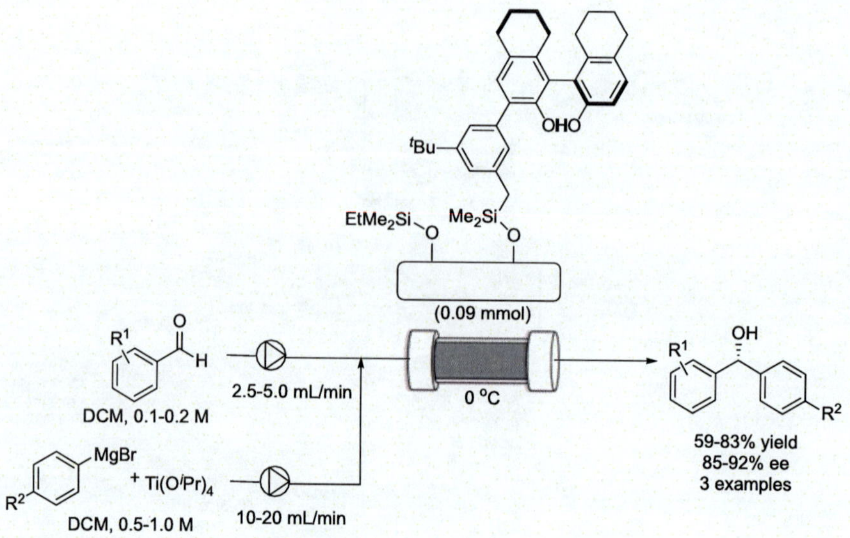

Scheme 43 Silica-immobilized H_8-BINOL ligand

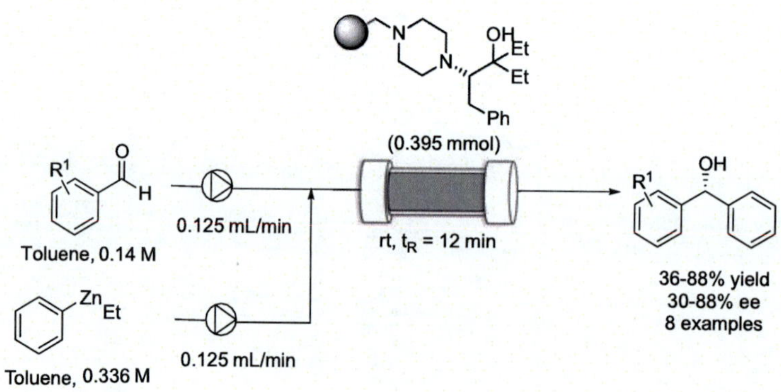

Scheme 44 Heterogeneous chiral ligands for the enantioselective 1,2-addition of aryl zinc to aldehyde

enantioselectivity, and with excellent TON and TOF values. The catalyst was stable for at least 8 h in the flow reaction.

In 2018, Pastre et al. developed another type of heterogeneous chiral ligand for the enantioselective 1,2-addition of aryl zinc to aldehydes (Scheme 44) [50]. Ligand screening under homogeneous conditions revealed that phenyl alanine derived amino alcohol ligand was identified as the best ligand in terms of enantioselectivity. The immobilization of the chiral ligand was performed by the alkylation of secondary amine with Merrifield resin. The heterogeneous ligand showed comparable activity and selectivity compared with the homogeneous ligand, and it was employed

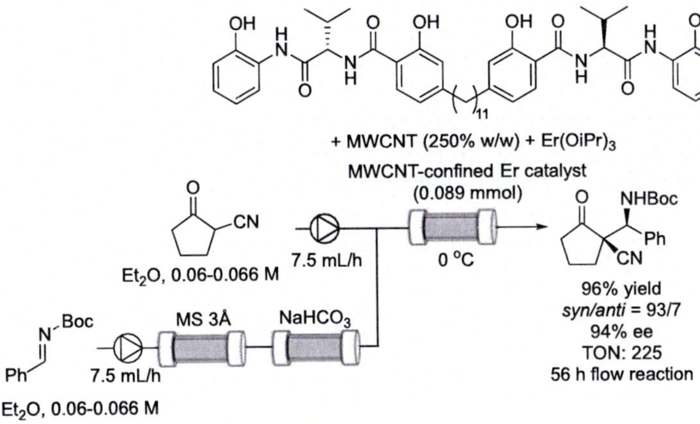

Scheme 45 MWCNT-supposed chiral Er catalyst.

for the flow reaction to give the target product in 86% yield with 80% ee within a residence time of 12 min.

In 2015, Kumagai and Shibasaki et al. developed a multiwalled carbon nanotube (MWCNT) supported chiral Er catalyst for enantioselective Mannich reaction of cyanoketone. The authors previously reported the noncovalent immobilization of Na/Nd bimetallic catalyst on MWCNT via noncovalent immobilization (Scheme 45) [51]. Although the immobilization method is simple and gave highly active, selective, and robust catalysts, it was only applicable for insoluble metal salts. To improve the versatility of this method, the authors prepared linked chiral amide ligands to enhance the formation of insoluble material and applied this technique for their immobilization method. The linked chiral ligands were prepared by the double functionalization of dialkyne with a variety of linker length and evaluated under batch conditions. It was revealed that longer linkers gave higher diastereoselectivity and enantioselectivity, and that linkers with an aromatic ring decreased the selectivity, which may suggest the negative influence of closely located active sites. The heterogeneous Er catalyst could be successfully incorporated with MWCNT simply by mixing during complexation. The MWCNT-confined catalyst not only showed almost the same activity and selectivity but also showed recyclability and reusability under batch conditions. Finally, it was applied under continuous-flow conditions to give the target product in 96% yield with excellent stereoselectivity and the total TON reached up to 225 during 56 h of flow reaction.

In 2017, the same group performed a detailed study on heterogeneous Nd/Na catalyst for the synthesis of the key intermediate of AZD7594, which is a therapeutic candidate for asthma and chronic obstructive pulmonary disease (Scheme 46) [52]. The heterogeneous catalysts were prepared by mixing $NdCl_3$, NaO^tBu, chiral amide ligand, and MWCNT. The structures of the chiral ligand and solvent were examined in detail for the asymmetric Henry reaction under batch conditions. As a result, an isoleucine derived ligand in 2-Me-THF gave the best yield with the best stereoselectivity. With the best conditions, the flow reaction was performed. The

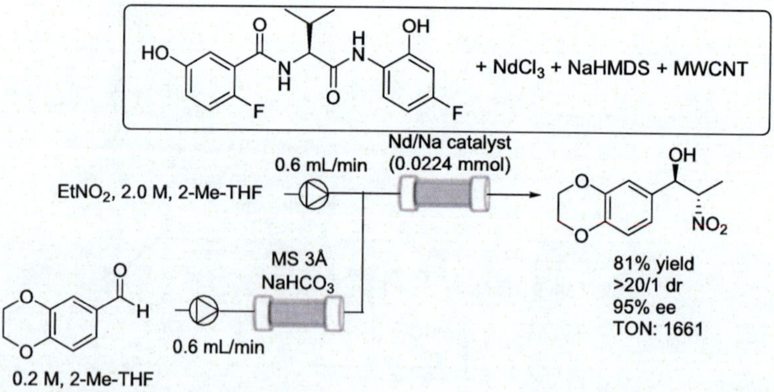

Scheme 46 Synthesis of the key intermediate of AZD7594

Scheme 47 Enantioselective Henry reaction with α-ketoester

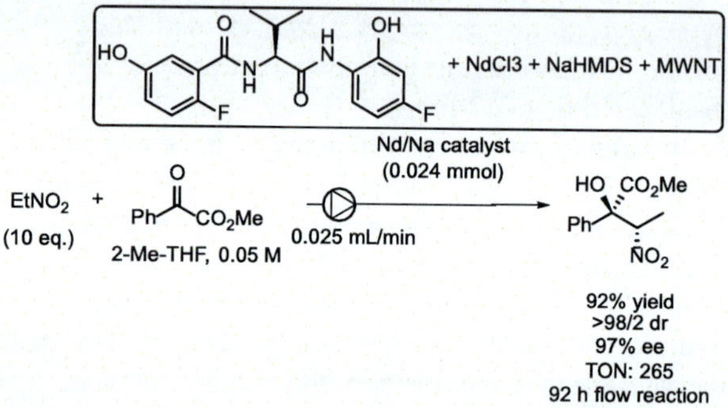

aldehyde solution was passed through the columns packed with MS 3A and $NaHCO_3$ to remove trace amounts of water and acidic impurities before flowing into the catalyst column. The catalyst retained its activity and selectivity over a 398 h reaction to afford the target product in 81% yield with >20:1 dr and 95% ee. The TON reached up to 1,661.

In 2018, the same group applied MWCNT supported Nd/Na catalyst for an enantioselective Henry reaction with α-ketoesters (Scheme 47) [53]. The catalyst was prepared in the same manner as previous reports and evaluated under batch conditions. During an investigation of the substrate scope, it was revealed that the solvent has a strong effect on both reactivity and stereoselectivity. Especially, the use of 2-Me-THF gave a much better result than the use of THF despite the similarity of their structures. After various control experiments, it was concluded that one of the enantiomers of 2-Me-THF interacted preferentially with a heterogeneous catalyst to create an efficient reaction environment to give high activity and selectivity. The catalyst could be applicable for a flow reaction of nitroethane and α-ketoester. After

92 h of flow reaction, the target product was obtained in 92% yield with >98:2 dr and 97% ee and the TON value reached up to 265.

3.5 Biocatalysis

In 2015, Jeong et al. developed a Novozyme 435 catalyzed continuous-flow kinetic resolution of 1-amino-2-indanol (Scheme 48) [54]. Novozyme 435 is a commercially available immobilized lipase on acrylic resin and one of the most frequently used heterogeneous biocatalysts. The authors investigated the kinetic resolution of amino alcohol for the synthesis of a key intermediate of the HIV protease inhibitor. After the optimization of acyl donor and solvent under batch conditions, flow reaction was investigated. The excellent enantioselectivity was maintained, even increasing the conversion and reaction time; finally, 50% conversion was achieved with an E value of >200. Furthermore, the catalyst retained activity even after 21 days, which demonstrated the robustness of Novozyme 435.

In 2017, Debecker et al. developed another type of kinetic resolution with an immobilized biocatalyst under continuous-flow conditions (Scheme 49) [55]. They focused on enzymatic enantioselective transamination of α-methyl benzylamine. The ω-transaminase ATA-50 was immobilized on a silica monolith called Si (HIPE) by three different methods. The first immobilization was simple adsorption on the surface of a support. The second method utilized ionic interaction between negatively charged enzyme and positively charged surface-functionalized Si(HIPE). The third method used covalent bonds by the formation of di-imine between surface amines of the support and lysine residue of the enzyme with glutaraldehyde as an

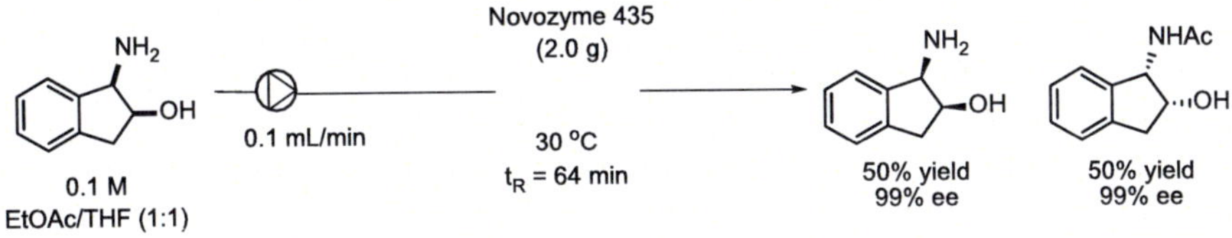

Scheme 48 Novozyme 435 catalyzed continuous-flow kinetic resolution

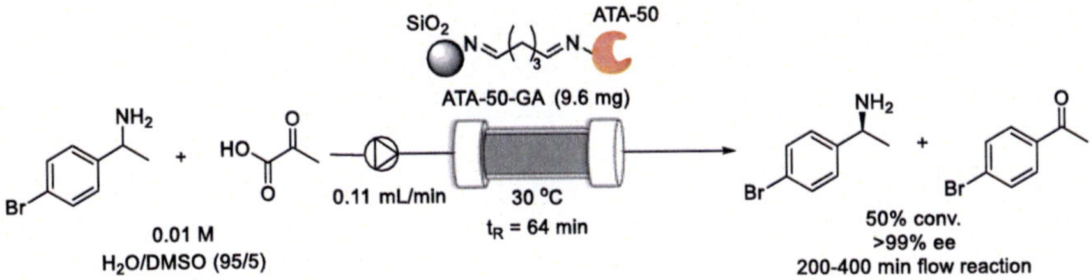

Scheme 49 Kinetic resolution with immobilized biocatalyst

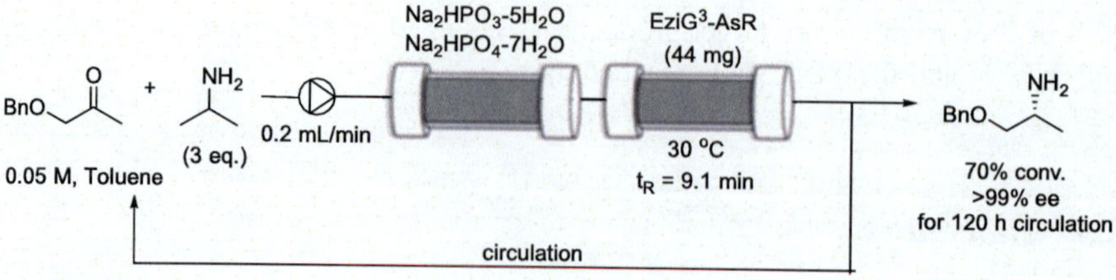

Scheme 50 Transamination of ketone in organic solvent

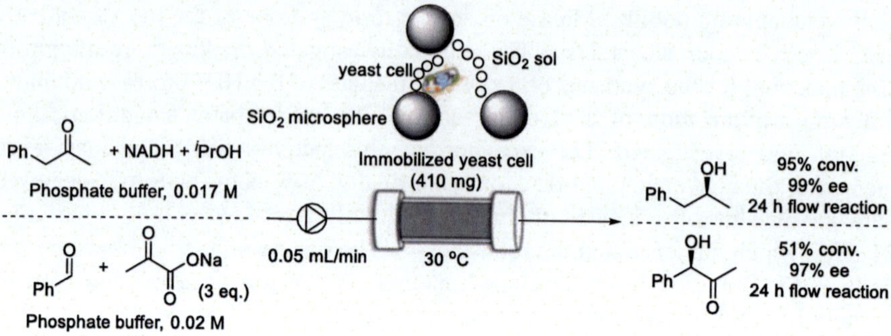

Scheme 51 Whole-cell immobilization catalyst

anchoring reagent. Three different catalysts were evaluated under flow conditions, and a covalently immobilized catalyst showed the highest activity and durability. After the optimization of conditions, 50% conversion was achieved for over 200 min of flow reaction.

In 2020, Mutti et al. developed transamination of ketone in an organic solvent under continuous flow with novel immobilized transaminases (Scheme 50) [56]. Immobilization of transaminase was performed by simple adsorption onto porous glass carrier material (EZiG) in buffer solution. With immobilized catalysts in hand, the catalyst activity was evaluated under batch conditions. The authors realized that the water content of the immobilized catalyst affected the catalyst activity significantly, and that it was precisely controlled by the pretreatment of hydrophilic organic solvent with hydrated salt. With this knowledge in hand, the circulated flow reaction was performed by packing the immobilized catalyst into a column reactor. The key to the success was to use a precolumn filled with hydrated salt to introduce a suitable amount of water to the solvent. After 120 h of reaction, the conversion reached 90% with >99% enantioselectivity.

In 2019, Paizs and Poppe et al. developed an immobilization of whole yeast cell onto silica sol-gel matrix for the enantioselective reduction of ketones and acyloin condensation (Scheme 51) [57]. Whole-cell immobilization catalysts offer a number of advantages such as easy handling, no need for enzyme purification, and improved stability. The yeast cell was first mixed with silica microsphere MAT540 to facilitate surface adsorption followed by the addition of silica sol to stabilize the structure and

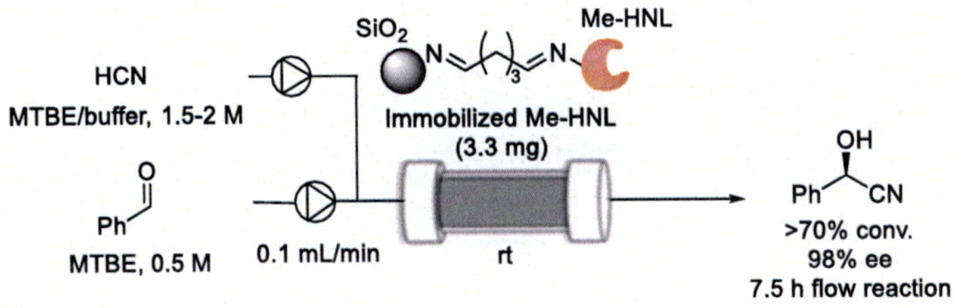

Scheme 52 Covalently immobilized hydroxynitrile lyase catalyst

prevent leaching. The solid material was washed with buffer to remove homogenous cells and used for flow enantioselective reactions. To demonstrate the versatility and compatibility with different kinds of reactions, ketone reduction and acyloin condensation were performed using the same catalyst column. First, reduction of phenylacetone with NADH was performed for 24 h to give stable conversion and enantioselectivity. The column was then washed with buffer, and acyloin condensation of benzaldehyde with sodium pyruvate was performed for 24 h. Stable results were also obtained for this reaction. Finally, the column was again washed with buffer and the reduction of phenylacetone was performed to give results that were identical to those obtained in the first reaction. These results suggest the high stability of heterogeneous catalysts.

Szymanska and Hnefeld et al. developed covalently immobilized hydroxynitrile lyase catalysts for the enantioselective hydrocyanation reactions (Scheme 52) [58]. They focused on the siliceous monolithic reactor due to the rapid and efficient mixing and developed surface immobilized enzyme catalysts for flow reaction. Two types of supports were chosen for the covalent immobilization. The first had aldehyde functionalized mesostructured cellular foam, which was expected to form an imine with the surface lysine of the enzyme. The second had epoxide-functionalized MCF, which was also expected to react with lysine on the enzyme. The efficiency of immobilization was determined, and aldehyde functionalized MCF showed superior immobilization efficiency. Flow enantioselective hydrocyanation was performed with this catalyst and the catalyst maintained activity for 7.5 h of flow reaction to give >70% conversion with 98% ee at the best. However, catalyst deactivation started after 7.5 h and both conversion and enantioselectivity decreased. Under the best conditions, STY was 1,229 g L^{-1} h^{-1}.

In 2019, Niemeyer et al. investigated the effect of support on the enzymatic reduction of ketone under continuous-flow conditions to improve the space-time yield (Scheme 53) [59]. Five kinds of reactors with different supports were evaluated based on the STY. Among the tested supported enzymes, including a physisorbed catalyst, covalently immobilized catalyst, biofilm-supported catalyst, and particle supported catalyst, hydrogels consisting of a polydimethylsiloxane supported catalyst showed the best activity and STY. Moreover, the catalyst supported on PDMS showed high stability and >50% conversion was observed after 150 h reaction. Under the best conditions, STY of >450 was achieved.

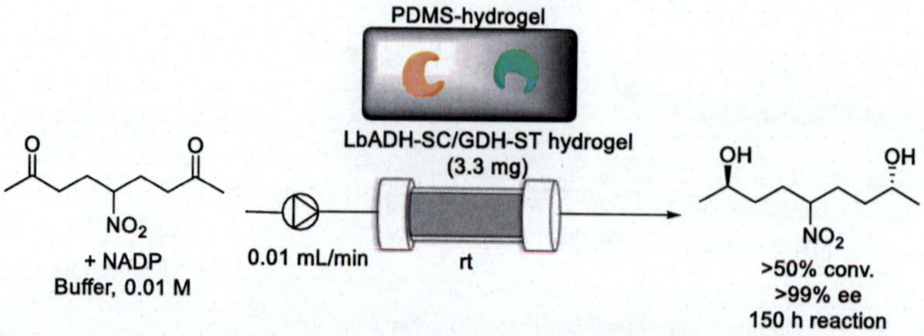

Scheme 53 Hydrogel consisting of polydimethylsiloxane supported catalyst

4 Application for Multistep Synthesis of a Complex Molecule

Multistep continuous-flow synthesis is a highly attractive method for the synthesis of complex molecules such as APIs, because tedious workup for each single reaction and purification of each intermediate can be avoided, with concomitant reduction in the amount of waste. However, multistep flow synthesis is far more challenging than single-step reactions because each reaction should not produce any catalyst poison for the subsequent reactions. Therefore, an overall synthetic route should be carefully designed to minimize the formation of by-products and side products, and each reaction should be completed just using a stoichiometric amount of starting materials. Given these restrictions, successful examples are still limited to date. On the other hand, the following examples offer proof that even optically active complex molecules could be synthesized under multistep continuous-flow conditions.

In 2015, Kobayashi et al. achieved a continuous-flow synthesis of Rolipram by using four different kinds of heterogeneous catalysts (Scheme 54) [60]. In the first column, an amine-functionalized silica catalyzed condensation between nitromethane and aldehyde was performed to give nitroalkene. The nitroalkene was subjected to enantioselective 1,4-addition reaction with malonate. In this reaction, polystyrene-immobilized chiral Ca-Pybox was employed as a heterogeneous catalyst. The obtained nitro compound was reduced to the corresponding amine by a heterogeneous Pd catalyst with H_2, and it was cyclized to form the lactam structure. Finally, the ester moiety was hydrolyzed by carboxylic acid functionalized silica, and the target compound was obtained by subsequent decarboxylation. It should be emphasized that only heterogeneous catalysts were used in all chemical transformations in this report, and throughout all transformations, generated by-products are only water, CO_2, and methanol. Such highly atom-economical synthesis is one of the keys to accomplishing multistep flow synthesis.

Ley et al. reported the combination of enzymatic flow reactions for the synthesis of cyanated compounds (Scheme 55) [61]. In the first step, the hydrolysis of ethyl cyanoformate was performed using commercially available Novozyme 435 as a heterogeneous catalyst to form hazardous HCN in situ. The generated HCN solution was mixed with aldehydes, and enantioselective hydrocyanation was performed with

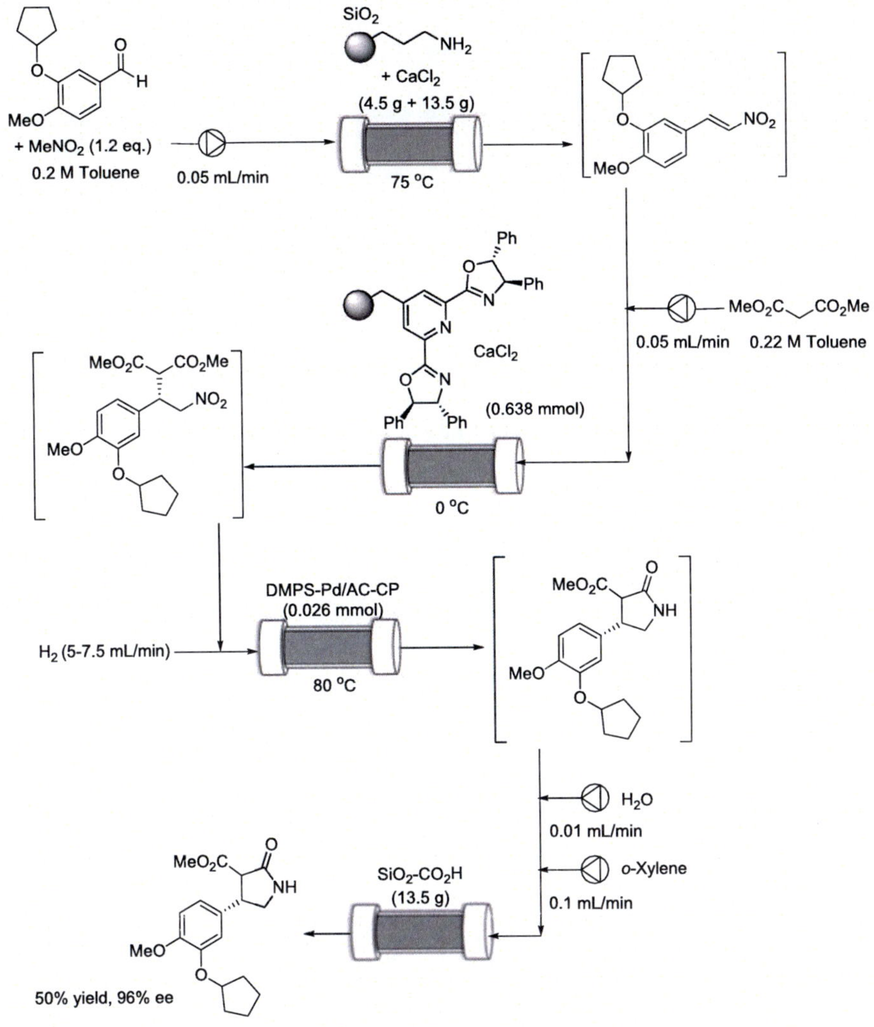

Scheme 54 Continuous-flow synthesis of Rolipram

a whole-cell immobilized E-coli-AtHNL heterogeneous catalyst. The obtained alcohol was trapped by Ac_2O in pyridine to give chiral cyanated compounds in good yields with good enantioselectivity. This is a nice example of the in-line formation of a hazardous reagent.

Otvos and Kappe et al. achieved the flow synthesis of a key intermediate of (−)-paroxetine in 2019 (Scheme 56) [62]. In the first reaction, an enantioselective 1,4-addition of malonate to unsaturated aldehyde was performed with a heterogeneous prolinol silyl ether catalyst. Interestingly, this reaction proceeded under solvent-free conditions, and the target product was obtained in excellent yield and selectivity. After purification of this intermediate, it was dissolved in 2-Me-THF and

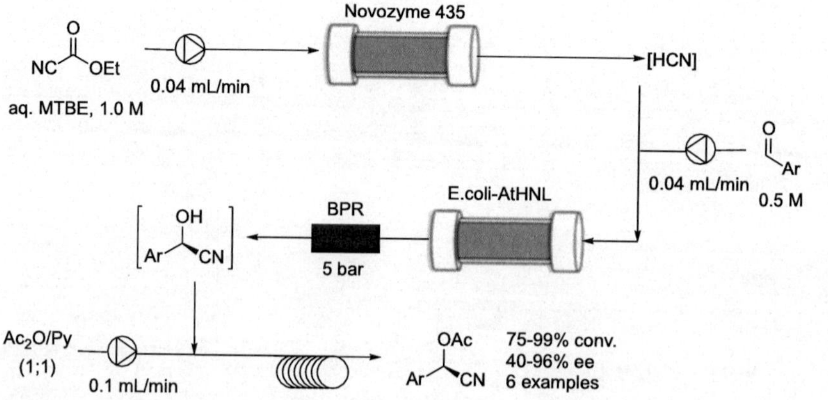

Scheme 55 Enzymatic flow reaction for the synthesis of cyanated compound

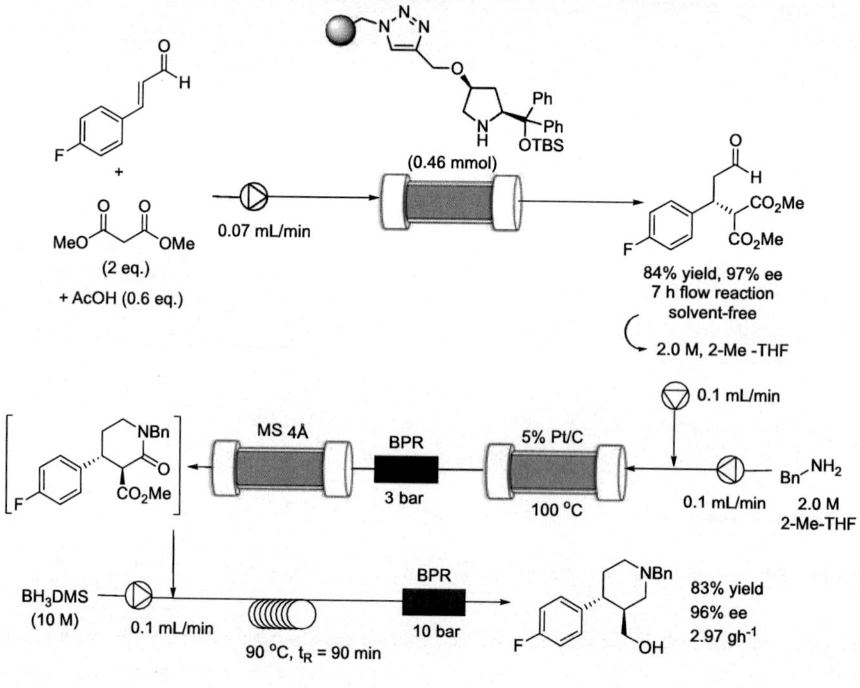

Scheme 56 Flow synthesis of a key intermediate of (−)-paroxetine

the sequential flow reaction was started. The first step was reductive amination with benzylamine with H_2 catalyzed by Pt/C. Both the fluorine substituent and benzyl groups were tolerated, and the cyclic intermediate was obtained after lactam formation. MS 4 Å column was introduced to remove water for the final step of BH_3

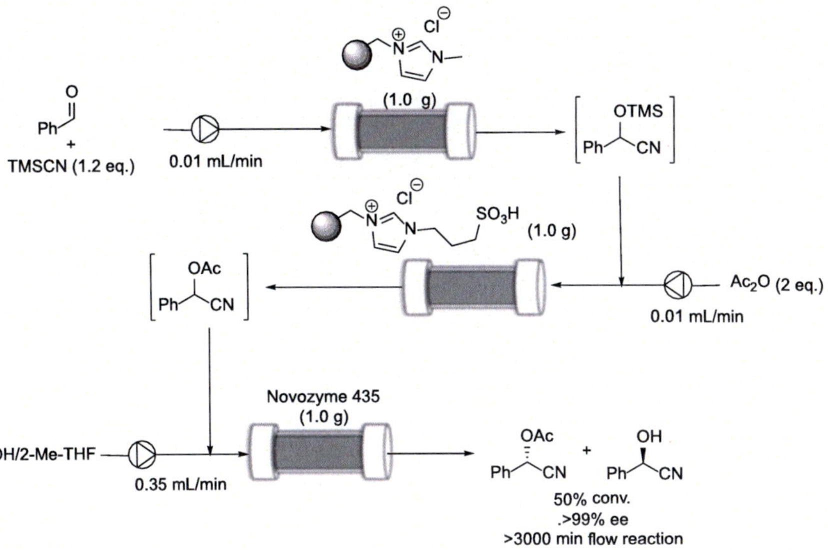

Scheme 57 Three-step heterogeneous flow catalysis for the preparation of a chiral cyano compound

reduction. Both the lactam and the ester were reduced to the corresponding amine and alcohol, respectively, and the sequential synthesis was achieved.

In 2019, Verdugo, Luis et al. developed a three-step heterogeneous flow catalysis for the preparation of a chiral cyano compound by the combination of imidazolium catalysts and enzymatic reaction (Scheme 57) [63]. Benzaldehyde was reacted with TMSCN in the presence of a heterogeneous imidazolium catalyst to form cyanated silyl ether in the first step. The silyl ether was cleaved and acetylated in the presence of another type of heterogeneous imidazolium catalyst having sulfonic acid functionality, to give racemic acetate. Finally, a deacylative kinetic resolution was performed with immobilized enzyme Novozyme435 to give enantiopure compounds. Notably, the system remained stable for over 3,000 min of flow reaction.

In 2019, Ishitani and Kobayashi et al. developed mesoporous silica-supported Ni-diamine catalysts for 1,4-addition reactions (Scheme 58) [64]. The catalyst was prepared by simply mixing NiI_2, diamine ligand, and mesoporous silica. Detailed characterization of this material revealed a 1:1 complex of Ni-diamine ligand, which is reported to be unstable in solution, that was immobilized by the stabilization of the Ni-O-Si bond in the mesopore. The effect of the support was highlighted in the reaction of 1,4-addition of malonate and nitroalkene. MCM-41, which is a common mesoporous silica support, showed much higher activity compared with a catalyst supported on other materials. Interestingly, the heterogeneous catalyst showed much higher activity than the original homogeneous catalyst.

In 2020, Ishitani and Kobayashi et al. developed the sequential-flow synthesis of a Baclofen precursor in a three-step transformation with heterogeneous catalysts (Scheme 59) [65]. The synthetic route was similar to that of Rolipram, which was

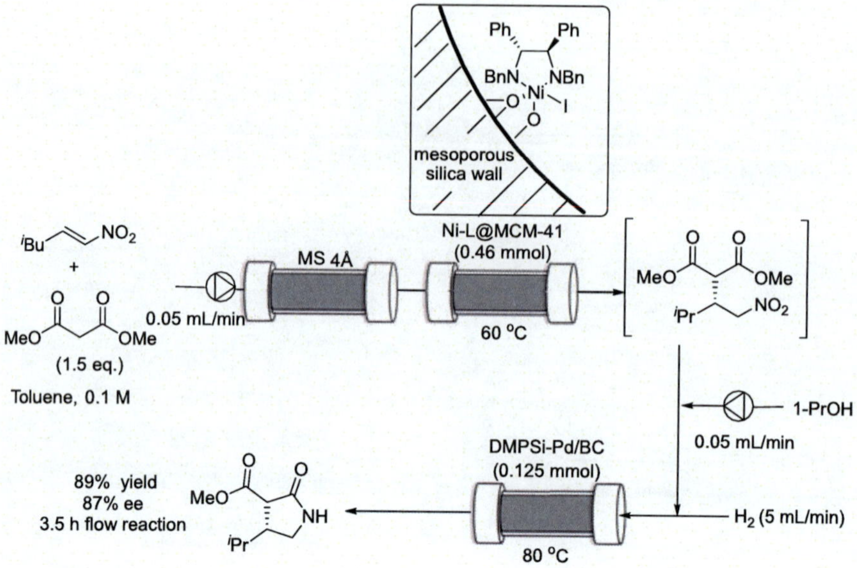

Scheme 58 Mesoporous silica-supported Ni-diamine catalysts for 1,4-addition reaction

reported by the same group. Nitroalkene was obtained by the condensation between aldehyde and nitromethane by an amine-functionalized silica catalyst. Enantioselective 1,4-addition of malonate was accomplished by the use of a polystyrene-immobilized Ca-Pybox catalyst. The catalyst was modified from Pd to Pt for the reduction of the NO_2 group for compatibility with the chloride substituent. Through the three-step sequential flow reaction, the Baclofen precursor was synthesized in excellent yield with 92% ee for a 69 h reaction.

5 Perspective

From the viewpoint of environmental compatibility, efficiency, and safety, continuous-flow reactions with heterogeneous catalysts can be regarded as ideal reactions for multistep continuous-flow synthesis of fine chemicals including pharmaceuticals [66]. At this moment, examples of these organic transformations are relatively scarce, and this is very different from batch systems. Among many continuous-flow transformations, hydrogenations with heterogeneous catalysts are the most advanced and promising, and high yield, selectivity, productivity, and catalyst efficiency have been attained. On the other hand, asymmetric hydrogenation with chiral heterogeneous catalysts is limited. For example, asymmetric hydrogenations of ketimines and direct asymmetric reductive amination of ketones provide efficient methods for the preparation of chiral amines. During the decades, highly active and selective chiral homogeneous catalysts have been developed to realize

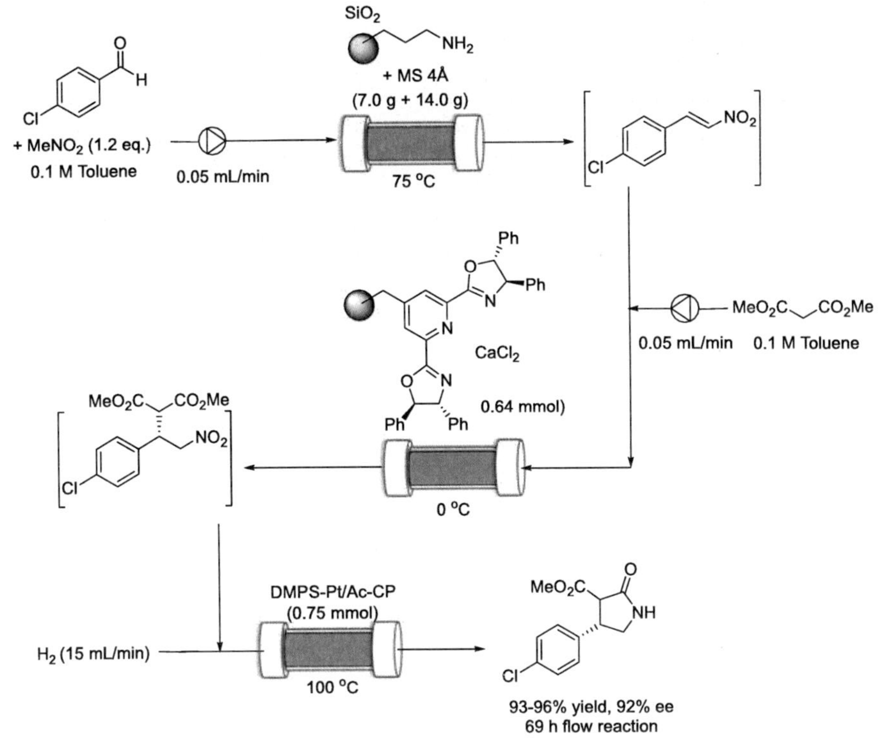

Scheme 59 Sequential-flow synthesis of Baclofen precursor

excellent TONs and TOFs. In these cases, the recovery and reuse of catalysts become less important from the economic viewpoints. However, the current status may be changed from a standpoint of continuous production, because flow reactions with heterogeneous catalysts provide powerful tools for the continuous synthesis of fine chemicals. Furthermore, for catalytic asymmetric hydrogenations with homogeneous catalysts, high reaction temperature and high hydrogen pressure are often required. If these issues could be addressed, practical continuous-flow asymmetric hydrogenations to prepare chiral amines could be realized. This is quite useful for production as well as medicinal chemistry and drug discovery. The development of heterogeneous catalysts for asymmetric hydrogenations toward continuous production will be an important research theme in the future.

There have been limited efficient C–C bond-forming reactions with heterogeneous catalysts. Chiral heterogeneous catalysts for asymmetric C–C bond-formation reactions are often based on immobilized organocatalysts, and some chiral heterogeneous metal catalysts have also been reported. This is very different from asymmetric catalysis with homogeneous catalysts. Another important aspect is that TON and TOF, productivity, and lifetime of catalysts in asymmetric C–C bond-formation reactions are inferior to those of asymmetric hydrogenations in many cases.

Therefore, further improvements toward more powerful chiral heterogeneous organocatalysts, as well as the development of truly efficient chiral heterogeneous metal catalysts, will be needed.

Other atom-economical bond-forming reactions, such as cycloadditions, rearrangements, and especially clean oxidation reactions (with oxygen gas or hydrogen peroxide), are crucial in multistep flow synthesis enabling the creation of more diverse sets of complex organic compounds. Other recent exciting developments in atom-economical transformations are photoinduced C–H and decarboxylative addition and cross-coupling reactions. To apply these processes to multistep continuous-flow synthesis, the development of efficient and robust heterogeneous photocatalysts is required.

For a long time, organic synthesis has been carried out in batch, and reactions and catalysts have been developed under batch conditions. However, reaction environments, such as molarity and frequency of contact of catalysts and substrates, are fundamentally different between batch and flow systems even at the molecular level. Therefore, unique reactivity and selectivity are often observed in flow compared with in batch, and even new reactions that proceed exclusively in flow systems are possible. When developed heterogeneous catalysts, various factors, such as size of particles, surface area, and dispersion state of the catalytic species on the support, must be well tuned for obtaining high efficiency in continuous-flow reactions. Heterogenous catalysts that were originally optimized in batch systems may not be suitable for flow reactions, and in those cases, novel heterogeneous catalysts for flow synthesis are required.

Continuous-flow synthesis will become one of the key technologies in chemical synthesis, and the science and technology will move society forward toward a green sustainable world. The sequential flow system shown in Scheme 54 is one of the final goals for future production of fine chemicals including pharmaceuticals, and to obtain this goal, we expect further developments in synthetic organic chemistry.

Compliance with Ethical Standards

Funding: This work was supported in part by a Grant-in-Aid for Scientific Research from Japan Society for the Promotion of Science (JSPS KAKENHI, Grant No. 19K15557) and Japan Science and Technology Agency (JST).

Ethical Approval: This chapter does not contain any studies with human participants or animals performed by any of the authors.

Informed Consent: Informed consent was obtained from all individual participants included in the study.

References

1. O'Neal EJ, Lee CH, Brathwaite J, Jensen KF (2015) ACS Catal 5(4):2615–2622
2. Touge T, Kuwana M, Komatsuki Y, Tanaka S, Nara H, Matsumura K, Sayo N, Kashibuchi Y, Saito T (2019) Org Process Res Dev 23(4):452–461

3. Kluson P, Stavarek P, Penkavova V, Vychodilova H, Hejda S, Jaklova N, Curinova P (2019) J Flow Chem 9(4):221–230
4. Kluson P, Stavarek P, Penkavova V, Vychodilova H, Hejda S, Bendova M, Došek M (2020) Chem Eng Res Design 153:537–546
5. Abrams ML, Buser JY, Calvin JR, Johnson MD, Jones BR, Lambertus G, Landis CR, Martinelli JR, May SA, McFarland AD, Stout JR (2016) Org Process Res Dev 20(5):901–910
6. Dai W, Mi Y, Lv Y, Chen B, Li G, Chen G, Gao S (2016) Adv Synth Catal 358(4):667–671
7. Fleming GS, Beeler AB (2017) Org Lett 19(19):5268–5271
8. Wang YF, Jiang ZH, Chu MM, Qi SS, Yin H, te Han H, Xu DQ (2020) Org Biomol Chem 18 (26):4927–4931
9. de Angelis S, de Renzo M, Carlucci C, Degennaro L, Luisi R (2016) Org Biomol Chem 14 (18):4304–4311
10. Kisszekelyi P, Alammar A, Kupai J, Huszthy P, Barabas J, Holtzl T, Szente L, Bawn C, Adams R, Szekely G (2019) J Catal 371:255–261
11. Schober L, Ratnam S, Yamashita Y, Adebar N, Pieper M, Berkessel A, Hessel V, Gröger H (2019) Synthesis 51(5):1178–1184
12. Kondo M, Wathsala HDP, Sako M, Hanatani Y, Ishikawa K, Hara S, Takaai T, Washio T, Takizawa S, Sasai H (2020) Chem Commun 56(8):1259–1262
13. Tang XF, Zhao JN, Wu YF, Zheng ZH, Feng SH, Yu ZY, Liu GZ, Meng QW (2019) Org Biomol Chem 17(34):7938–7942
14. Tang XF, Zhao JN, Wu YF, Feng SH, Yang F, Yu ZY, Meng QW (2019) Adv Synth Catal 361 (22):5245–5252
15. Cunillera A, Blanco C, Gual A, Marinkovic JM, Garcia-Suarez EJ, Riisager A, Claver C, Ruiz A, Godard C (2019) ChemCatChem 11(8):2195–2205
16. Zhang Z, Franciò G, Leitner W (2015) ChemCatChem 7(13):1961–1965
17. Geier D, Schmitz P, Walkowiak J, Leitner W, Franciò G (2018) ACS Catalysis 8(4):3297–3303
18. Amara Z, Poliakoff M, Duque R, Geier D, Franciò G, Gordon CM, Meadows RE, Woodward R, Leitner W (2016) Org Process Res Dev 20(7):1321–1327
19. Madarász J, Nánási B, Kovács J, Balogh S, Farkas G, Bakos J (2018) Monatsh Chem 149 (1):19–25
20. Saito Y, Kobayashi S (2020) J Am Chem Soc:jacs.0c08109
21. Yasukawa T, Masuda R, Kobayashi S (2019) Nat Catal 2(12):1088–1092
22. Kawakami Y, Borissova A, Chapman MR, Goltz G, Koltsova E, Mitrichev I, Blacker AJ (2019) Eur J Org Chem 2019(45):7499–7505
23. Gaunt MJ, Johansson CCC, McNally A, Vo NT (2007) Drug Discov Today 12(1–2):8–27
24. Ötvös SB, Szloszár A, Mándity IM, Fülöp F (2015) Adv Synth Catal 357(16–17):3671–3680
25. Izquierdo J, Ayats C, Henseler AH, Pericàs MA (2015) Org Biomol Chem 13(14):4204–4209
26. Sagamanova I, Rodríguez-Escrich C, Molnár IG, Sayalero S, Gilmour R, Pericàs MA (2015) ACS Catalysis 5(11):6241–6248
27. Llanes P, Rodríguez-Escrich C, Sayalero S, Pericàs MA (2016) Org Lett 18(24):6292–6295
28. Lai J, Sayalero S, Ferrali A, Osorio-Planes L, Bravo F, Rodríguez-Escrich C, Pericàs MA (2018) Adv Synth Catal 360(15):2914–2924
29. Ochiai H, Nishiyama A, Haraguchi N, Itsuno S (2020) Org Process Res Dev 24(10):2228–2233
30. de La Torre AF, Scatena GS, Valdés O, Rivera DG, Paixão MW (2019) Beilstein J Org Chem 15:1210–1216
31. Canellas S, Ayats C, Henseler AH, Pericàs MA (2017) ACS Catalysis 7(2):1383–1391
32. Clot-Almenara L, Rodríguez-Escrich C, Osorio-Planes L, Pericas MA (2016) ACS Catal 6 (11):7647–7651
33. Chen X, Jiang H, Hou B, Gong W, Liu Y, Cui Y (2017) J Am Chem Soc 139(38):13476–13482
34. Izquierdo J, Pericàs MA (2016) A recyclable, immobilized analogue of benzotetramisole for catalytic enantioselective domino Michael addition/cyclization reactions in batch and flow. ACS Catalysis 6(1):348–356

35. Porta R, Benaglia M, Annunziata R, Puglisi A, Celentano G (2017) Adv Synth Catal 359 (14):2375–2382
36. Kasaplar P, Ozkal E, Rodríguez-Escrich C, Pericàs MA (2015) Green Chem 17(5):3122–3129
37. Osorio-Planes L, Rodríguez-Escrich C, Pericàs MA (2016) Cat Sci Technol 6(13):4686–4689
38. Ragno D, di Carmine G, Brandolese A, Bortolini O, Giovannini PP, Massi A (2017) ACS Catalysis 7(9):6365–6375
39. Zhao L, Bao X, Hu Q, Wang B, Lu AH (2018) ChemCatChem 10(6):1248–1252
40. Neyyappadath RM, Chisholm R, Greenhalgh MD, Rodríguez-Escrich C, Pericàs MA, Hähner G, Smith AD (2018) Catalysis 8(2):1067–1075
41. Guha NR, Neyyappadath RM, Greenhalgh MD, Chisholm R, Smith SM, McEvoy ML, Young CM, Rodríguez-Escrich C, Pericàs MA, Hähner G, Smith AD (2018) Green Chem 20 (19):4537–4546
42. Ciogli A, Capitani D, di Iorio N, Crotti S, Bencivenni G, Donzello MP, Villani C (2019) Eur J Org Chem 2019(10):2020–2028
43. Rodríguez-Rodríguez M, Maestro A, Andrés JM, Pedrosa R (2020) Adv Synth Catal 362 (13):2744–2754
44. Lai J, Neyyappadath RM, Smith AD, Pericàs MA (2020) Adv Synth Catal 362(6):1370–1377
45. Liu H, Feng J, Zhang J, Miller PW, Chen L, Su CY (2015) Chem Sci 6(4):2292–2296
46. Shen G, Osako T, Nagaosa M, Uozumi Y (2018) J Org Chem 83(14):7380–7387
47. Crowley DC, Lynch D, Maguire AR (2018) J Org Chem 83(7):3794–3805
48. Yoo C-J, Rackl D, Liu W, Hoyt CB, Pimentel B, Lively RP, Davies HML, Jones CW (2018) Angew Chem 130(34):11089–11093
49. Watanabe S, Nakaya N, Akai J, Kanaori K, Harada T (2018) Org Lett 20(9):2737–2740
50. Forni JA, Novaes LFT, Galaverna R, Pastre JC (2018) Catal Today 308:86–93
51. Hashimoto K, Kumagai N, Shibasaki M (2015) Chem A Eur J 21(11):4262–4266
52. Nonoyama A, Kumagai N, Shibasaki M (2017) Tetrahedron 73(11):1517–1521
53. Karasawa T, Oriez R, Kumagai N, Shibasaki M (2018) J Am Chem Soc 140(38):12290–12295
54. Kim YJ, Choi YS, Yang S, Yang WR, Jeong JH (2015) Synlett 26(14):1981–1984
55. van den Biggelaar L, Soumillion P, Debecker DP (2017) Catalysts 7(2):54
56. Böhmer W, Volkov A, Engelmark Cassimjee K, Mutti FG (2020) Adv Synth Catal 362 (9):1858–1867
57. Nagy-Györ L, Lăcătuş M, Balogh-Weiser D, Csuka P, Bódai V, Erdélyi B, Molnár Z, Hornyánszky G, Paizs C, Poppe L (2019) ACS Sustain Chem Eng 7(24):19375–19383
58. van der Helm MP, Bracco P, Busch H, Szymańska K, Jarzębski AB, Hanefeld U (2019) Cat Sci Technol 9(5):1189–1200
59. Peschke T, Bitterwolf P, Hansen S, Gasmi J, Rabe KS, Niemeyer CM (2019) Catalysts 9(2):164
60. Tsubogo T, Oyamada H, Kobayashi S (2015) Nature 520(7547):329–332
61. Brahma A, Musio B, Ismayilova U, Nikbin N, Kamptmann SB, Siegert P, Jeromin GE, Ley SV, Pohl M (2016) Synlett 27(2):262–266
62. Ötvös SB, Pericàs MA, Kappe CO (2019) Chem Sci 10(48):11141–11146
63. Peris E, Porcar R, Burguete MI, García-Verdugo E, Luis SV (2019) ChemCatChem 11 (7):1955–1962
64. Ishitani H, Kanai K, Yoo WJ, Yoshida T, Kobayashi S (2019) Angew Chem Int Ed 58 (38):13313–13317
65. Ishitani H, Furiya Y, Kobayashi S (2020) Chem Asian J 15(11):1688–1691
66. Yoo WJ, Ishitani H, Saito Y, Laroche B, Kobayashi S (2020) J Org Chem 85:5132–5145

Top Med Chem (2021) 38: 223–240
https://doi.org/10.1007/7355_2021_113

Published online: 16 April 2021

Multiple Organolithium Reactions for Drug Discovery Using Flash Chemistry

Yosuke Ashikari, Takashi Tamaki, Masahiro Takumi, and Aiichiro Nagaki

Contents

Abstract The high reactivity of organolithium reagents often renders them too unstable to be used, thereby limiting their application in organic synthesis. This review highlights our approach to various synthetic multiple reactions for drug discovery mediated by organolithium reagents based on flash chemistry.

Keywords Flash chemistry, Flow microreactor, Organolithium, Reaction integration

Y. Ashikari, T. Tamaki, M. Takumi, and A. Nagaki (✉)
Department of Synthetic Chemistry and Biological Chemistry, Graduate School of Engineering, Kyoto University, Kyoto, Japan
e-mail: ashikari.yousuke.5r@kyoto-u.ac.jp; tamaki.takashi.4r@kyoto-u.ac.jp; takumi.masahiro.6n@kyoto-u.ac.jp; anagaki@sbchem.kyoto-u.ac.jp

1 Introduction

One advantage of synthetic organic chemistry is the myriad of useful organic compounds that are created, including biologically active compounds. Numerous pharmaceuticals, potential drug candidates, and their precursors have been synthesized, mass produced, and supplied. To meet future demands, enhancing the speed and applicability of organic syntheses is desirable. Thus, new methods implementing reaction integration have gained significant attention.

Traditional organic syntheses, including drug synthesis, have comprised stepwise formations of individual bonds in the target molecule. A more efficient method would be to form several bonds in a single sequence without needing to isolate intermediates. Indeed, conventional step-by-step syntheses are being supplemented with integrated syntheses, where multiple components are combined by a single operation, such as one-pot or single-flow processes [1]. In particular, one-pot sequential reactions effectively carry out numerous transformations and form several bonds in a single vessel. Furthermore, the reaction sequence can be modified by varying the order in which the reaction components are added. However, it can be difficult to use highly reactive and short-lived intermediates in a consecutive sequence because, in practice, the time taken to add the next reaction components often substantially exceeds the lifetimes of such intermediates.

Extensive efforts have been made in recent years to develop a new synthetic method denoted "space integration." This is a continuous-flow method in which a sequence of reactions is conducted by adding components at different points along the flow [2]. This is accomplished in a flow reactor by controlling the residence time, which is similar to the reaction time in a batch reactor. The residence time of a solution is defined as the time that elapses from when the reaction components are mixed to when the quenching agent is added. The reaction time can be controlled by changing the distance (or volume) between the two mixing points in the flow reactor (Fig. 1).

In flow microreactors, the residence time is precisely controlled and greatly decreased by reducing the volume of the channel or tube. This enables highly reactive and short-lived intermediates that were generated in the system to be used in a flow microreactor, which consequently allows reactions that are very difficult or impossible to achieve by conventional means to be accomplished. Two types of space integration methods have been developed: linear and convergent, both of

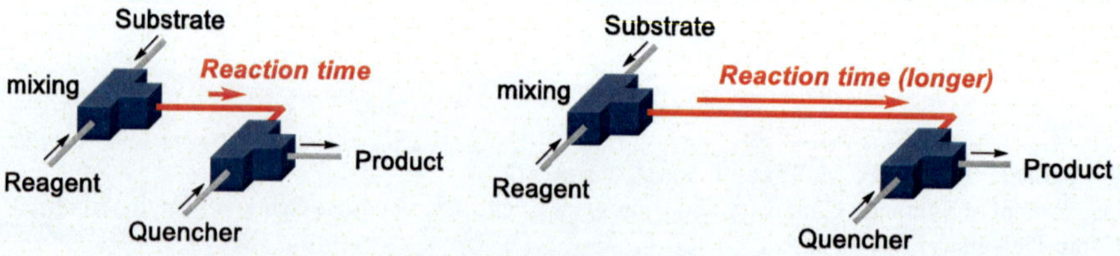

Fig. 1 Reaction time controlled by changing the distance between the inlets of the reagent and the inlet of the quencher in the flow reactor

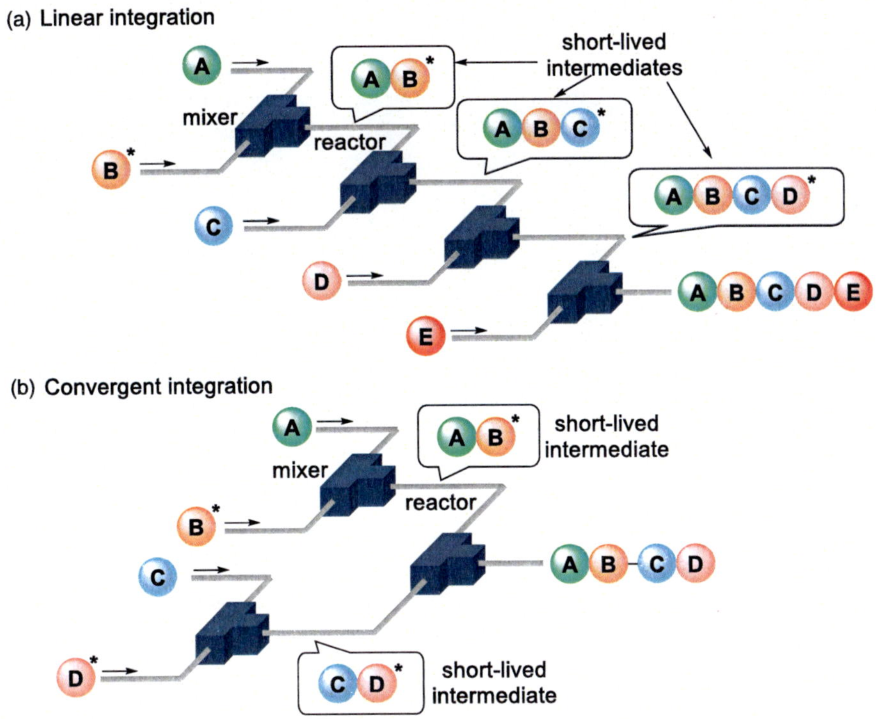

Fig. 2 Two types of space integration reactions involving several unstable intermediates: (**a**) linear and (**b**) convergent. Asterisk mark means reactive species

which employ a single series of reactions involving several short-lived intermediates (Fig. 2). In linear integration (Fig. 2a), two components can be reacted together to generate a short-lived intermediate, which can be followed by subsequent generation of other short-lived intermediates upon adding more reaction components. This linear procedure can be repeated within a single-flow process to obtain the target molecule. Moreover, this sequence of reactions can be easily regulated by varying reaction component combinations. In convergent integration (Fig. 2b), several different short-lived intermediates can be generated separately and then combined to create the target molecule.

Organometallic intermediates are crucial in organic synthesis because they serve as carbanion equivalents that can be used for carbon–carbon bond formation to construct the carbon skeletons of organic compounds [3]. The properties of organometallics are highly dictated by the nature of their metal. In general, greater differences between the electronegativities of the metal and carbon cause greater degrees of polarization, which consequently leads to a greater reactivity of the organometallic compound. Therefore, of the organometallics commonly used in organic synthesis, the most reactive intermediates are organolithiums. Unfortunately, the high reactivity of organolithiums often renders them too unstable to use, thereby limiting their synthetic applications. In fact, extremely low temperatures, such as −78°C or below, are often required in conventional batch reactors to avoid organolithium decomposition [4]. Moreover, as is often the case with batch chemistry, it is nearly impossible to control the reactions, even at such an extremely low temperatures. Flow microreactors can solve such problems by implementing space integration. Consequently, chemical reactions using highly reactive and short-lived organolithiums that are otherwise difficult or even impossible to perform in batch processes can be accomplished in flow microreactors. This article details the reactions we developed that employ space integration; particularly, novel reactions involving several highly reactive and short-lived organolithium intermediates. Some of these reactions produce pharmaceuticals and their candidates.

2 Flash Chemistry

2.1 Reactions Mediated by Decomposable Intermediates

Before introducing reaction integration, the development of flash chemistry should be mentioned. In this chemistry, the super-short-lived intermediates are generated and used up in seconds to enable transformations that could not be performed in batch reactors. Such reactions mediated by highly unstable intermediates can be carried out in flow reactors, since they can precisely control reaction time, as shown in Fig. 1, and instantly make the reaction mixture homogeneous. In this section, we discuss some reactions mediated by reactive intermediates from the viewpoint of discovering and developing new drugs.

Halogen–lithium exchange reactions of bromobenzenes to generate aryllithiums, followed by their reaction with electrophiles, are one of the most straightforward approaches to introduce a lithium substituent onto the benzene ring. The high reactivity of aryllithiums enhances the reaction efficiency, but it could also cause problems regarding functional group tolerability. Functional groups are indispensable for synthesizing highly functionalized complex organic molecules. Since they can undergo further transformations, they are interesting candidates for linear integration. Therefore, the use of functional organolithiums has been extensively investigated. Organolithiums are known to be incompatible with many electrophilic functional groups in conventional batch reactions. Consequently, these electrophilic functional groups often require protection prior to their introduction to the organolithium compound. Space-integration via a flow microreactor can be used to conduct these sensitive organolithium reactions without requiring electrophilic functional group protection. This is possible because of the short residence times in the generation step.

For example, aryllithiums bearing ketone carbonyl groups have been generated from aryl iodides and mesityllithium (2,4,6-trimethylphenyl lithium) via iodine–lithium exchange where the residence time was reduced to 3 ms or less [5]. The resulting aryllithiums have successfully reacted with various electrophiles in a flow

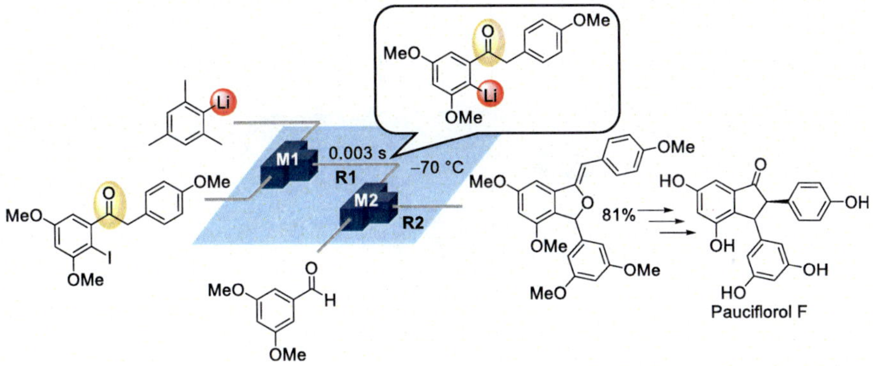

Fig. 3 Flash generation of aryllithium bearing a ketone carbonyl group for the formal synthesis of Pauciflorol F

microreactor without their ketone carbonyl groups being affected. Furthermore, this method has been successfully applied to the formal synthesis of Pauciflorol F, a natural product isolated from stem bark (Fig. 3). Pauciflorol F belongs to a family of compounds that have gained significant attention owing to their multi-functional bioactivities, including cytotoxic, antibacterial, and anti-HIV effects [6]. The I–Li exchange reaction of the ketone-substituted aryliodide with mesityllithium followed by a reaction with 3,5-dimethoxybenzaldehyde was carried out using a flow microreactor system consisting of an integrated device, where micromixers M1 and M2 and microtube reactor R1 were integrated into a single device. The greatly reduced residence time of 3 ms in R1 enables the product to be obtained in 81% isolated yield after subsequent cyclization with the unchanged ketone carbonyl group. Notably, the present method exhibits relatively high productivity (1.06 g from 5 min of operation) and provides an efficient way of producing useful pharmaceutical compounds in quantities sufficient for screening and clinical studies. Treating the product with HCl/*i*-PrOH in the presence of O_2 in a batch reactor gave the cyclopentenone derivative, which can be converted to Pauciflorol F via one-pot hydrogenation and *epimerization* followed by deprotection. According to the similar strategy, the generation and reaction of aryllithiums bearing alkoxycarbonyl- [7, 8], nitro- [9], and cyano groups [10] were also achieved.

Flash chemistry also permits the generation and reaction of functionalized alkyllithiums, which are known to be more unstable than aryllithiums because of their high reactivity and may react with electrophilic functionalities, such as an epoxy group, much faster. Figure 4 shows the flow system for the generation and reaction of the alkyllithiums bearing an epoxide group [11]. Using lithium naphthalenide (LiNp), the alkyllithium can be generated in a reductive lithiation manner and then reacted with electrophiles after 6.7 ms. With this system, the synthetic utility of alkyllithiums bearing chloro, alkoxy, vinyl, silyl, cyano, and alkoxycarbonyl groups has also been established [11, 12]. Moreover, since alkyllithiums can act as initiators for living anionic polymerizations, a

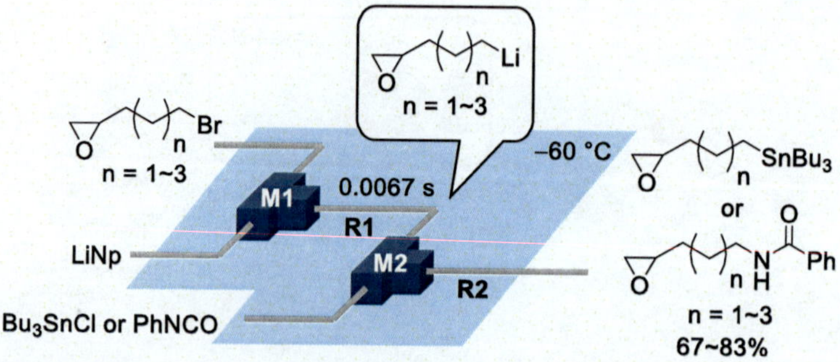

Fig. 4 Flash generation of alkyllithiums bearing electrophilic functionalities

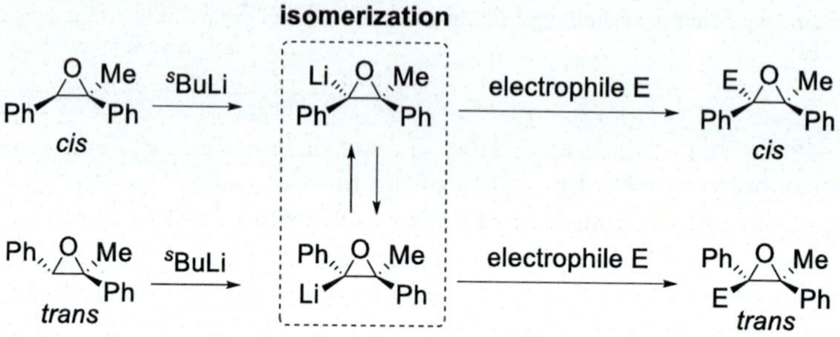

Fig. 5 Isomerizable oxyranyl lithiums and their reactions with an electrophile

heterotelechelic polymer synthesis initiated by those alkyllithiums has also been reported [12].

2.2 *Reactions Mediated by Isomerizable Intermediates*

Flash chemistry can control reactions mediated by intermediates that can easily lose their stereochemistry. Oxyranyl anions, which are generated by the deprotonation of epoxides by *sec*-butyl lithium, are typical examples of such isomerizable intermediates. The oxyranyl lithium species generated from trisubstituted epoxides have two types of stereochemistry (*cis* and *trans*) that tend to convert back and forth (Fig. 5). In addition, these oxyranyllithiums also undergo decomposition reactions. Thus, these reactions are usually operated at very low temperatures, such as −100°C, in conventional batch apparatuses [13, 14].

The flow system depicted in Fig. 6 allows the oxyranyl anions to react with electrophiles before they epimerize. The oxyranyl anions were generated at −48°C and reacted with iodomethane after 23.8 s to produce the desired tetrasubstituted epoxides in a quantitative yield with a high stereoselectivity [15, 16]. A similar method gives aziridinyllithiums from *N*-butylsulfonyl- [17] and *N*-tosyl-aziridines

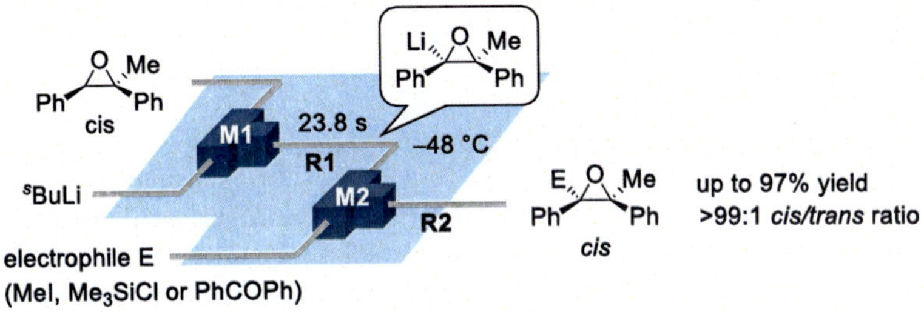

Fig. 6 Flash generation of oxyranyl lithium species and their reactions with electrophiles

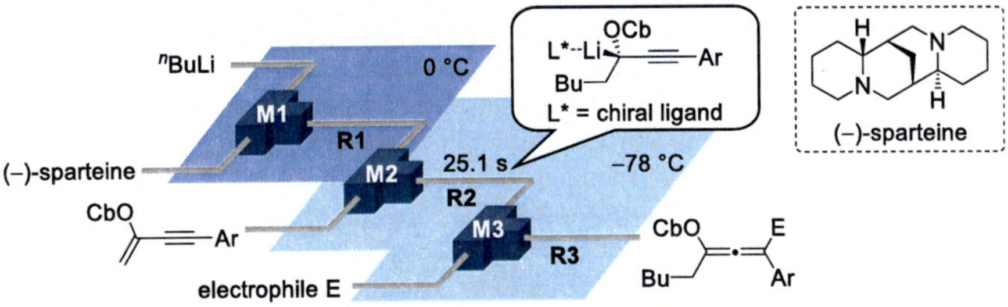

Fig. 7 Flash generation of optically active oganolithium species via the carbolithiation of conjugated enynes and their reaction prior to racemization

[18], which can efficiently react with electrophiles to synthesize functionalized aziridines and 1,2,3,4-tetrahydroisoquinolines (THIQs) [19].

2.3 *Reactions Mediated by Racemizable Intermediates*

In the field of asymmetric synthesis, flash chemistry plays a crucial role. Even if the starting material (or the intermediate itself) is highly enantio-enriched, unstable chiral-intermediates often undergo rapid racemization that the batch reactors cannot control, resulting in enantio-poor products. However, flash chemistry allows optically active intermediates to react before they epimerize. The following carbolithiation of conjugated enynes (Fig. 7) is a typical example [20]. In the presence of (−)-sparteine, an optically active amine, *n*-butyllithium can react with a conjugated enyne bearing a directing group (*N*,*N*-diisopropylcarbamoyloxy, OCb) to generate the enantio-enriched intermediate, which easily becomes a racemic mixture. The flow system allows the intermediate to react with the electrophiles by controlling the residence time in R2 (Fig. 7).

3 Reaction Integration

3.1 Linear Integration

When simply conducted in a flask, the halogen–lithium exchange reaction of *o*-dibromobenzene to generate *o*-bromophenyllithium [21, 22] is usually performed at −110°C. This is because the elimination of LiBr to form benzyne is very fast, even at −78°C. In contrast, flash chemistry in a microreactor allows the reaction to be performed at −78°C with a residence time of approximately 0.82 s. This is possible because *o*-bromophenyllithium is effectively trapped with an electrophile [23]. This even enables the sequential introduction of two lithium substituents onto the benzene ring by the linear integration of two consecutive halogen–lithium exchange reactions of dibromobenzene. This can be performed in a microreactor system comprising four micromixers (M1, M2, M3, and M4) and four microtube reactors (R1, R2, R3, and R4) (Fig. 8). In M1 and R1, a bromine–lithium exchange of the dibromobenzene is conducted to generate the corresponding bromophenyllithium, which is immediately trapped with an electrophile in M2 and R2. The second bromine–lithium exchange occurs next in M3 and R3, generating another substituted aryllithium that undergoes subsequent electrophilic substitution in M4 and R4 to give the corresponding disubstituted benzene. The linear integration method can also be used to synthesize disubstituted arenes from *m*- or *p*-dibromobenzenes [24], dibromopyridines [25], and dibromo biaryls [26].

TAC-101, 4-[3,5-bis(trimethylsilyl)benzamido]benzoic acid, is a synthetic retinoid that manifests differentiation-inducing activity on the human promyelocytic leukemia cell line HL-60 [27]. TAC-101 and its analogs can be efficiently synthesized via the versatile linear integration of three consecutive halogen–lithium exchange reactions of tribromobenzene. Three consecutive Br–Li exchange reactions of 1,3,5-tribromobenzene, each generating a different aryllithium that can be reacted with an electrophile, have been space-integrated. The integrated flow

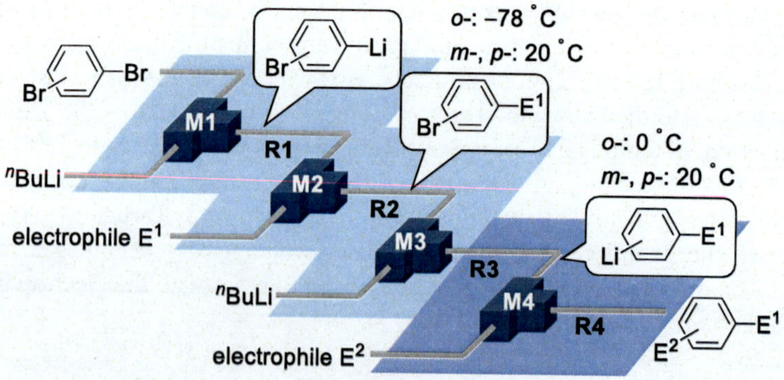

Fig. 8 Linear integration of two consecutive halogen–lithium exchange reactions of dibromobenzene in an integrated flow microreactor

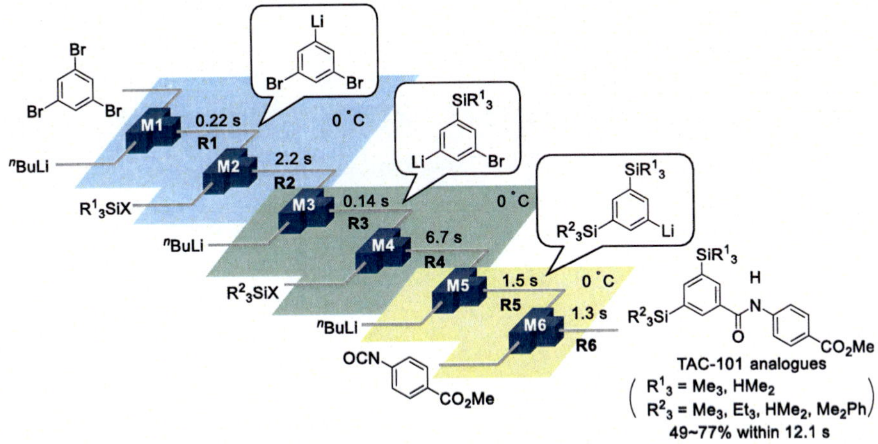

Fig. 9 A straightforward synthesis of TAC-101 and its analogues by the linear integration of three, consecutive halogen–lithium exchange reactions of 1,3,5-tribromobenzene in an integrated flow microreactor

microreactor is composed of six micromixers (M1, M2, M3, M4, M5, and M6) and six microtube reactors (R1, R2, R3, R4, R5, and R6), as shown in Fig. 9. By optimizing and adjusting the residence time in each reactor, the sequential introduction of two identical or different trimethylsilyl groups, followed by the introduction of an amide functionality, can be accomplished at 0°C in a single-flow process to yield TAC-101 methyl esters in good yield. Various TAC-101 methyl ester analogs with two different silyl groups have been synthesized from 1,3,5-tribromobenzene in a single-flow process. Notably, the total residence time was equal to the whole reaction time (12.2 s) and the productivity ranged from 132 mg to 194 mg per minute depending on the nature of the silyl groups. Hydrolysis of the methyl esters with NaOH in ethanol gave the corresponding carboxylic acids.

Deprotonation together with halogen–lithium exchange reactions has also become a useful method for generating short-lived organolithiums. Linear integration is an effective method for accomplishing this and can be used for reactions involving several short-lived organolithiums. The linear integration of two consecutive deprotonations of fluoroiodomethane, which is readily available, also demonstrates the precise control of the temperature and residence time that can be obtained in an integrated flow microreactor system consisting of four micromixers (M1, M2, M3, and M4) and four microtube reactors (R1, R2, R3, and R4), as depicted in Fig. 10 [28]. Fluoroiodomethyllithium is first generated by a deprotonation of fluoroiodomethane with lithium diisopropylamide (LDA) in M1 and R1. It is subsequently trapped with chlorotrialkylstannanes in M2 and R2, and the resulting fluoroiodomethylstannane undergoes a second deprotonation with LDA in M3 and R3. This generates an unusual and very unstable organolithium bearing two metal atoms (Li and Sn) and two halogen atoms (F and I) on the same carbon. The generated organolithium requires a very short residence time of 8.1 ms in R3.

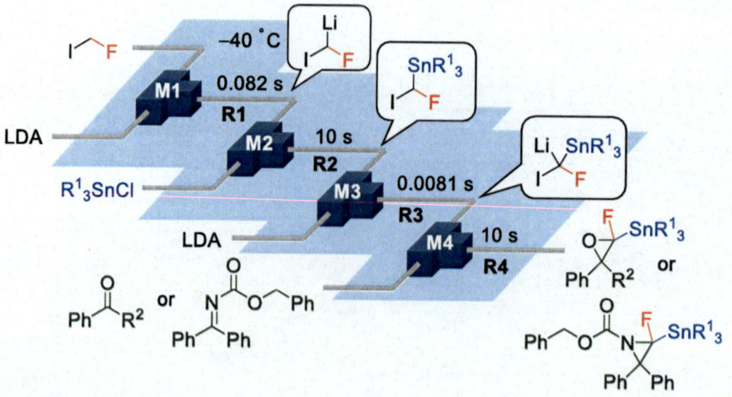

Fig. 10 Flash generation of iodofluoromethyl lithium and its space-integrated reactions with ketones and imines

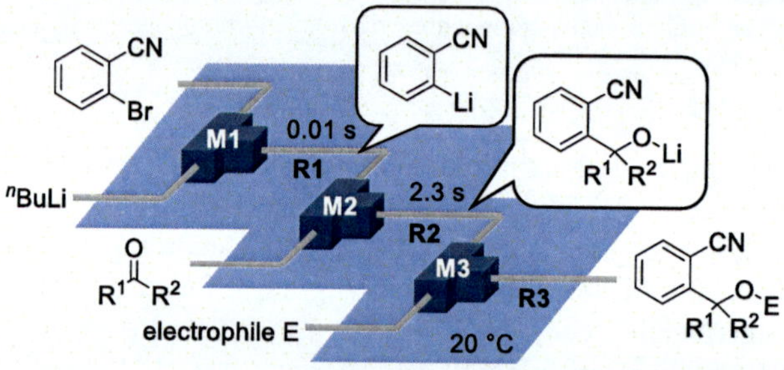

Fig. 11 Linear integration of a halogen–lithium exchange reaction, the reaction of a carbonyl compound with *o*-lithiobenzonitrile, and electrophilic addition to give alkoxylithiums

Subsequent effective-trapping with electrophiles (in M4 and R4), such as ketones and imines, provides tetrasubstituted epoxides and tetrasubstituted aziridines in good yields.

As mentioned in Sect. 2.1, various electrophilic functional groups are compatible with organolithium reactions if flash chemistry is used. Therefore, linear integration via the generation of electrophilic-functionalized aryllithiums in flow microreactor systems can facilitate further chemical transformations. For example, an integrated flow microreactor comprising three micromixers and three microtube reactors was used to generate alkoxylithiums by reacting carbonyl compounds with *o*-lithiobenzonitrile. The alkoxylithiums underwent subsequent direct and sequential trapping reactions upon electrophilic addition to produce three-component coupled products (Fig. 11) [10].

The generation of electrophilic-functionalized organolithiums can also be linearly integrated with transition-metal catalyzed coupling reactions. For example, homogeneous Pd-catalyzed Murahashi coupling [29] of an aryllithium with an aryl bromide can be achieved using integrated flow microreactor systems composed of

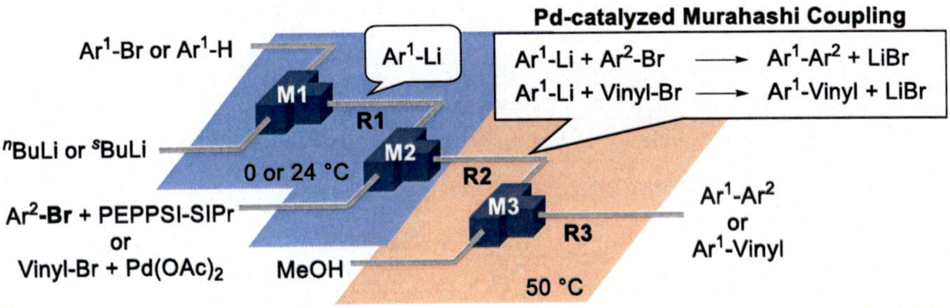

Fig. 12 Linear integration of a lithiation reaction and a homogeneous, Pd-catalyzed Murahashi coupling

three micromixers and three microtube reactors [30]. Aryllithiums can be generated by halogen–lithium exchange or deprotonation. Then, a highly reactive, homogeneous Pd catalyst (PEPPSI™-SIPr for aryl bromides or $Pd(OAc)_2$ for vinyl bromides) can be used for faster cross-coupling before the organolithiums decompose, which gives various cross-coupling products in good to high yields (Fig. 12). The integrated lithiation and Murahashi cross-coupling sequence can be completed within a total residence time of 1 min.

Aryllithiums with electrophilic functionalities can also be converted to boronic esters, which can be engaged in Suzuki-Miyaura cross-coupling reactions. Integrated lithiation and borylation-coupling processes using homogeneous palladium catalysts produce a variety of biphenyls [31], but Suzuki-Miyaura cross-coupling reactions using heterogeneous Pd catalysts are more environmentally friendly. Thus, Suzuki-Miyaura coupling reactions using supported Pd catalysts have been developed for the flow process. Pd(0) catalysts supported on a polymer monolith with pores of two different diameters were prepared and packed in a high-performance liquid chromatography (HPLC) column [32]. The catalyst was incorporated into an integrated flow microreactor to sequentially perform lithiation, borylation, and heterogeneous Suzuki-Miyaura coupling reactions in one flow.

The productivity of the coupling reaction was effectively improved by connecting multiple Pd monolith catalysts with larger pore sizes in series [33]. The flow microreactor system using a Pd monolith packed in an HPLC column (Fig. 13) was used for the gram-scale synthesis of adapalene, which has been used for psoriasis, photoaging, and acne treatments [34]. The integrated flow microreactor was assembled with two micromixers, two microtube reactors, and five Pd monolith reactors connected in series. First, the aryl bromide underwent lithiation, and the resulting aryllithium underwent borylation with trimethoxyborane at 0°C to give the corresponding lithium arylborate. The lithium arylborate solution was then mixed with another aryl halide and passed through a Pd monolith reactor at 120°C using a plunger pump. No additional base was required to facilitate the cross-coupling reaction by the direct use of in situ generated lithium boronate. After 4 h, 1.49 g of the desired methyl ester was obtained, making the overall productivity 370 mg/h. Subsequent hydrolysis with NaOH gave adapalene in 89% yield. Recently, an

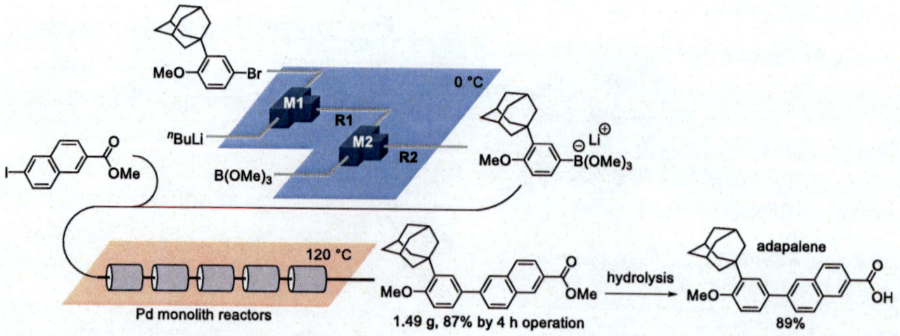

Fig. 13 Linear integration of lithiation, borylation, and heterogeneous Pd-catalyzed Suzuki-Miyaura coupling for adapalene synthesis

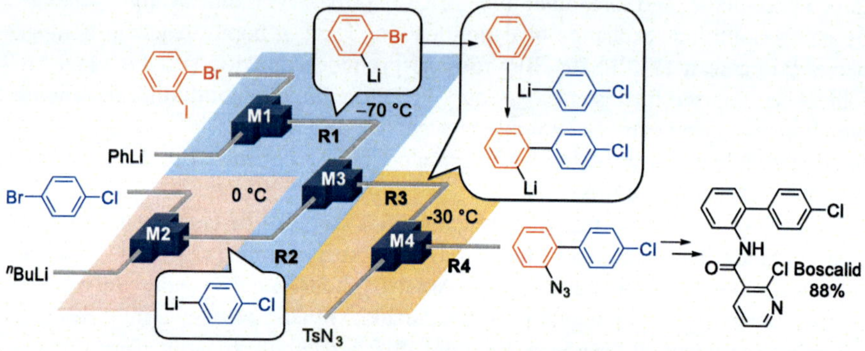

Fig. 14 Convergent integration via several organolithiums for the three-component coupling of benzyne, functionalized aryllithiums, and an electrophile for Boscalid synthesis

adapalene production of 5.1 g/h was achieved using a hierarchical, bimodal porous silica gel supported palladium column reactor having a low pressure drop [35].

3.2 *Convergent Integration*

Carbolithiation of a benzyne with functionalized aryllithiums followed by reactions with various electrophiles is an example of convergent-integrated organolithium reactions involving several lithium intermediates that proceed in a flow-integrated microreactor system (Fig. 14) [36]. Two different organollithiums, such as 2-bromophenyllithium in M1 and R1 and a functional aryllithium in M2 and R2, can be separately generated from 1-bromo-2-iodobenzene and the corresponding aryl halide, respectively, and then integrated at −70°C in M3 and R3. In R3, 2-bromophenyllithium decomposes at −30°C to generate a benzyne without affecting the functional aryllithium. This process can be followed by the spontaneous carbolithiation of the benzyne with the aryllithium. The resulting functional biaryllithium can then be reacted with an electrophile in reactor R4 to yield the corresponding three-component coupling product. The precise optimization of the reaction conditions using temperature–residence time mapping is responsible for the success of this three-component coupling. Furthermore, this method has been successfully applied to the synthesis of Boscalid (Fig. 14) [37], an important fungicide belonging to a class of succinate dehydrogenase inhibitors.

3.3 *Three-Component Coupling Based on Convergent and Linear Integration*

Convergent and linear integration using the flow microreactor method can also be applied to chemoselective three-component couplings. A central issue in organic synthesis is chemoselectivity, which refers to the preferential reaction of a chemical reagent or reactive species with one of two or more different functional groups. In general, chemoselective nucleophilic reactions of difunctional electrophiles are simple if the reactivity of one functional group is higher than that of the other. However, this is not true if the reaction is very fast. In fact, the reaction of 4-benzoylbenzaldehyde with an equivalent amount of phenyllithium in a batch reactor leads to the formation of a mixture of three products (Fig. 15a). These are the alcohol derived from the reaction of the aldehyde (aldehyde-adduct), the alcohol from the ketone (ketone-adduct), and the diol (double adduct), despite aldehydes generally being more reactive than ketones. Conversely, remarkable chemoselectivity was achieved using fast micromixing, affording the desired compound with high selectivity at high flow rates (Fig. 15b) [38].

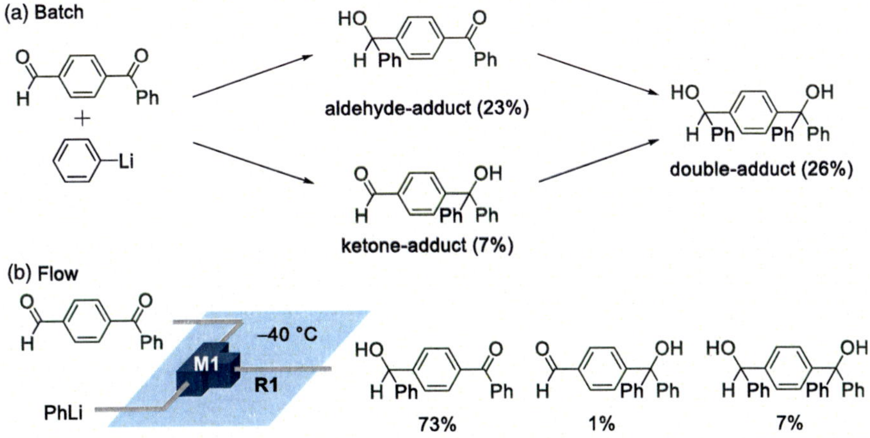

Fig. 15 Chemoselectivity for the reaction of 4-benzoylbenzaldehyde with phenyllithium in (**a**) batch and (**b**) flow reactions

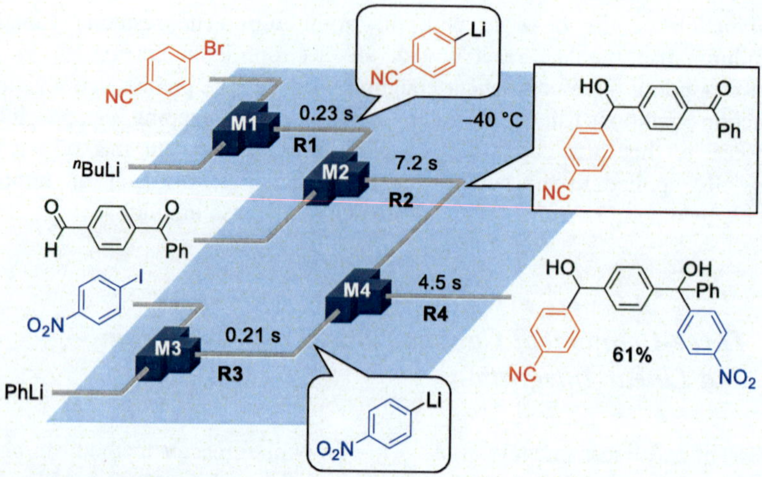

Fig. 16 Chemoselective three-component coupling reaction of difunctional electrophiles with functionalized aryllithiums in a linear and convergent-integrated flow microreactor

With this high chemoselectivity, subsequent reaction of this product with 4-cyanophenyllithium at the aldehyde carbonyl group, followed by reaction with 4-nitrophenyllithium at the ketone carbonyl group, successfully afforded the chemoselective three-component coupling product in 61% yield (Fig. 16) [38].

Reaction integration allows not only the integration of the bond-forming reactions, but also the catalyst-generating reactions. When palladium(II) acetate mixed with tri-*tert*-butylphosphine, a highly reactive palladium species, which can be converted to its less-reactive state within sub-seconds, is generated [39]. To take full advantage of such a reactive catalyst, the catalyst solution should be mixed with coupling reagents soon after generation. The flow system depicted in Fig. 17 integrates the lithiation-borylation sequence with the Suzuki-Miyaura cross-coupling reaction, as well as the catalyst generation process [40]. The aryl bromide bearing an electrophilic functionality is lithiated and borylated to provide the corresponding aryl borate. After adding water, the borate was mixed with another aryl bromide and the reactive catalyst, the latter of which was generated in a residence time of 1 s. As a result, this integrated sequence of lithiation, borylation, catalyst generation, and coupling affords biphenyls bearing two different electrophilic functionalities in good to high yields.

With this combination of convergent and linear integration, a precursor of the histone deacetylase (HDAC) inhibitor was synthesized in quantitative yield (Fig. 18) [41].

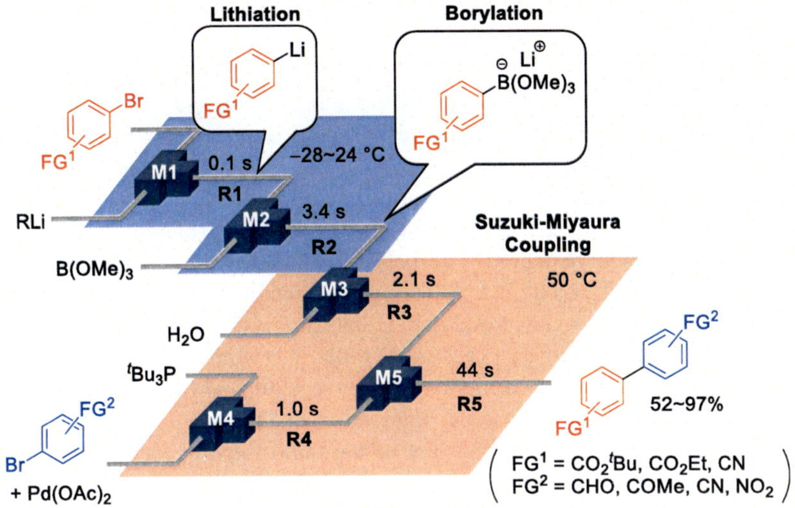

Fig. 17 Three-component (aryl borate, aryl bromide, and reactive Pd catalyst) coupling reaction synthesizing biphenyls bearing two different electrophilic functionalities

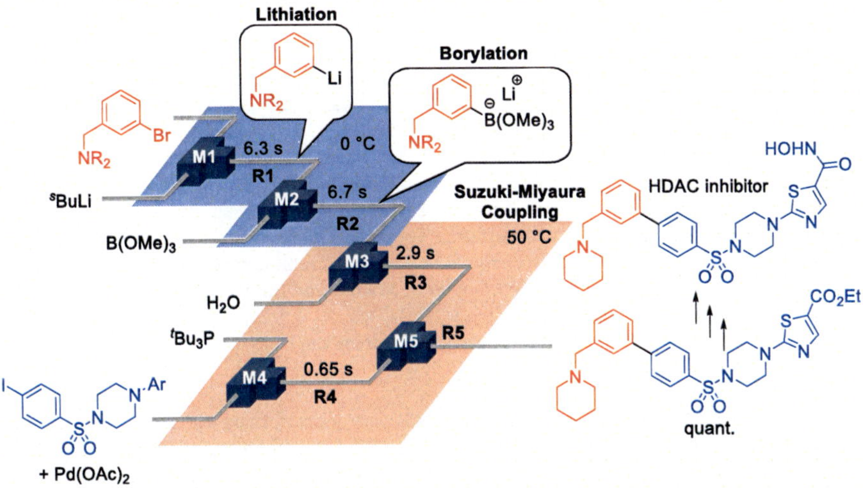

Fig. 18 Three-component coupling sequence for synthesizing a histone deacetylase (HDAC) inhibitor precursor (NR_2 = 1-piperidinyl)

4 Conclusion

Due to their high reactivity, organolithium reagents are considered unsuitable for organic syntheses, especially those fabricating pharmaceuticals because they require accurate reactions. To avoid side reactions, chemists conducting drug synthesis prefer mild and controllable reagents. The aforementioned examples demonstrate that organolithium reagents can be used for synthesizing compounds with complex structures because their high reactivity enhances the reaction efficiency.

Furthermore, these examples and many others show that flash chemistry and space integration serve as powerful tools for efficient organic syntheses. Flash chemistry enables reactions mediated by unstable, isomerizable, and racemizable intermediates to successfully proceed before the intermediates decompose. Reaction integration enables serial and combinatorial syntheses, where a variety of compounds are generated in a sequential manner using a single-flow reactor. These integration methods serve not only as efficient methods for synthesizing compounds with biological activities, but also as an environmentally benign way of producing pharmaceuticals. As the U.S. Food and Drug Administration (FDA) announced, reaction integration in a flow process could provide several advantages to pharmaceutical manufacturing, such as highly productive, safe, and reproducible reactions [42]. Moreover, the use of reaction integration enables on-demand and on-site chemical syntheses, which leads to less energy consumption for product storage and transportation.

Flash chemistry and reaction integration will be indispensable for pharmaceutical research and production. Various methods in these fields will be developed to help meet the ever increasing expectations of drug discovery and production.

Compliance with Ethical Standards

Funding: This work was partially supported by JSPS KAKENHI Grant Numbers JP15H05849 (Grant-in-Aid for Scientific Research on Innovative Areas 2707 Middle molecular strategy), JP26288049 (Grant-in-Aid for Scientific Research (B)), JP26220804 (Grant-in-Aid for Scientific Research (S)), JP25220913 (Grant-in-Aid for Scientific Research (S)), JP17865428 (Grant-in-Aid for Scientific Research (C)), and JP20K15276 (Grant-in-Aid for Early-Career Scientists). This work was also partially supported by Japan Agency for Medical Research and Development (AMED, JP19ak0101090), the New Energy and Industrial Technology Development Organization (NEDO, JPNP19004), the Core Research for Evolutional Science and Technology (CREST, JPMJCR18R1) and the Ogasawara Foundation for the Promotion of Science and Engineering.

Ethical Approval: All procedures in this manuscript were not performed with human participants, nor any other animals.

Informed Consent: All procedures in this manuscript were not performed with human participants.

References

1. Kobayashi S (2016) Chem Asian J 11:425–436
2. Yoshida J (2015) Basics of flow microreactor synthesis. Springer, Tokyo
3. Knochel P (2005) Handbook of functionalized organometallics: applications in synthesis. Wiley-VCH, Weinheim
4. Stanetty P, Mihovilovic MD (1997) J Org Chem 62:1514–1515
5. Kim H, Nagaki A, Yoshida J (2011) Nat Commun 2:264
6. Ito T, Tanaka T, Iinuma M, Nakaya K-i, Takahashi Y, Sawa R, Murata J, Darnaedi D (2004) J Nat Prod 67:932–937
7. Nagaki A, Kim H, Yoshida J (2008) Angew Chem Int Ed 47:7833–7836

8. Nagaki A, Kim H, Moriwaki Y, Matsuo C, Yoshida J (2010) Chem Eur J 16:11167–11177
9. Nagaki A, Kim H, Yoshida J (2009) Angew Chem Int Ed 48:8063–8065
10. Nagaki A, Kim H, Usutani H, Matsuo C, Yoshida J (2010) Org Biomol Chem 8:1212–1217
11. Nagaki A, Yamashita H, Hirose K, Tsuchihashi Y, Yoshida J (2019) Angew Chem Int Ed 58:4027–4030
12. Nagaki A, Yamashita H, Tsuchihashi Y, Hirose K, Takumi M, Yoshida J (2019) Chem Eur J 25:13719–13727
13. Eisch JJ, Galle JE (1990) J Org Chem 55:4835–4840
14. Capriati V, Florio S, Luisi R, Salomone A (2002) Org Lett 4:2445–2448
15. Nagaki A, Takizawa E, Yoshida J (2009) J Am Chem Soc 131:1654–1655
16. Nagaki A, Takizawa E, Yoshida J (2010) Chem Eur J 16:14149–14158
17. Nagaki A, Takizawa E, Yoshida J (2009) Chem Lett 38:1060–1061
18. Takizawa E, Nagaki A, Yoshida J (2012) Tetrahedron Lett 53:1397–1400
19. Giovine A, Musio B, Degennaro L, Falcicchio A, Nagaki A, Yoshida J, Luisi R (2013) Chem Eur J 19:1872–1876
20. Tomida Y, Nagaki A, Yoshida J (2011) J Am Chem Soc 133:3744–3747
21. Chen LS, Chen GJ, Tamborski C (1980) J Organomet Chem 193:283–292
22. Bettinger HF, Filthaus M (2007) J Org Chem 72:9750–9752
23. Usutani H, Tomida Y, Nagaki A, Okamoto H, Nokami T, Yoshida J (2007) J Am Chem Soc 129:3046–3047
24. Nagaki A, Tomida Y, Usutani H, Kim H, Takabayashi N, Nokami T, Okamoto H, Yoshida J (2007) Chem Asian J 2:1513–1523
25. Nagaki A, Yamada S, Doi M, Tomida Y, Takabayashi N, Yoshida J (2011) Green Chem 13:1110–1113
26. Nagaki A, Takabayashi N, Tomida Y, Yoshida J (2008) Org Lett 10:3937–3940
27. Nagaki A, Imai K, Kim H, Yoshida J (2011) RSC Adv 1:758–760
28. Colella M, Tota A, Takahashi Y, Higuma R, Ishikawa S, Degennaro L, Luisi R, Nagaki A (2020) Angew Chem Int Ed 59:10924–10928
29. Murahashi S, Yamamura M, Yanagisawa K, Mita N, Kondo K (1979) J Org Chem 44:2408–2417
30. Nagaki A, Kenmoku A, Moriwaki Y, Hayashi A, Yoshida J (2010) Angew Chem Int Ed 49:7543–7547
31. Shu W, Pellegatti L, Oberli MA, Buchwald SL (2011) Angew Chem Int Ed 50:10665–10669
32. Nagaki A, Hirose K, Moriwaki Y, Mitamura K, Matsukawa K, Ishizuka N, Yoshida J (2016) Catal Sci Tech 6:4690–4694
33. Nagaki A, Hirose K, Moriwaki Y, Takumi M, Takahashi Y, Mitamura K, Matsukawa K, Ishizuka N, Yoshida J (2019) Catalysts 9:300
34. Mukherjee S, Date A, Patravale V, Korting HC, Roeder A, Weindl G (2006) Clin Interv Aging 1:327–348
35. Ashikari Y, Maekawa K, Takumi M, Tomiyasu N, Fujita C, Matsuyama K, Miyamoto R, Bai H, Nagaki A (2020) Catal Today. https://doi.org/10.1016/j.cattod.2020.07.014
36. Nagaki A, Ichinari D, Yoshida J (2014) J Am Chem Soc 136:12245–12248
37. Krämer W, Schirmer U (2007) Modern crop protection compounds. Wiley, Weinheim
38. Nagaki A, Imai K, Ishiuchi S, Yoshida J (2015) Angew Chem Int Ed 54:1914–1918
39. Nagaki A, Takabayashi N, Moriwaki Y, Yoshida J (2012) Chem Eur J 18:11871–11875
40. Nagaki A, Moriwaki Y, Yoshida J (2012) Chem Commun 48:11211–11213
41. Takahashi Y, Ashikari Y, Takumi M, Shimizu Y, Jiang Y, Higuma R, Ishikawa S, Sakaue H, Shite I, Maekawa K, Aizawa Y, Yamashita H, Yonekura Y, Colella M, Luisi R, Takegawa T, Fujita C, Nagaki A (2020) Eur J Org Chem:618–622
42. Lee SL, O'Connor TF, Yang XC, Cruz CN, Chatterjee S, Madurawe RD, Moore CMV, Yu LX, Woodcock J (2015) J Pharm Innov 10:191–199

Top Med Chem (2021) 38: 241–274
https://doi.org/10.1007/7355_2021_119

Published online: 7 April 2021

Organocatalysis in Continuous Flow for Drug Discovery

Laura Amenós, Esther Alza, and Miquel A. Pericàs

Contents

Abstract Although organometallic chemistry has been the predominant category for the preferred catalytic transformations applied in pharmaceutical industry since the nineteenth century, during the last 20 years, organocatalytic processes are becoming more common and of wide use in drug discovery field. In particular, enantioselective organocatalysis is emerging as one of the main tools and a greener protocol for the synthesis of chiral molecules. As for today, the main applications of chiral enantiopure compounds are in the pharma and fine chemicals industries, in crop protection, and in the biotechnology area.

L. Amenós, E. Alza (✉), and M. A. Pericàs (✉)
Institute of Chemical Research of Catalonia, ICIQ, Tarragona, Spain
e-mail: lamenos@iciq.es; ealza@iciq.es; mapericas@iciq.es

Organocatalytic flow processes are of practical application in drug discovery and in the manufacturing of Active Pharmaceutical Ingredients (APIs). In this chapter, we will describe the most recent examples of continuous flow organocatalytic processes reported in the literature for the synthesis of molecules with pharmaceutical interest.

Keywords Achiral catalysis, Chiral catalysis, Drug synthesis, Flow chemistry, Organocatalysis

1 Introduction

Continuous flow processes have recently emerged as a powerful technology for performing chemical transformations since they ensure some important advantages over traditional batch procedures.

Although most industries have employed continuous manufacturing processes for decades, the pharmaceutical industry has remained committed to batch operations, despite the advantages in cost, time-length of the process, and quality characteristics of final products obtained by continuous flow methodologies. However, this trend is changing due to fast growing use of more personalized medicines and therapies involving complex compounds that renders continuous processes competitive with respect to traditional batch synthesis.

Although organometallic chemistry has been the predominant category for the preferred catalytic transformations applied in pharmaceutical industry since the nineteenth century, during the last 20 years, organocatalytic processes [1–8] are becoming more common and of wide use in drug discovery field. In particular, enantioselective organocatalysis is emerging as one of the main tools and a greener protocol for the synthesis of chiral molecules. As for today, the main applications of chiral enantiopure compounds are in the pharma and fine chemicals industries, in crop protection, and in the biotechnology area [9].

By definition, *organocatalysis* is the acceleration of chemical reactions with a substoichiometric amount of an organic compound in absence of a metal element [10].

The fundamental advantages of organocatalysis are mainly related with the easier experimental procedures required with mild reaction conditions. That involves low levels of chemical waste resulting in savings in time and energy and also in the overall cost of the process because usually organocatalysts are relatively inexpensive, compared with organometallic species.

Usually, organocatalysts are more robust and not prone to decomposition during reaction, in contrast with traditional metal catalysts. Moreover, many organocatalysts have been modified to allow their immobilization onto different supports (from polymeric resins to nanoparticles or ionic liquids), for their use as heterogeneous catalysis in flow synthesis. The immobilization of organocatalysts is long-established as an approach that allows the recovery and reuse of these usually expensive molecules. The heterogeneous organocatalyst is used in continuous flow usually by placing it into a packed bed reactor as a swellable microporous resin, as a monolithic macroporous column, as a covalent attachment to the reactor internal walls of the reactor, or even as a 3D printed material [11–13].

Organocatalysis represents an effective methodology to perform metal-free catalytic processes avoiding the contamination of final products with potentially toxic metal specie. In fact, in the case of stereoselective reactions, the use of organocatalysts allows the achievement of high levels of enantioselectivity *in continuo* [14–17]. However, one of the main limitations of these processes is the propensity to deactivation of organocatalysts, a fact that reduces their efficiency for some transformations and hampers their large-scale application [18–20]. The identification of the mechanisms that cause deactivation is an important challenge to address in order to implement corrective measures, thus allowing a wider use of organocatalysis in pharmaceuticals and fine chemicals industries.

Many research groups have focused their attention in the organocatalysis field and all these efforts have demonstrated that organocatalytic flow processes are of practical application in the manufacturing of Active Pharmaceutical Ingredients (APIs). In this chapter, we will describe the most recent examples of continuous flow organocatalytic processes reported in the literature for the synthesis of molecules with pharmaceutical interest.

The application of organocatalysis in continuous flow has been approached in two ways (Fig. 1):

(a) mixing a soluble homogeneous catalyst in the solution being pumped and
(b) using organocatalysts immobilized onto a solid support in packed bed reactors.

Both strategies have a lot of potential, but the use of an heterogeneous catalyst has additional benefits: the product is not contaminated with catalyst and, if this proves robust enough, the effective catalyst loading becomes a function of operation time, leading to extremely high turnover numbers (TONs) for the overall process even when small-size devices are operated for long periods of time, whereas at any given time the substrates are exposed to (super)stoichiometric amounts of catalyst inside the reactor and this allows high conversions in short contact times.

In this chapter, we have decided to divide the described organocatalytic processes in flow by the use of achiral (Sect. 2) and chiral (Sect. 3) organocatalysts (also depending on their nature), in both cases including homogeneous and heterogeneous species.

2 Achiral Organocatalysts in Continuous Flow

2.1 Carbene Catalyzed Reactions

The group of Monbaliu developed, in 2016, a versatile methodology for the generation of *N*-heterocyclic carbenes (NHCs) in continuous flow [21]. Once the synthesis

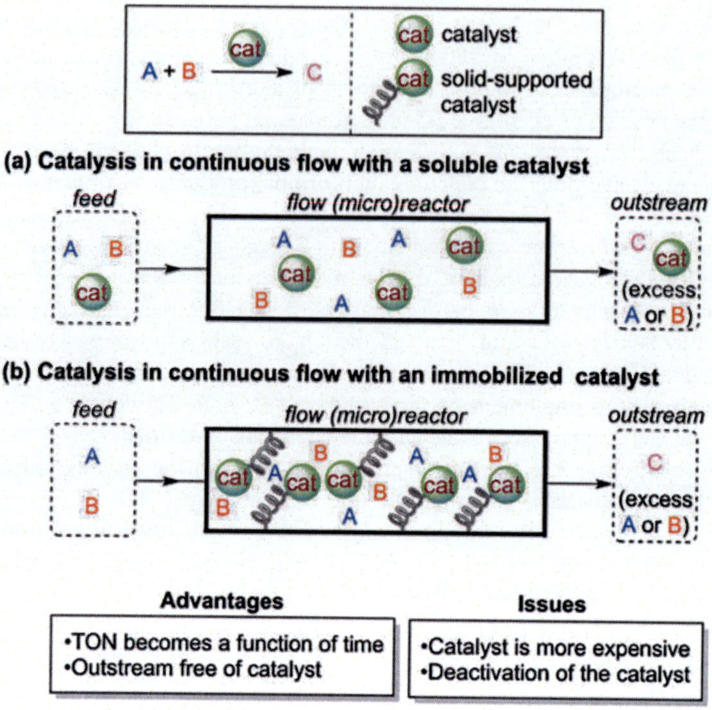

Fig. 1 Continuous flow catalysis with (**a**) Homogeneous or (**b**) Heterogeneous catalysts

of NHCs in flow was optimized, the sequence was telescoped with the organocatalytic transesterification of vinyl acetate and benzyl alcohol (Fig. 2). The microfluidic setup enabled the fast study of different reaction parameters and screening of various NHC precursors to determine the optimal conditions. In-line reaction monitoring, downstream quench, and liquid–liquid separation techniques were also implemented in the sequence.

A similar strategy was applied for the amidation reaction of *N*-Boc-glycine methyl ester with ethanolamine catalyzed by IMes. This example showed the direct telescoping of three synthetic steps, first the generation of the carbene, then the transesterification, and finally an intramolecular *O*-to-*N* acyl shift. Both organocatalytic transformations proceeded with total conversion and excellent yields were achieved, representing the first examples of organocatalysis with NHCs in continuous flow.

A continuous flow procedure for the umpolung of aromatic α-diketones was described by Massi in 2016 [22]. 2-*tert*-butylimino-2-diethylamino-1,3-dimethylperhydro-1,3,2-diazaphosphorine on polystyrene (PS-BEMP) was found to be the optimal base for the chemoselective synthesis of benzoin and Stetter-like products. The supported base was capable to initiate the two electron-transfer process from the carbamoyl anion of *N*,*N*-dimethylformamide to the α-diketone

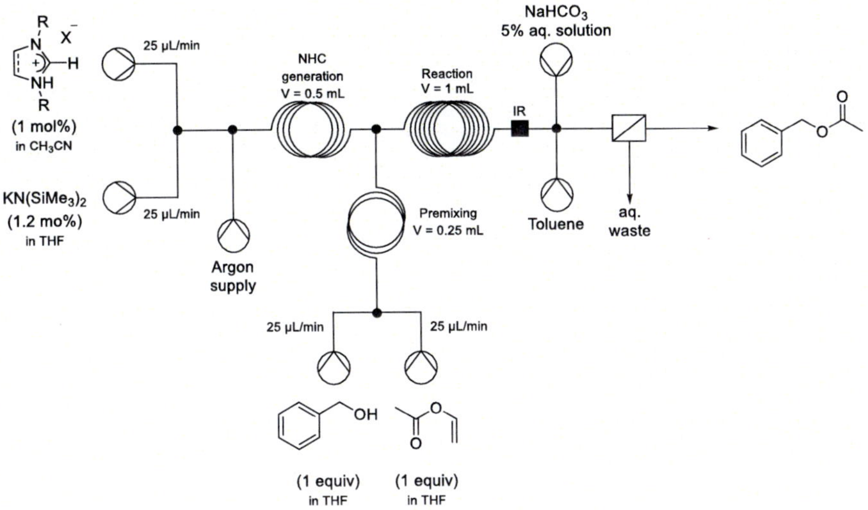

Fig. 2 Sequential NHCs synthesis and organocatalytic transesterification

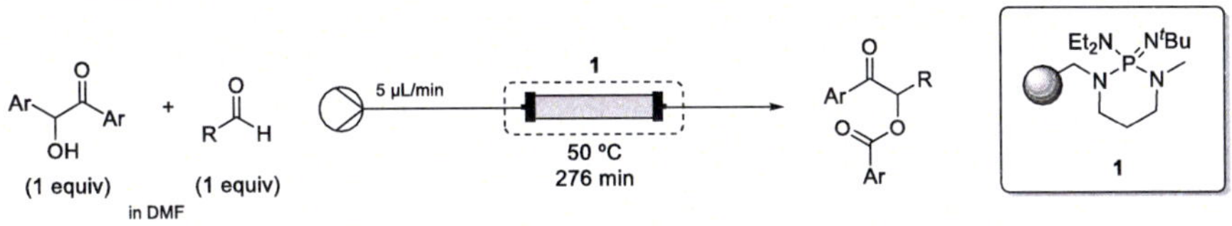

Fig. 3 Continuous flow Benzoin-like reaction

and generates the corresponding enediolate active species. The continuous flow setup consisted of a fixed-bed reactor (pressure-resistant stainless-steel column) packed with the polymer-supported base at 50°C with a residence time of 276 min (Fig. 3). Under these conditions, the benzoylated benzoin product was isolated in pure form by simple evaporation of the solvent. Together with the ease of product/catalyst separation, an important benefit of the continuous flow process was the long-term stability of the packing bed (ca. 5 five days on streams).

The application of continuous flow *N*-heterocyclic carbene (NHC)-catalysis into the valorization of renewable chemicals was reported also by the group of Massi in 2018 [23]. The synthesis of supported azolium salt precatalyst **2** was developed to produce monoesters of glycerol by oxidative NHC-catalysis under continuous flow conditions. The esterification of glycerol with naphthaldehyde (Fig. 4) was performed as a benchmark reaction. The setup consisted of a stainless-steel column packed with active polystyrene-supported triazolium salt precatalyst. Two feed solutions were pumped into the packed bed reactor allowing the efficient monoesterification of glycerol. Moreover, the same group also developed a novel catalytic procedure for the synthesis of 5-hydroxymethyl-2-furancarboxylic acid,

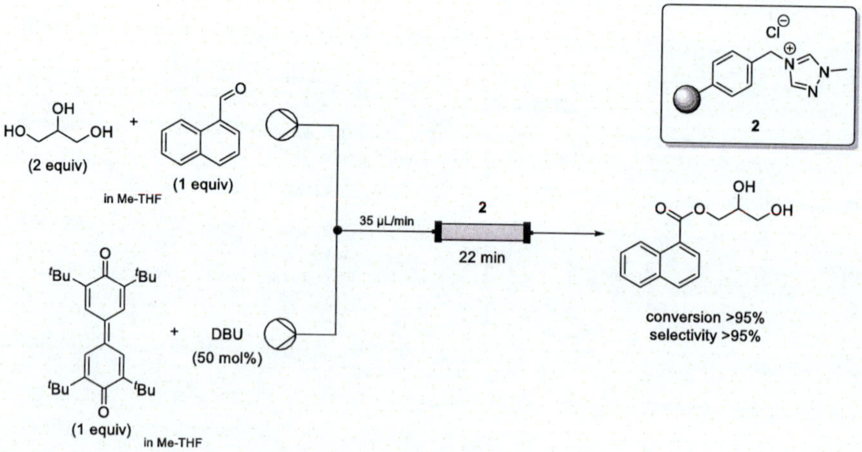

Fig. 4 Monoesterification of glycerol with naphthaldehyde

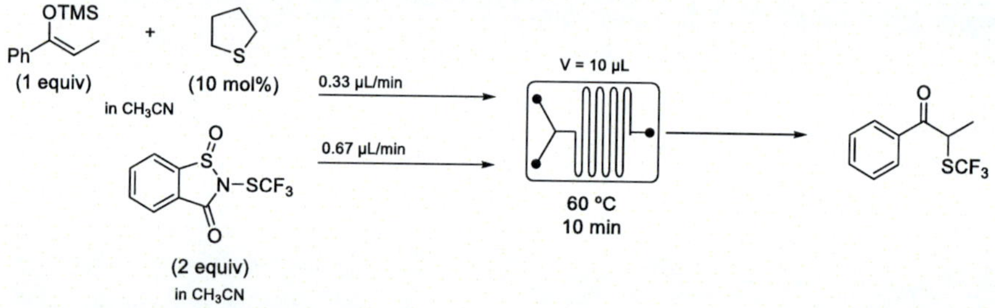

Fig. 5 Catalytic α-trifluoromethylthiolation

catalyzed by polystyrene-supported triazolylidene via oxidative esterification of 5-hydroxymethyl-2-furan under continuous flow conditions [24].

2.2 *Lewis Base Catalyzed Reactions*

The organocatalytic α-trifluoromethylthiolation of ketone-derived enoxysilanes promoted by Lewis bases was reported in continuous flow by Benaglia and Rossi in 2018 [25]. The use of *N*-(trifluoromethylthio)saccharin, activated by the presence of catalytic amounts of a Lewis base, as trifluoromethylthiolating reagent was successfully developed under continuous flow conditions (Fig. 5). The setup used consisted of a glass microreactor that was fed by two different solutions: the mixture of the silylenol ether combined with tetrahydrothiophene and the trifluoromethylthiolating agent. Shorter reaction time and higher productivity were observed when the reaction was operated in the flow system, offering clear advantages compared to the batch approach.

2.3 *Lewis Acid Catalyzed Reactions*

A novel method for the diastereoselective synthesis of *cis*-4-aminobenzodihydropyran derivatives using an efficient organocatalyst under mild conditions was reported in 2017 [26]. By using a low loading of tritylium cation $TrBF_4$ (1 mol%), pyranobenzodihydropyrans and furanobenzodihydropyrans were obtained via the interrupted Povarov reaction between salicylaldimines and alkenes. After optimization in batch, a flow system for the catalytic three-component condensation of salicylaldehydes, aniline, and 2,3-dihydrofuran was developed via Lewis acidic catalysis (Fig. 6). The flow setup involved a first microreactor for the non-catalytic imine formation, fed by a salicylaldehyde solution and an aniline solution. Then, the resulting salicylaldimine solution passed through a water absorption tube, and the water-free salicylaldimine was combined with 2,3-dihydrofuran solution containing tritylium tetrafluoroborate and pumped into the second microreactor. Gratifyingly, the pyranobenzodihydropyran was efficiently formed at high flow rate resulting in a residence time of 1 min. This flow process allowed a streamlined preparation of the pyranobenzodihydropyrans in a practical manner with high yield and short residence time. Interestingly, the process transfer from batch to flow resulted in comparable yields while avoided preparation and separation steps of preformed salicylaldimines, thus representing an efficient and scalable approach.

The same year, the group of Nguyen presented for the first time the use of tropylium salts as organic Lewis Acid catalysts to facilitate the acetalization and transacetalization of a wide range of aldehydes [27]. This metal-free method was studied in both batch and flow conditions, offering an interesting protocol to mask aldehydes in organic chemistry. The flow setup consisted of a heated tubular reactor coil fed by the reaction mixture solution (Fig. 7). The optimal conditions under continuous flow allowed an efficient multigram synthesis of acyclic acetals.

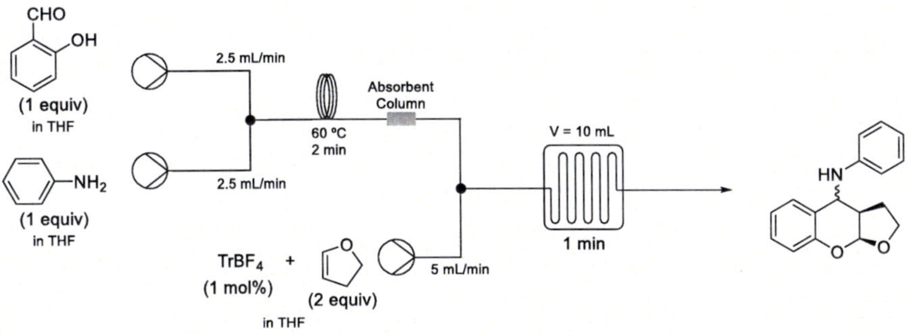

Fig. 6 Diastereoselective synthesis of *cis*-4-aminobenzodihydropyran

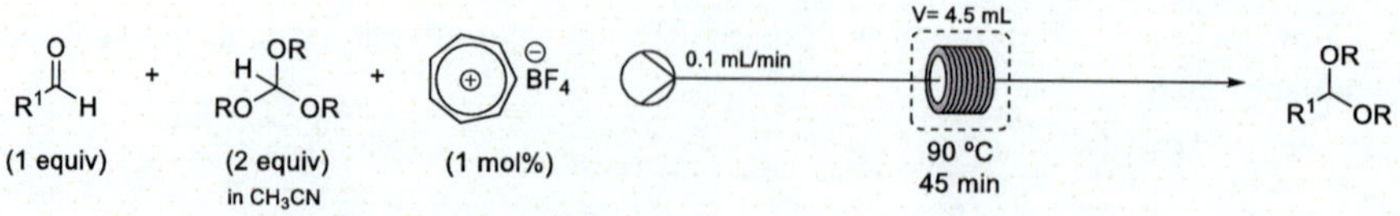

Fig. 7 Tropylium catalyzed acetalization under continuous flow conditions

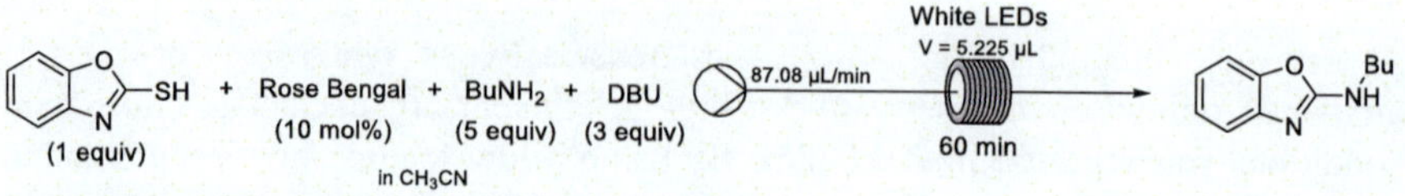

Fig. 8 Photocatalyzed amination of heterocyclic thiols

Moreover, the tropylium catalyst could be isolated and recycled, as it remained mostly unchanged after the reaction.

2.4 *Photocatalytic Reactions*

An efficient amination of heterocyclic thiols by photocatalytic reaction under continuous flow conditions was reported by the group of Wacharasindhu [28]. Rose Bengal was employed as photocatalyst and commercial LEDs as light source in the S_NAr of heterocyclic thiols (Fig. 8). Different functional groups were well tolerated allowing the synthesis of diverse amino-substituted heterocycles such as 2-aminobenzoxazole and 4-aminoquinazoline derivatives in moderate to excellent yields. The novel use of thiols as alternative substrates in an S_NAr-type reaction leading to C – N bond formation represented a mild, cost-effective, and metal-free alternative to conventional amination. The flow setup consisted of a transparent perfluoroalkoxy alkane (PFA) tube, which was surrounded by a visible-light source. A mixture of the starting material, DBU, *n*-$BuNH_2$, and Rose Bengal was simply flowed through the irradiated tube. The suitability of the mild conditions involved in flow with the amination reaction allowed the efficient formation of pharmaceutical intermediates and demonstrated the high potential of the flow approach for industrial scale up.

Melchiorre and coworkers developed a photochemical S_N2 process to generate open-shell intermediates from electrophilic substrates that would be incompatible with traditional radical-generating strategies [29]. The method required an organic catalyst **3** and occurred under visible-light irradiation. This strategy was applied to the synthesis of an intermediate of Tolmetin, a marketed nonsteroidal anti-inflammatory drug. The reaction was carried out under continuous flow conditions, in a photoreactor irradiated with Blue LEDs to efficiently afford the pyrrole derivative (Fig. 9). The broad substrate scope highlighted the potential of this transformation, which involved the use of a low-cost catalyst.

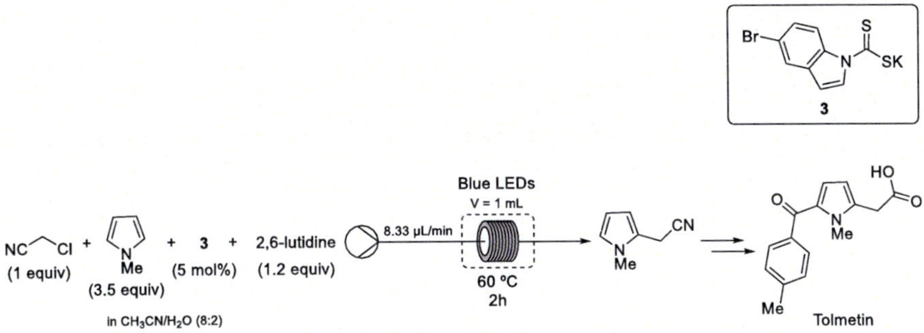

Fig. 9 Photocatalyzed S_N2 process under continuous flow conditions

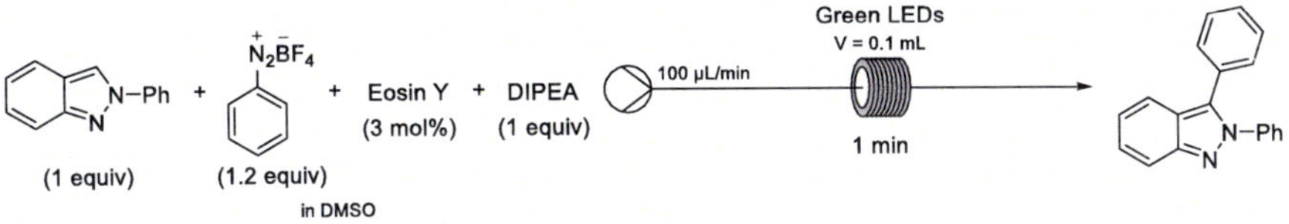

Fig. 10 Photocatalyzed C3 arylation of 2H-indazoles with aryldiazonium salts

The group of Kim reported in 2019 a metal-free, direct C3 arylation of 2*H*-indazoles with aryldiazonium salts at room temperature by continuous flow organo-photoredox catalysis involving the use of green light [30]. They applied the optimal conditions previously established in batch to a continuous flow microfluidic reactor (Fig. 10). The reactor was a highly transparent PFA coil irradiated with LED light. The efficient light irradiation achieved under these conditions and the efficient mixing under continuous flow conditions had a very positive impact on the reaction allowing a time reduction from 18 h in batch to 1 min residence time in the flow process. The visible light promoted an efficient cross coupling in the presence of Eosin Y as the photocatalyst. Besides, this approach represented an interesting alternative to the known transition-metal-catalyzed strategies for C – H arylation of 2*H*-indazoles.

2.5 *Phosphine Catalyzed Reactions*

A phosphine catalyzed "anti-Michael addition" on alkynes was developed in continuous flow heterogeneous conditions by Sharma and Van der Eycken [31]. This nucleophilic catalysis enabled the synthesis of a wide range of spiroindolines and spiroindolenines, that are interesting compounds due to their presence in various natural products and biologically relevant molecules. Prior to the continuous flow process, the catalytic performance was investigated in a batch mode. Then, a reusable catalytic-system was described using polymer-supported triphenylphosphine under continuous flow conditions. The system consisted of a

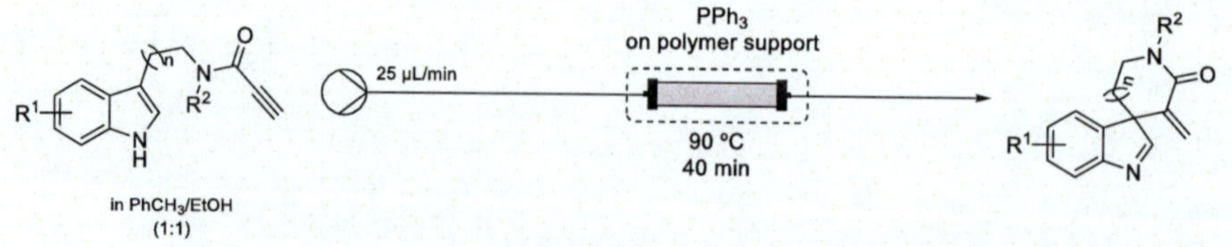

Fig. 11 Synthesis of spiroindolines and spiroindolenines in continuous flow

packed bed reactor, filled with the supported catalyst, fed by a solution of the corresponding alkyne in a mixture of Toluene/EtOH (1:1). The reaction worked very efficiently affording the desired products in up to 98% yield. Moreover, it was demonstrated the scalability of the process performing a gram-scale reaction that gave 79% isolated yield of 1′-benzyl-3′-methylenespiro[indole-3,4′-piperidin]-2′-one (Fig. 11).

3 Chiral Organocatalysis in Continuous Flow

Biologically active molecules are in most cases chiral. For this reason, enantioselective catalytic transformations are crucial in the synthesis of such molecules. In that sense, the use of chiral organocatalysts in flow chemistry promotes the effective and sustainable production of enantiomerically pure compounds providing a very powerful tool for pharmaceutical industry.

Although there were some drawbacks associated to high catalytic loading, low turnover numbers of impossibility of recovery and recycling the organocatalysts, nowadays there has been a huge advance in the field and these problems have been overcome by the immobilization of the catalysts. This strategy allows the production of chiral compounds in a rapid and purest way, making asymmetric organocatalysis a real powerful tool for drug discovery and drug synthesis, even in large scale.

3.1 Pyrrolidine-Derived Catalysts

In 2015, a novel polymer-supported fluorinated organocatalyst was developed by the group of Pericàs [32]. The catalytic resin **4** demonstrated great potential for the Michael addition of aldehydes to nitroalkenes in batch and flow (Fig. 12). This enantioselective transformation was selected as a benchmark to evaluate the performance and recyclability of this novel catalyst. The study constituted the first report of its application in the enamine activation mode. A highly efficient continuous flow process was designed, allowing either the multigram synthesis of a single Michael adduct or the sequential generation of a library of enantiopure Michael adducts from different combinations of substrates. The setup used consisted of resin **4** packed in a

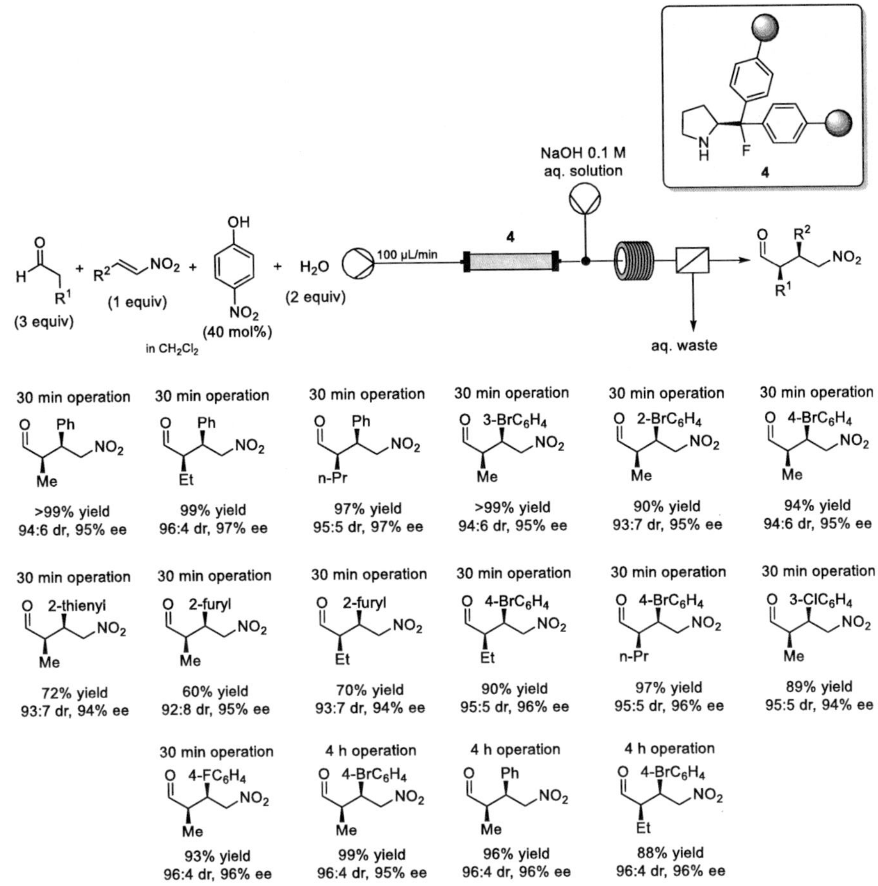

Fig. 12 Enantioselective Michael addition of aldehydes to nitroalkenes

glass column and fed by the reaction mixture. Additional in-line workup allowed to simplify the isolation of the adduct. Indeed, pure product was obtained evaporating the organic outstream, in most cases. Catalyst **4** was proved to be extremely active, displaying excellent selectivity with a wide variety of substrates. Key to the success was a polymer design that did not affect the active site.

A wide library of compounds was synthesized using this continuous flow setup. To demonstrate the robustness of the catalyst, this flow process was run with different reagent combinations. Each pair of starting materials was circulated through the system for 30 min, rinsing with solvent for 1 h afterward. In total, 16 consecutive runs were carried out, with high yields and stereoselectivities observed in all cases. Remarkably, the overall TON for this set of experiments was 72.

Supported diarylprolinols are very useful catalysts for different transformations. The use of a solid-supported organocatalyst of this type for the enantioselective

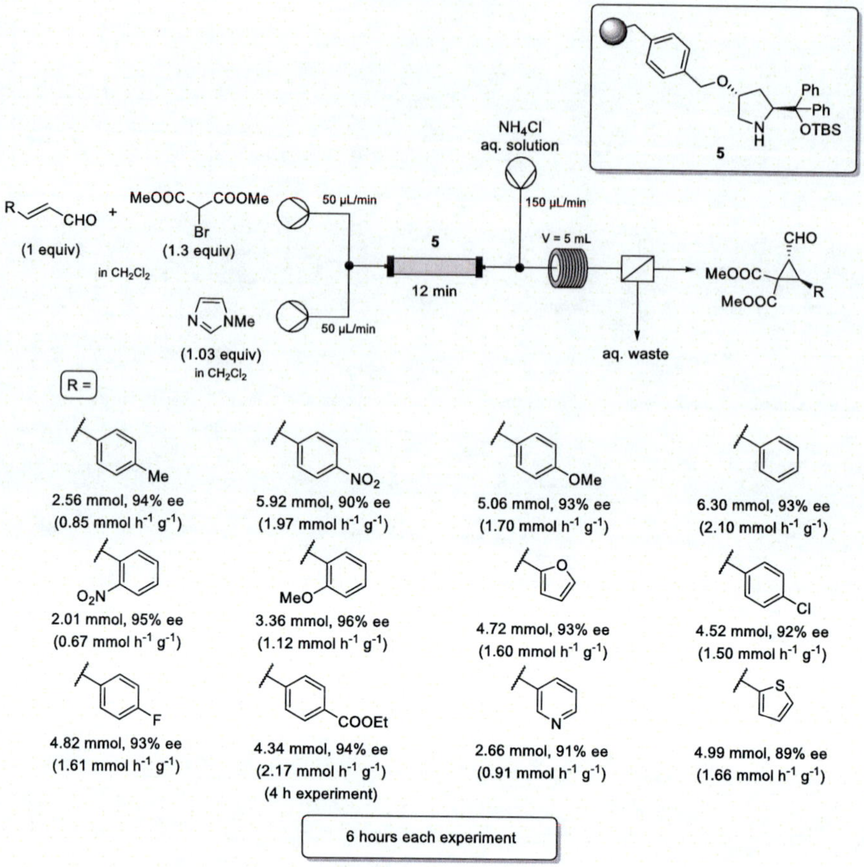

Fig. 13 Enantioselective cyclopropanation reaction

cyclopropanation of α,β-unsaturated aldehydes in flow was reported in 2016 by the group of Pericàs (Fig. 13) [33]. After testing a set of polymeric catalysts, the use of resin **5** in the flow asymmetric cyclopropanation of enals with bromomalonates was found to be the best option. This work included the evaluation of both the solid phase (microporous vs. macroporous polystyrene) and the anchoring strategy.

The flow setup consisted of two feeding streams that were combined right before a glass Omnifit® column packed with the catalytic resin **5**. *N*-methylimidazole was used as a base, and a downstream process was required to prevent a secondary reaction involving ring-opening of the cyclopropane. To this end, a third inlet with ammonium chloride and a liquid–liquid separator were placed after the packed bed reactor to remove the base from the outgoing flow. This process allowed to perform an in-line workup, with only evaporation and chromatographic purification remaining to obtain the final product.

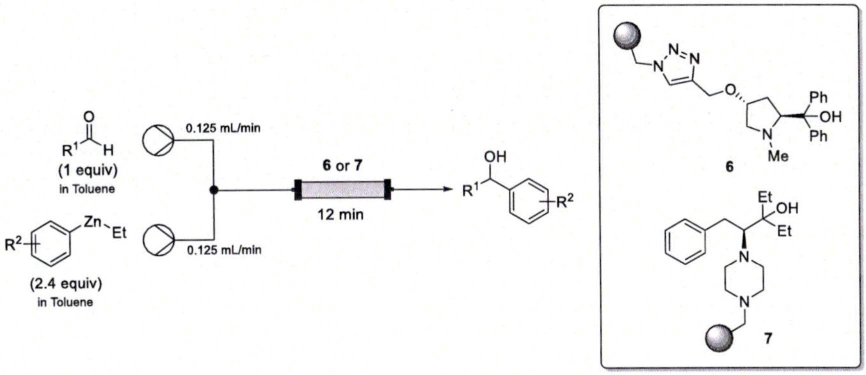

Fig. 14 Enantioselective arylation reaction

Both electron-rich and electron-poor cinnamaldehydes were well tolerated in this process, and even heteroaromatic enals produced the desired cyclopropanes with high enantiomeric purity. Each flow experiment was run for 6 h under the optimal conditions. After every example, the column was simply rinsed with CH_2Cl_2 to be used with the next combination of substrates. With this sequential approach, up to 12 different analogues were prepared with excellent *ee*'s and diastereoselectivities. The catalytic resin **5** proved to be extremely robust, with productivities up to 2.1 mmol h^{-1} $g_{resin}{}^{-1}$, which amounted to a total TON of 94.0 (up to 76 h running).

Pastre and coworkers developed, in 2017, different immobilized amino acids derived ligands for the enantioselective arylation of aldehydes with organozinc reagents in flow [34]. Supported ligands derived from (*S*)-proline, (2*S*,4*R*)-4-hydroxyproline, (*S*)-tyrosine, and (*S*)-phenylalanine on the Merrifield resin were synthesized and first, tested in batch. The best results were obtained with resins **6** and **7** (Fig. 14) and these immobilized ligands were evaluated under continuous flow conditions, with the main goal of improving the reaction time and enantioselectivity of the process. The designed setup consisted of a glass Omnifit® column packed with the functionalized resin and fed with the stock solutions of the arylzinc reagent and aldehyde. Under optimal conditions, full conversion was obtained with only 1.5 min of residence time affording the desired product with high enantiomeric excess. Moreover, the stability of the ligands was evaluated, and it was found out that **7** could be used for a longer period of time than **6**. In fact, no significant change in reaction yield was observed, but the enantioselectivity slightly decreased, after a 3 days experiment.

In 2018, a new family of polystyrene-supported *cis*-4-substituted diarylprolinols was described by the group of Pericàs [35]. A tandem sequence consisting of aza-Michael addition catalyzed by diarylprolinol **8** followed by hemiacetalization was developed to afford chiral 5-hydroxyisoxazolidine (Fig. 15). Encouraged by the robustness of the supported catalyst, a family of α,β-unsaturated aldehydes was submitted to the flow process. The system consisted of a packed bed reactor filled with catalyst **8** and fed by a solution of the aldehyde and a mixture of *N*-protected

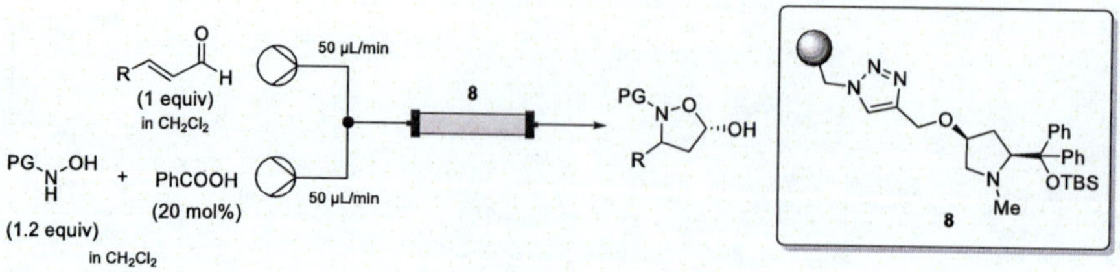

Fig. 15 Enantioselective Aza-Michael addition reaction

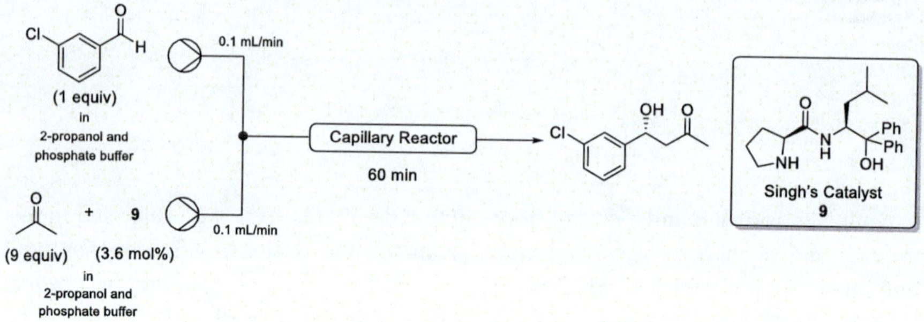

Fig. 16 Asymmetric organocatalytic aldol reaction promoted by Singh's catalyst

hydroxylamine and benzoic acid. Overall, the accumulated TON in these flow processes was of 134. Moreover, the synthetic versatility of the products was demonstrated by a sequence consisting of oxidation and continuous flow hydrogenation that allowed the preparation of β-amino acids.

Asymmetric organocatalytic aldol reaction with hydrophobic substrates in water under continuous flow conditions was reported by Gröger and coworkers in 2019 [36]. The optimal conditions were found using 3.6 mol% catalyst loading (Singh's catalyst), residence time of 60 min at room temperature in 2-propanol/phosphate buffer (pH 7) as a reaction medium. The continuous flow system consisted of a capillary type tube reactor and two feed streams (Fig. 16). First feed solution contained 3-chlorobenzaldehyde, 2-propanol, and phosphate buffer, whereas the second one consisted of Singh's catalyst (**9**), 2-propanol, acetone, and phosphate buffer. The outlet stream was quenched using a mixture of dichloromethane and 2.0 M aqueous hydrochloric acid solution (1:1 ratio) at 0°C. The desired product, (*R*)-4-(3-Chlorophenyl)-4-hydroxybutan-2-one, was obtained with 74% conversion and 89% *ee*.

3.2 *TRIP-Derived Catalysts*

In 2016, Pericàs and coworkers reported the use of new catalytic resin based on polystyrene-supported TRIP (**10**) for the asymmetric allylboration of aldehyde

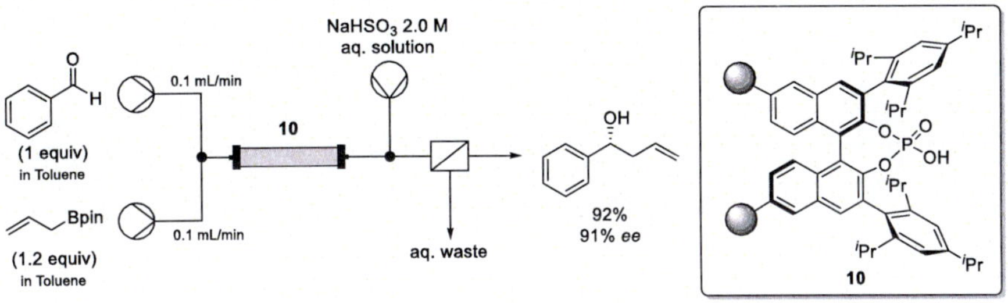

Fig. 17 TRIP catalyzed enantioselective allylation of aldehydes

[37]. Encouraged by the excellent results observed in batch, they extended the enantioselective synthesis of homoallylic alcohols to the corresponding flow process. The designed system consisted of a packed bed reactor containing the functional resin **10** (Fig. 17) that was fed with two solutions containing the aldehyde and the allylboronic ester. The authors observed a background reaction taking place in the collecting flask between the excess of allylboronic ester and small amounts of remaining aldehyde which was detrimental to the enantiomeric purity of the final product. To avoid that, a downstream process was required at the end of the column, the addition of an aqueous solution of $NaHSO_3$, which scavenged any unreacted aldehyde.

Under the optimal operation conditions, (*R*)-1-phenylbut-3-en-1-ol was obtained in 92% yield and 91% *ee*, with a TON of 282 and a productivity of 2.22 mmol h^{-1} g_{resin}^{-1}, after 28 h running. Noteworthy, no detectable decrease in the catalytic activity of the resin was observed. This flow process allowed the production of enantiopure homoallylic alcohol in short periods of time with highly reduced energy and material costs, under safety conditions.

3.3 *Thiourea-Derived Catalysts*

In 2015, the Pericàs laboratory reported the enantioselective α-amination of 1,3-dicarbonyl compounds with azodicarboxylates catalyzed by the immobilized thiourea organocatalyst **11** (Fig. 18) [38]. After the optimization in batch and proving the recyclability of the catalyst, the continuous process was developed. The flow reactor consisted of a packed bed reactor with **11** that was fed with a mixture of both reagents. A second pump was used to circulate a solution of Et_3N to wash the column periodically, as this process was required to preserve the catalytic activity of the resin.

Overall, this experiment led to the desired product in 71% isolated yield and 93% *ee* with a productivity of 4.88 mmol $mmol_{cat}^{-1}$ h^{-1} and a TON of 37. This represents an interesting alternative to the previously reported immobilized squaramides, as it offers the versatility of thiourea catalysts and the modularity of this simplified approach avoiding the use of linkers/spacers.

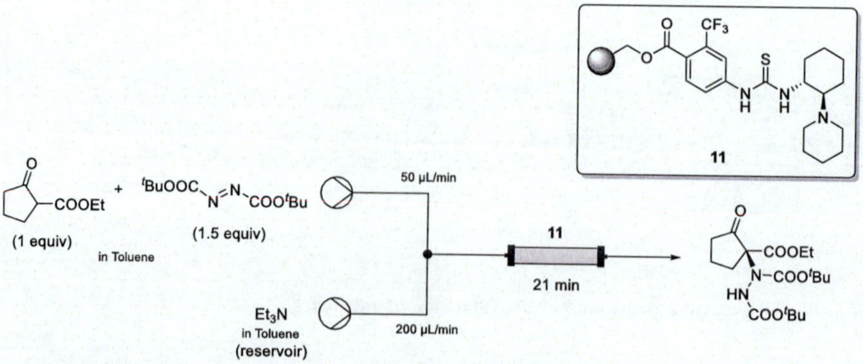

Fig. 18 Enantioselective α-amination of 1,3-dicarbonyl compounds

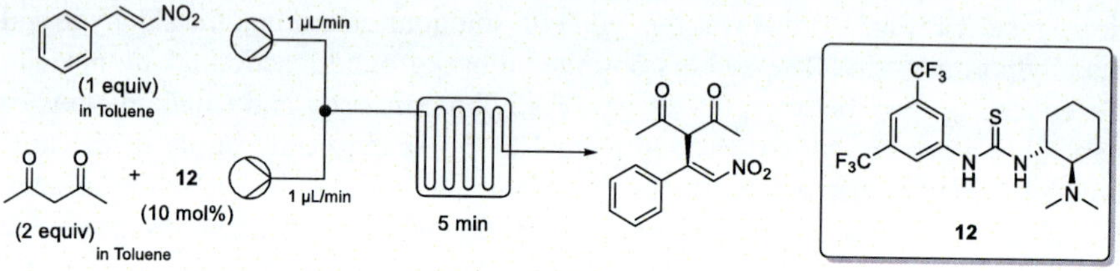

Fig. 19 Metal-free Michael addition of acetylacetone to β-nitrostyrene

In the same year, the metal-free Michael addition of acetylacetone to β-nitrostyrene using bifunctional, *(R,R)-trans*-1-[3,5-bis(trifluoromethyl)phenyl]-3-[2-(*N,N*-dimethylamino)cyclohexyl]thiourea as a catalyst was reported by Maggini and coworkers [39]. The use of a flow process resulted more efficient, versatile, and sustainable compared to batch. The continuous flow system used in this work was the Labtrix® Start Standard system equipped with a 10-μL glass microreactor. The reactants were supplied in two feeds, one contained β-nitrostyrene in toluene whereas the second one contained acetylacetone and the bifunctional catalyst (Fig. 19). The use of microreactor technology significantly accelerated the reaction providing the Michael adduct in high yield and enantioselectivity, up to 85% *ee*, in only 5 min residence time. In the same report, the authors, also presented successful synthesis of an advanced intermediate for the preparation of the $GABA_B$ receptor agonist Baclofen.

A novel polystyrene-supported isothiourea organocatalyst **13** was reported by Pericàs and coworkers in 2016 based on the enantiopure benzotetramisole structure [40]. Its catalytic activity was demonstrated in a domino Michael addition/cyclization reaction leading to dihydropyridinones, displaying excellent results. Encouraged by the promising results of recyclability, a continuous flow protocol was developed comprising the sequential preactivation of the carboxylic acid and the organocatalytic domino transformation (Fig. 20). The flow setup consisted of a simple tubular reactor, for the formation of the mixed anhydride, connected to a

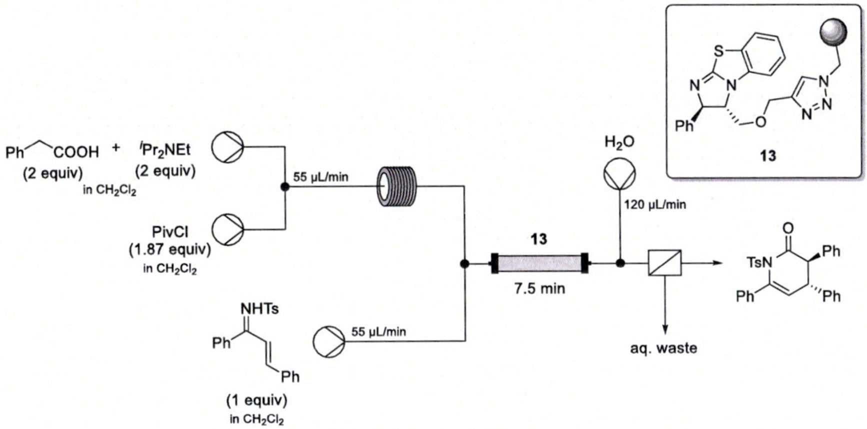

Fig. 20 Enantioselective domino Michael addition/cyclization

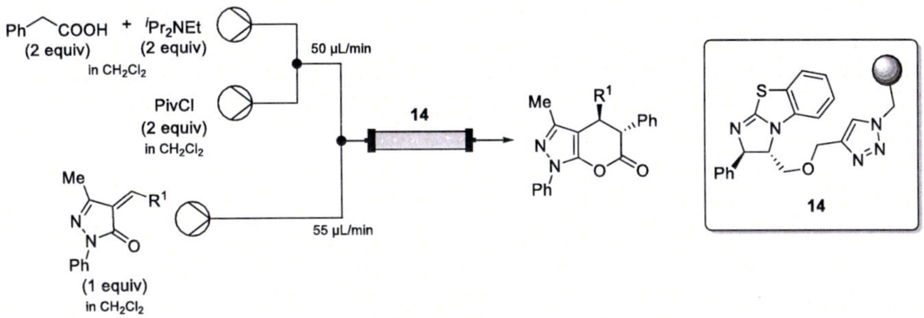

Fig. 21 Enantioselective [4 + 2] cycloaddition reaction

packed bed reactor containing **13** for the asymmetric domino reaction. The integrated in-line liquid–liquid separator allowed the continuous collection of the organic phase, from which the desired dihydropyridinone product was obtained by simple evaporation. The system was operated for 11 h with very high conversion and enantioselectivity, with a TON of 22.5.

The next year (2017), the same group reported the first enantioselective formal [4 + 2] cycloaddition reaction between unsaturated heterocycles and in situ activated arylacetic acids in a flow process catalyzed by an immobilized isothiourea catalyst [41]. The flow system consisted of a Omnifit® glass chromatography column loaded with polystyrene-supported isothiourea **14** (Fig. 21). The reactor was fed with a mixture of ^{i}Pr$_2$NEt and phenylacetic acid, previously combined with pivaloyl chloride, and with a solution of alkylidene pyrazolones. The desired product was obtained as a single diastereoisomer in 67% yield and 99% *ee*. Remarkably, the accumulated TON for 24 h run experiment was 76.8.

Remarkably, also using the same organocatalyst **14**, the first synthesis of optically pure spirocyclic oxindole-pyranopyrazolone scaffolds was developed, when

N-substituted isatin-derived pyrazolones were used. It was performed the cascade [4 + 2] reaction leading to the corresponding spirocyclic adducts in good yields with high *ee*'s. This approach allowed to assemble a range of optically active substituted spiropyranopyrazolones under the established optimal reaction conditions.

3.4 Cinchona-Derived Catalysts

9-amino(9-deoxy)epi cinchona alkaloids immobilized onto polystyrene resins were proved to be highly efficient organocatalysts promoting asymmetric Michael additions under batch and flow conditions [42]. The high catalytic activity exhibited by **15** and its robustness converted this catalytic resin into a suitable candidate for the activation of enones towards the addition of *C*-nucleophiles (Fig. 22). The flow reactor used to test this reaction involved resin **15** packed into a simple Teflon® tube between two plugs of glass wool. The packed bed reactor was heated to 30°C while a solution mixture containing the two reactants and benzoic acid was pumped into the system. Notably, 3.6 g of desired product were collected in 21 h of operation, corresponding to a TON of 32. Moreover, the versatility of the PS-supported cinchona aminocatalyst **15** allowed the sequential preparation of a small library of enantiopure adducts using the optimized continuous flow process. The complete

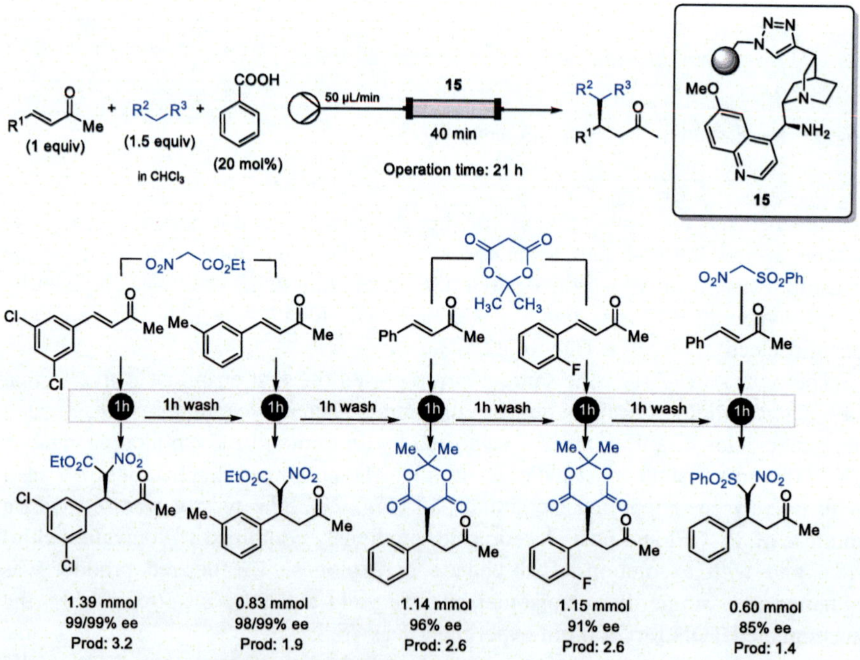

Fig. 22 Enantioselective Michael addition

process took place with remarkable productivity (TOF = 1.4–3.2 $mmol_{product}$/$mmol_{resin}$ h) and afforded a variety of Michael adducts with high enantioselectivity.

Taking advantage of the versatility of the PS-supported cinchona aminocatalyst **15**, a library of enantioenriched Michael adducts was sequentially prepared. This flow setup allowed the preparation of five Michael adducts by combination of four different enones with three nucleophiles. Each solution containing the enone, nucleophile, and the benzoic acid, was run through the system for 2 h (first hour was required for stabilization of the flow system whilst only the effluent corresponding to the second hour was collected). In order to avoid cross-contamination, the column was washed with $CHCl_3$ for 1 h between the preparation of two consecutive adducts. The complete process took place with remarkable productivity and afforded a library of Michael adducts with higher enantioselectivity than the one obtained in batch processes.

The synergistic combination of asymmetric organocatalysis with photochemical reactivity has resulted in a very useful tool to access to new reaction pathways, and thus, to enable the synthesis of novel compounds [43]. The use of light to promote reactions is non-hazardous, environmentally friendly, very efficient (high temperatures or harsh conditions are often non-required) and with broad and easy access. Probably for these reasons, this combination methodology has gained attention in the pharmaceutical industry for drug discovery and development [44, 45]. However, only a few publications on asymmetric photocatalysis in continuous flow using organocatalysis to control the absolute configuration of the reaction products have been reported [46, 47].

Very recently, the Meng's group reported the enantioselective photooxygenation of β-dicarbonyl compounds in continuous flow [48] using C-2'modified cinchonine-derived phase-transfer organocatalysts with excellent yields (up to 97%) and high enantioselectivities (up to 90% *ee*). In fact, α-Hydroxy-β-dicarbonyl moiety is an important structural functionality present in a variety of natural products and pharmaceuticals. α-Hydroxy-β-dicarbonyl compounds are also useful synthetic building blocks for the preparation of a variety of biologically active substances, including naturally occurring heterocycles and carbocycles [49, 50].

The authors developed a visible-light driven heterogeneous gas–liquid–liquid asymmetric aerobic oxidation reaction taking place in a flow photomicroreactor. After reaction conditions optimization in batch, they found that the best organocatalyst **16** could be used at 10 mol% loading and in combination with a 0.01% of phthalocyanine (Pc) as a photosensitizer. In order to perform the reaction in continuous flow, they chose the CORNING Advanced-Flow Reactor, for photochemistry (Fig. 23). They did not apply the same optimized conditions used in batch since, for instance, the low temperature needed in batch (−15°C) caused precipitation and expansion of water and the strongly basic media generated upon oxidation damaged the surface of the glass reactor. The organic phase was prepared by solubilizing the corresponding substrates, PTC **16** (10 mol%) and Pc (1 mol%) in toluene. The aqueous phase was a K_2HPO_4 solution (10%) acting as a buffer. The organic phase and aqueous phase were pumped separately at a flow rate of 1.5 mL min^{-1}. Oxygen was introduced through a mass flow controller at a flow

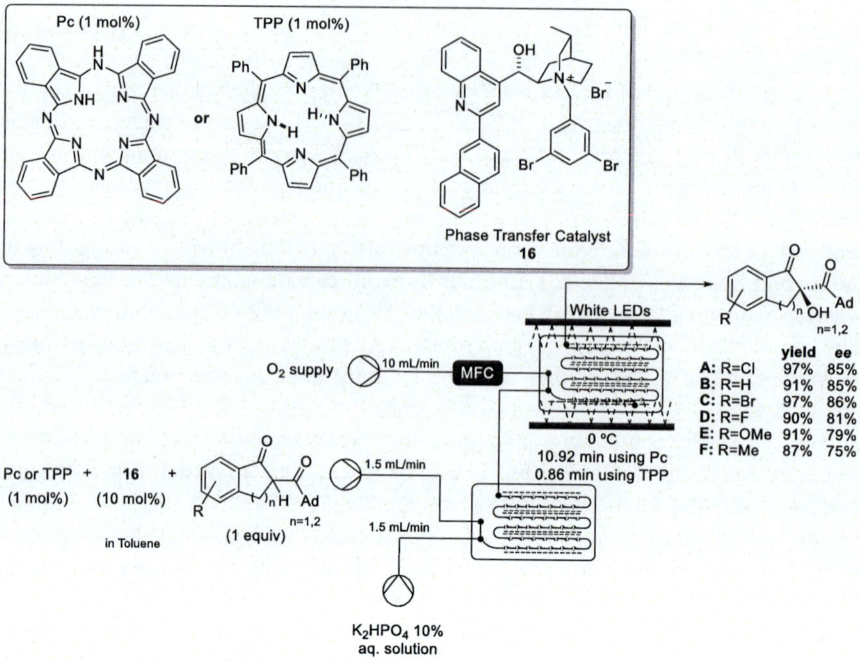

Fig. 23 Asymmetric photocatalytic oxidation of β-dicarbonyl substrates in continuous flow

rate of 10 mL min^{-1}, with a reaction pressure of 3 atm and reaction temperature of 20°C.

The reaction time could be shortened from 8 h in batch to a residence time of 10.92 min in the flow reactor, high yields and enantioselectivities being preserved. Surprisingly, when TPP (tetraphenylporphyrin) was used as the photosensitizer, the residence time was drastically shortened to 0.89 min without major variations in yield or enantiomerical excess, in contrast with results obtained in batch. It was also demonstrated that stereoselectivity was greatly decreased if the catalyst loading was reduced, and that reducing the TPP loading to 0.1 mol% was not affecting the performance of the reaction. With these results, they were able to perform a gram-scale reaction in flow to obtain 3 mmol (1.03 g) of product A, without any decrease in yield or enantioselectivity.

Benaglia and coworkers reported developed in 2016 the development of a variety of immobilized chiral picolinamides, derived from cinchona alkaloid primary amines, supported on silica and polystyrene [51]. They investigated the stereoselective reduction of imines with trichlorosilane catalyzed by these family of catalysts. Encouraged by the good results obtained and the preliminary recyclability studies of picolinamide **17**, a continuous flow process was designed for this transformation (Fig. 24). The system consisted of an Omnifit® glass column filled with catalyst **17** and fed by a solution of imine and another of trichlorosilane. The continuous flow process was run for 3 h and the corresponding chiral amine was

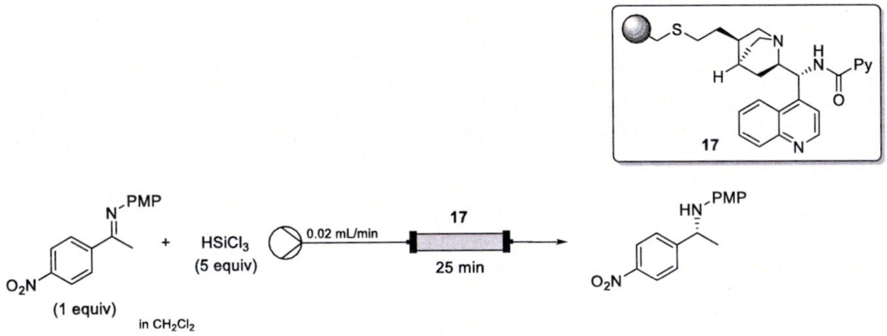

Fig. 24 Stereoselective reduction of imines catalyzed by supported chiral picolinamides

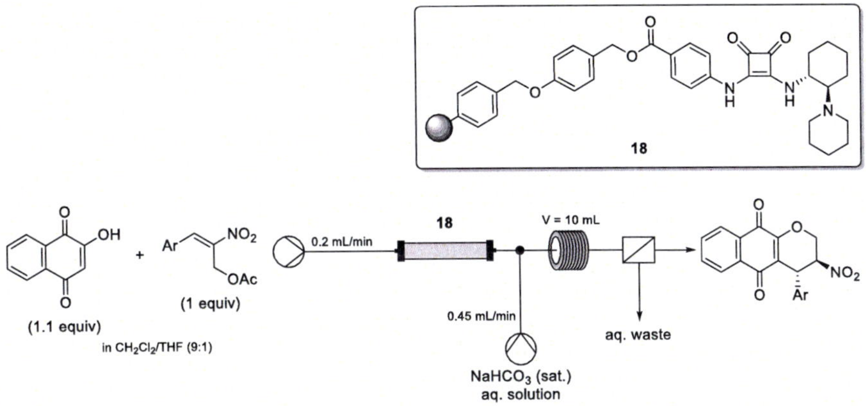

Fig. 25 Enantioselective Oxa-Michael reaction

continuously obtained in very high yield. However, the enantioselectivity of the process was quite low and it dropped over time.

3.5 *Squaramide-Derived Catalysts*

A new, simple, and low-cost immobilized chiral squaramide **18** was developed by the group of Pericàs to expand the applicability of its predecessors [52]. This new resin was applied in a benchmark reaction, and gratifyingly, the reaction time was reduced by half and the enantioselectivity was maintained in comparison with previous explored catalysts. The target transformation comprised two sequential reactions: the organocatalyzed Michael addition and the base promoted cyclization. Regarding the enantioselective reaction, the experimental setup consisted of a packed bed reactor with PS-Squaramide **18** (Fig. 25). Then, the outlet of the column was coupled with a T-junction connecting the base solution and the coil reactor.

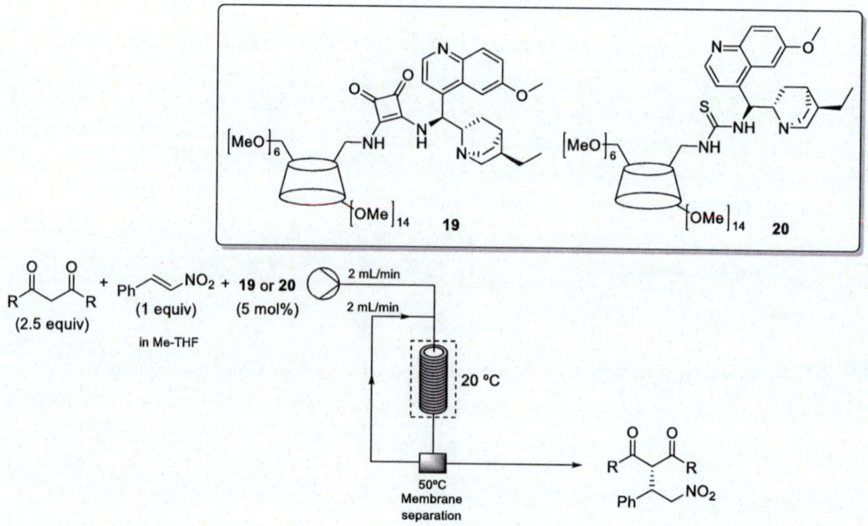

Fig. 26 Schematic process scheme for the continuous synthesis–separation platform

Remarkably, the total residence time of the process including the two reactions and in-line workup was only 30 min. Pericàs and coworkers demonstrated with this study the importance of the proper linker for the optimal performance of the heterogenized catalytic species. Besides, this polystyrene-supported resin was implemented for the production of a library of highly enantioenriched pyranonaphthoquinones by a sequential two-step synthesis.

Kupai and Szekely have recently reported the anchoring of organocatalytic species onto cyclodextrin (CD) as a methodology to facilitate the recovery of covalently bonded organocatalysts [53]. Cinchona-squaramide **19** and cinchona-thiourea **20** were the selected catalysts for the asymmetric Michael reaction of 1,3-diketones and trans-β-nitrostyrene (Fig. 26). Prior to the continuous flow process, the catalytic performance was optimized in batch mode. Then, continuous flow organocatalysis was performed in a coiled tube flow reactor integrated with a membrane separation unit. Under the optimal operation conditions, the resulting adducts were obtained through a continuous flow process achieving 80 g L^{-1} h^{-1} productivity, 98% final product purity, up to 99% enantiomeric excess and 100% catalyst recovery.

The cyclodextrin anchor was found to have two key roles. First, to allow the complete recovery of the catalyst due to its size and second, the improvement of the catalytic performance due to the favourable conformation of the anchored cinchona-squaramide catalyst **19**. The separation process was carried out through the coupling of an in-line membrane separation unit to the flow reactor. The nanofiltration module contained a DM900 membrane and the cross-flow nanofiltration unit was kept at 50°C to avoid precipitation of the product. The system reported was very efficient for the online recovery of the catalyst, 100% present in the retentate stream. In addition, the separation unit was also able to recycle up to 50% of the reaction solvent.

3.6 Diamine-Derived Catalysts

The polystyrene-supported chiral vicinal diamine **21** (Fig. 27) was developed combining the high selectivity of the homogeneous version with the advantages from its

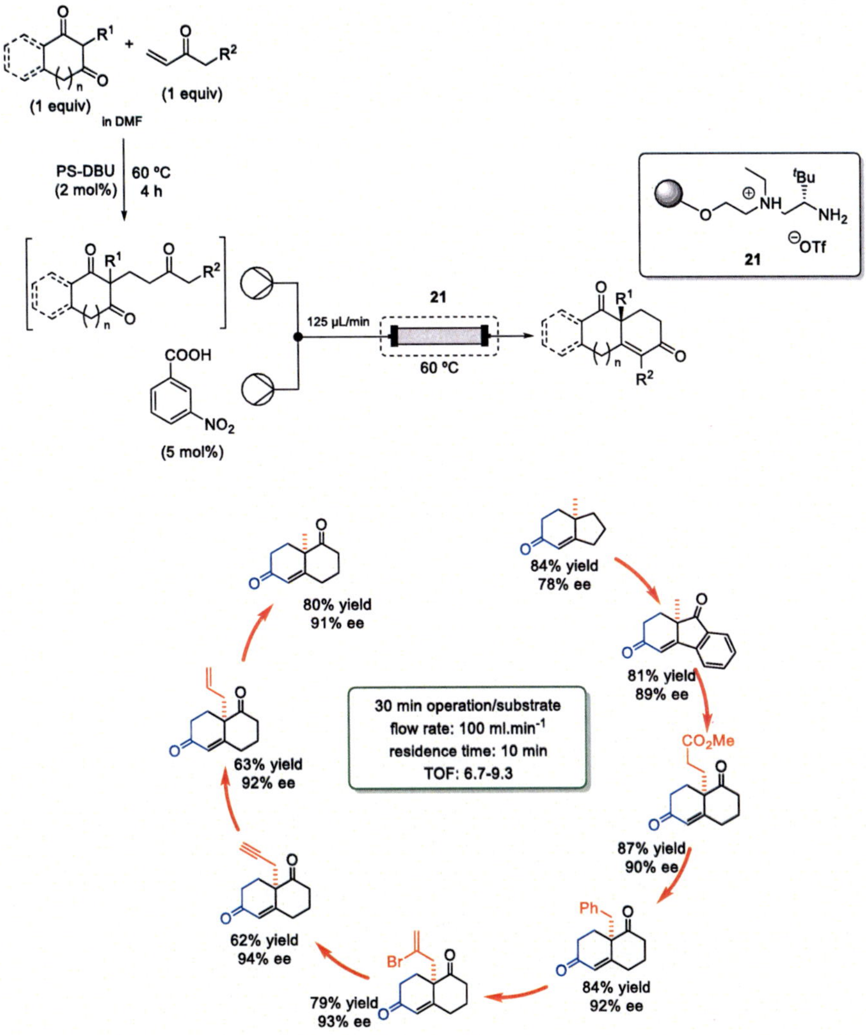

Fig. 27 Enantioselective Robinson annulation

polymeric nature [54]. The great catalytic activity showed by immobilized catalyst **21** in batch encourages Pericàs and coworkers to convert the enantioselective Robinson annulation catalyzed by the chiral vicinal-diamine **21** into a continuous flow process. A preformed solution of the Michael adduct was combined with *m*-$NO_2C_6H_4CO_2H$ (5 mol%) and pumped through a MPLC jacketed glass column packed with **21** and heated at 60°C. In this manner, the Wieland–Miescher ketone (R^1 = CH_3, R^2 = H) was obtained in high enantiomeric purity (91% ee) with a productivity of 2.7 g h^{-1} and TON of 117, after 24 h operation of the flow system.

Catalytic asymmetric flow processes based on immobilized catalysts are very useful for preparing libraries of complex compounds, since different reactions can be performed in a sequential manner. Taking advantage of the versatility of the supported catalyst **21**, a diverse library of enantioenriched Robinson annulation products was sequentially prepared using the designed setup for the flow system (Fig. 27). In this way, the diversity-oriented continuous flow process was successfully carried out with very high productivities (TOF = 6.7–9.3 $mmol_{prod}mmol_{resin}^{-1}\ h^{-1}$) easily affording a library of annulation products.

3.7 *Imidazolidinone-Derived Catalysts*

In 2017, a new class of solid-supported chiral imidazolidinones organocatalysts for the catalytic reduction of imines was reported by Benaglia and coworkers [55]. Polystyrene-supported catalyst **22** was employed for the fabrication of packed bed reactors. An efficient and convenient continuous flow process was developed where the immobilized imidazolidinone **22** was packed into a stainless-steel HPLC column and fed by the imine and the trichlorosilane (Fig. 28). The flow system was stable for 7 h, producing the corresponding chiral amine in excellent yield and good enantiomeric excess. The stereoselectivity of the process slightly decreased over time, probably due to the degradation of the imidazolidinone ring of the catalyst.

In homogeneous version, a new class of chiral Lewis bases (picolinamides based on chiral imidazolidinones) was developed by the same group for the

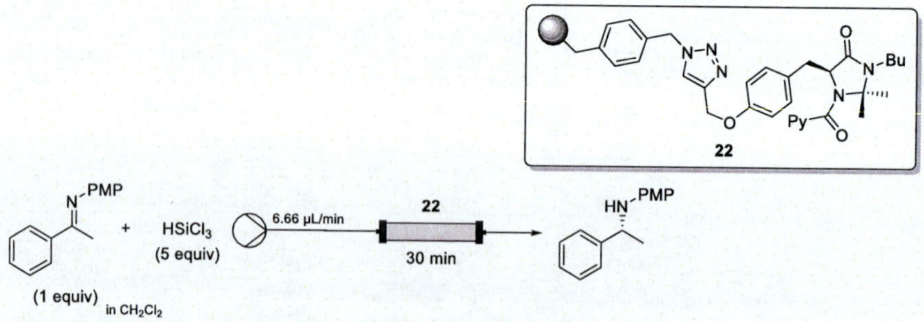

Fig. 28 Asymmetric reduction of imines catalyzed by supported chiral imidazolidinones

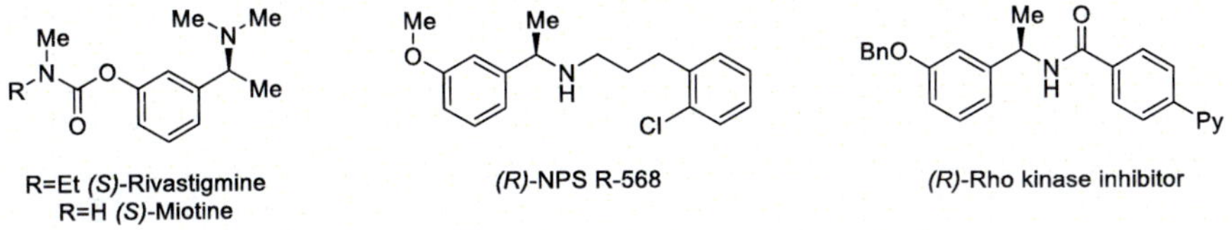

Fig. 29 Chiral amines as valuable APIs

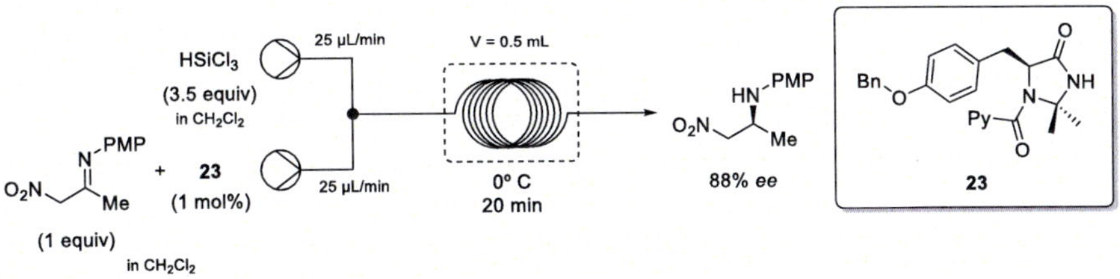

Fig. 30 Organocatalytic enantioselective reduction of imines

enantioselective, metal-free imine reduction, mediated by $HSiCl_3$ [56]. The use of $HSiCl_3$ is very convenient due to its high chemoselectivity to reduce C=N bonds and affordability, being an extremely cheap reagent available in large quantities.

After catalyst structure optimization and reaction optimization in batch, Benaglia and coworkers identified the active organocatalyst **23** that was able to promote the reduction of a large variety of functionalized substrates in high yields with enantioselectivities (> 90%) using only 0.1–1 mol% of catalyst loading. They demonstrated the synthetic potential of that approach by the preparation of advanced intermediates of important active pharmaceutical ingredients. For example, the reduction of 3-alkoxy-substituted acetophenone imines, either *N*-benzyl protected or *N*-(3-phenylpropyl) substituted is an efficient step for the synthesis of enantiomerically pure compounds, used in the treatment of Alzheimer and Parkinson diseases, hyperparathyroidism, neuropathic pain, and neurological disorders (Fig. 29).

The authors applied the optimized catalytic methodology in continuous flow for the synthesis of chiral α-nitro amines, that can be easily reduced with Ni-Raney to achieve chiral 1,2-diamines (a functionality displaying a broad spectrum of biological activity) in only two steps (Fig. 30).

The simple experimental procedure, the low cost of the reagents, the mild reaction conditions, and the straightforward isolation of the product make the methodology attractive for large-scale applications.

3.8 Chiral Carbene-Derived Catalyst

The immobilization of the Rovis catalyst, a triazolium carbene precursor, onto polystyrene and silica supports was described by the group of Massi [57]. The synthesis of optically active chromanones via stereoselective intramolecular Stetter reaction was selected as benchmark reaction to test their catalytic activity. Encouraged by the excellent activity displayed in batch by **24**, a continuous flow process was designed and further optimized, thus allowing the production of the desired chromanone with a satisfactory instant conversion and excellent enantioselectivity. The system consisted of a microreactor based on a stainless-steel column charged with **24** (Fig. 31). The recyclability of the heterogeneous Rovis catalyst **24** was confirmed by a long run experiment of 48 h, with unchanged values of conversion and stereoselectivity. Significantly, 120 h operation determined a loss of only half the initial productivity. These results corresponded to a total TON of 132. Moreover, an important increase of productivity was observed from batch to flow conditions.

4 Direct Application of Organocatalysis in Continuous Flow for Drug Synthesis

As it becomes evident from the previous sections in this chapter, organocatalysis has been extensively used under continuous flow conditions. However, despite being a powerful tool, there are still rather few examples reported where organocatalysis in flow is applied in the key step of the synthesis of APIs or drugs. It is even harder to find examples where multistep processes using this approach are involved.

In 2015, the groups of Puglisi and Benaglia reported the synthesis in multigram scale of an *(S)*-Pregabalin precursor and *(S)*-Warfarin in continuous flow with high enantioselectivity in an experimentally very simple procedure [58]. (*S*)-Pregabalin is a widely administered anticonvulsant and antiepileptic drug, and (*S*)-Warfarin is an oral anticoagulant. In the first case, the authors use the chiral thiourea organocatalyst

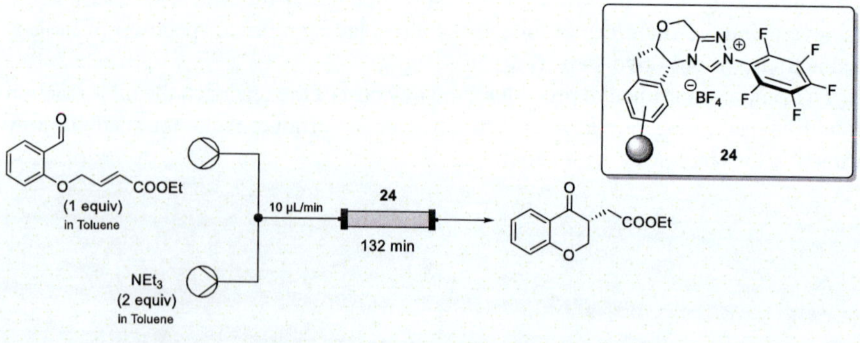

Fig. 31 Enantioselective Stetter reaction

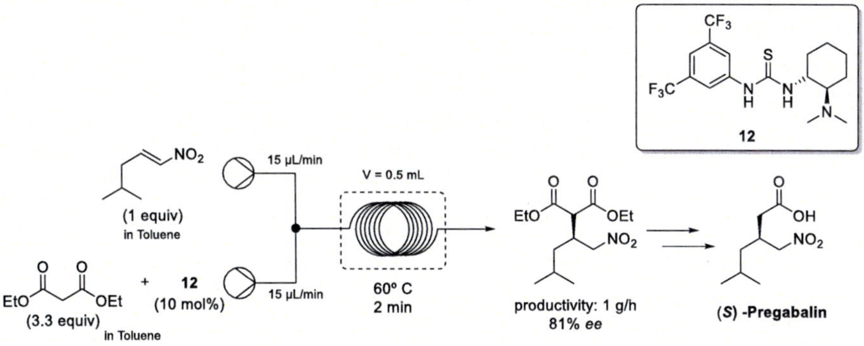

Fig. 32 Asymmetric organocatalytic mediated synthesis of (*S*)-pregabalin key intermediate

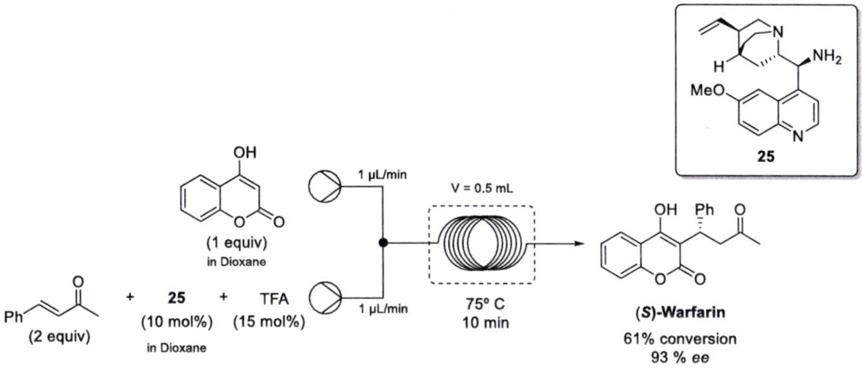

Fig. 33 Asymmetric synthesis under continuous flow conditions of (*S*)-warfarin

12 to perform the stereoselective nucleophilic addition in flow of diethyl malonate to the appropriate nitroalkene in order to obtain a key chiral intermediate that, after reduction of the nitro moiety followed by hydrolysis and decarboxylation, lead to the formation of (*S*)-Pregabalin (Fig. 32).

Under the best reaction flow conditions (60°C, 2 min residence time) they were able to produce 1 g of desired intermediate in 81% *ee* in only 1 h using a very simple flow, lab size setup.

They also were able to synthesize (*S*)-Warfarin through an organocatalytic approach by the nucleophilic addition of 4-hydroxycoumarin to benzalacetone catalyzed by the cinchona-derived primary amine **26** catalyst as the most straightforward methodology (Fig. 33). The optimal reaction conditions found involved performing the reaction at 75°C with 10 min residence time, leading to the formation of the desired product with 61% conversion and 93% *ee*, using 10 mol% of catalyst.

Another outstanding proof of concept, which demonstrates the potential of flow chemistry for sustainable pharmaceutical manufacturing, has been recently reported by Kappe and Pericàs: a multistep continuous flow process leading to the synthesis of the key intermediate of (−)-Paroxetine [59].

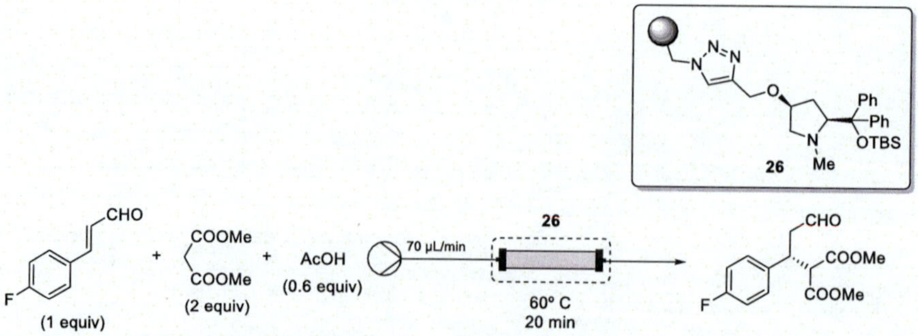

Fig. 34 Asymmetric Michael reaction under continuous flow conditions

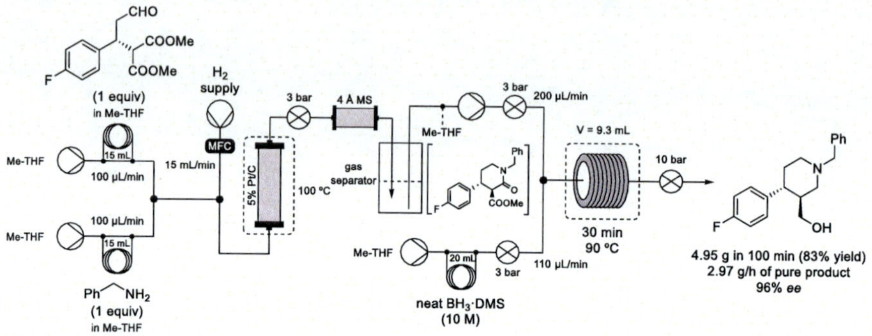

Fig. 35 Multistep continuous flow synthesis of the intermediate of (−)-paroxetine

The chiral aldehyde (Fig. 34) was obtained via organocatalytic asymmetric Michael reaction of *p*-fluorocinnamaldehyde and dimethyl malonate promoted by the heterogeneous polystyrene-supported *cis*-4-hydroxydiphenylprolinol catalyst designed by Pericàs. The use of the catalyst was previously reported by Pericàs and coworkers in 2018 [35].

The critical step of the process is presented in Fig. 34. The chiral aldehyde was obtained in a solvent-free enantioselective conjugate addition. Under optimal conditions, involving 20 min residence time at 60 °C, 93% conversion with complete chemoselectivity and 98.5% enantioselectivity was achieved. Moreover, the stability of the system was evaluated in a 7 h long continuous flow experiment leading to a productivity of 2.47 g h^{-1} of pure product (17.26 g, 84% isolated yield, 97% *ee*). It should be emphasized that the catalyst proved to be very robust, practically constant selectivity (95–98% *ee* and 100% chemoselectivity), and only a small decrease in catalytic activity (93–85% conversion), under the neat conditions applied. Overall, the data generated result in an effective catalyst loading of 0.6 mol%, TON of 132, STY of 1.76 kg L^{-1} h^{-1} as well as very low waste formation confirmed by an E-factor of only 0.7.

The chiral aldehyde thus obtained was used in a telescoped process (Fig. 35) that consisted of a reductive amination/lactamization/amide-ester reduction sequence

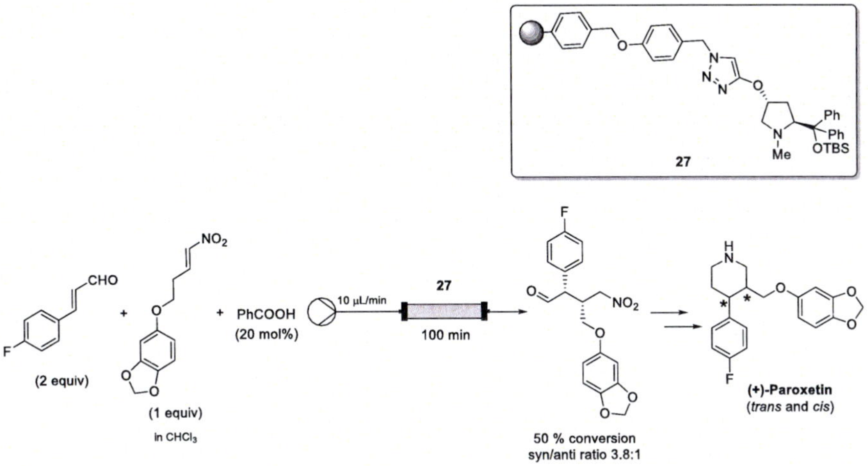

Fig. 36 Asymmetric Michael reaction for the synthesis of (+)-paroxetine key intermediate

leading to the catalytic enantioselective synthesis of the chiral key intermediate, ((3*S*,4*R*)-4-(4-fluorophenyl)piperidin-3-yl)methanol, of (−)-paroxetine. Overall, the telescoped process proved stable and afforded the desired product in 83% isolated yield and 96% *ee* with a productivity of 2.97 g h^{-1}.

This solvent-free/highly concentrated multistep process in combination with the remarkably robust catalysts enabled a significant process intensification, leading to a productivity of ((3*S*,4*R*)-4-(4-fluorophenyl)piperidin-3-yl)methanol on multigram per hour scale. In addition, the process offered high chemo- and stereoselectivity while generating minimal amounts of waste, as confirmed by a cumulative E-factor of 6.

An analogous *Jørgensen*–Hayashi catalyst supported onto a Wang resin was also applied by Mlynarski and coworkers in the continuous flow organocatalytic Michael addition to prepare the key chiral intermediate in the total asymmetric synthesis of (+)-Paroxetine and (+)-Femoxetine (Fig. 36) [60].

In this case although the best results for the asymmetric Michael addition was using a flow rate of 0.005 mL min^{-1}. Complete conversion to the desired product was achieved with this flow rate with *syn*/*anti* ratio of 3.2:1. However more than 16 h were required to pump 5 mL of the starting material through the system. In contrast, when a flow rate of 0.5 mL min^{-1}, although the *syn*/*anti* ratio observed was higher (4.5:1) the conversion dropped to 17%. The authors decided to apply a compromise between residence time and productivity, so the flow rate was set to 0.01 mL min^{-1} to achieving a maximum conversion of 50% with a residence time of 100 min.

Using the same supported catalyst **27** developed by the Pericàs group, Ötvos, Kappe, and coworkers have also reported a two-step asymmetric synthesis in flow of chiral γ-nitrobutyricacids as key intermediates of the GABA analogues Baclofen, Phenibut, and Fluorophenibut on a multigram scale [61]. The telescoped process comprises an enantioselective Michael-type addition facilitated by a PS-supported

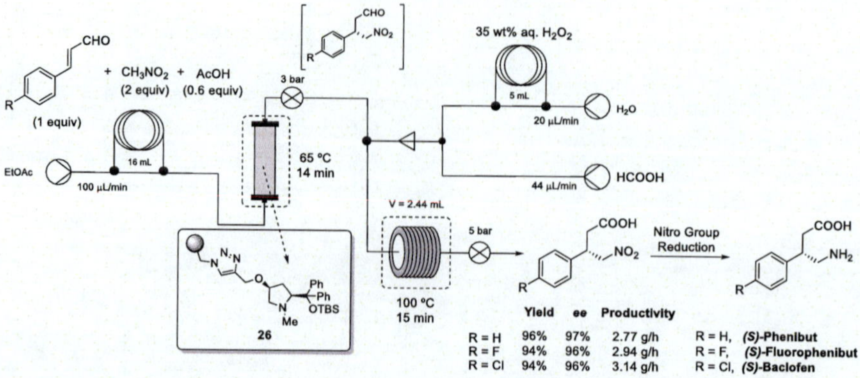

Fig. 37 Telescoped organocatalytic conjugate addition–aldehyde oxidation sequence for the synthesis in continuous flow of chiral GABA intermediates

organocatalyst under neat conditions followed by the oxidation of the resulting aldehyde with in situ-generated performic acid. Chiral GABA derivatives are of great interest for pharmaceutical applications. For example, Baclofen is an antispastic drug applied as a muscle relaxant, while Phenibut is used to treat anxiety and insomnia.

The overall process for the synthesis of these derivatives involves the asymmetric Michael-type addition of nitromethane to α,β-unsaturated aldehydes to obtain γ-nitro-aldehydes key intermediate which can be transformed into the desired GABA analogues via subsequent oxidation and reduction steps (Fig. 37).

The key step in the telescoped flow sequence is the organocatalytic asymmetric Michael addition of nitromethane to the corresponding cinnamaldehyde. This process is performed under neat conditions. The reactants pass through a column heated at 65°C and filled with catalyst **27**. Then, neat formic acid and an aqueous H_2O_2 solution are introduced at flow rates that correspond to 1 equiv. of H_2O_2 and 5 equiv. of formic acid with respect to the aldehyde. The reaction mixture coming from the packed bed reactor, containing the γ-nitroaldehyde, is mixed then with the combined formic acid/H_2O_2 stream, and the resulting mixture solution is passed through a coil reactor at 100°C where performic acid generation and aldehyde oxidation take simultaneously place. The desired γ-nitrobutyric acids are obtained in high yields and excellent enantioselectivities, the processes being characterized by a remarkably high productivity. Notably, no chromatography purification is required for products obtained at the multigram scale.

Other important transformation is the continuous flow procedure for the transformation of biobased glycerol into high value-added oxiranes that was reported by Monbaliu in 2019 [62]. The flow procedure consisted of a hydrochlorination/dechlorination sequence and presented economically and environmentally favourable conditions involving an organocatalyst and aqueous solutions of hydrochloric acid and sodium hydroxide (Fig. 38). First, the chemoselective hydrochlorination of crude, biobased glycerol to 1,3-dichloro-2-propanol with

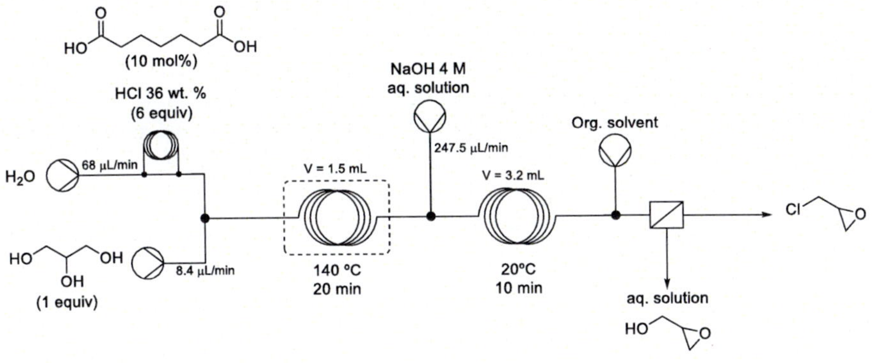

Fig. 38 Hydrochlorination/dehydrochlorination sequence for the synthesis of oxiranes

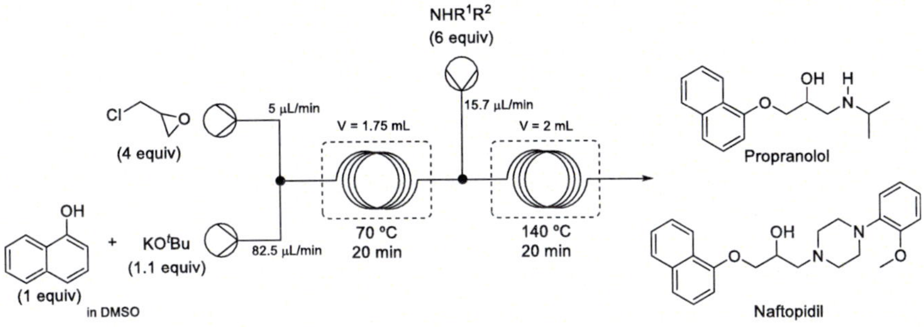

Fig. 39 Sequential flow process for the production of APIs

36 wt. % aqueous HCl was catalyzed by pimelic acid at 140°C. Sequentially, a dehydrochlorination/hydrolysis step was performed with concentrated aqueous sodium hydroxide. Finally, an in-line liquid–liquid separation unit in the downstream process allowed the separation of epichlorohydrin (organic solution) and glycidol (aqueous solution).

Moreover, the so-prepared biobased epichlorohydrin was subsequently used for the continuous flow preparation of active pharmaceutical ingredients including propranolol (hypertension, WHO essential) and naftopidil (prostatic hyperplasia) via formation of the glycidyl ether intermediate and sequential aminolysis reaction (Fig. 39) [62].

5 Conclusions

In this chapter, key organocatalytic processes in continuous flow for the synthesis of important Active Pharmaceutical Ingredients (APIs) and building blocks in drug discovery have been discussed. Both achiral and chiral organocatalysts, including homogeneous and heterogeneous species have been presented. The combination of the advantages of continuous processes and organocatalysis represents an excellent approach to a more sustainable and smart production of fine chemicals. Usually, organocatalytic methodologies require easier experimental procedures and milder reaction conditions, resulting in savings in time and energy and also in the overall cost of the process, compared with the use of organometallic species. Furthermore, shorter reaction time and higher productivity are observed when the organocatalytic reaction is performed in continuous flow, offering clear advantages compared to the batch approach.

Additionally, the application of supported organocatalysts in continuous flow allows the recovery and reuse of these molecules. However, the identification of the deactivation mechanisms associated to the organocatalysts represents an important challenge to be addressed, in order to allow a wider use of organocatalytic process in flow in pharma and fine chemical industries, in particular for a large-scale application.

Compliance with Ethical Standards

Funding: There is no funding related to the chapter.

Ethical Approval: This manuscript is a review of previously published accounts, as such, no animal or human studies were performed.

Informed Consent: No patients were studied in this chapter.

References

1. Melchiorre P, Marigo M, Carlone A, Bartoli G (2008) Angew Chem Int Ed 47:6138
2. Bertelsen S, Jørgensen KA (2009) Chem Soc Rev 38:2178–2189
3. Giacalone F, Gruttadauria M, Agrigento P, Noto R (2012) Chem Soc Rev 41:2406–2447
4. Alemán J, Cabrera S (2013) Chem Soc Rev 42:774–793
5. Jimeno C (2016) Org Biomol Chem 14:6147–6164
6. Quin Y, Zhu L, Luo S (2017) Chem Rev 117:9433–9520
7. Chanda T, Zhao JC-G (2018) Adv Synth Catal 360:2–79
8. Xiang S, Tan B (2020) Nat Commun 11:3786
9. Krištofíková D, Modrocká V, Mečiarová M, Šebesta R (2020) Chem Sus Chem 13:2828–2858
10. Dalko PI, Moisan L (2004) Angew Chem Int Ed 43:5138–5175
11. Tanimu A, Jaenicke S, Alhooshani K (2017) Chem Engineering J 327:792–821
12. Colella M, Carlucci C, Luisi R (2018) Top Curr Chem 376:46
13. de Oliveira PHR, da BM, Santos S, Leão RAC, Miranda LSM, San Gil RAS, de Souza ROMA, Finelli FG (2019) Chem Cat Chem 11:5553–5561
14. Rodríguez-Escrich C, Pericàs MA (2015) Eur J Org Chem:1173–1188
15. Atodiresei I, Vila C, Rueping M (2015) ACS Catal 5:1972–1985
16. Finelli FG, Miranda LSM, de Souza ROMA (2015) Chem Commun 51:3708–3722
17. De Risi C, Bortolini O, Brandolese A, Di Carmine G, Ragno D, Massi A (2020) React Chem Eng 5:1017–1052
18. Maity C, Trausel F, Eelkema R (2018) Chem Sci 9:5999–6005
19. Burès J, Armstrong A, Blackmond DG (2014) Angew Chem Int Ed 53:8700–8704
20. Burès J, Armstrong A, Blackmond DG (2013) Pure App Chem 85:1919–1934

21. Di Marco L, Hans M, Delaude L, Monbaliu J-CM (2016) Chem Eur J 22:4508–4514
22. Zaghi A, Ragno D, Di Carmine G, De Risi C, Bortolini O, Giovannini PP, Fantin G, Massi A (2016) Beilstein J Org Chem 12:2719–2730
23. Ragno D, Brandolese A, Urbani D, Di Carmine G, De Risi C, Bortolini O, Giovannini PP, Massi A (2018) React Chem Eng 3:816–825
24. Brandolese A, Ragno D, Di Carmine G, Bernardi T, Bortolini O, Giovannini PP, Ginoble Pandoli O, Altomarec A, Massi A (2019) Org Biomol Chem 16:8955–8964
25. Abubakar SS, Benaglia M, Rossi S, Annunziata R (2018) Catal Today 308:94–101
26. Liu J, Xu J, Li Z, Huang Y, Wang H, Gao Y, Guo T, Ouyang P, Guo K (2017) Eur J Org Chem:3996–4003
27. Lyons DJM, Crocker RD, Enders D, Nguyen TV (2017) Green Chem 19:3993–3996
28. Rattanangkool E, Sukwattanasinitt M, Wacharasindhu S (2017) J Org Chem 82:13256–13262
29. Schweitzer-Chaput B, Horwitz MA, Beato EP, Melchiorre P (2019) Nat Chem 11:129–135
30. Vidyacharan S, Ramanjaneyulu BT, Jang S, Kim D-P (2019) ChemSusChem 12:2581–2586
31. Ranjan P, Ojeda GM, Sharma UK, Van der Eycken EV (2019) Chem Eur J 25:2442–2446
32. Sagamanova I, Rodríguez-Escrich C, Molnár IG, Sayalero S, Gilmour R, Pericàs MA (2015) ACS Catal 5:6241–6248
33. Llanes P, Rodríguez-Escrich C, Sayalero S, Pericàs MA (2016) Org Lett 18:6292–6295
34. Forni JA, Novaes LFT, Galaverna R, Pastre JC (2018) Catal Today 308:86–93
35. Lai J, Sayalero S, Ferrali A, Osorio-Planes L, Bravo F, Rodríguez-Escrich C, Pericàs MA (2018) Adv Synth Catal 360:2914–2924
36. Schober L, Ratnam S, Yamashita Y, Adebar N, Pieper M, Berkessel A, Hessel V, Gröger H (2019) Synthesis 51:1178–1184
37. Clot-Almenara L, Rodríguez-Escrich C, Osorio-Planes L, Pericàs MA (2016) ACS Catal 6:7647–7651
38. Kasaplar P, Ozkal E, Rodríguez-Escrich C, Pericàs MA (2015) Green Chem 17:3122–3129
39. Rossi S, Benaglia M, Puglisi A, De Filippo CC, Maggini M (2015) J Flow Chem 5:17–21
40. Izquierdo J, Pericàs MA (2016) ACS Catal 6:348–356
41. Wang S, Izquierdo J, Rodríguez-Escrich C, Pericàs MA (2017) ACS Catal 7:2780–2785
42. Izquierdo J, Ayats C, Henseler AH, Pericàs MA (2015) Org Biomol Chem 13:4204–4209
43. Silvi M, Melchiorre P (2018) Nature 554:41–49
44. Blakemore DC, Castro L, Churcher I, Rees DC, Thomas AW, Wilson DM, Wood A (2018) Nat Chem 10:383–394
45. Bogdos MK, Pinard E, Murphy JA, Beilstein J (2018) Org Chem 14:2035–2064
46. Sakeda K, Wakabayashi K, Matsushita Y, Ichimura T, Suzuki T, Wada T, Inoue Y (2007) J Photochem Photobiol A 192:166–171
47. Neumann M, Zeitler K (2012) Org Lett 14:2658–2661
48. Tang X-F, Zhao J-N, Wu Y-F, Zheng Z-H, Feng S-H, Yu Z-Y, Liu G-Z, Meng Q-W (2019) Org Biomol Chem 17:7938–7942
49. Christoffers J, Baro A, Werner T (2004) Adv Synth Catal 346:143–151
50. Reddy DS, Shibata N, Nagai J, Nakamura S, Toru T (2009) Angew Chem Int Ed 48:803–806
51. Fernandes SD, Porta R, Barrulas PC, Puglisi A, Burke AJ, Benaglia M (2016) Molecules 21:1182
52. Osorio-Planes L, Rodríguez-Escrich C, Pericàs MA (2016) Cat Sci Technol 6:4686–4689
53. Kisszekelyi P, Alammar A, Kupai J, Huszthy P, Barabas J, Holtzl T, Szente L, Bawn C, Adams R, Szekely G (2019) J Catal 371(255):261
54. Cañellas S, Ayats C, Henseler AH, Pericàs MA (2017) ACS Catal 7:1383–1391
55. Porta R, Benaglia M, Annunziata R, Puglisi A, Celentano G (2017) Adv Synth Catal 359:2375–2382
56. Brenna D, Porta R, Massolo E, Raimondi L, Benaglia M (2017) ChemCatChem 9:941–945
57. Ragno D, Carmine GD, Brandolese A, Bortolini O, Giovannini PP, Massi A (2017) ACS Catal 7:6365–6375
58. Porta R, Benaglia M, Coccia F, Rossi S, Puglisi A (2015) Symmetry 7:1395–1409

59. Ötvös SB, Pericàs MA, Kappe CO (2019) Chem Sci 10:11141–11146
60. Szcześniak P, Buda S, Lefevre L, Staszewska-Krajewska O, Mlynarski J (2019) Eur J Org Chem:6973–6982
61. Ötvös SB, Llanes P, Pericàs MA, Kappe CO (2020) Org Lett 22:8122–8126
62. Morodo R, Gérardy R, Petit G, Monbaliu J-CM (2019) Green Chem 21:4422–4433

Top Med Chem (2021) 38: 275–316
https://doi.org/10.1007/7355_2021_116

Published online: 25 April 2021

Biocatalysis in Flow for Drug Discovery

Itziar Peñafiel and Sebastian C. Cosgrove

Contents

I. Peñafiel (✉)
Future Biomanufacturing Research Hub, Manchester Institute of Biotechnology, University of Manchester, Manchester, UK
e-mail: itziar.penafiel@manchester.ac.uk

S. C. Cosgrove (✉)
Future Biomanufacturing Research Hub, Manchester Institute of Biotechnology, University of Manchester, Manchester, UK

Lennard-Jones Laboratories, School of Chemical and Physical Sciences, Keele University, Staffordshire, UK
e-mail: s.cosgrove@keele.ac.uk

Abstract The use of enzymes in organic synthesis has seen a significant rise in the past 20 years. This is due to the increased application of directed evolution, allowing enzymes to be improved to the point of being the best catalyst for a specific function. In relation to drug manufacture, this has seen engineered biocatalysts used in several manufacturing processes. But in early-stage drug discovery, there is much less direct application. This has been attributed to slow pace of biocatalysts not matching up to the speed required in drug discovery. In this context, it seems flow chemistry can offer a route into discovery programmes for biocatalysis. This chapter will discuss some recent advances of biocatalysis in continuous flow and consider how this can impact on the use of enzymes in drug discovery programmes.

Keywords Biocatalysis, Bioprocessing, Continuous biocatalysis, Enzyme immobilisation

1 Introduction

1.1 Biocatalysis in Drug Development

Biocatalysis has seen extraordinary advances over the last 30 years since the discovery and development of directed evolution methods pioneered by Arnold and others [1]. Wild-type enzymes typically operate under mild conditions (room temperature & pressure, aqueous environments), making them attractive propositions as sustainable catalysts for organic synthesis [2, 3]. The chemistry of enzymes is often considered to be limited to natural substrates, but directed evolution has pushed their capability beyond this, with non-natural chemistries being reported more frequently now than ever before [1, 4].

Nevertheless, the way in which enzymes are used practically, much like traditional organic synthesis, has not really changed in the last 30 years. Most industrial applications of enzymes are performed as batch operations. A recent trend has seen the application of continuous flow applied to enzymes as a means to improving their performance as biocatalysts [5, 6]. Some of the key reasons for transferring biocatalysts into continuous flow reactors include:

- Improved productivity through immobilisation and application in packed-bed reactors
- Ability to apply non-ambient conditions to enable rate enhancement

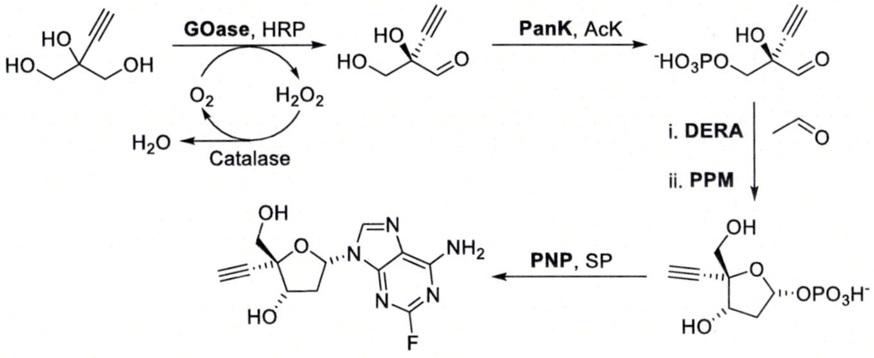

Fig. 1 Biocatalytic synthesis of islatravir, with bold enzymes representing enzymes that were engineered in the study [9]. *GOase* galactose oxidase, *HRP* horseradish peroxidase, *PanK* pantothenate kinase, *AcK* acetate kinase, *DERA* deoxyribose 5-phosphate aldolase, *PPM* phosphopentamutase, *PNP* purine nucleoside phosphorylase, *SP* sucrose phosphorylase

– Modularisation to permit ease of combination with other enzymes or chemical reactions

Biocatalyst application in industrial settings to date has primarily been as specialist catalysts used for large-scale preparation of active pharmaceutical ingredients (APIs) [7]. One famous example is the engineering of a transaminase for the synthesis of sitagliptin which is used as a treatment for type two diabetes [8]. A protein engineering campaign delivered a final enzyme that was tolerant of organic solvent (50% DMSO), high substrate loadings (>100 g L^{-1}) and high concentration of the amine donor, isopropylamine (1 M). More recently, the same group at Merck engineered five enzymes to achieve a one-pot synthesis of islatravir **1**, an experimental HIV drug [9]. All five enzymes were engineered to work on specific substrates, delivering the nucleoside analogue in an impressive 51% overall yield (Fig. 1).

These examples are very impressive and have changed the way chemists think about using enzymes for the manufacture of APIs. The approach to these manufacturing problems is to evolve a biocatalyst to a point where it is the best available catalyst to do a specific transformation. Having this ability to fine-tune the enzymes through directed evolution presents many opportunities for process chemists who want highly active, selective catalysts that operate under ambient conditions.

1.2 Enzymes in Early-Stage Drug Discovery Approaches

Nevertheless, the perception of enzymes in organic synthesis is often that they only work with specific substrates, which limits their scope in the discovery of new molecules [10]. Indeed, as discussed, the enzymes that are applied in the manufacture of APIs are often only able to work that effectively with the substrate they have been evolved towards. Some early-stage drug discovery approaches prefer low levels of stereoselectivity in the discovery of new bioactive molecules as there is less knowledge with regard to which stereoisomer of a molecule will interact most effectively with a protein target [11]. This can hinder biocatalyst application due to the fact they possess often perfect stereochemical control. Therefore, the application of enzymes in early-stage drug discovery is somewhat less.

There has been one application of enzymes to diversity-oriented synthesis (DOS) [12], by the groups of Arnold and Reisman [13]. An enzymatic cyclopropanation with a vinyl boronate ester **2** and α-diazo esters **3** generated boryl cyclopropanes **4** which were then derivatised via a Suzuki reaction to give a small library of cyclopropanated aromatics **5** (Fig. 2).

This being the single example of biocatalysis for a synthesis-oriented drug discovery approach highlights an absence of biocatalysts from the portfolios of many pharmaceutical companies' early-stage drug discovery programmes. Several reports and reviews have discussed the benefits that biocatalysis could bring to this part of drug discovery, however this does not seem to have changed the pathway as of yet [10, 14]. This is where flow chemistry could play a role in enabling enzymes to help in the search for new bioactive molecules.

We recently made the case for enzymes application in early-stage drug discovery to be increased, specifically with respect to fragment-based drug discovery [15]. We argued as well as the necessity for new biological tools such as protein engineering, there is also the opportunity for reaction engineering to play a role in the enablement of biocatalysis in different aspects of early-stage drug discovery. Process design has a significant effect on how a biocatalyst performs, so the importance of designing reactions in an efficient manner and applying the correct tools can vastly improve the output of enzymes in organic synthesis [16].

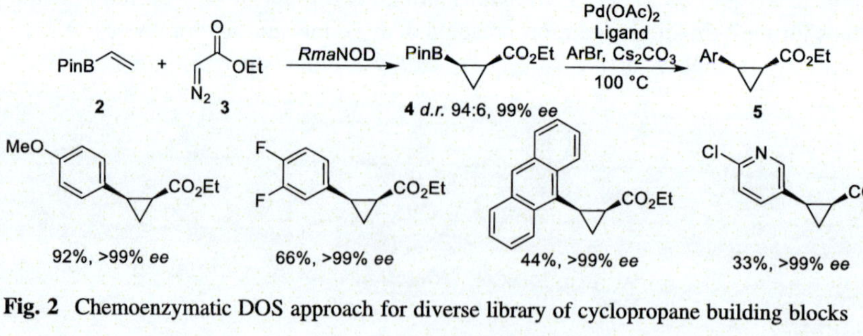

Fig. 2 Chemoenzymatic DOS approach for diverse library of cyclopropane building blocks

1.3 Flow Chemistry as an Enabling Tool for Enzymes

The benefits of flow chemistry to enzymes are well documented and have been covered in several prior reports by ourselves and others [5, 6, 17]. Due to the high volume of previous reviews, we will mainly focus on the last few years (2016-), although there are many additional examples of continuous biocatalysis reported prior to this point. This chapter will focus on new reactor designs and novel syntheses that could play a part in drug discovery.

1.4 To Immobilise or Not to Immobilise?

Most examples of continuous biocatalysis use immobilised enzymes. This is attributed to the many benefits that immobilisation can afford to a protein, mainly increased stability and reuse, as well as the ease of applying immobilised catalysts in packed-bed reactors [6, 17]. The stability benefits from immobilising enzymes are advantageous not just to flow biocatalysis, but also to the application of enzymes more widely used in synthesis, with immobilised enzymes having played important roles in industrial processes for decades [4]. Immobilised enzymes can be easily recovered, therefore improving overall costs of a bioprocess.

This has been applied in batch for several important industrial processes. For example, the biocatalytic cleavage of penicillin G affords the intermediate substrate 6-aminopenicillanic acid which is used to produce a host of semi-synthetic penicillin derivatives [18]. This is performed with penicillin G acylase on a multi-tonne scale every year, with an immobilised preparation used to reduce costs. In addition, the isomerisation of glucose to fructose is carried out by glucose isomerase which is a high-cost enzyme. Immobilisation of the isomerase permits its reuse and makes the process economically feasible [19].

Ultimately, immobilised biocatalysts provide higher economy to bioprocesses. They are also amenable to flow chemistry as they can be easily applied in packed-bed reactors. In this chapter we will discuss applications of both immobilised and soluble enzymes, however the latter has much lower representation for the reasons mentioned.

2 Transaminase

Transaminase enzymes catalyse the transfer of an amino group between two molecules via pyridoxal-5′-phosphate (PLP). They are classed as a transferase and have been used by synthetic chemists to produce chiral amines. The use of an engineered transaminase (TA) for the industrial production of sitagliptin spurred application of this class of enzyme in organic synthesis [8, 20], including multiple applications of

TA in continuous flow reactors. Some of the general challenges with using TA include the requirement for the exogenous addition of PLP as a cofactor and the use of costly amine donors in large amounts due to the equilibrium state of the TA reactions.

2.1 *Transaminase in Continuous Flow*

An early example of TA in flow came from the Jamison group, where whole *E. coli* cells expressing the TA from *Arthobacter* were immobilised on methacrylate beads [21]. The authors described the continuous amination of four α-alkoxy/aryloxyl ketones including the precursor **6** to mexiletine **8**, where the system was run continuously for 5 days at 50 °C with *ee* consistently at >99% and conversion maintained at approximately 70–80% for the duration, with isopropylamine **7** used as amine donor (Fig. 3). The PLP cofactor was also immobilised with the *E. coli* and impressively the reaction was run entirely in MTBE, meaning the PLP did not leach from the immobilised cells due to its preference for aqueous environments. The ease of production of whole-cell preparations offers a quick method for screening potential enzymes using this type of continuous reactor.

The Paradisi group has spearheaded the use of immobilised TA in continuous flow reactors over the past 5 years. An early such example exploited the His-tag technology to apply the isolated TA from *Halomonas elongata* (HEWT) in the bioamination of several aldehydes [22]. The stability of the enzyme was significantly enhanced upon immobilisation and could be stored for a month in the fridge with less than a 10% drop in activity over that time. The paper described the production of several amines including cinnamylamine, vanillylamine and alanine from their respective carbonyl precursors. Perhaps due to the removal of the cell, this system was able to produce all the amines in much shorter timeframes than the previous example (Fig. 3), with the amination of pyruvate proceeding at >99% conversion with only 1 min residence time. Importantly, the immobilisation procedure could be applied directly to the crude cell extract, exploiting the selectivity for the His-tags and avoiding the need to purify the biocatalyst.

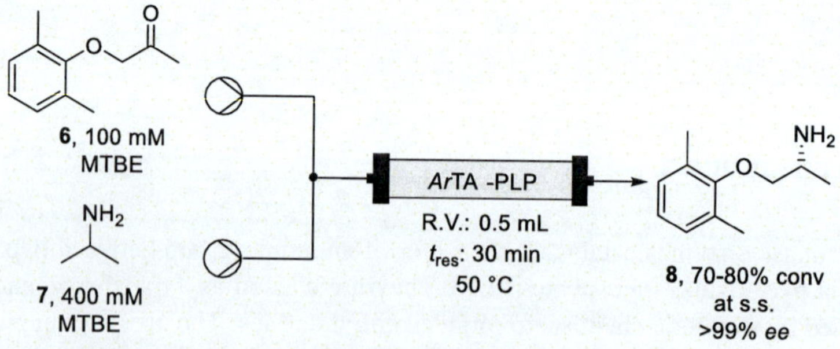

Fig. 3 Continuous production of mexiletine using immobilised whole-cell TA

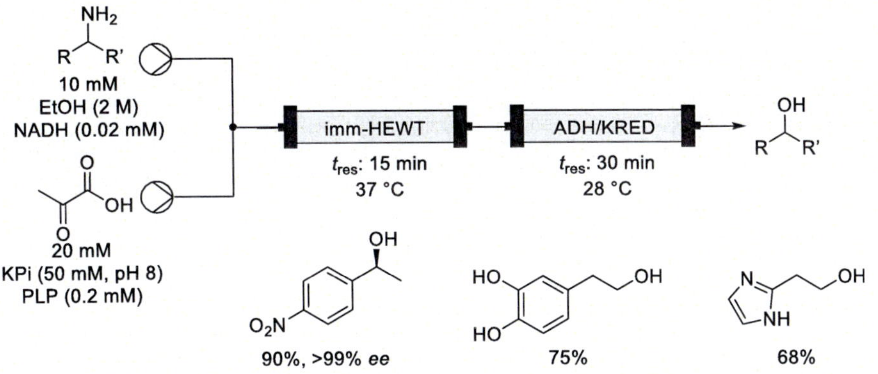

Fig. 4 Two-stage deamination then reduction for continuous production of biogenic alcohols

The same group used the same enzyme however this time in reverse, using the reversible nature of the transferases [23]. A packed-bed reactor containing the immobilised HEWT was used to continuously deaminate a range of amines, which were directly pumped into a column containing a ketoreductase (KRED) to deliver the alcohol derivatives. The biogenic alcohols included compounds derived from dopamine, cinnamaldehyde and histamine (Fig. 4). The modularity of the described reactors was further demonstrated with several in-line product recovery steps that also extended to an elegant system that permitted closed-loop recycling of the costly cofactors that were necessary for the biocatalysts (NADH and PLP). The closed-loop system also removed the gluconic acid by-product which was generated by the GDH oxidation of glucose.

The same group has used a similar strategy with the same TA enzyme to generate biogenic aldehydes, rather than the fully reduced alcohol products as shown above. The first report detailed the synthesis of a panel of 10 benzaldehyde derivatives from the respective benzylamines [24]. Using almost identical conditions to that described above (Fig. 4), impressive residence times as low as 3 min meant the reactor had high efficiency, but it was also coupled to a continuous extraction tool, using the Zaiput extraction membrane. Use of this permitted the removal of buffer which contained unreacted starting material and cofactor, whilst producing a stream of aldehyde in organic solvent being continuously removed from the reactor, allowing for potential combination with further compatible chemical reactions (Fig. 5).

The authors applied the same system to extend the scope to the deamination of a panel of 2-hydroxyarylethylamine derivatives to generate several alkyl biogenic aldehydes as well [25].

Another study from Semproli et al. studied the immobilisation of a different TA on a glyoxyl-agarose support for the continuous production of an amino-pyrimidine **10** used in the synthesis of JAK2 kinase inhibitor AZD1480 **11** [26]. The authors discussed an important implication of immobilisation, namely the loss of activity at the cost of stabilisation. Using the agarose support, the method for immobilisation was through transient imine formation between aldehyde groups on the support and

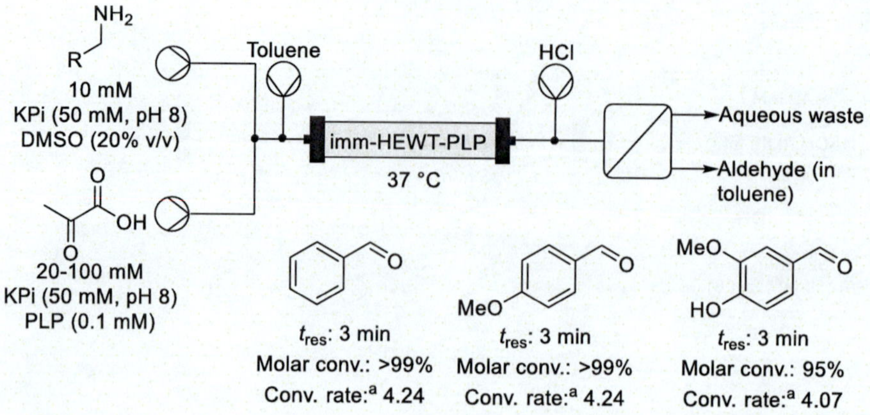

Fig. 5 Continuous deamination to produce aldehydes [a]Conversion rate in units of μmol min^{-1} $g_{Biocatalyst}^{-1}$

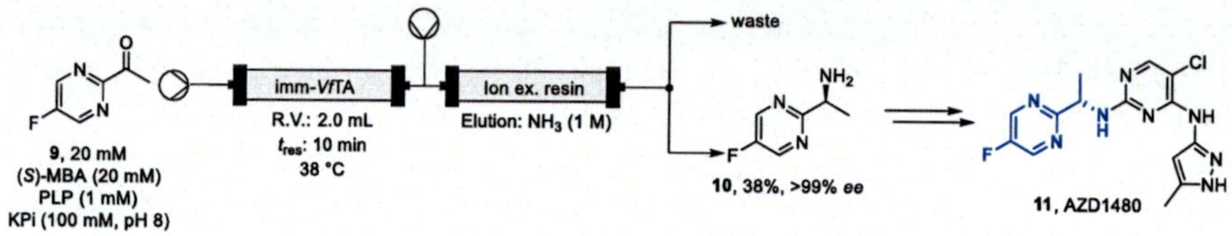

Fig. 6 Synthesis of intermediate for API for experimental API AZD1480 **11**

surface amines of the protein. To reduce leaching, reduction of the imines was implemented, however in all cases <10% activity recovery was obtained. Use of glycerol, often used for protein preservation, and reduction of the temperature of the immobilisation increased the recovered activity to around 30% (Fig. 6).

Using the TA from *Vibrio fluvialis* (*Vf*TA), the product was continuously produced and recovered in-line to allow for maximum efficiency. Under the optimised conditions, 100 mL of reaction mixture was passed through and collected.

Using another TA and a different resin, Mutti and co-workers also used an affinity resin for the immobilisation of two stereocomplementary TAs, namely the (*R*)-selective TA from *Arthobacter* (*Ar*TA) and the (*S*)-selective TA from *Chromobacterium violaceum* (*Cv*TA) [27]. The authors used an affinity resin called EziG, which comes in three formats [28], deciding on the carrier named EziG[3] or amber. The benefit of immobilisation, as mentioned already in this chapter, is that the enzymes are stabilised which can allow non-aqueous conditions to be used. The authors described using hydrate salt pairs to carefully control the water content of different organic solvents to ensure the immobilised biocatalyst did not lose activity during continuous use. The closed-loop flow system was run with 20 mL of reaction mixture in toluene, with a packed-bed column containing a mixture of hydrated phosphate salts, which was then run directly into the immobilised TA column and recycled. The reaction was run for five sets of 24 h, achieving a productivity of

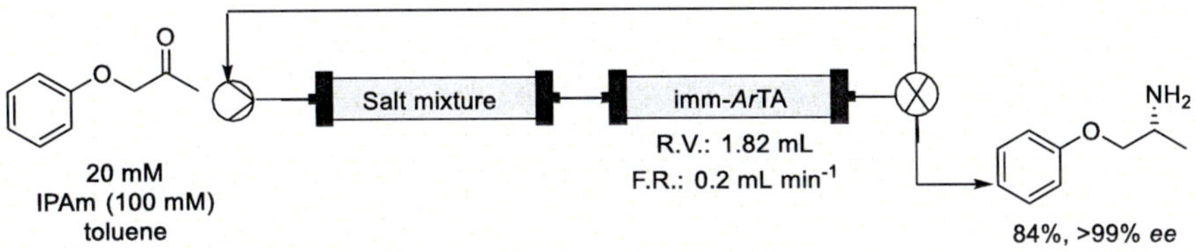

Fig. 7 Closed-loop transamination for aryloxyamine synthesis

around 32 mg d^{-1} and a space-time yield of 1.99 g L^{-1} h^{-1}. No additional PLP was required (Fig. 7).

Several groups have recently applied new approaches to the application of TA enzymes in continuous flow. In 2017, the groups of Sans and O'Reilly used 3D printed bioreactors for the expedient immobilisation and applications of TA enzymes [29]. The cartridges were printed from nylon fibres then chemically etched with acid and functionalised with glutaraldehyde to allow for enzyme immobilisation. This study was more focussed on the design of the reactor, with a simple deamination of α-methylbenzylamine (MBA), the only synthesis demonstrated. Nevertheless, the speed with which these new reactors could be designed gives a glimpse into what could be possible in terms of rapid prototyping of biocatalytic reactions.

Debecker and co-workers have also reported the same reaction (with the *p*-Br derivative) with fabricated silica monolithic reactors [30, 31]. The reactors were fabricated using a sol-gel method, with the resultant reactors functionalised to allow glutaraldehyde cross-linking with proteins. The monolith reactors were applied to the resolution of the MBA derivative and importantly were stabilised over time. As with the previous example, this needs to be extended to additional substrates and enzyme classes to showcase its full potential, however having access to a variety of reactor types and immobilisation methods will help to broaden the chemistry attainable with biocatalysts.

In a similar process, Molnár et al. described the immobilisation of whole *E. coli* cells in a silica monolith structure [32]. The difference here to that described by Debecker was that during the polymer fabrication a solution of the cells expressing the TA was incubated to allow encapsulation within the reactor structure. The structures stabilised the cells so much that even after storage of several months the authors reported that full activity was effectively retained. They demonstrated the utility of the encapsulated enzyme complexes through the continuous kinetic resolution of four amines (Fig. 8). Either enantiomer of the amine could be obtained through selection of the appropriate TA, with the (*S*)-selective TA from *Arthobacter* (*Ar*(S)TA) or the (*R*)-selective TA from *Aspergillus terreus* (*At*TA), the two TA found to be the most effective in this study.

Transaminase has also been combined with a transketolase (TK) enzyme by Szita and co-workers to synthesise (2*S*,3*R*)-2-amino-1,3,4-butanetriol (ABT) **12**, an amino triol that is used in the synthesis of several important molecules [33]. Transketolases mediate aldol reactions at the expense of CO_2, with the erythrulose product of this reaction a substrate for the TA reaction. Whilst only demonstrated on this one

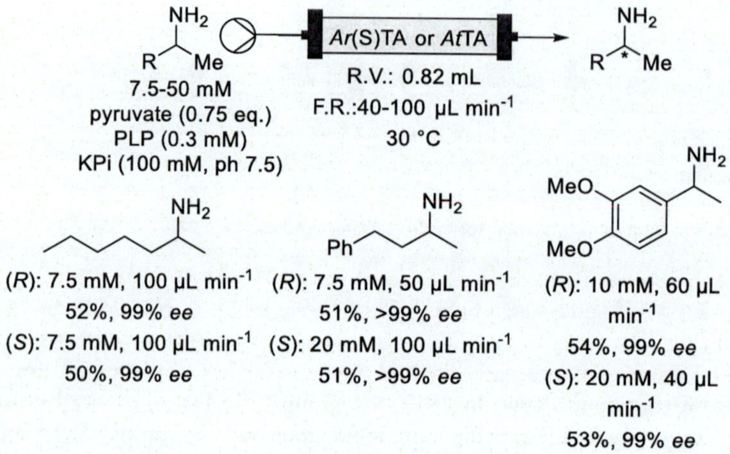

Fig. 8 Monolith encapsulated whole cells expressing TA for amine kinetic resolution

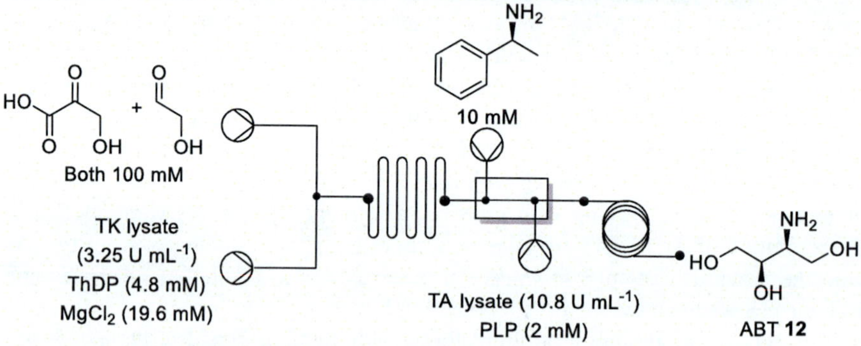

Fig. 9 Continuous transketolase/transaminase synthesis of ABT. *ThDP* thiamine diphosphate

substrate, the authors discussed at length issues with previous batch syntheses that this flow system overcame. The final optimised conditions showed the TK step could be completed in under 10 min, and this fed directly into the TA reaction without need for purification. Batch efforts from others had shown this had inhibitory effects on the TA (Fig. 9).

2.2 *Outlook for Transaminase in Continuous Flow*

Transaminase is without doubt one of the most important biocatalysts available to synthetic chemists, underlined by its ever expanding role in the production of important molecules [20]. This has also seen it applied in continuous flow more than many other enzymes. Despite this, the same challenges relating to equilibrium that plague TA use in batch also apply to flow. Some of the smart solutions involving continuous product removal have helped to alleviate these issues somewhat. There is still yet to be a significant application for early-stage drug discovery. This will change as flow technology improves, as with all enzymes. What will certainly help is the number of transaminase enzymes that are available, both natural and engineered. Having such a broad substrate scope allows a greater volume of chemical space to be accessed, an essential part of early-stage drug discovery.

3 Oxidoreductases

Oxidoreductase enzymes in general catalyse the oxidation and reduction of molecules. The subclasses include dehydrogenase, oxidase, reductase, oxygenase, peroxidase and hydroxylase. The most common classes are dehydrogenase and reductase, consequently meaning there are more reports of them being applied in synthesis and continuous flow.

3.1 *Ketoreductase and Alcohol Dehydrogenase*

Ketoreductase (KRED) and alcohol dehydrogenase (ADH) enzymes catalyse the reversible reduction of carbonyl compounds to the corresponding alcohols [34, 35]. They are one of the biocatalysts that has been applied most within industrial settings as they have broad activity and deliver chiral alcohols in perfect stereocontrol.

Several examples of KRED batch processes have been scaled up to meet the requirements for large-scale carbonyl bio-reduction. One such example is the synthesis of key intermediate for the cholesterol-lowering blockbuster drug atorvastatin (Fig. 10) [36]. The described process demonstrated how KREDs could be applied to the production of important APIs, with the final process running at very high substrate loadings on a 2,000 L scale, delivering a final catalyst productivity of 117 $g_{product}\ g_{enzyme}^{-1}$.

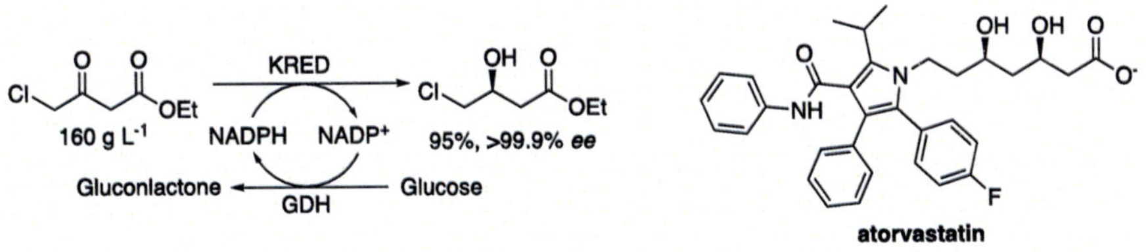

Fig. 10 KRED-catalysed reduction of ethyl-4-chloroacetate for API used in synthesis of atorvastatin

3.2 Application of KRED/ADH in Continuous Flow

An early example of a continuous KRED process was reported by Li et al., where covalent immobilisation of a commercially available biocatalyst on an epoxy resin allowed the continuous production of a chiral benzyl alcohol **14** for 7 days [37]. The alcohol product **14** was an intermediate in the synthesis of Emend **15**, an NK1 receptor antagonist for chemotherapy-induced emesis, previously synthesised by asymmetric transfer hydrogenation [38]. Screening of the selected KRED with several supports in batch demonstrated stability in high levels of organic solvent (up to 90% isopropyl alcohol (IPA)), which also permitted high substrate loadings (up to 50 g L^{-1}) to further improve the overall process. This process was then transferred to a packed-bed reactor (PBR) where it was run continuously for 7 days, producing the product **14** at 98% conversion with perfect stereoselectivity (Fig. 11). The authors commented that the tight binding of the cofactor by the enzyme meant there was no requirement of exogenous addition of NADPH after the enzyme had been immobilised. The regeneration occurred through oxidation of the IPA solvent which was in a large excess.

The continuous process clearly required improvement, with a 10-h residence time necessary to obtain near full conversion. But with high substrate loading and perfect stereoselectivity this example shows that KREDs can be used in continuous reactors under generally challenging conditions.

The groups of Tamborini and Serra demonstrated an elegant example of using an immobilised KRED in a packed-bed reactor to selectively reduce several interesting molecules that could hold promise as bioactive molecule precursors [39]. The KRED from *Pichia glucozyma* (*Pglu*KRED) was co-immobilised on an agarose gel with glucose dehydrogenase from *Bacillus megaterium* (*Bm*GDH). Using 4-nitroacetophenone as a test substrate, the authors found the immobilised biocatalysts had increased solvent tolerance (up to 20% DMSO) and could be easily recycled in batch. The enzyme also, impressively, retained around 70% of the initial activity after 6 months in the reduction of 4-nitroacetophenone. The continuous system was then applied to the reduction of several diketones shown below (Fig. 12). The small reactor was run continuously for 15 days for all three of the substrates, producing mmol of product with quantities of enzyme that could only produce μmol amounts in batch.

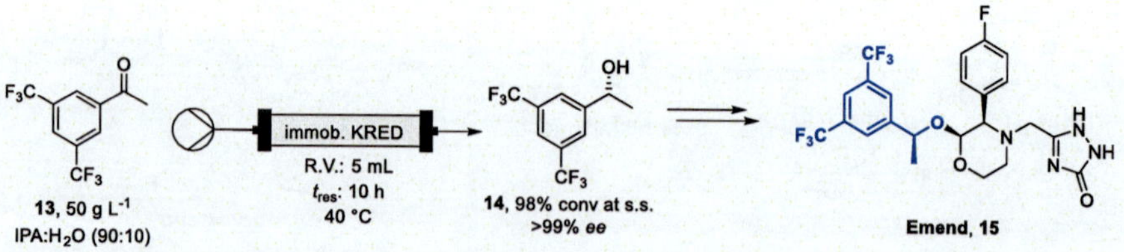

Fig. 11 Continuous KRED-catalysed production of 2,5-bis-trifluoromethylbenzyl alcohol **14**

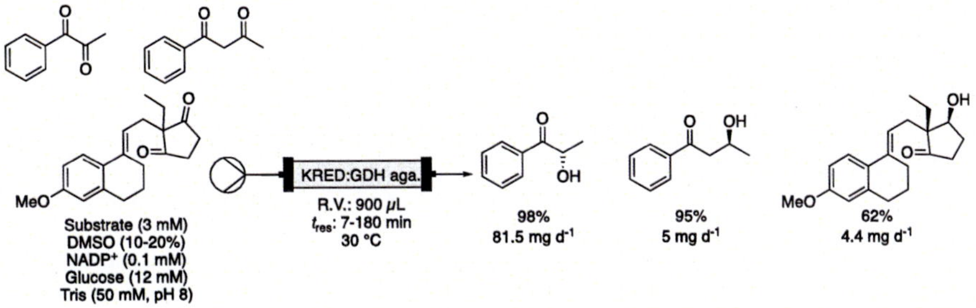

Fig. 12 Diketones reduced using immobilised *Pglu*KRED. *aga.* agarose

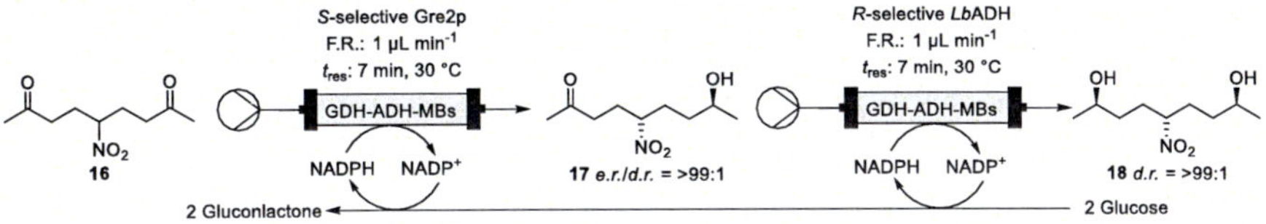

Fig. 13 Dual enzymatic system for stereoselective reduction of NDK **16**

The divergent nature of this application could be applied to the generation of libraries of compounds, particularly when considering the potential for keto-alcohols in organic synthesis.

The Niemeyer group has demonstrated the reduction of 5-nitrononane-2,8-dione **16** (NDK) as model substrate across multiple flow systems. The interesting aspect of this molecule is that reduction of both ketones generates three stereocentres, therefore producing up to eight isomers. The Niemeyer group has shown that careful selection of ADH can deliver high diastereo- and enantioselectivities [40]. The group described the use of streptavidin protein for specific binding of the KRED. A small binding peptide is coded with the protein of interest, which binds to streptavidin with high affinity. The streptavidin protein was immobilised on magnetic beads, which after immobilisation of the KRED could be isolated by magnetic separation. The process took 30 min and allowed immobilisation of the protein directly from the crude lysate with a simple recovery of the final immobilisate (Fig. 13).

The system was extremely small, with reactor volumes ranging from 7–10 μL, so this has a long way to go before it could be applied in large quantities. But it could be envisaged that this system be applied to general use biocatalysts for the rapid screening of substrates for activity. The Niemeyer group has also applied systems that use isopropyl alcohol and another magnetically immobilised ADH to turn over the nicotinamide cofactor [41] and 3D printed reactor cartridges for thermostable enzymes [42].

Very recently the Vincent group demonstrated that lactate dehydrogenase (LDH) could be immobilised by adsorption to carbon black nanoparticles [43]. This readily available support was able to immobilise both the LDH and the recycling enzymes formate dehydrogenase (FDH) together, to allow for in situ cofactor recycling. Full conversions were achieved in only 10 min in some cases. This simple system that operates via adsorption could offer quick ways to test new biocatalysts in flow systems as the carbon supports are low cost.

3.3 *Outlook for KRED/ADH in Drug Discovery*

The potential for KREDs and ADHs in drug discovery lies in the exquisite control they deliver in the products of their reactions and the utility a lot of these products could find in the synthesis of bioactive molecules. The speed with which several of these enzymes have been shown to work underlines their potential for use in small molecule or library generation. As with TA, there are many KREDs/ADHs to choose from, giving access to a lot of chemicals space. Having multiple biocatalysts also offers different starting point for directed evolution campaigns, further increasing potential chemical space that can be accessed.

3.4 *Other Oxidoreductases in Continuous Flow*

3.4.1 Imine Reductase in Continuous Flow

Imine reductases (IREDs) catalyse the stereoselective reduction of imines to afford chiral amines [44]. As with KRED/ADH, they are nicotinamide dependent. They are attractive to chemists as the ability to reduce imines stereoselectively in water is inherently challenging due to their instability in aqueous media. They were only discovery within the last 10 years, but there has even been a sub-class that was disclosed by Turner and co-workers in 2017 that catalysed intermolecular reductive amination of carbonyl and amines compounds, termed reductive aminases (RedAms) [45]. This significant discovery has already been taken up by industry, with GSK using an engineered RedAm for the production of an API [46].

Recently, Mangas-Sanchez et al. disclosed new RedAms that could accept ammonium salts as the amine source to produce primary amines [47]. The benefit of using this over TA is the use of cheap ammonium salts as the amine donor, in this case ammonium chloride or ammonium hydroxide. As the reaction is nicotinamide dependent there are also no equilibrium issues to overcome. The reaction proceeded in good conversions on analytical batch scale with a panel of carbonyl compounds, however the requirement for high levels of ammonium (up to 1 M) destabilised the enzyme on scale leading to poor preparative conversions. The enzyme was therefore immobilised using the EziG system and applied in continuous flow for the continuous amination of 2-hexanone. Whilst the system maintained >99% conversion with perfect stereocontrol for up to 10 h, the high ammonium concentration led to leeching of the enzyme from the support. This proof-of-concept study shows however that enzymes that use cheap and readily available amine sources could challenge TA in the future.

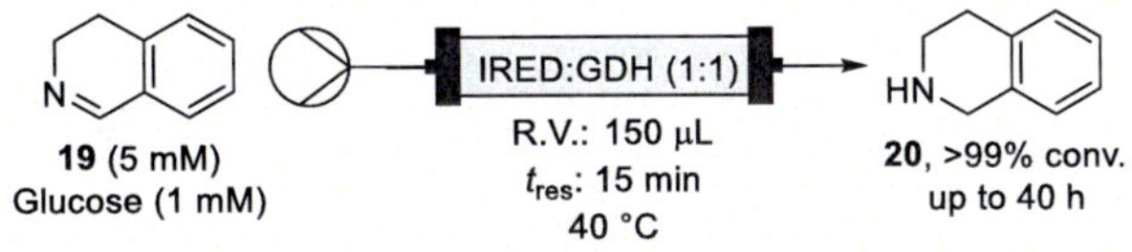

Fig. 14 Fused hydrogel of IRED and GDH via SpyCatcher system for continuous imine reduction

The Turner group also used a computational approach to optimise a continuous RedAm process [48]. The RedAm from *Ajellomyces dermatitidis* (*Ad*RedAm) was immobilised on EziG with *Bs*GDH. Using a mechanistic model, a set of optimal conditions were predicted based on the Michaelis kinetic parameters of the biocatalysts. The experimental results did not align exactly with the predictions, so the authors constructed an empirical model using a design of experiment, with the mechanistic model prediction as a starting point for optimisation. Only one reaction was studied, the reductive amination of hydrocinnamaldehyde and allylamine, but the modelling approach allowed the authors to tune the conditions to allow for high STY or low waste. The application of these computational tools significantly reduced the time it would have taken to optimise using a one factor at a time approach.

Niemeyer and co-workers have also shown that IREDs can be used in a continuous system [49]. They were specifically applied in a hydrogel system, formed exclusively from the proteins of interest, something they had already applied on the di-ketone system described previously (see Fig. 13) [50]. The enzymes self-assembled in pairs as they were encoded to contain the complementary SpyCatcher and SpyTag domains. Simply mixing the tagged proteins caused them to polymerise and form porous nanostructures that were used in packed-bed reactors. The IRED GF3546 from *Streptomyces* sp. was combined with *Bs*GDH to allow for a self-sufficient system that only required catalytic nicotinamide. This was applied to the reduction of dihydroisoquinoline **19** in a packed-bed reactor continuously for 40 h (Fig. 14).

The hydrogel system holds much promise for continuous biocatalysis. Being able to tag any nicotinamide-dependent biocatalyst to the GDH to allow for recycling is a powerful tool. In addition, this system required no additional support to be added, just the cost of the biocatalyst. Innovation such as this will help to drive down the cost of implementing enzymes in continuous flow and synthesis more widely.

3.4.2 Amine/Amino Acid Dehydrogenase

Amino acid dehydrogenase (AADH) enzymes are involved in the metabolic production of amino acids [51]. There is also a sub-class called amine dehydrogenase (AmDH) containing both engineered and natural biocatalysts that accept carbonyl substrates without alpha carboxyl groups [51, 52]. These enzymes are similar in activity to IREDs, being nicotinamide-dependent, and are often coupled to recycling enzymes for cofactor regeneration. A common strategy for AmDH is to use

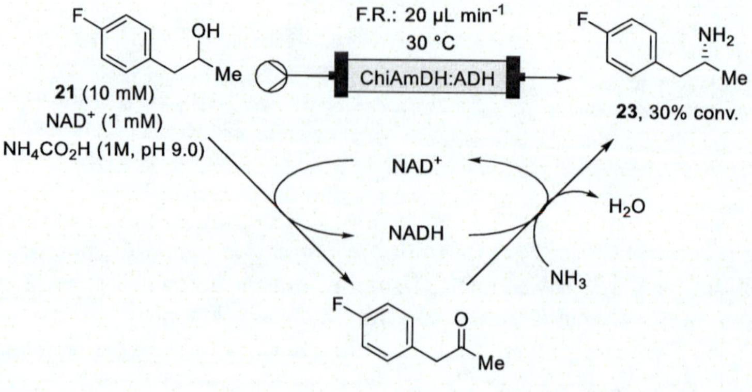

Fig. 15 Continuous biocatalytic borrowing hydrogen cascade for chiral amine synthesis

ammonium formate as buffer in concert with formate dehydrogenase (FDH); this provides both an amine source in ammonium and sacrificial hydride in formate [53].

There have been several applications of AADH and AmDH in flow reactors. Thompson et al. showed that a chimeric AmDH (ChiAmDH) could be immobilised via the His-tags on the EziG carrier and applied in continuous flow [54]. First, they showed that in situ cofactor recycling could be achieved with co-immobilisation of FDH for the continuous amination of 4-fluorophenyl acetone **22**. A conversion of 68% was maintained for 3 h, however this rapidly deteriorated after 6 h. The authors reasoned this was due to oxidation of cysteine residues on the FDH leading to deactivation, however it may also have been due to leeching from the high ammonium concentrations within the buffer (1 M). In the same study, the FDH was switched out for an engineered form of the ADH from *Thermoanaerobacter ethanolicus* (*Tes*ADH W110A/G198D) to enable continuous biocatalytic borrowing hydrogen. Using the alcohol precursor to **21**, a closed system whereby NAD^+/NADH continuously regenerate through oxidation/reduction of the cascade products delivered the same amine **23** continuously (Fig. 15) [55].

Franklin et al. also showed a single substrate for AmDH and FDH working with in situ cofactor recycling [56]. The authors reported the use of a different affinity resin, namely the Nuvia IMAC Ni-NTA from Bio-Rad, which works in the same way as the EziG discussed already. The authors described a lengthy optimisation for the continuous production of (*R*)-5-methyl-2-aminohexane from the precursor ketone, with a maximum conversion of 48% obtained with an 11-min residence time. Importantly, the enzymes remained stable for up to 5 days, much longer than the previous report which used EziG. However, as with the previous report, low conversions will need to be addressed to allow more use in target synthesis.

AmDHs were engineered from amino acid dehydrogenase (AADH), naturally occurring enzymes from amino acid biosynthesis pathways. López-Gallego and co-workers have conducted much work into different immobilisation systems and recently disclosed a two-enzyme system (AADH and FDH) that could be used for

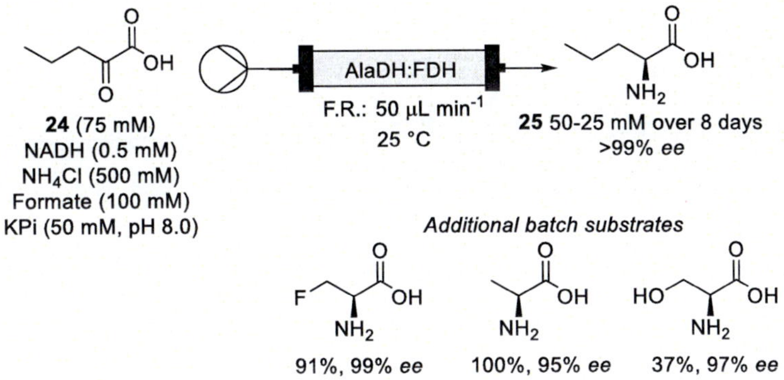

Fig. 16 Continuous synthesis of norvaline **25** and batch synthesis of amino acids using immobilised AlaDH

the continuous production of amino acids [57]. The alanine dehydrogenase from *Bacillus subtilis* (*Bs*AlaDH) was used in combination with FDH whilst immobilised on a hierarchical architecture in which *Bs*AlaDH was immobilised first, followed by the FDH after some mild chemical modifications. The study demonstrated a panel of amino acids, both natural and non-natural, could be synthesised under the batch conditions. They also reported a continuous production of non-natural amino acid L-norvaline **25** (Fig. 16).

Paradisi and co-workers showed that two dehydrogenases could be used in tandem to convert lysine **26** to pipecolic acid **28** [58]. They initially reported HEWT for the deamination of ε-amino group which then cyclised to the α-amino group which was reduced by a pyrroline-5-carboxylate reductase from *Halomonas elongate* (HeP5C). This cascade could only reach around 50% conversion and required coupling to FDH for the NADH recycling. To address these limitations, an NAD-dependent lysine dehydrogenase from *Geobacillus stearothermophilus* (Gs-Lys6DH) was combined with HeP5C for a redox neutral system that managed to reach >99% conversion with a 30-min residence time (Fig. 17).

These promising results hold promise for early-stage drug discovery due to the volume of available AADH enzymes. Being able to access amino acids quickly could allow access to new scaffolds for peptide-based pharmaceuticals.

3.4.3 Hydrogenase

Hydrogenase enzymes can catalyse the reduction of nicotinamide cofactors in the oxidised form to afford the reduced form that is essential for many hydride dependent bioprocesses [59]. They are attractive as they can use hydrogen gas as the reductant. This leads to much better atom economy when compared to GDH or FDH processes, which produce gluconolactone and carbon dioxide, respectively. There are obvious processing issues around using hydrogen gas, so the control that is

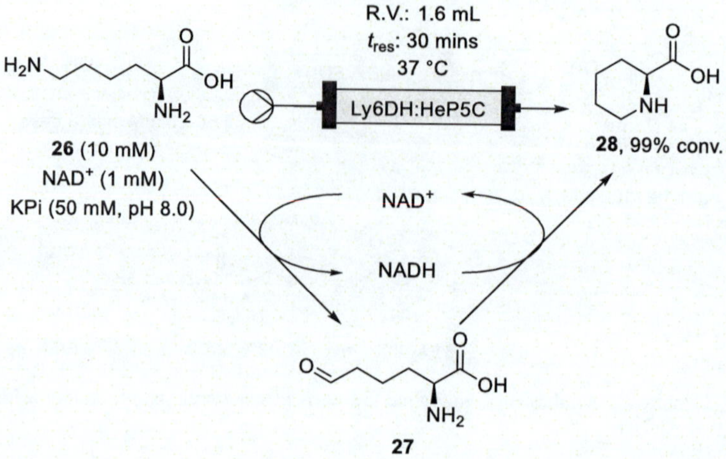

Fig. 17 Continuous conversion of lysine **26** to pipecolic acid **28** by a two-enzyme cascade reaction

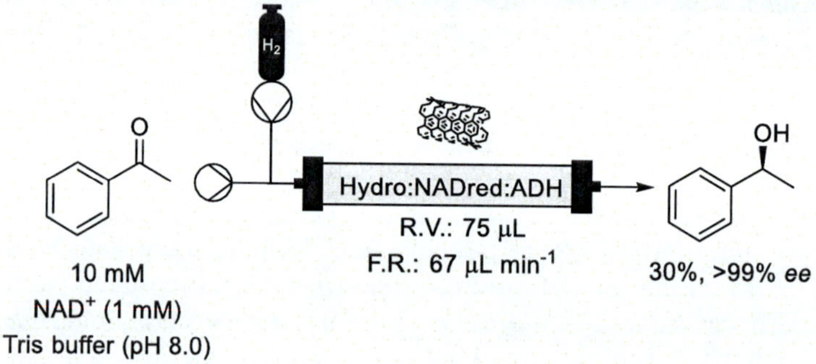

Fig. 18 Carbon nanotube supported three-enzyme cascade that uses H_2 gas as reductant

achievable with flow systems has allowed better application of these enzymes for synthesis.

The Vincent group has pioneered the application of H_2-driven biocatalysis, coupling hydrogenases to several nicotinamide-dependent biocatalysts. The group originally demonstrated that using graphite as an adsorptive scaffold for immobilisation could provide stable biocatalysts that operated in a cascade format that begins with H_2-driven reduction of $NAD(P)^+$ [60]. This involved immobilising three enzymes: the hydrogenase, NAD^+ reductase, and either an ADH or AADH. The hydrogenase provided electrons as a co-substrate for the NAD^+ reduction, which then provided the hydride equivalents for the enzyme of interest. The initial paper demonstrated the ADH-mediated reduction of acetophenone and the L-alanine dehydrogenase mediated production of alanine (Fig. 18). This impressive initial result was limited by the activity of the recycling system. When moved to a single-pass, fully continuous system, the conversion dropped to only 8% of the alcohol.

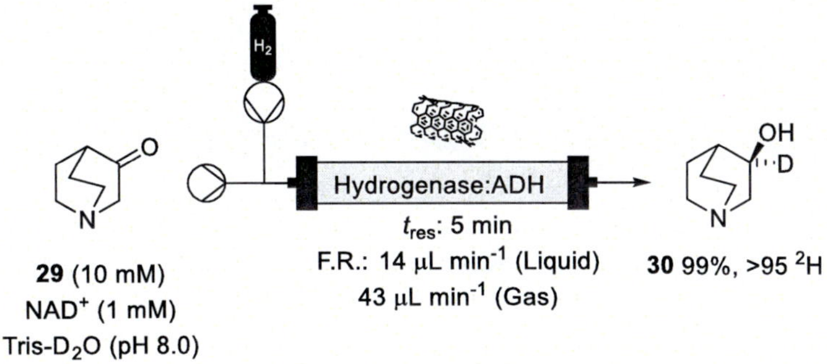

Fig. 19 Continuous deuteration of 3-quinuclidone **29** by NAD^{2}H production in continuous flow

The Vincent group has since improved and extended the application of hydrogenases in continuous flow to several systems that could have implications in early-stage drug discovery. A particularly important extension of this methodology was reported recently, where Thompson et al. demonstrated that biocatalytic deuteration could be achieved by switching to heavy water for the buffer [61]. A different hydrogenase was used which directly reduced NAD$^+$, obviating the need for the NAD reductase. The authors showed that NAD^{2}H could be continuously obtained at >99% conversion with a 132 s t_{res} using a single hydrogenase enzyme. They reported significant improvements on their initial system, including not having to the run system as a closed loop in a semi-continuous manner. In addition, this hydrogenase enzyme was coupled to an ADH to continuously reduce 3-quinuclidone **29** at >99% conversion and 95% incorporation of ^{2}H at the C_3 position (Fig. 19).

As with many examples in biocatalysis, the substrate scope has not been robustly tested. Nevertheless, this straightforward technology could play a very important role in future studies that require expedient deuterium incorporation.

3.4.4 Oxidases in Continuous Flow

Oxidase enzymes catalyse the irreversible oxidation of compounds using molecular oxygen as a co-substrate [62]. These powerful classes of enzymes have been used for the oxidation of alcohols [63], amines [64], as well as for the oxidation of sugar molecules with sugar alcohol oxidases [65]. Their synthetic appeal lies in the fact that they are often free from the requirement for expensive stoichiometric cofactors such as nicotinamide or PLP. They often only rely on oxygen; however, this brings with its other problems centred around oxygen supply [66]. The maximum solubility of oxygen in aqueous systems is approximately 250 µM [67], despite the fact the estimated kinetic requirements for engineered biocatalysts being much higher than this, sometimes as high as 5 mM [68]. This means under standard conditions many oxidase enzymes can be left operating under severe oxygen-limitation. Using flow chemistry, several solutions have been demonstrated that aid in the improvement of oxidase process performance through improved oxygen supply.

Carbohydrate Oxidases

Utilising the fact that biocatalysts can easily operate under identical conditions, Chapman et al. described the use of an in situ biocatalytic decomposition of hydrogen peroxide to generate higher concentrations of oxygen [69]. A new reactor, termed a multipoint injection reactor (MPIR), was used to build up reservoirs of hydrogen peroxide which were instantly degraded to oxygen and water by catalase, and extremely fast-acting and cheap enzyme. The study used an engineered form of galactose oxidase (GOase), termed GOase $M_{3–5}$, which was evolved from the original sugar oxidation enzyme to have a broader substrate scope by Turner and co-workers [70]. The increased soluble oxygen resulted in much faster reactions than when compared to the equivalent batch reactions. A set of benzyl alcohols were converted to their carbonyl counterparts with residence time of around 10 min, delivering at least 88% conversion for the 15 substrates that were tested (Fig. 20). The speed with which these products can be generated via this method presents an opportunity in potential library generation for early-stage drug discovery, with some of the aldehydes shown in the work medicinally relevant such as pyridine and halogen containing benzaldehydes.

A follow-up study to this from the same authors showed that different variants of GOase (M_1 and F_2) could also be significantly enhanced to produce large amounts of oxidised sugars, which hold promise as potential precursors to biobased materials [71]. The in situ oxygen supply allowed >200-fold improvement in the STY, and a sevenfold improvement in the biocatalyst productivity with respect to the GOase M_1 catalysed oxidation of lactose (Fig. 21). Being able to incorporate sugars into drugs is integral to finding new activity, however the synthesis of pure sugar products remains a challenge due to purification issues. Methods such as this, which deliver the oxy-sugars with full selectivity (no over-oxidation) and in 100% yield mean rare sugars could potentially be functionalised with ease to allow incorporation into new drugs.

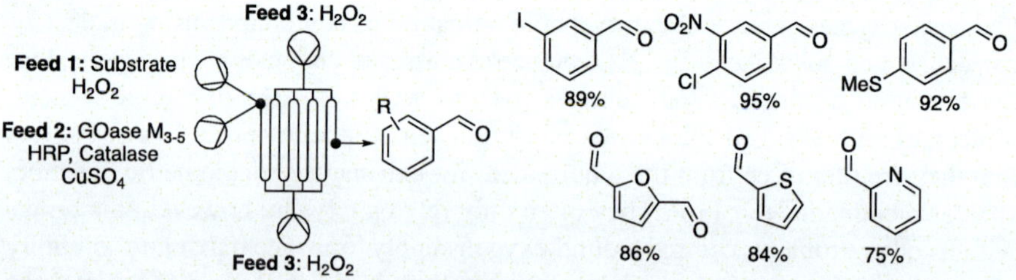

Fig. 20 MPIR reactor used to enhance GOase activity. Feed 1: Substrate at 30 mM, H_2O_2 at 30 mM; Feed 2: GOase CFE at 6.5 mg mL^{-1}, HRP at 0.1 mg mL^{-1}, Catalase at 0.13 mg mL^{-1}; Feed 3: H_2O_2 at 60 mM

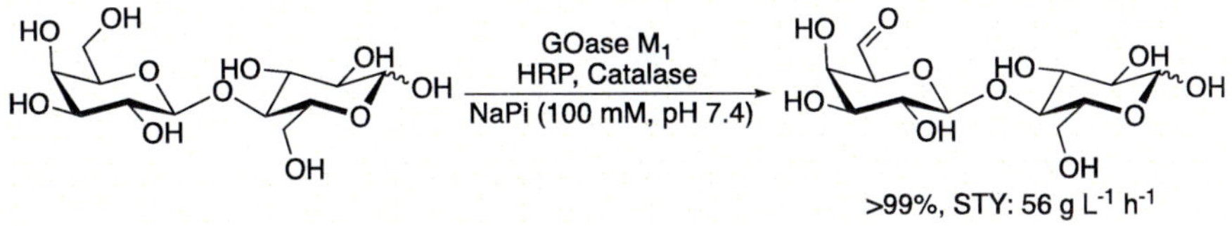

Fig. 21 The GOase M_1 catalysed oxidation of lactose using an MPIR

The unusual nature of this work is the use of soluble enzymes. Whilst the work shows a marked improvement over equivalent batch versions of the reactions, the long-term feasibility for larger scale applications is limited using soluble, non-recoverable enzymes. A system that could use the in situ oxygen-generation principle with immobilised enzymes is desirable and would represent a significant step forward.

There have been different approaches to improve oxidase enzymes in flow using improved gas supply to the reactor. Despite showing better metrics for oxidase enzymes, most specially designed reactors have yet to be applied to the synthesis of synthetically interesting molecules. Many systems employ glucose oxidase (GOx) as a proof of principle to demonstrate the benefits of the reactor. These include a falling film reactor which drops a stream of reaction mixture next to a gas flow moving in the opposite direction [72]. The authors calculated the oxygen transfer coefficient (k_La), which is independent for every reactor, to be 20,590 h^{-1}, which permits the system to saturate with oxygen in only 6 s. Comparing this to a normal beaker with 200 rpm stirring, which the authors calculated to have a k_La value of 1.13 h^{-1}, underlines how much of an increase in rate can be obtained with specialist reactor designs. This allowed the oxidation of glucose to proceed in a matter of seconds, with the same reaction taking 142 min in a beaker.

Several groups have also applied the Coflor agitated cell reactor to improve the rates of reactions of an enzymatic transformation. Woodley and co-workers improved the GOx reaction metrics, measuring a k_La value of 344 h^{-1} and allowing the full conversion of 100 mL of a 100 mM solution of glucose in 45 min [73]. The same system was also applied by Gaspirini et al. to enhance the rate of reaction of an amino acid oxidase, permitting a continuous resolution of 1 M solutions of racemic alanine [74]. The reactions on a 10 L scale were complete in only 8 h, showing how scalable the technology is.

While these such examples have minimal impact from a synthetic product perspective, the proof-of-concept articles demonstrate the potential flow chemistry has to improve the rate of reactions of oxygen-dependent enzymes and help with their uptake as general use catalysts.

Alcohol Oxidases

Alcohol oxidases act in a similar manner to that of carbohydrate oxidases, but on non-carbohydrate alcohol molecules rather than specifically on sugars [75]. As with

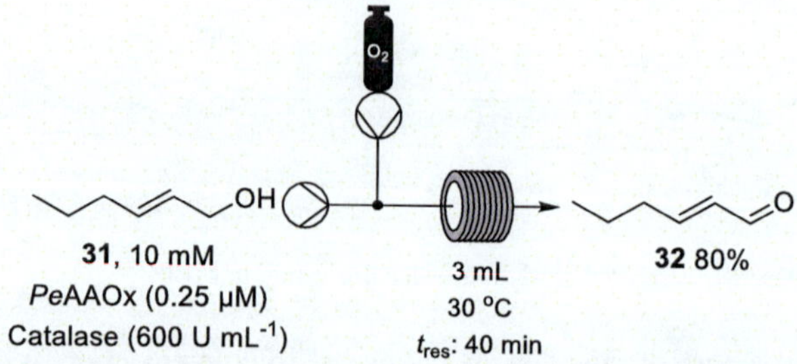

Fig. 22 Plug flow reactor used for *Pe*AAOx oxidation of *trans*-hex-2-en-1-ol **31**

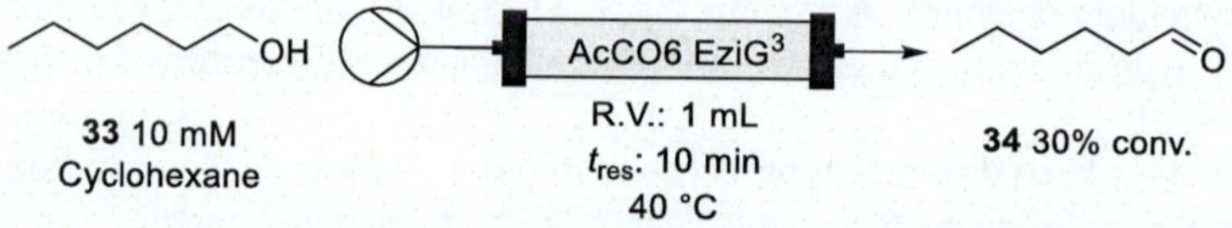

Fig. 23 Continuous oxidation of hexanol in pure cyclohexane using an engineered choline oxidase

other oxidases, they often depend only on flavin cofactors and oxygen as a co-substrate which leads to attractive propositions as catalysts, despite the fact they operate under sub-optimal conditions.

Hollman and Noel worked together using an alcohol oxidase from *Pleurotus eryngii* (*Pe*AAOx) in the continuous oxidation of *trans*-hex-2-en-1-ol **31** [76]. After determining apparent kinetic values ($K_M = 1$ mM, $k_{cat} = 22\ s^{-1}$), a slug flow reactor with phases of reaction mixture and gas was used to apply the enzyme to the oxidation of the allylic alcohol **31**. The oxygen transfer rate for the reactor was determined to be 0.25 mM min^{-1} at steady state, which led to turnover frequencies of up to 38 s^{-1}, higher than the apparent k_{cat} value in the literature and therefore underlining the importance of oxygen supply (Fig. 22).

Thompson et al. used an engineered choline oxidase for the continuous oxidation of hexanol **33** [54]. The enzyme was immobilised on EziG (discussed earlier), via the His-tags used for purification, and used in a packed-bed reactor. During the first run, the authors found that rather than selective oxidation to hexanal they obtained only hexanoic acid, likely from oxidation of the aldehyde hydrate which forms under aqueous conditions. The benefit of immobilisation was that the increased stability of the enzyme allowed use of cyclohexane as solvent instead of aqueous buffer. Under these conditions, the enzyme was completely selective for the aldehyde **34**, reaching a steady state conversion of 3 mM which was maintained for 120 column volumes with residence time of 10 min. The 3 mM concentration did however only represent a 30% conversion, once again demonstrating the need for improved oxygen supply for these classes of enzymes (Fig. 23).

Amine oxidases

In the same multipoint injection reactor (MPIR) article as that reported earlier (Fig. 20), the authors also used whole *E. coli* cells expressing monoamine oxidase D9, an enzyme that is capable of the selective oxidation of amines to imines

[69, 77]. Impressively, despite the viscous nature of the solution and the metabolic need for oxygen within the cells, the MPIR still managed to enhance the oxidation of tetrahydroisoquinoline to go to full conversion in a matter of minutes, with the equivalent batch reaction usually proceeding in a matter of hours. As with many examples of biocatalysis in flow, there is minimal substrate scope reported despite the promise reactors such as these have shown.

3.5 *Outlook for Oxidoreductases in Continuous Flow*

Many of these enzymes have only recently been discovered, and even their application under standard batch conditions is dwarfed by the synthetic equivalents. Nevertheless, the pace of application is fast, as evidenced by the increasing number of industrial syntheses that have been disclosed in the last few years [9, 46]. As new reactor technologies are discovered more applications will be reported; this will ultimately lead to increased usage for the discovery of novel bioactive molecules. The exquisite redox control imparted by oxidoreductases will be essential to reducing our need for unsustainable chemical reagents and catalysts. Flow chemistry has shown that it can provide a vehicle to enable them in synthesis.

4 Chemoenzymatic Catalysis in Flow for Drug Discovery

Chemoenzymatic reactions are catalytic cascade processes as a result of coupling chemical- and biocatalytic steps [78]. The main value of this integration is the reduction in the number of processing steps (8th principle of Green Chemistry). The use of biocatalysts decreases the number of synthetic steps (thanks to their high specificity). In addition, the combination of both avoids cumbersome separation steps as no isolation is necessary. As a result, the overall process is conferred by an impressive reduction of waste and energy; thus, new and improved synthetic routes can be developed for the generation of molecular diversity.

However, the integration of chemo- and biocatalysis presents intrinsic incompatibility derived from the specific reaction conditions needed in each case. Usually, organometallic catalysts work at high temperatures, high pressures and in organic solvents while enzymes most frequently operate at room temperature, atmospheric pressure and use aqueous buffers as reaction media. As a result of the efforts made for the integration of both catalytic techniques, several chemoenzymatic approaches have been described [79], and with some even applied in the synthesis of natural products [80–83].

Biocatalytic steps have been integrated into traditional synthetic processes [84]. In 2008, a chemoenzymatic process for the scalable manufacture of (*S*)-3-(aminomethyl)-5-methylhexanoic acid (Pregabalin **37**) was developed and optimised by Pfizer. Pregabalin is a lipophilic GABA (γ-aminobutyric acid) analogue which is used for the treatment of several central nervous system disorders

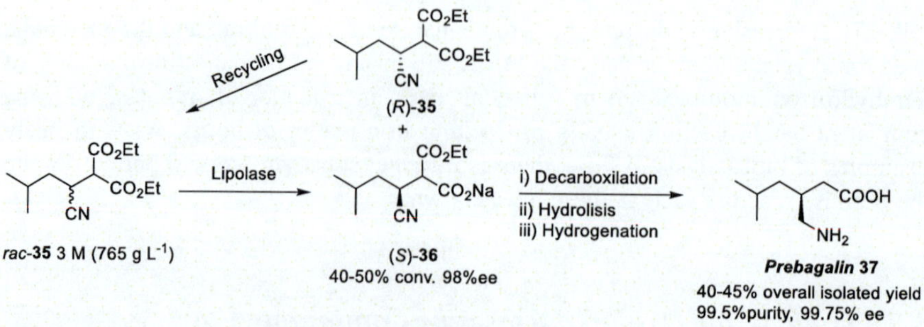

Fig. 24 Chemoenzymatic Pregabalin **37** Synthesis by Pfizer

including epilepsy, neuropathic pain, anxiety and social phobia [85]. The process uses a commercially available lipase from *Thermomyces lanuginosus* (Lipolase) to efficiently resolve *rac*-2-carboxyethyl-3-cyano-5-methylhexanoic acid ethyl ester *rac*-**35** to form the key enantiopure intermediate 2-carboxyethyl-3-cyano-5-methylhexanoic acid (*S*)-**36**. A heat-promoted decarboxylation of *rac*-**35** generates (*S*)-3-cyano-5-methylhexanoic acid ethyl ester (*S*)-**36**, a precursor of Pregabalin. The chemoenzymatic process resulted in dramatically improved yields of pregabalin (40–45% after one recycle of (*R*)-**35**) compared to the previous chemocatalyzed route. It also demonstrated the application of green chemistry by reducing the organic solvent usage in a mostly aqueous process: two isolation steps were eliminated, and the total reduction of waste resulted in an impressive fivefold decrease in the E factor from 86 to 17 (Fig. 24).

Although this optimised synthetic route demonstrated the clear advantages associated with chemo- and biocatalysis integration, the number of chemoenzymatic routes for industrially relevant molecules is minimal. The intrinsic compatibility issues challenge the combination of them in one-pot processes. It is at this point where continuous flow emerges as a key tool. Benefits imparted by using flow, such as separation of the catalysts, change of reaction conditions between steps or the possibility to run sequential reaction steps, can help in the combination of the two disciplines. This is consistent with the millions of efficient biotransformations that take place in a cell in continuous mode which chemists constantly attempt to mimic.

4.1 Chemoenzymatic Systems in Continuous Flow

In this section, the focus will be on chemoenzymatic reactions carried out in continuous flow during the last 5 years, describing high efficiencies, novel approaches and extended molecular diversity generation. This will not be extended to the coupling of biotransformations and stoichiometric steps such as (de)protection reactions or nucleophilic additions under continuous flow, which have been recently reviewed elsewhere [86–89]. Also, for a recent extensive review of biocatalytic

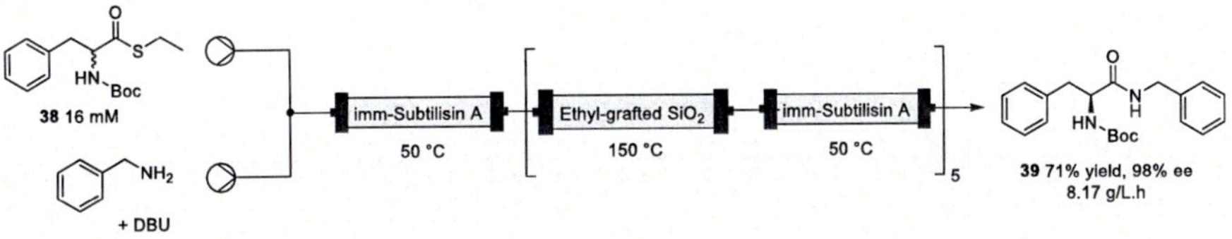

Fig. 25 Continuous Dynamic Kinetic Resolution of *N*-Boc-phenylalanine ethyl thioester by alternating packed-bed reactors containing chemo- and biocatalysts

cascade reactions, including chemoenzymatic reactions in batch, see Kroutil and co-workers [90].

One of the most frequently studied chemoenzymatic reactions is the Dynamic Kinetic Resolution (DKR). It consists of the generation of enantiopure products through the combination of enantioselective biotransformations with chemocatalytic racemisation. However, it is only recently that the first continuous example was published by the Poppe research group. They reported a continuous DKR for the generation of chiral amines from racemic *N*-Boc-phenylalanine thioethyl ester **38** [91]. The amide functional group is chemically stable, neutral and it can be both a hydrogen acceptor and donor. These desirable properties maybe explain its presence in around 25–30% of all pharmaceutical drugs [92].

The authors independently optimised both bio- and chemocatalytic steps. *Tert*-Amyl alcohol was identified as a common solvent and the main incompatibility (significantly difference reaction temperatures, 50°C for KR and 150°C for racemisation) was overcome by alternating a packed-bed enzyme reactor and a packed-bed base reactor at the selected temperatures (11 columns overall) (Fig. 25). The kinetic resolution (KR) was catalysed by an immobilised Alcalase (Subtilisin A). The enzyme was hydrophobically adsorbed onto ethyl-grafted macroporous silica gels. Secondly, the base (attached to an ethyl-grafted silica gel) was packed into a different reactor to carry out the racemisation step. The desired product (*S*)-*N*-Boc-phenylalanine benzylamide **39** was generated with a space-time yield of 8.17 g L^{-1} h^{-1}, in high yield and enantioselectivity (79%, 98% *ee*).

The same research group also reported a different approach to operate the continuous chemoenzymatic DKR of six benzylic amines using a mixed-bed DKR unit (a column filled with CAL-B/TDP10 and Pd/AMP-KG) (Fig. 26) [93]. In most of the examples, it was necessary to link just one column for KR (containing the immobilised biocatalyst) before a resolution unit. A six-membered library of chiral amides was generated in a fully continuous process, with isolated yields from 57 to 96% and enantioselectivities up to 98.8% for 48 h.

Another example taking advantage of continuous flow techniques to circumvent bio- and chemocatalyst incompatibility issues was reported by de Souza and co-workers [94], a continuous DKR of 1-phenylethanol **40** to give enantiopure (*R*)-esters with excellent conversions and high productivities.

In this case, integration of chemo- (heterogeneous catalyst $VOSO_4$) and biocatalyst (*Candida antarctica* lipase B, CAL-B, Novozym-435) was facilitated by finding common reaction conditions for both catalytic steps (reaction was carried

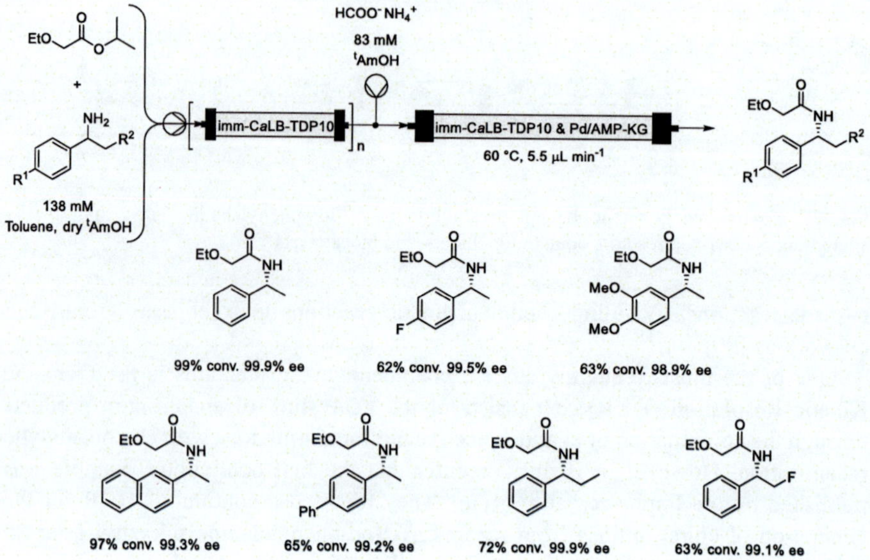

Fig. 26 Continuous chemoenzymatic dynamic kinetic resolution of amines mediated by a mixed-bed DKR unit

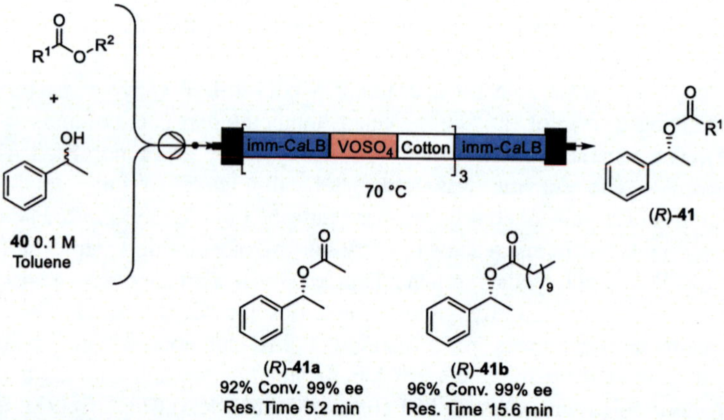

Fig. 27 Continuous CAL-B and $VOSO_4$ mediated chemoenzymatic dynamic kinetic resolution of 1-phenylethanol **40**

out at 70°C using toluene as solvent) (Fig. 27). To overcome the inhibition of the enzyme by the heterogeneous vanadium catalyst, the authors decided to alternate layers of CALB and $VOSO_4$, physically separated by thin cotton partitions in a single packed-bed reactor. This approach allowed the generation of the desired product with conversions of 92% and 96% for vinyl acetate and vinyl decanoate as acylating agents, respectively. In the case of employing vinyl decanoate, it was possible to increase substrate concentration 10-fold to generate (*R*)-phenylethyl decanoate (*R*)-**41b** with overall 82% yield (2.72 g in 2 h) and 90% *ee*.

The use of non-conventional solvents has also been described to successfully link chemo- and biocatalytic reactions under continuous flow mode. These solvents have been mixed with water to overcome well-known drawbacks such as high toxicity and viscosity. Although these examples involved the synthesis of just a singular product, we have decided to highlight them because they can serve as a proof-of-concept. They also bring to light the necessity of carefully designing chemical catalysts to work under aqueous conditions.

In 2018, Porcar et al. reported the use of supported ionic liquid like phases (SILLP) for the continuous synthesis of chiral 1,2-amino alcohols [95]. To enable the chemoenzymatic process, three different compartmentalised catalytic modules were assembled: The first was a biotransformation catalysed by CALB-SILLP-dec-NTf_2, where the alkene epoxidation occurred after organic peroxyacid generation. The second step consisted of the chemocatalysed epoxide ring opening, mediated by SIILP-SO_3-$Sc(OTf)_2$ with an amine as the nucleophile. In the third and final step, the KR was biocatalysed by Novozyme-435. The desired chiral product was generated in 92% conversion with 99% *ee* and a productivity of 0.96 g h^{-1}. Fundamental to the success of this reaction was dual role of dimethyl carbonate (DMC), as solvent and reagent for two of the three synthetic steps (precursor of the organic peroxyacid and acylating agent for the KR).

In an innovative approach using deep eutectic solvent (DES) with water, Grabner et al. developed an efficient integrated flow process combining an enzymatic decarboxylation and subsequent Heck coupling via a heterogeneous Pd-catalyst. Both catalytic steps took place in independent packed-bed reactors linked in line [96]. In this example, immobilised phenolic acid decarboxylase from *Bacillus subtilis* (*Bs*PAD) decarboxylated *p*-coumaric acid. Subsequently an aryl halide was added for the Pd-catalysed Heck cross-coupling to synthesise (*E*)-4-hydroxystilbene. The DES demonstrated an outstanding potential to overcome compatibility issues by increasing substrate solubility (substrate concentration went from 5 to 20 mM), maintaining enzymatic activity and stability and reducing chemocatalyst-mediated hydrolysis. The temperature incompatibility issue (30°C was required to maintain enzyme stability but the ideal reaction temperature for the palladium catalysed step was 85°C) was solved by compartmentalising the chemo- and biocatalyst in different packed-bed reactors. Full conversion was achieved for 16 h in continuous flow mode.

Recently, Chrobok and co-workers reported the first continuous flow chemoenzymatic Baeyer–Villiger oxidation mediated by CAL-B supported on commercially available, unmodified multiwalled carbon nanotubes (MWCNTs, Nanocyl NC7000) [97]. The nanobiocatalyst mediated the oxidation of 2-methylcyclohexanone to 6-methyl-ε-caprolactone in ethyl acetate, with 30% aq. hydrogen peroxide employed as primary oxidant (Fig. 28). There were six lactones generated in high conversions (83–99%) under mild reaction conditions (40°C) and short reaction times (5 min). Moreover, the nanobiocatalyst was still

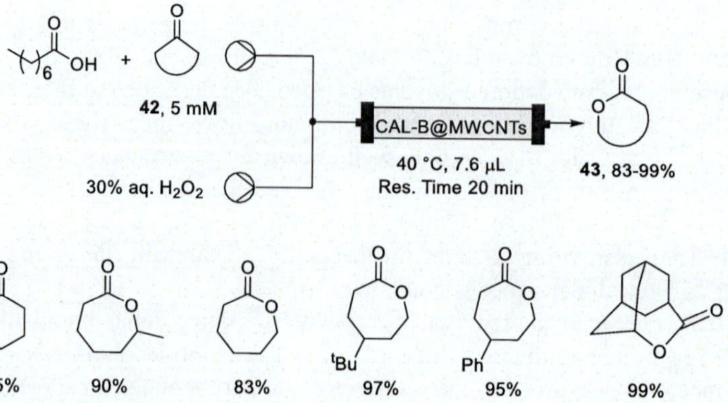

Fig. 28 Continuous CAL-B mediated chemoenzymatic Baeyer–Villiger oxidation

stable after 8 h on stream. Due to the instability of peracids, this approach provides a safer opportunity for this kind of transformation.

4.2 Outlook for Continuous Chemoenzymatic Processes for Drug Discovery

Continuous flow, which has shown potential for drug discovery [98], has emerged as a key tool to overcome incompatibility issues for the integration of bio- and chemocatalysis. Clearly, this is the beginnings of the field where efforts should be addressed to combine organocatalysts and enzymes, and to design chemical catalysts that are able to work in aqueous conditions. Some of the technological innovations, such as non-conventional media, provide a way to compartmentalise and even combine multiple types of catalyst. Some of the inventive ways that different reactions have been combined in batch, such as membrane separation or implementation of surfactants [78, 99, 100], will also undoubtedly have a role to play in the future of continuous chemoenzymatic synthesis.

5 Microreactors and Biocatalysis for Library Compounds Generation

Microreactors are defined as being flow reactors that are below 1 mm in channel size. This allows rapid mass and heat transfer, small reaction volumes and short diffusion. The application of biocatalysis in microreactors offers opportunities and challenges for faster and reliable implementation in industry [101]. Highlighted here are some recent examples where the use of this technology has enabled the quick generation of

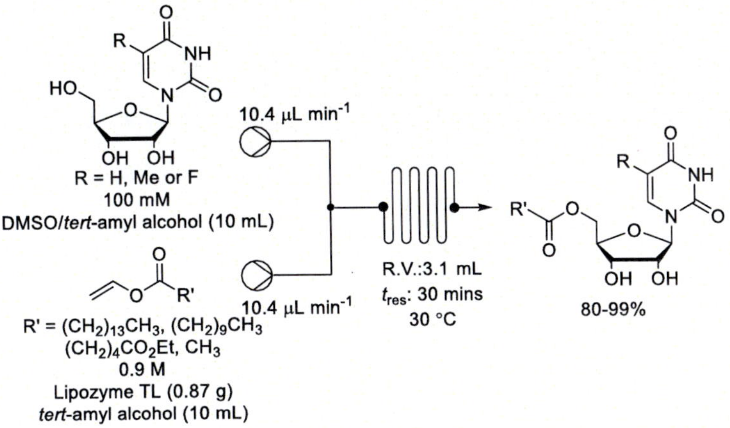

Fig. 29 Lipozyme-mediated acylation of nucleoside analogues

libraries of compounds without the use of protecting groups. This useful feature could help turn biocatalysis into an essential tool for drug discovery.

The Luo research group has successfully employed Lipase TL IM from *Thermomyces lanuginosus* for different transformations generating libraries from 12 to 20 products.

In 2018, they reported the continuous enzymatic acylation of the primary hydroxyl group of uridine derivatives with excellent conversions and regioselectivities [102]. Combination of different uridine derivatives and vinyl esters led to a substrate scope of 12 final products with conversions from 74–99%. The increasing importance of nucleoside analogues in the treatment of diseases, and the challenge of synthesising them with traditional methods, means a biocatalytic approach is timely. The biocatalytic synthesis presents a series of advantages, such as mild reaction conditions (30°C), short reaction time (30 min) and reduction of DMSO content (Fig. 29).

They used the same strategy to quickly build a thioester library through the lipase-catalysed transesterification reaction of thiols with vinyl carboxyl esters in a continuous flow microreactor [103]. The substrate scope was tested by varying the thiols and vinyl esters to generate 12 thioesters with conversions between 62 and 96%. After reaction optimisation and a thorough comparison of batch and flow methodology, a maximum conversion of 96% was obtained at 50°C for about 30 min using 4-methylbenzyl mercaptan:vinyl esters as substrates in a molar ratio of 1:2.

They employed the same catalytic system for an aza-Michael addition of imidazoles **44** or benzimidazoles **45** to acrylates **46** under continuous flow to generate potentially bioactive *N*-substituted imidazole **47** and benzimidazoles **48** [104] (Fig. 30). The authors reported an exhaustive comparative of the same biotransformation in batch and flow with analysis of reaction medium, temperature, substrate molar ratio, time/flow rate and the effect of substrate structure on the reaction. Here again the protocol for the synthesis of 12 *N*-substituted imidazole analogues under

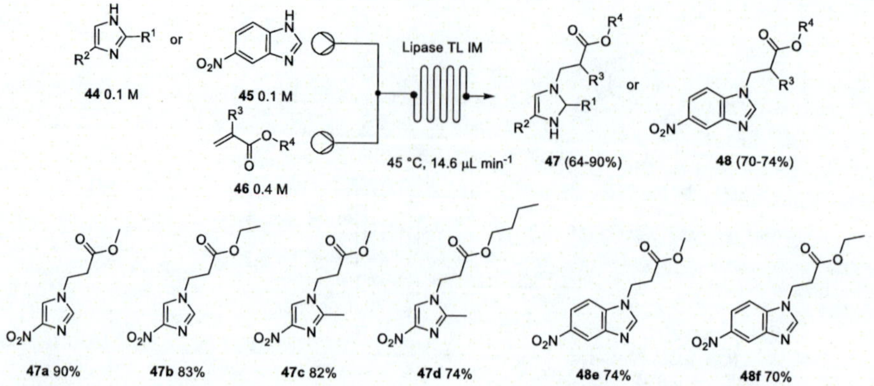

Fig. 30 Continuous Michael addition mediated by Lipase TL IM in a flow microreactor

continuous flow resulted in yields from 64 to 90% and carried associated benefits such as shorter reaction time (35 min) and mild reaction conditions (45°C).

More recently, Luo and co-workers have gone a step further and they have reported an enzyme–enzyme microfluidic reaction cascade for the preparation of sugar-containing coumarin derivatives **51** [105]. The same enzyme was employed in both biotransformations. After extensive optimisation, the authors reported that a small groups such as methyl in position R^4 of the diester gave the best conversion due to steric hindrance. Therefore, dimethyl malonate **50** was used for the substrate exploration study (Fig. 31).

Five salicylaldehyde derivatives **49** and four sugars were subjected to the reaction conditions using both shaker reactors and two-step tandem continuous flow microreactors (Fig. 31). A wide range of substrates were accepted by the enzyme in both cases. It is remarkable that for batch experiments reaction time needed to be about 40 h, while in flow 18 compounds were synthesised in parallel in a single experiment with a 50 min residence time and a final yield of 43–75%. The flow approach avoids intermediate product separation and purification, improving reaction efficiency. These promising results may encourage scientists to expand this research field [106].

Recently Heinzler et al. demonstrated the use of microreactors for automated glycan synthesis [107]. The authors described a series of glycosyltransferases that were immobilised via the His-tags on to magnetic beads. A six-enzyme cascade was then used in a compartmentalised fashion to synthesise non-sulfated human natural killer cell-1 (HNK-1) epitope, which has been identified as an important trisaccharide in several metabolic processes. The reactor consisted of three modules which produced the precursors and a final module that assembled them into the desired HNK-1 molecule. The first module contained the galactokinase (GalK) and the UDP-sugar pyrophosphorylase (USP) to produce UDP-galactose. Combination of *N*-acteylglucosamine was mediated by a second module containing β1,4-galactosyltransferase (GalT) and delivered *N*-acetyl-D-lactosamine. The third module contained UDP-glucose-dehydrogenase (UGDH), generating uridine

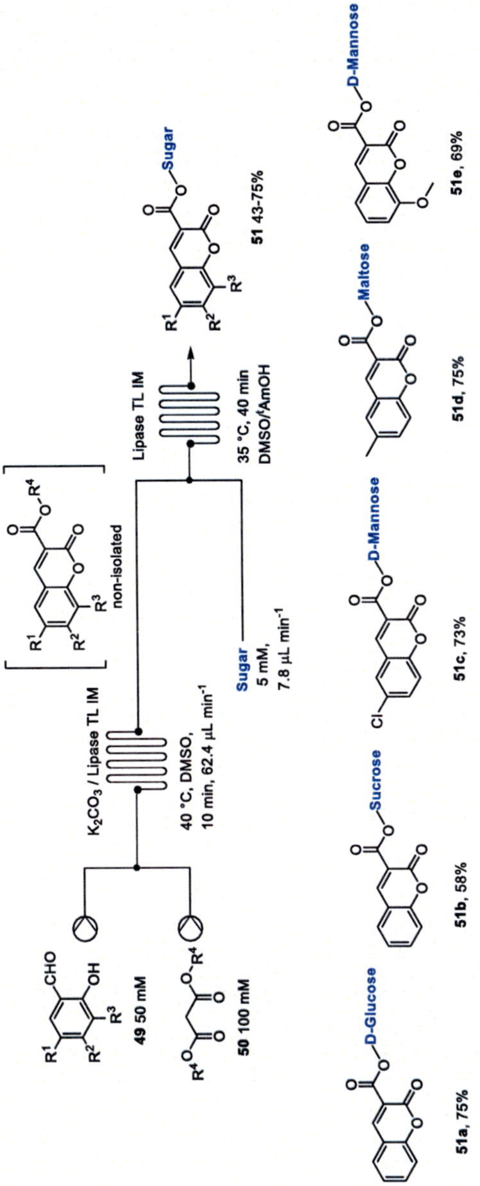

Fig. 31 Continuous tandem synthesis of sugar-containing coumarin derivatives

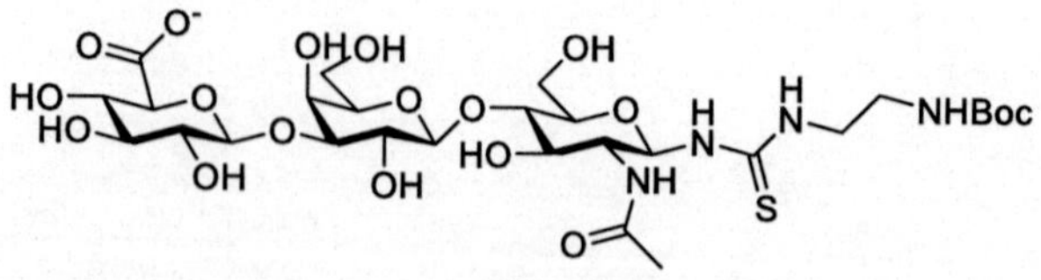

52

Fig. 32 Structure of HNK-1 **52**

5′-diphosphate glucuronic acid, which combined with the lactosamine in the final module to produce HNK-1 **52** (Fig. 32). The authors stated that the use of their immobilised enzyme flow reactor allowed the yield to increase to 96% of the desired product, which is 40% higher than they had achieved with the same enzymes in soluble form. They also calculated the reactor to have a space-time yield of 17.6 g L^{-1} d^{-1}.

This important biomolecule proves extremely difficult to synthesise under other conditions. It is hoped that this modular approach could be applied to other biocatalysts to enable the synthesis of other challenging molecules.

5.1 The Future of Microreactor Technology in Biocatalysis

Microreactors clearly have a role to play in enabling biocatalysis for early-stage drug discovery. As per the examples discussed above, library generation from common intermediates could provide significant numbers of target molecules on more than suitable scales for assay-based screening. This could see application in areas such as diversity-oriented synthesis or lead-oriented synthesis, where diversity introducing reactions are significant parts of the process.

6 Miscellaneous Reactions

The following section will deal with reports of biocatalytic transformations in flow that were deemed relevant enough to include but that did not fit into the categories previously described. They provide solutions and new biological chemistry that could in theory advance how continuous biocatalysis is performed.

6.1 Pickering Emulsion

Usually, enzymes are sensitive to the presence of organic solvents. In this context, a conceptually novel method for continuous flow liquid–liquid interface catalysis based on a water-in-oil pickering emulsion was presented by the group of Yang.

They exploited the immiscibility of water and organic solvents to generate what they called a flow pickering emulsion (FPE). FPE consists of compartamentalisation of the biocatalyst in water droplets, with an organic phase at the interface of the droplet [108]. They reported the successful application of FPE methodology to the CAL-B-catalysed hydrolysis kinetic resolution of four racemic acetates (4-methyl-2-pentanol acetate, 1-octen-3-ol acetate, 2-octanol acetate and 1-phenylethyl acetate). Using this method increased the efficiency of the biocatalyst 10-fold in comparison with the conventional batch conditions. The reactions were run in some cases continuously for 2,000 h, with no observable drop in enantioselectivity. This methodology is operationally simple and efficient and obviates the need for continuous agitation and intermittent separation of product with catalyst.

6.2 *Esterifications*

The generation of small libraries of chiral esters has been effectively catalysed by immobilised CAL-B under continuous flow with high conversion values. In 2016 Buchmeiser and co-workers employed ionic liquids as enzyme stabilisers. Enzyme-containing ionic liquids (ILs) were immobilised in cellulose-2,5-acetate microbead particles embedded in a porous monolithic polyurethane matrix. The catalytic system mediated the enantioselective transesterification of *rac*-1-phenylethanol with vinyl butyrate and vinyl acetate and the esterification of (+/−)-2-isopropyl-5-methylcyclohexanol with propionic anhydride. The authors reported equally high or higher enantioselectivities when compared with batch conditions, as well as high turnover numbers (TONs) of up to 5.1×10^6 and space-time yields (STYs) up to 28 g L^{-1} h^{-1} (which in comparison with batch methodology imply improvement factors up to 3,100 and 40, respectively) [109].

Another effective continuous esterification using CAL-B, in this case whilst immobilised on a polyacrylic resin as hydrophobic carrier (Novozyme 435), was reported by Salvi et al. [110]. In this study, geraniol was esterified with five different acids with conversions from 84 to 97%, the longer the alkyl chain, the higher the conversion.

A different approach was carried out by Haumann and co-workers for the continuous gas-phase transesterification of vinyl propionate and 2-propanol using supported ionic liquid phase (SILP) technology to immobilise CAL-B within a hybrid monolith. Total turnover numbers of 2.4×10^8 mol_{subst} mol_{CALB} $^{-1}$ were obtained and the supported CAL-B showed high stability for at least 700 h in stream at 65°C [111].

In 2019, the Luo research group reported fast, continuous esterification and transesterification of phytosterol, mediated by a commercially available lipase immobilised on cellulose [112]. The authors reported an exhaustive analysis of the lipase (AYS) immobilisation on a skeleton of cellulose acetate (CA) monolith (CA-MN). The supported biocatalytic system carried out the esterification and transesterification of phytosterol with six different acyl donors, including free fatty acids and triglycerides, with high conversion above 90% in just 10 min for 200 h. Phytosterol esters were produced at a rate of 442 g $g_{monolith}^{-1}$. Impressively, the kinetic parameter *Km*/*Vm* of the supported enzyme (AYS@CA-MN) under continuous flow conditions increased >66-fold in comparison with the conventional batch process. It is another remarkable example of quick molecular library generation which might extend the frontiers of continuous flow bioreactor applications for early-stage drug discovery.

6.3 Carboligation

Giovannini et al. reported the remarkable biocatalytic synthesis of chiral tertiary α-hydroxy ketones **54** using immobilised acetylacetoin synthase (AAS) [113]. Immobilisation of AAS from *Bacillus licheniformis* (ThDP dependent enzyme) on silica enhanced its stability and permitted the generation of four chiral tertiary alcohols (Fig. 33). The chiral products are a key feature of a range of natural products and antibiotics and due to the quaternary stereocentre present high synthetic utility for the synthesis of diols or amino alcohols. The union of biocatalysis and flow permitted long-term operation up to 15 days.

6.4 Lyases for Drug Synthesis

The Paradisi research group has recently reported the enzymatic enantioselective hydrolysis of the naproxen butyl ester (a non-steroidal anti-inflammatory drug) under continuous flow [114]. BS2mT4L1 (a fusion protein of BS2m and T4 lysozyme) was immobilised on an epoxy derivatised methacrylate support to provide the chiral final product with a molar conversion of 24% and 80% *ee*. The reaction took place under mild conditions and a surfactant (Triton®X-100) was employed to

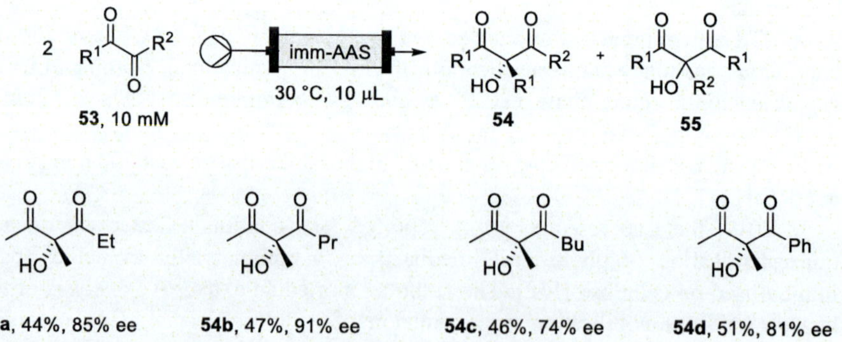

Fig. 33 Continuous production of tertiary alcohols **54** mediated by AAS

overcome substrate solubility issues. Moreover, the catalytic bed exhibited a good operational stability after 6 h, with 57% of retained activity and a remarkable long-term stability (more than 15 days on stream).

6.5 Cyanation

There are different contributions in the literature for the continuous enzymatic generation of cyanohydrins where hydroxynitrile lyases are immobilised on silica [115] or Celite [116]. In 2016, Pohl and co-authors reported an orthogonal approach by coupling two bioreactors in line [87]. Firstly, Novozyme 435 mediated the hydrolysis of ethyl cyanoformate **56** for the in situ safe generation of toxic HCN. In a second bed packed reactor, Celite supported hydroxynitrile lyase (*At*HNC) catalysed the addition of HCN to different aryl aldehydes. As the resulting cyanohydrins are unstable at neutral to alkaline pH and at room temperature, in situ chemical acylation took place in the third step. As a result of the three in-line reactions, a library of six chiral *O*-acetylcyanohydrins **57** was formed with conversions from 75 to 99% and enantioselectivities from 40 to 98% (Fig. 34). This flow protocol offered ease of operation and shorter reaction time when compared with the conventional batch methodology (40 min vs 345 min), opening access to reactions which have not been considered due to safety concerns.

6.6 Protein Modification

Sortase-mediated ligation (sortagging) is a powerful strategy for protein modification [117]. In 2014, Pentelute and co-authors reported the first flow-based sortagging platform flow for the ligation of two peptides at low nucleophile concentration [118]. With this aim sortase A was easily immobilised via a simple His-tag and catalytic activity was retained. This approach allowed access to protein

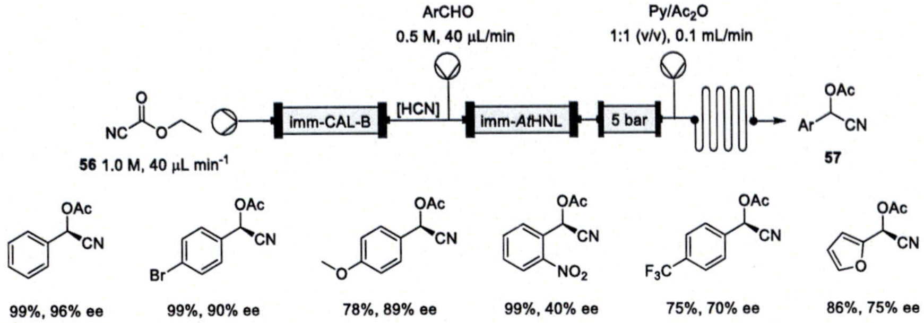

Fig. 34 Multistep continuous preparation of chiral *O*-acetylcyanohydrins **57**

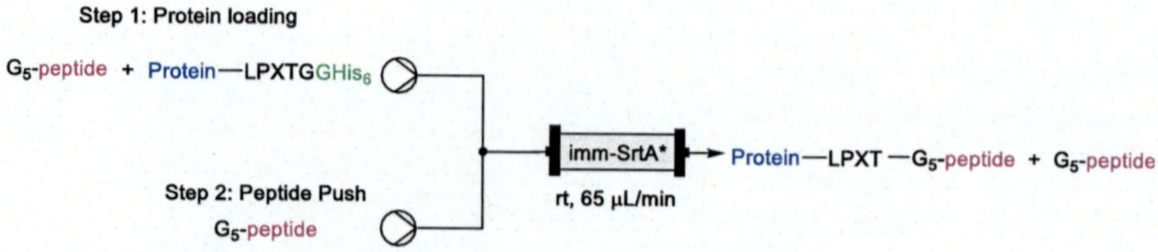

Fig. 35 Biocatalytic protein modification using sortase-mediated ligation

bioconjugates, usually inaccessible by solution-phase batch sortagging (Fig. 35). Sortagging benefits from continuous flow methodology features: (1) immediate release of the transpeptidation product (minimising contact time between the LPXTGG containing protein and SrtA* and by-product formation); (2) to hold nucleophile concentration at a low, fixed concentration (overcoming the decrease in nucleophile concentration typical from conventional batch sortagging). This strategy could potentially enhance the value of sortagging reactions for difficult substrates prone to side reactions such as hydrolysis or dimerisation.

Formylglycine generating enzymes (FGEs) possess the ability to install site-specific aldehyde functionalities in proteins of interest [119]. This approach to protein modification has recently had growing interest in the biotechnological and pharmaceutical areas. Very recently, Li et al. reported continuous-flow aldehyde tag conversion using immobilised FGE [120]. This strategy overcame some limitations of traditional batch-mode catalysis such as low efficiency and stability and enhanced productivity by 10 times compared with the batch methodology. A variant of *Tc*FGE (C187A/Y273F, from *Thermomonospora curvata*) was immobilised on epoxy activated Sepharose beads via amine residues (89% immobilisation yield and 47% activity recovery). Moreover, the authors developed an in situ strategy for copper cofactor reconstitution that avoided the need for cofactor regeneration while supported FGE can be reused consecutively. This work may establish a novel platform for site-specific protein modification by aldehyde tag technology.

6.7 *Cofactor Regeneration*

Cofactors and co-substrates play a decisive role in the cost of a biocatalytic reaction. In batch, the use of stoichiometric amounts of the expensive redox equivalent nicotinamide adenine dinucleotide [phosphate] (NAD[P](H)) is already a challenge. Under continuous flow, the challenge is greater, since usually non-immobilised cofactors are washed away. Among different approaches developed to address this challenge [121–123], a general protein engineering strategy recently reported by Hartley et al. is a distinct, innovative solution [124]. The authors developed three specifically designed biocatalysts. Each biocatalyst was constituted by a genetically encoded multi-enzyme fusion protein and a tethered, modified cofactor which was retained and regenerated. There were three protein modules: a cofactor-dependent

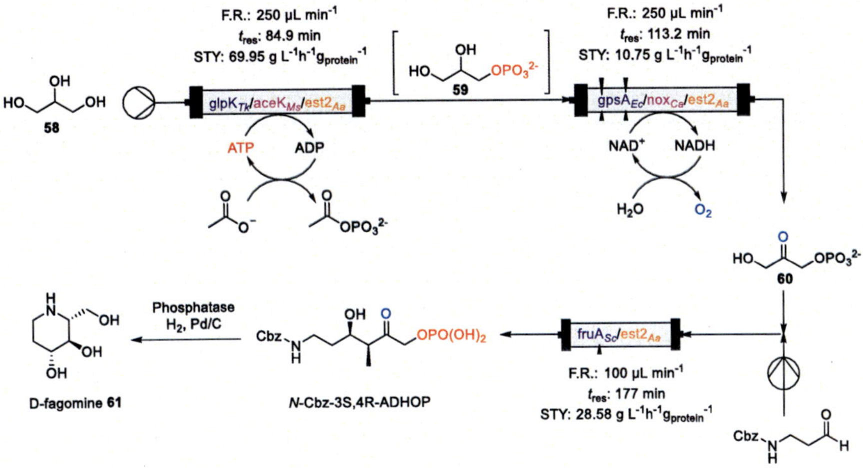

Fig. 36 Engineered enzymes for cofactor regeneration in flow

catalytic enzyme module (purple), a cofactor-recycling module (pink) and a conjugation module (orange) separated by short amino acid spacers. The conjugation module allowed immobilisation of the biocatalyst in trifluoroketone-activated agarose. The three packed-bed modules were linked for the continuous generation of the antidiabetic drug D-fagomine **61** (Fig. 36).

Each reactor carried out a different reaction mediated by a specifically designed biocatalyst. For glycerol **58** phosphorylation to glycerol-3-phosphate **59**, glycerol kinase from *Thermococcus kodakarensis* (GlpK*Tk*) and an acetate kinase from *Mycobacterium smegmatis* (AceK*Ms*) were employed. Secondly, *E. coli* glycerol-3-phosphate dehydrogenase (G3PD*Ec*) and the water-forming NADH oxidase from *Clostridium aminovalericum* (NOX*Ca*) were selected for the NAD^+-dependent production of DHAP **60** from glycerol-3-phosphate **59**. The third reaction step was a cofactor-independent aldolase-catalysed aldol addition mediated by a monomeric fructose aldolase (FruA) homologue from *Staphylococcus carnosus* and was selected from a panel of five potential aldolases. Conversions were maintained between 85% and 90% at 23°C for more than 7 h. The TONs for the cofactors exceeded 10,000 (ca. 11,000 for the NAD^+-dependent oxidation reactor and ca. 17,000 for the ATP-dependent phosphorylation reactor).

Complex multistep biochemical pathways can be built using molecular modularity with both serial and parallel reactor compartments. This strategy could be extended to the development of artificial metabolic networks. Over the next few years, this generalisable this generalisable approach will develop to become an important research field in the area of continuously operated biocatalysis.

7 The Future of Flow Biocatalysis for Early-Stage Drug Discovery

It is without doubt that there is massive potential for biocatalysis in early-stage drug discovery. Many of the bottlenecks associated with their application are starting to be overcome, such as substrate scope and pace of reactions, with directed evolution techniques. Even so, the improvement of reaction engineering is essential to ensuring the engineered biocatalysts that are produced are used in an appropriate manner. Here is where flow can really make the difference. The examples demonstrated in this chapter show that flow technologies can significantly enhance how enzymes are applied. There are even benefits that flow brings, such as integration of gases or overcoming equilibrium issues, that are improving the way biocatalysts are used in synthesis. As more of these examples are published, and as more researchers adopt flow methodology the application in early-stage drug discovery will start to increase rapidly.

Computational tools, such as the modelling approach highlighted from the Turner group [48], will undoubtedly have a role to play in the future. Recently, the same group has developed a biocatalytic retrosynthetic planning tool, named RetroBioCat [125]. Having access to large datasets such as this will allow all chemists to understand the broad applications of enzymes, helping to integrate them more easily into their synthetic sequences.

It is certain that a union of chemical and biocatalysis will lead to a more successful exploration of chemical space. Whilst enzymes have met the challenge of mild redox chemistry and esterification, the power of challenging bond-formation reactions remains the forte of transition metal catalysts. The issues that arise with combining the two disciplines could be best addressed using continuous flow it seems. Compartmentalisation and solvent switching are some of the long-standing solutions flow can offer to organic synthesis and are integral to uniting chemo- and biocatalysis. Immobilisation has a key role to play in this area: one could envisage a single material that can be used for the immobilisation of both types of catalyst in future perhaps. Such a material would enable off-the-shelf catalyst modules to push the boundaries of what chemists can really achieve in continuous chemoenzymatic synthesis for early-stage drug discovery.

Compliance with Ethical Standards

Funding: Itziar Peñafiel and Sebastian C. Cosgrove. Both acknowledge The UK Catalysis Hub for funding. Sebastian C. Cosgrove also acknowledges Keele University for funding.

Informed Consent: All procedures in this manuscript were not performed with human participants.

Ethical Approval: All procedures in this manuscript were not performed with human participants, nor any other animals.

References

1. Chen K, Arnold FH (2020) Nat Catal 3:203–213
2. Sheldon RA, Woodley JM (2018) Chem Rev 118:801–838
3. Sheldon RA, Brady D, Bode ML (2020) Chem Sci 11:2587–2605
4. Sheldon RA, Brady D (2018) Chem Commun 54:6088–6104
5. Britton J, Weiss GA, Britton J (2018) Chem Soc Rev 47:5891–5918
6. Thompson MP, Peñafiel I, Cosgrove SC, Turner NJ (2019) Org Process Res Dev 23:9–18
7. Wu S, Snajdrova R, Moore JC, Baldenius K, Bornscheuer UT (2021) Angew Chem Int Ed 60:88–119
8. Savile CK, Janey JM, Mundorff EC, Moore JC, Tam S, Jarvis WR, Colbeck JC, Krebber A, Fleitz FJ, Brands J, Devine PN, Huisman GW, Hughes GJ (2010) Science 329:305–309
9. Huffman MA, Fryszkowska A, Alvizo O, Borra-Garske M, Campos KR, Canada KA, Devine PN, Duan D, Forstater JH, Grosser ST, Halsey HM, Hughes GJ, Jo J, Joyce LA, Kolev JN, Liang J, Maloney KM, Mann BF, Marshall NM, McLaughlin M, Moore JC, Murphy GS, Nawrat CC, Nazor J, Novick S, Patel NR, Rodriguez-Granillo A, Robaire SA, Sherer EC, Truppo MD, Whittaker AM, Verma D, Xiao L, Xu Y, Yang H (2019) Science 366:1255–1259
10. Devine PN, Howard RM, Kumar R, Thompson MP, Truppo MD, Turner NJ (2018) Nat Rev Chem 2:409–421
11. Grygorenko OO, Volochnyuk DM, Ryabukhin SV, Judd DB (2019) Chem A Eur J 26:1196–1237
12. Trabocchi A (2013) Diversity-oriented synthesis: basics and applications in organic synthesis drug discovery and chemical biology. Wiley, Hoboken
13. Wittmann BJ, Knight AM, Hofstra J, Reisman SE, Kan SBJ, Arnold FH (2020) ACS Catal 10:7112–7116
14. Fryszkowska A, Devine PN (2020) Curr Opin Chem Biol 55:151–160
15. Ramsden JI, Cosgrove SC, Turner NJ (2020) Chem Sci 11:11104–11112
16. Ringborg RH, Woodley JM (2016) React Chem Eng 1:10–22
17. Romero-Fernández M, Paradisi F (2020) Curr Opin Chem Biol 55:1–8
18. Chandel AK, Rao LV, Narasu ML, Singh OV (2008) Enzyme Microb Technol 42:199–207
19. Parker K, Salas M, Nwosu VC (2006) Biotechnol Mol Biol Rev 5:71–78
20. Slabu I, Galman JL, Lloyd RC, Turner NJ (2017) ACS Catal 7:8263–8284
21. Andrade LH, Kroutil W, Jamison TF (2014) Org Lett 16:6092–6095
22. Planchestainer M, Contente ML, Cassidy J, Molinari F, Tamborini L, Paradisi F (2017) Green Chem 19:372–375
23. Contente ML, Paradisi F (2018) Nat Catal 1:452–459
24. Contente ML, Dall'Oglio F, Tamborini L, Molinari F, Paradisi F (2017) ChemCatChem 9:3843–3848
25. Contente ML, Paradisi F (2019) Chembiochem 20:2830–2833
26. Semproli R, Vaccaro G, Ferrandi EE, Vanoni M, Bavaro T, Marrubini G, Annunziata F, Conti P, Speranza G, Monti D, Tamborini L, Ubiali D (2020) ChemCatChem 12:1359–1367
27. Böhmer W, Volkov A, Engelmark Cassimjee K, Mutti FG (2020) Adv Synth Catal 362:1858–1867
28. Engelmark Cassimjee K, Federsel H-J (2018) Biocatalisys an industrial perspective. The Royal Society of Chemistry, Cambridge, pp 345–362
29. Peris E, Okafor O, Kulcinskaja E, Goodridge R, Luis SV, Garcia-Verdugo E, O'Reilly E, Sans V (2017) Green Chem 19:5345–5349
30. Biggelaar L, Soumillion P, Debecker D (2017) Catalysts 7:54
31. Van Den Biggelaar L, Soumillion P, Debecker DP (2019) RSC Adv 9:18538–18546
32. Molnár Z, Farkas E, Lakó Á, Erdélyi B, Kroutil W, Vértessy BG, Paizs C, Poppe L (2019) Catalysts 9:438
33. Gruber P, Carvalho F, Marques MPC, O'Sullivan B, Subrizi F, Dobrijevic D, Ward J, Hailes HC, Fernandes P, Wohlgemuth R, Baganz F, Szita N (2018) Biotechnol Bioeng 115:586–596

34. Hollmann F, Opperman DJ, Paul CE Angew Chem Int Ed:anie.202001876. https://doi.org/10.1002/anie.202001876(2020
35. Adams JP, Brown MJB, Diaz-Rodriguez A, Lloyd RC, Roiban GD (2019) Adv Synth Catal 361:2421–2432
36. Ma SK, Gruber J, Davis C, Newman L, Gray D, Wang A, Grate J, Huisman GW, Sheldon RA (2010) Green Chem 12:81–86
37. Li H, Moncecchi J, Truppo MD (2015) Org Process Res Dev 19:695–700
38. Brands KMJ, Payack JF, Rosen JD, Nelson TD, Candelario A, Huffman MA, Zhao MM, Li J, Craig B, Song ZJ, Tschaen DM, Hansen K, Devine PN, Pye PJ, Rossen K, Dormer PG, Reamer RA, Welch CJ, Mathre DJ, Tsou NN, McNamara JM, Reider PJ (2003) J Am Chem Soc 125:2129–2135
39. Dall'Oglio F, Contente ML, Conti P, Molinari F, Monfredi D, Pinto A, Romano D, Ubiali D, Tamborini L, Serra I (2017) Catal Commun 93:29–32
40. Peschke T, Skoupi M, Burgahn T, Gallus S, Ahmed I, Rabe KS, Niemeyer CM (2017) ACS Catal 7:7866–7872
41. Peschke T, Bitterwolf P, Rabe KS, Niemeyer CM (2019) Chem Eng Technol 42:2009–2017
42. Maier M, Radtke CP, Hubbuch J, Niemeyer CM, Rabe KS (2018) Angew Chem Int Ed 57:5539–5543
43. Poznansky B, Thompson LA, Warren SA, Reeve HA, Vincent KA (2020) Org Process Res Dev 24:2281–2287
44. Mangas-Sanchez J, France SP, Montgomery SL, Aleku GA, Man H, Sharma M, Ramsden JI, Grogan G, Turner NJ (2017) Curr Opin Chem Biol 37:19–25
45. Aleku GA, France SP, Man H, Mangas-Sanchez J, Montgomery SL, Sharma M, Leipold F, Hussain S, Grogan G, Turner NJ (2017) Nat Chem 9:961–969
46. Schober M, MacDermaid C, Ollis AA, Chang S, Khan D, Hosford J, Latham J, Ihnken LAF, Brown MJB, Fuerst D, Sanganee MJ, Roiban GD (2019) Nat Catal 2:909–915
47. Mangas-Sanchez J, Sharma M, Cosgrove SC, Ramsden JI, Marshall JR, Thorpe TW, Palmer RB, Grogan G, Turner NJ (2020) Chem Sci 11:5052–5057
48. Finnigan W, Citoler J, Cosgrove SC, Turner NJ (2020) Org Process Res Dev 24:1969–1977
49. Bitterwolf P, Ott F, Rabe KS, Niemeyer CM (2019) Micromachines 10:783
50. Peschke T, Bitterwolf P, Gallus S, Hu Y, Oelschlaeger C, Willenbacher N, Rabe KS, Niemeyer CM (2018) Angew Chem Int Ed 57:17028–17032
51. Au SK, Groover J, Feske BD, Bommarius AS (2016) Organic syntesis using biocatalysis. Elsevier Inc., Amsterdam, pp 187–212
52. Mayol O, Bastard K, Beloti L, Frese A, Turkenburg JP, Petit JL, Mariage A, Debard A, Pellouin V, Perret A, de Berardinis V, Zaparucha A, Grogan G, Vergne-Vaxelaire C (2019) Nat Catal 2:324–333
53. Knaus T, Böhmer W, Mutti FG (2017) Green Chem 19:453–463
54. Thompson MP, Derrington SR, Heath RS, Porter JL, Mangas-Sanchez J, Devine PN, Truppo MD, Turner NJ (2019) Tetrahedron 75:327–334
55. Mutti FG, Knaus T, Scrutton NS, Breuer M, Turner NJ (2015) Science 349:1525–1529
56. Franklin RD, Whitley JA, Caparco AA, Bommarius BR, Champion JA, Bommarius AS (2020) Chem Eng J 127065. https://doi.org/10.1016/j.cej.2020.127065
57. Velasco-Lozano S, da Silva ES, Llop J, López-Gallego F (2018) Chembiochem 19:395–403
58. Roura Padrosa D, Benítez-Mateos AI, Calvey L, Paradisi F (2020) Green Chem 22:5310–5316
59. Greiner L, Schröder I, Müller DH, Liese A (2003) Green Chem 5:697–700
60. Zor C, Reeve HA, Quinson J, Thompson LA, Lonsdale TH, Dillon F, Grobert N, Vincent KA (2017) Chem Commun 53:9839–9841
61. Thompson LA, Rowbotham JS, Nicholson JH, Ramirez MA, Zor C, Reeve HA, Grobert N, Vincent KA (2020) ChemCatChem 12:3913–3918
62. Turner NJ (2011) Chem Rev 111:4073–4087
63. Pickl M, Fuchs M, Glueck SM, Faber K (2015) Appl Microbiol Biotechnol 99:6617–6642

64. Cosgrove SC, Brzezniak A, France SP, Ramsden JI, Mangas-Sanchez J, Montgomery SL, Heath RS, Turner NJ (2018) Methods enzymol, vol 608. Academic Press Inc., San Diego, pp 131–149
65. Parikka K, Master E, Tenkanen M (2015) J Mol Catal B: Enzym 120:47–59
66. Lindeque RM, Woodley JM (2020) Org Process Res Dev 24:2055–2063
67. Hoschek A, Schmid A, Bühler B (2018) ChemCatChem 10:5366–5371
68. Ringborg RH, Toftgaard Pedersen A, Woodley JM (2017) ChemCatChem 9:3285–3288
69. Chapman MR, Cosgrove SC, Turner NJ, Kapur N, Blacker AJ (2018) Angew Chem Int Ed 57:10535–10539
70. Escalettes F, Turner NJ (2008) Chembiochem 9:857–860
71. Cosgrove SC, Mattey AP, Riese M, Chapman MR, Birmingham WR, Blacker AJ, Kapur N, Turner NJ, Flitsch SL (2019) ACS Catal 9:11658–11662
72. Illner S, Hofmann C, Löb P, Kragl U (2014) ChemCatChem 6:1748–1754
73. Toftgaard Pedersen A, de Carvalho TM, Sutherland E, Rehn G, Ashe R, Woodley JM (2017) Biotechnol Bioeng 114:1222–1230
74. Gasparini G, Archer I, Jones E, Ashe R (2012) Org Process Res Dev 16:1013–1016
75. Hollmann F, Arends IWCE, Buehler K, Schallmey A, Bühler B (2011) Green Chem 13:226–265
76. van Schie MMCH, Pedroso de Almeida T, Laudadio G, Tieves F, Fernández-Fueyo E, Noël T, Arends IWCE, Hollmann F (2018) Beilstein J Org Chem 14:697–703
77. Ghislieri D, Turner NJ (2014) Top Catal 57:284–300
78. Rudroff F, Mihovilovic MD, Gröger H, Snajdrova R, Iding H, Bornscheuer UT (2018) Nat Catal 1:12–22
79. Dumeignil F, Guehl M, Gimbernat A, Capron M, Ferreira NL, Froidevaux R, Girardon JS, Wojcieszak R, Dhulster P, Delcroix D (2018) Cat Sci Technol 8:5708–5734
80. Li J, Zhang X, Renata H (2019) Angew Chem Int Ed 58:11657–11660
81. Pyser JB, Baker Dockrey SA, Benítez AR, Joyce LA, Wiscons RA, Smith JL, Narayan ARH (2019) J Am Chem Soc 141:18551–18559
82. Doyon TJ, Perkins JC, Baker Dockrey SA, Romero EO, Skinner KC, Zimmerman PM, Narayan ARH (2019) J Am Chem Soc 141:20269–20277
83. Zwick CR, Renata H (2018) J Org Chem 83:7407–7415
84. de María PD, de Gonzalo G, Alcántara AR (2019) Catalysts 9:802
85. Martinez CA, Hu S, Dumond Y, Tao J, Kelleher P, Tully L (2008) Org Process Res Dev 12:392–398
86. Delville MME, Koch K, Van Hest JCM, Rutjes FPJT (2015) Org Biomol Chem 13:1634–1638
87. Brahma A, Musio B, Ismayilova U, Nikbin N, Kamptmann SB, Siegert P, Jeromin GE, Ley SV, Pohl M (2016) Synlett 27:262–266
88. Döbber J, Pohl M, Ley SV, Musio B (2018) React Chem Eng 3:8–12
89. De Vitis V, Dall'Oglio F, Pinto A, De Micheli C, Molinari F, Conti P, Romano D, Tamborini L (2017) ChemistryOpen 6:668–673
90. Schrittwieser JH, Velikogne S, Hall M, Kroutil W (2018) Chem Rev 118:270–348
91. Falus P, Cerioli L, Bajnóczi G, Boros Z, Weiser D, Nagy J, Tessaro D, Servi S, Poppe L (2016) Adv Synth Catal 358:1608–1617
92. Henkel T, Brunne RM, Müller H, Reichel F (1999) Angew Chem Int Ed 38:643–647
93. Farkas E, Oláh M, Földi A, Kóti J, Éles J, Nagy J, Gal CA, Paizs C, Hornyánszky G, Poppe L (2018) Org Lett 20:8052–8056
94. De Miranda AS, De Silva MVM, Dias FC, De Souza SP, Leão RAC, De Souza ROMA (2017) React Chem Eng 2:375–381
95. Porcar R, Lozano P, Burguete MI, Garcia-Verdugo E, Luis SV (2018) React Chem Eng 3:572–578
96. Grabner B, Schweiger AK, Gavric K, Kourist R, Gruber-Woelfler H (2020) React Chem Eng 5:263–269

97. Szelwicka A, Zawadzki P, Sitko M, Boncel S, Czardybon W, Chrobok A (2019) Org Process Res Dev 23:1386–1395
98. Watts P, Haswell SJ (2003) Drug Discov Today 8:586–593
99. Cortes-Clerget M, Akporji N, Zhou J, Gao F, Guo P, Parmentier M, Gallou F, Berthon JY, Lipshutz BH (2019) Nat Commun 10:2169
100. Cosgrove SC, Thompson MP, Ahmed ST, Parmeggiani F, Turner NJ (2020) Angew Chem Int Ed 59:18156–18160
101. Žnidaršič-Plazl P (2019) Biotechnol J 14:1800580
102. Du LH, Shen JH, Dong Z, Zhou NN, Cheng BZ, Ou ZM, Luo XP (2018) RSC Adv 8:12614–12618
103. Zhou N, Shen L, Dong Z, Shen J, Du L, Luo X (2018) Catalysts 8:1–12
104. Du LH, Dong Z, Long RJ, Chen PF, Xue M, Luo XP (2019) Org Biomol Chem 17:807–812
105. Du LH, Chen PF, Long RJ, Xue M, Luo XP (2020) RSC Adv 10:13252–13259
106. Gruber P, Marques MPC, O'Sullivan B, Baganz F, Wohlgemuth R, Szita N (2017) Biotechnol J 12:1–13
107. Heinzler R, Fischöder T, Elling L, Franzreb M (2019) Adv Synth Catal 361:4506–4516
108. Zhang M, Wei L, Chen H, Du Z, Binks BP, Yang H (2016) J Am Chem Soc 138:10173–10183
109. Sandig B, Buchmeiser MR (2016) ChemSusChem 9:2917–2921
110. Salvi HM, Kamble MP, Yadav GD (2018) Appl Biochem Biotechnol 184:630–643
111. Lee C, Sandig B, Buchmeiser MR, Haumann M (2018) Cat Sci Technol 8:2460–2466
112. Xiao Y, Zheng M, Liu Z, Shi J, Huang F, Luo X (2019) ACS Sustain Chem Eng 7:2056–2063
113. Giovannini PP, Bortolini O, Cavazzini A, Greco R, Fantin G, Massi A (2014) Green Chem 16:3904–3915
114. Padrosa DR, De Vitis V, Contente ML, Molinari F, Paradisi F (2019) Catalysts 9:232
115. Van Der Helm MP, Bracco P, Busch H, Szymańska K, Jarzębski AB, Hanefeld U (2019) Cat Sci Technol 9:1189–1200
116. Coloma J, Guiavarc'h Y, Hagedoorn PL, Hanefeld U (2020) Cat Sci Technol 10:3613–3621
117. Li Y, Sun S, Fan L, Hu S, Huang Y, Zhang K, Nie Z, Yao S (2017) Angew Chem Int Ed 56:14888–14892
118. Policarpo RL, Kang H, Liao X, Rabideau AE, Simon MD, Pentelute BL (2014) Angew Chem Int Ed 53:9203–9208
119. Zang B, Ren J, Li D, Huang C, Ma H, Peng Q, Ji F, Han L, Jia L (2019) Org Biomol Chem 17:257–263
120. Peng Q, Zang B, Zhao W, Li D, Ren J, Ji F, Jia L (2020) Cat Sci Technol 10:484–492
121. Baumer B, Classen T, Pohl M, Pietruszka J (2020) Adv Synth Catal 362:2894–2901
122. De Santis P, Meyer L-E, Kara S (2020) React Chem Eng 5:2155–2184
123. Mordhorst S, Andexer JN (2020) Nat Prod Rep 37:1316–1333
124. Hartley CJ, Williams CC, Scoble JA, Churches QI, North A, French NG, Nebl T, Coia G, Warden AC, Simpson G, Frazer AR, Jensen CN, Turner NJ, Scott C (2019) Nat Catal 2:1006–1015
125. Finnigan W, Hepworth L, Turner N, Flitsch S (2021) Nat Catal. https://doi.org/10.1038/s41929-020-00556-z

Top Med Chem (2021) 38: 317–372
https://doi.org/10.1007/7355_2021_117

Published online: 7 April 2021

Improved Synthesis of Bioactive Molecules Through Flow Chemistry

Aline Aparecida Nunes de Souza, Elida Betania Ariza Paez, Francisco Fávaro de Assis, Timothy John Brocksom, and Kleber Thiago de Oliveira

Contents

Abstract This chapter presents the major impacts of continuous flow technologies on the development of syntheses of bioactive natural product molecules and active pharmaceutical ingredients (APIs). We emphasize scalable and robust protocols covering the literature from 2015 to mid-2020. In the case of continuous manufacturing of APIs we decided to cover examples with relevance considering the advanced stage for industrial production, the molecular complexity, and the importance of each API for Big Pharma.

Keywords Active pharmaceutical ingredients, Flow chemistry, Natural products, Synthesis

---

Aline Aparecida Nunes de Souza and Elida Betania Ariza Paez contributed equally to this work.

A. A. N. de Souza, E. B. A. Paez, T. J. Brocksom, and K. T. de Oliveira (✉)
Departamento de Química, Universidade Federal de São Carlos – UFSCar, São Carlos, SP, Brazil
e-mail: kleber.oliveira@ufscar.br

F. F. de Assis
Departamento de Química, Universidade Federal de Santa Catarina – UFSC, Florianópolis, SC, Brazil

1 Introduction

One of the greatest scientific successes of the last 200 years is certainly the development and production of legal drugs, obeying all the requirements of a modern society as far as efficacy, safety, and cost are concerned. Much of this science takes place in organic chemistry research laboratories and the corresponding fine chemical industrial environment of the pharmaceutical industry.

At the same time, the physical operation within these research and industrial laboratories does not seem to have changed very much over the last 100 years. The same round bottom flasks reign supreme, with their ground glass joints leading to the same addition funnels, condensers, stirrers, heating/cooling systems, and provision for the addition of solids and gases (inert or otherwise). This physical operation mode is denominated as the batch process.

However, since the beginning of this century, a quiet revolution is in progress with the introduction of continuous flow systems, and this new enabling technology has been rapidly accepted by both research and pharmaceutical laboratories. Although it is also true that chemical engineering processes in operation for at least the last 50 years have strong aspects of flow systems, the new technology brings a completely different approach to continuous flow. This substantial change of execution of the chemical reaction has also received a helping hand from the Food and Drug Administration (FDA), and other important regulatory agencies, over the last 10 years [1].

Initially, we should separate the two distinct phases of organic synthesis of bioactive molecules, on the way to APIs. The medicinal chemistry phase involves "academic" style organic syntheses but leading to some hundreds of grams of material appropriate for the first tests of suitable activities. In this part, the emphasis is on obtaining the desired molecule(s) as quickly as possible so that decisions can be made rapidly on the possible success or not. This analysis allows a relatively free choice of starting materials, reactions, reagents, solvents, and conditions, and can involve modifications of the synthetic route. Once the hits and then the leads have been produced, tested, and partially approved, the process chemistry syntheses become important with much more selective criteria on the choice of starting materials, reactions, reagents, solvents, and conditions. At this point, costs, sustainability, safety, environmental rules, and scalability become the dominant factors, especially as several hundreds of kilos may be required. This phase produces material for the last stages of the extensive testing necessary for final approval, and also in preparation for eventual commercial production.

This second stage also involves the testing of several different synthetic sequences, until the most efficient is determined, thus allowing the preparation of more than one protective patent. These comments are presented simply to clarify the relevance and importance of the enabling technologies being implanted in the last 20 years to attend these requirements.

Among these new enabling technologies, we should include the impact of artificial intelligence (AI), the computer in planning synthetic sequences (retrosynthetic analysis), robotics in the lab, high throughput experimentation, and machine learning. However, we must emphasize that continuous flow procedures have become the most relevant in academic, fine chemistry, and pharmaceutical industry labs.

The mechanics and equipment which are necessary for continuous flow are presently very well known, and more sophisticated versions have been commercially available for quite some time now. We should already emphasize that, in our perspective, continuous flow technologies will not substitute batch methods totally, and the two will co-exist depending upon their relative efficiencies. A second and also relevant comment is that many reactions can be conducted using home-made equipment or on-demand designed for continuous flow and at relatively low costs. Perhaps a very interesting secondary effect of all this scientific development is the striking interaction recently created between chemists and chemical engineers, with the latter demonstrating a certain leadership. We welcome this new relationship as being very positive both in research and in education.

The new continuous flow technology developed over the last 20 years has had a tremendous impact on our way of thinking and executing synthetic organic chemistry. The first objective is to determine optimized reaction conditions, which can be very time-consuming in the traditional batch system, whereas continuous flow permits rapid variations of solvents, concentrations, reaction temperatures, and reaction times, as well as the choice of appropriate reagents, acids/bases, and additives. The systematic investigation of structural variations of the substrates (scope) is also accelerated, and in all these experiments it is now acceptable to use NMR, IR, UV-Vis, and GC-MS analyses to determine the conversions and yields. The scalability is easily and quickly studied simply leaving the continuous flow operation in optimized activity for an extended period, with the facility of remote control linked to a cell phone. Currently, chemists and chemical engineers have proved it to be possible to think and design reaction protocols and processes on demand, taking into account many technological limitations before they appear, and highlighting the concept of flow-oriented design (FOD), recommending the development of the complete process in flow and aimed at flow systems since the beginning [2].

It is now relatively common to couple up sequences of reactions in end-to-end operation, without the necessity of transfers between distinct batch reactors. In principle, continuous flow can be transformed into scale-up and more frequently scale-out processes, where several systems can be operated side by side, with a common entry and exit to a simple unique isolation procedure. Reactions that require gases (inert or reactive) can be executed with the help of membrane reactors, and catalysis is performed with inline tubes and columns containing the solid catalyst.

In this chapter, we revise relevant (in our perspective) flow-assisted chemical syntheses of bioactive molecules suitable for drug acceptance, or active pharmaceutical ingredients (APIs), which have been effectively improved with continuous flow technologies. We then plan to discuss the syntheses of bioactive molecules and APIs, in a sequence based on their timeline publication, and with emphasis on the improvements promoted by continuous flow.

2 Synthesis of Bioactive Natural Products Under Continuous Flow Conditions

Natural products (secondary metabolites) have fascinated organic chemists for centuries, Nature being a captivating "natural laboratory" with improved skills and tactics for organic synthesis. Over the last 6–7 decades, organic synthesis has gained prominence by the many advances in terms of strategies and capabilities to offer solutions and real possibilities to prepare fine chemicals from milligrams to ton-scale having, in many cases, Nature as inspiration. Notably, Nature is an excellent example of efficiency and undeniably does all the processes in flow, giving us more inspiration on optimized technologies. We should emphasize that Nature knows best and biosynthesis is much more a continuous flow process than batch (sometimes called stop and go). In this regard, synthetic chemists have invested efforts in the last decade to apply continuous technologies in the synthesis of useful natural products, which we now highlight a selection of reported studies over the last 5 years.

The first endeavour on continuous natural product synthesis which we wish to present is the study from the Booker-Milburn group, reported in 2016. A short and scalable synthesis of (+)-Goniofufurone, a styryl-lactone natural product isolated from *Goniothalamus* trees in South East Asia, was reported (Fig. 1) [3]. The synthesis started from readily available *D*-isosorbide, and the key step involves a [2 + 2] Paternò-Büchi photocycloaddition for the construction of an intermediate containing an oxetane ring. In batch, the reaction delivered the oxetane product in 90% yield, albeit in very limited amounts due to the requirements of high dilution.

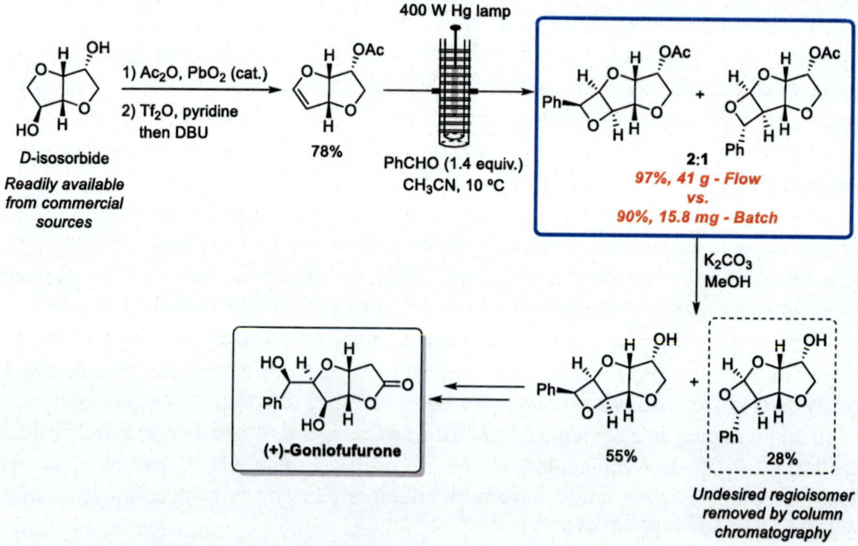

Fig. 1 Total synthesis of (+)-goniofufurone using a continuous flow photoreactor

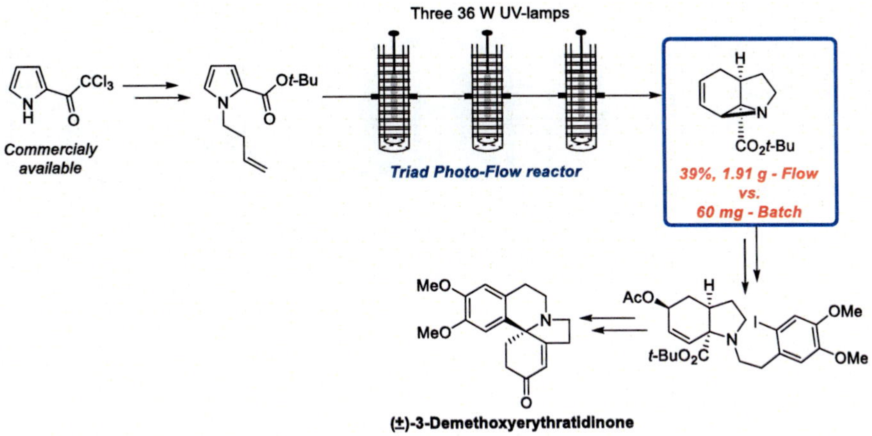

Fig. 2 Total synthesis of (±)-3-demethoxyerythratidinone using a triad photo-flow reactor

This limitation was overcome by using an FEP continuous flow reactor irradiated with a 400 W Hg lamp. Performing the reaction in a flow regime provided the desired oxetane product in nearly quantitative yield and on more than 40 g-scale. The oxetane intermediate was obtained as an inseparable mixture of regioisomers in 2:1 ratio for the desired one. After hydrolysis of the acetate group, the corresponding alcohols were separated by column chromatography, and the desired intermediate was converted into (+)-Goniofufurone after several steps in 9% overall yield.

Once the use of photoflow proved to be a promising alternative to scalability issues of photochemical reaction steps, Booker-Milburn used this again in the synthesis of another natural product, (±)-3-Demethoxyerythratidinone (Fig. 2) [4]. The commercially available 2-(trichloroacetyl)-pyrrole was converted to the *N*-buten-3-yl derivative which was subjected to a photochemical cyclization reaction. In batch conditions, the desired tricyclic product was obtained in up to 60 mg-scale. However, substituting the batch conditions by 3 consecutive continuous flow photoreactors, the key-intermediate was obtained in more than 1.9 g-scale in a single residence time run. Further steps involved the opening of the aziridine ring of the tricyclic intermediate, followed by an *N*-alkylation and decarboxylation/cyclization sequence, thus yielding (±)-3-Demethoxyerythratidinone in 15% overall yield (5 steps) from commercially available 2-(trichloroacetyl)-pyrrole (Fig. 2).

The same strategy of coupling together individual photo-flow reactors to scale up synthetic procedures was used by Beeler and co-workers [5]. In this case, the authors describe the use of a triad photo-flow reactor, analogous to the one used by Booker-Milburn, but immersed in a cooling system at 0°C, connected to a recirculating chiller (Fig. 3). This system was used to perform a 1,3-dipolar photocycloaddition reaction between 3-hydroxyflavones and cinnamic acid derivatives. The obtained cycloadducts can be used as intermediates in the syntheses of important natural products and analogues, and in the most significant example, this reaction set-up provided the desired cycloadduct in more than 12 g-scale (Fig. 3).

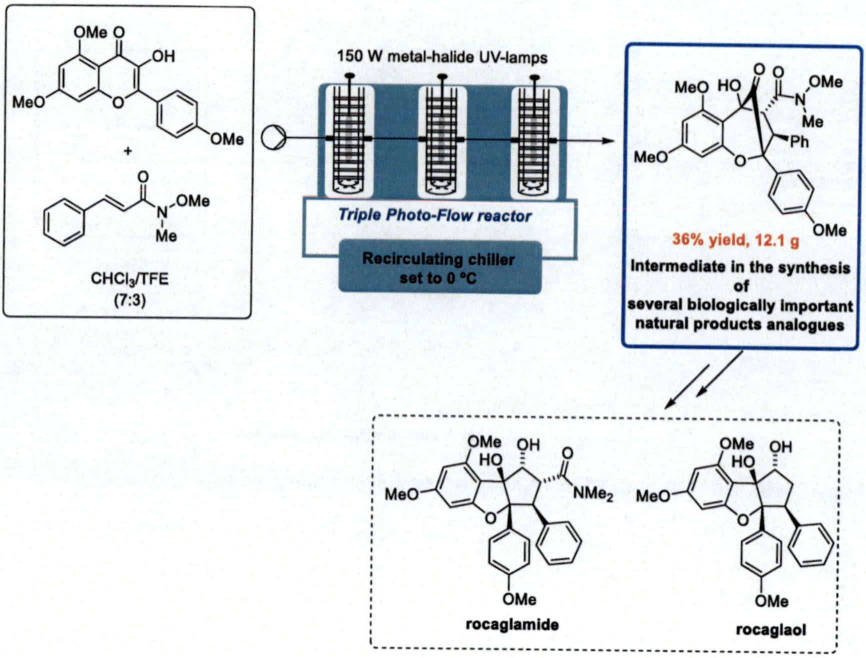

Fig. 3 [3 + 2] Photocycloaddition using a triad photo-flow reactor

In another impressive example of how photochemical reactions can benefit from continuous flow reaction set-ups, Hiemstra and co-workers reported the synthesis of an intermediate [6] for solanoeclepin A total synthesis (Fig. 4) [7]. To access this intermediate, the authors had to perform an intramolecular [2 + 2] photocycloaddition involving an enone and a vinyl boronate on the precursor. The photochemical reaction was carried out in a photo-flow apparatus, providing the intermediate in 87% yield and up to 18 g-scale. The cycloadduct product was further transformed (total of 18-steps) to yield solanoeclepin A (Fig. 4), showing the importance of flow chemistry not only in fully telescoped protocols but also in critical and otherwise poorly scalable steps.

The use of flow systems in the synthesis of natural products and analogues is not restricted to examples involving light-driven chemical transformations. In 2016, Fuse and co-workers reported the total synthesis of Feglymycin, a natural peptide with anti-HIV and antimicrobial properties, utilizing flow chemistry in the key steps [8]. The challenge associated with the synthesis of this compound concerns the manipulation of highly racemizable amino acid units. Using a set of syringe pumps and T-shaped mixers, the authors were able to develop a micro-flow strategy that allows peptide chain elongation without any signs of racemization (Fig. 5). Besides avoiding racemization of highly sensitive amino acids, this reaction set-up enabled the preparation of some small peptides in multigram-scale, such as in the case of the tripeptide shown in Fig. 5, which was obtained in more than 16 g with no

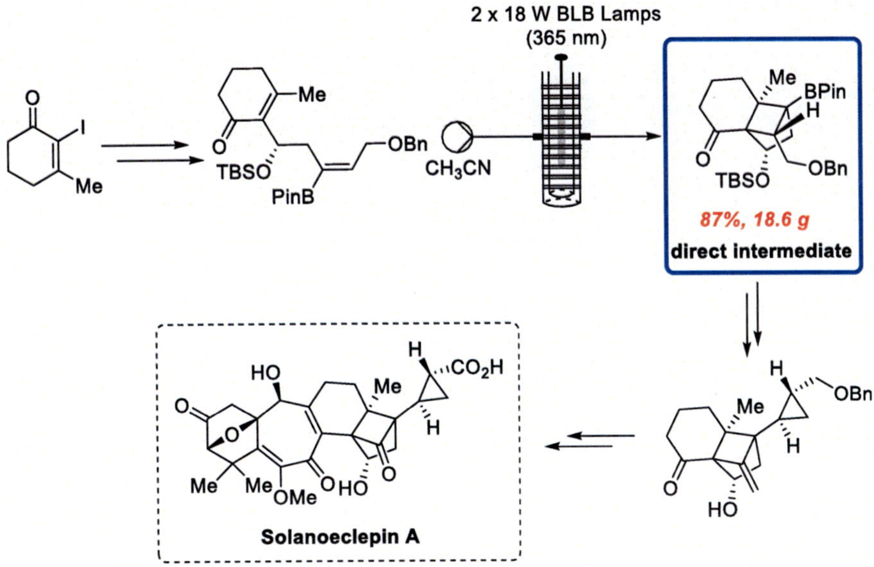

Fig. 4 Formal synthesis of solanoeclepin A

epimerization. Additionally, only 5 s, followed by simple liquid–liquid extraction were necessary for the success of this protocol.

In another example involving peptide chemistry, Wilson and co-workers demonstrate the use of a continuous flow system for the synthesis of some natural and unnatural cyclooligomeric depsipeptides, including beauvericin, bassianolide, and enniatin C (Fig. 6) [9]. The authors developed a protocol that allows the coupling of peptide chains in a 1:1 ratio and also provides an effective macrocyclization process, which is also a challenging step in the synthesis of such molecules. Comparisons between batch and continuous flow conditions were reported highlighting the advantages of flow protocols (Fig. 6). Besides improving the yields of the target compounds, the use of flow chemistry also allowed a reduction of the effort made by the chemists in the manipulation of the reaction set-up. However, no scale-up experiments were reported for the flow experiments (only mg-scale).

The use of flow set-ups is beneficial in both key-step or fully telescoped protocols, however, in some cases combining both batch and flow can be even better. This idea was demonstrated in 2017 by de Oliveira and co-workers in the synthesis of naturally occurring curcuminoids (Fig. 7) [10]. The authors evaluated three versions of the synthesis: (1) only batch, (2) hybrid batch-flow, (3) fully telescoped synthesis in flow. The whole synthetic route is comprised of three operations, and according to the authors, operation 1 is better performed in batch conditions, while operations 2 and 3 are more suitable to be carried out under continuous flow (Fig. 7). Three different curcuminoids were prepared by this protocol on multigram scales.

In 2017 Jamison and co-workers reported an end-to-end continuous protocol for the production of atropine, including the integration of separation/purification steps

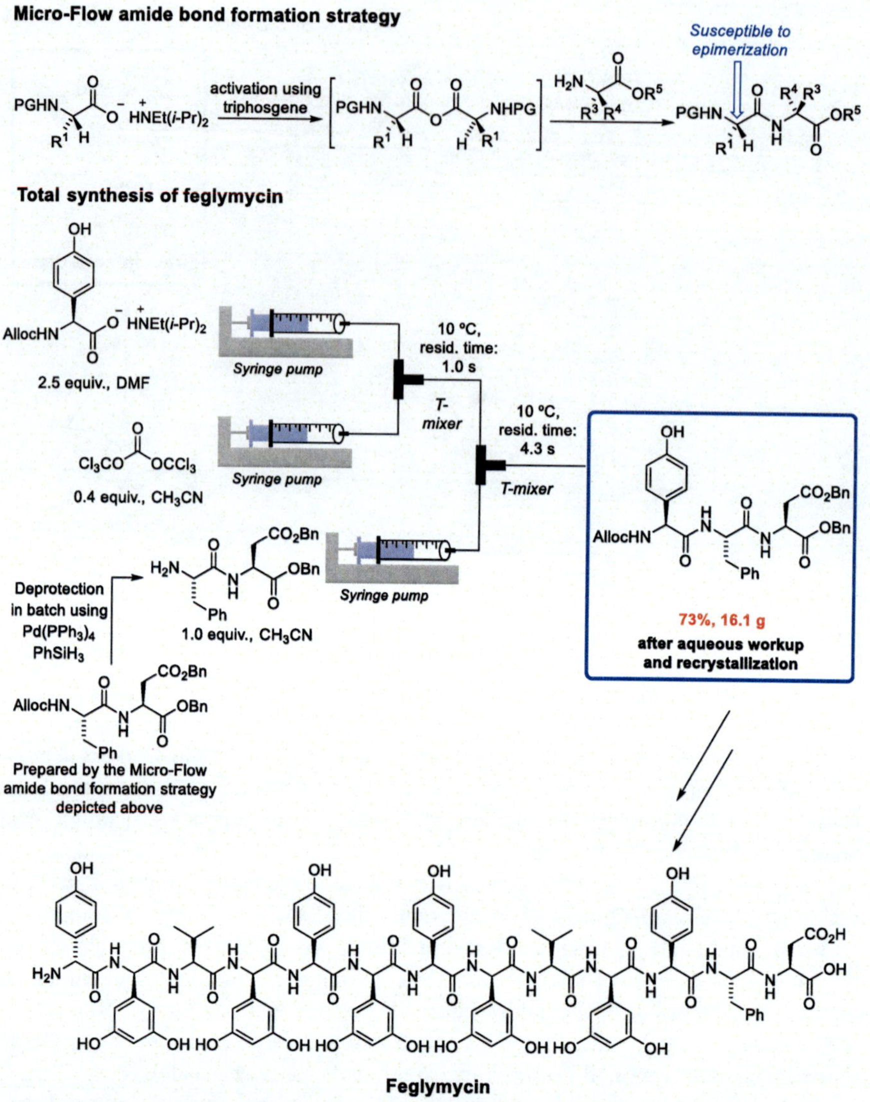

Fig. 5 Total synthesis of Feglymycin

[11]. This is an improved version of a continuous synthesis of atropine reported by the same group in 2015 [12]. The first step of the synthetic route involves the *O*-acylation of tropine with neat phenylacetyl chloride. The freshly generated acylated product is reacted in-line with aqueous NaOH and then formaldehyde, generating a mixture of atropine, apoatropine, and remaining acylated tropine (Fig. 8). By using a four-channel mixer, toluene and water are added in-line, and the resulting mixture then passes through a liquid–liquid in-line extractor, which separates the incoming

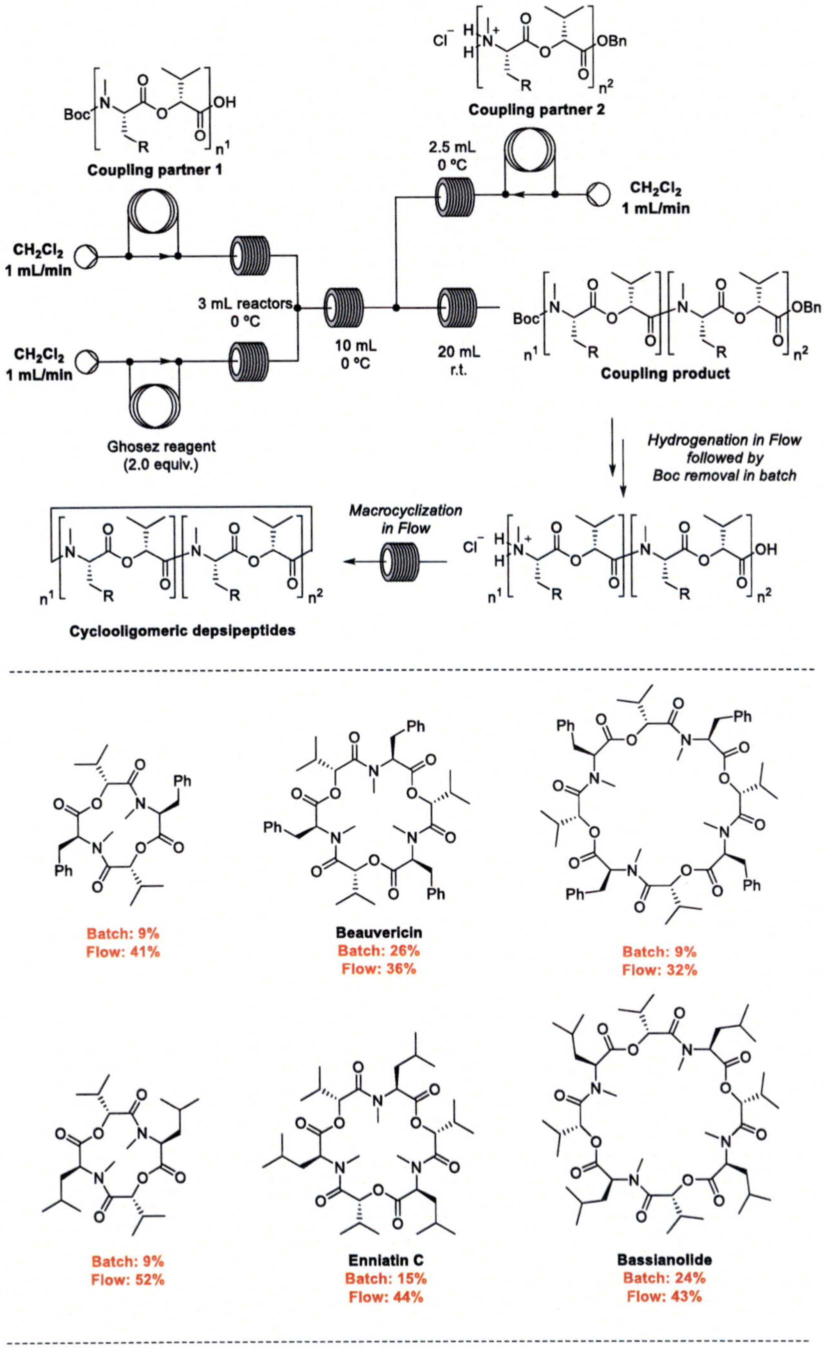

Fig. 6 Cyclooligomeric depsipeptides synthesis

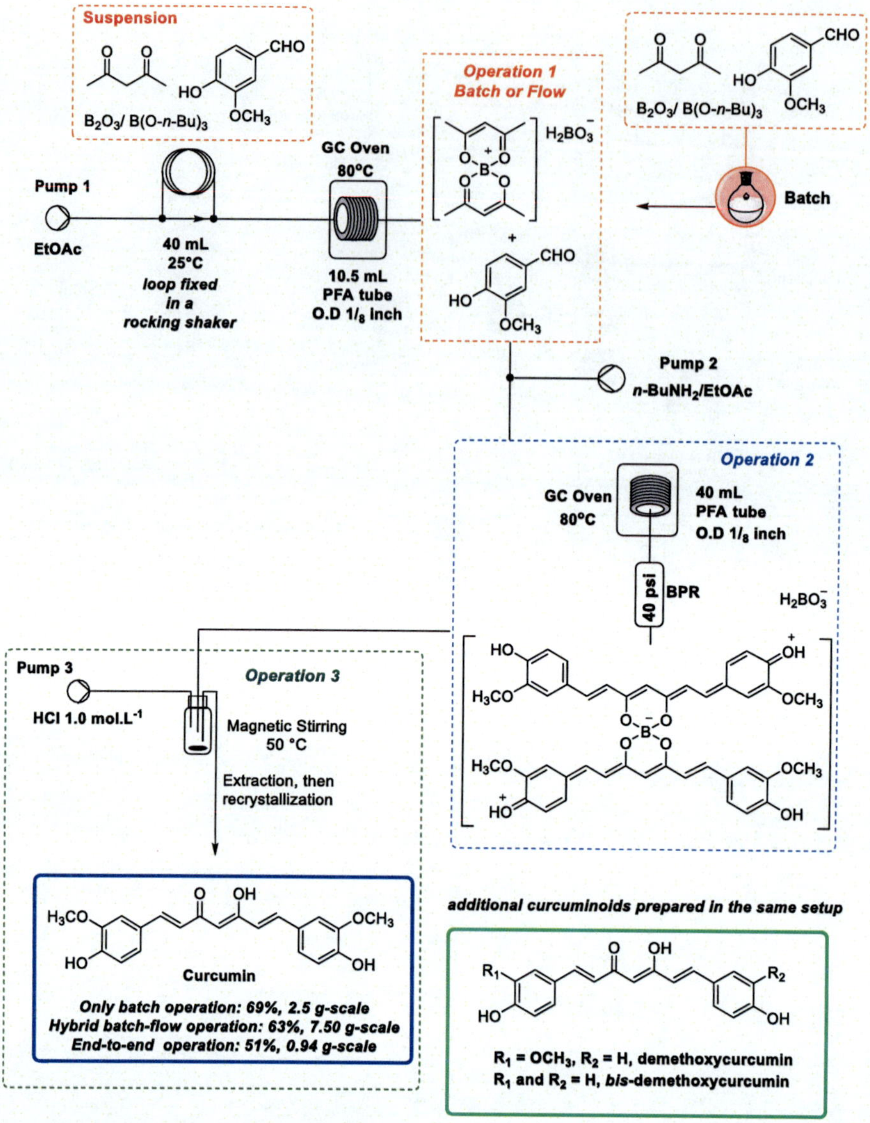

Fig. 7 Total synthesis of curcuminoids

stream into organic and aqueous streams. Apoatropine and acylated tropine are mostly concentrated in the organic stream, while the aqueous one contains the desired atropine, along with some impurities. Atropine is then obtained in 22% overall yield after a conventional extraction from H_2O/CH_2Cl_2 (up to 176 mg-scale). Despite the importance of this natural product, the authors did not report a better scale-up of this protocol (end-to-end gram-scale protocol) even with reported short residence times (<24 min).

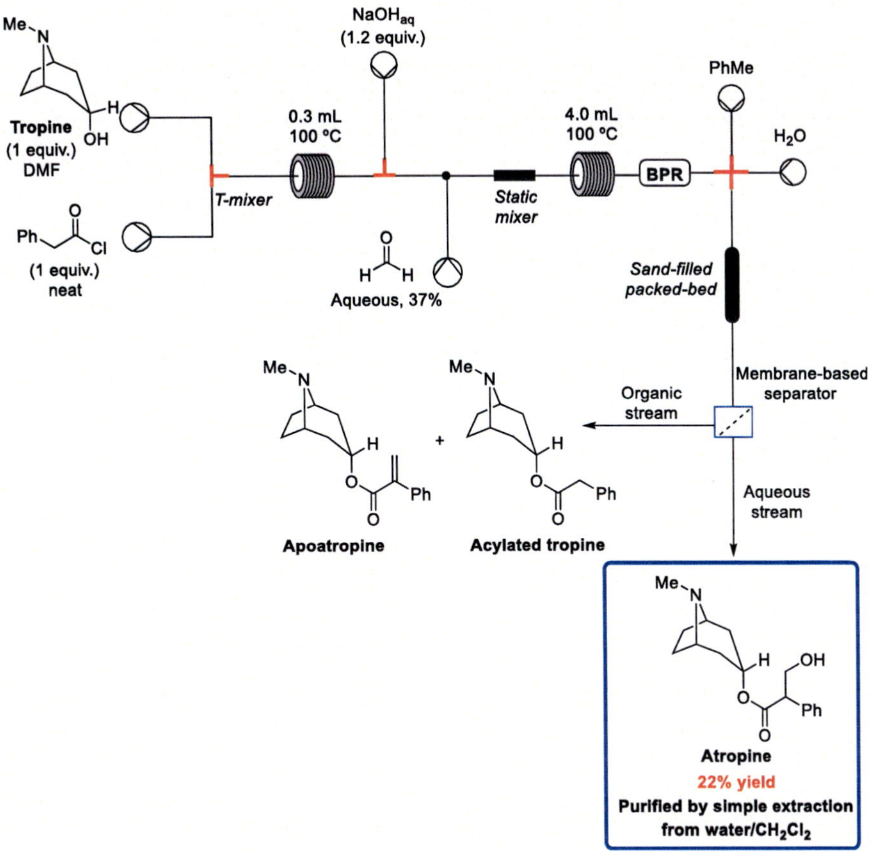

Fig. 8 End-to-end synthesis of atropine

Another relevant natural product which can be synthesized under continuous flow conditions is the antimalarial standard artemisinin. Many relevant approaches have been previously published by Seeberger, Gilmore, and co-workers in gram-scale [13]. The same authors recently described a very green approach starting from *Artemisia annua* extract using natural chlorophyll as photosensitizer in the photooxygenation step. However, this was developed with just NMR-yield determinations, and thus requires further development and scale-up [14].

One of the most recent (2015) gram-scale artemisinin intermediate syntheses was published by Kappe and co-workers, which demonstrates the conversion of artemisinic acid into dihydroartemisinic acid by the combination of oxygen and hydrazine under continuous flow regime (Fig. 9) [15]. This consists of pumping a solution of artemisinic acid and hydrazine hydrate along with a stream of O_2. The resulting segmented stream passes through a coil reactor at 60°C and the outgoing stream is then mixed in-line with more hydrazine hydrate and then passes again through a coil reactor at 60°C. This sequence is repeated twice more, providing a

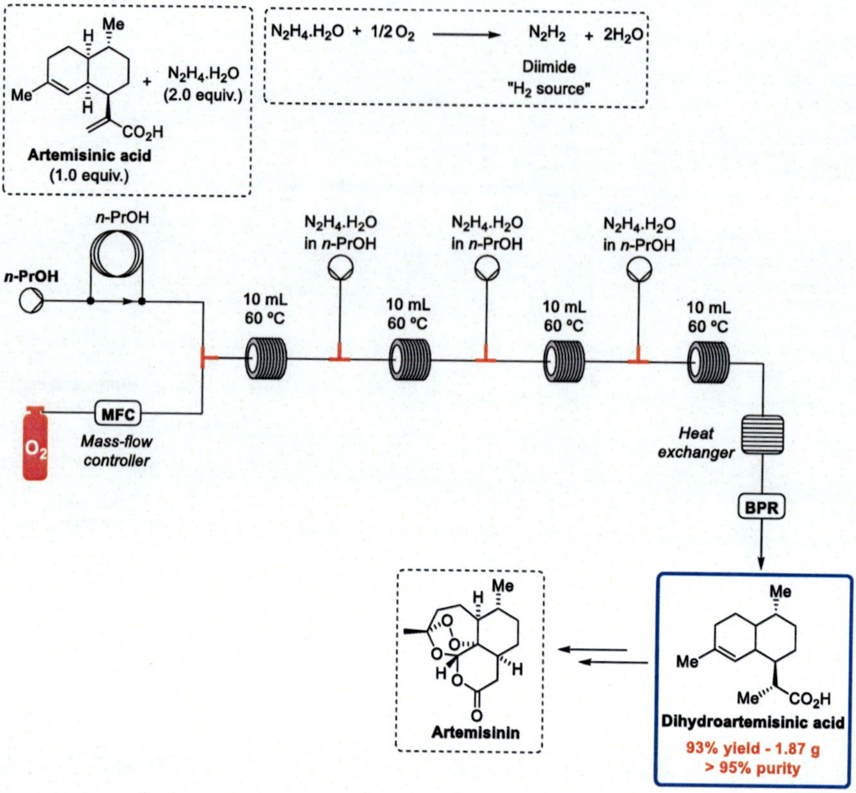

Fig. 9 Reduction of artemisinic acid for artemisinin synthesis

final mixture that contains dihydroartemisinic acid and some by-products. Dihydroartemisinic acid is then isolated by crystallization at controlled pH in 1.87 g-scale and 93% yield (Fig. 9). Further photooxygenation and rearrangements are well-known from dihydroartemisinic acid to artemisinin and are described in the literature [16, 17].

Thus, we can now envisage a completely continuous process from farnesol to artemisinin, by way of the Keasling group's biosynthetic engineering transformation into artemisinic acid, the Kappe group's reduction to dihydroartemisinic acid, and finally the Gilmore, Seeberger group's photo-oxidative transformation to artemisinin. This involves a Nature "flow" sequence, followed by two continuous flow processes, all amenable to extensive scale-up.

In the protocol reported by Kappe, oxygen is introduced into the reaction stream through the connection of a gas vessel to a T-shaped mixer. A more sophisticated and efficient way of working with gasses in flow set-ups is to use tube-in-tube devices (membrane reactors). This system is comprised of two tubes, one inside the other. The inner tube is permeable to gases, but not liquids. In this manner, the reaction stream may pass inside the inner tube while the outer tube is pressurized

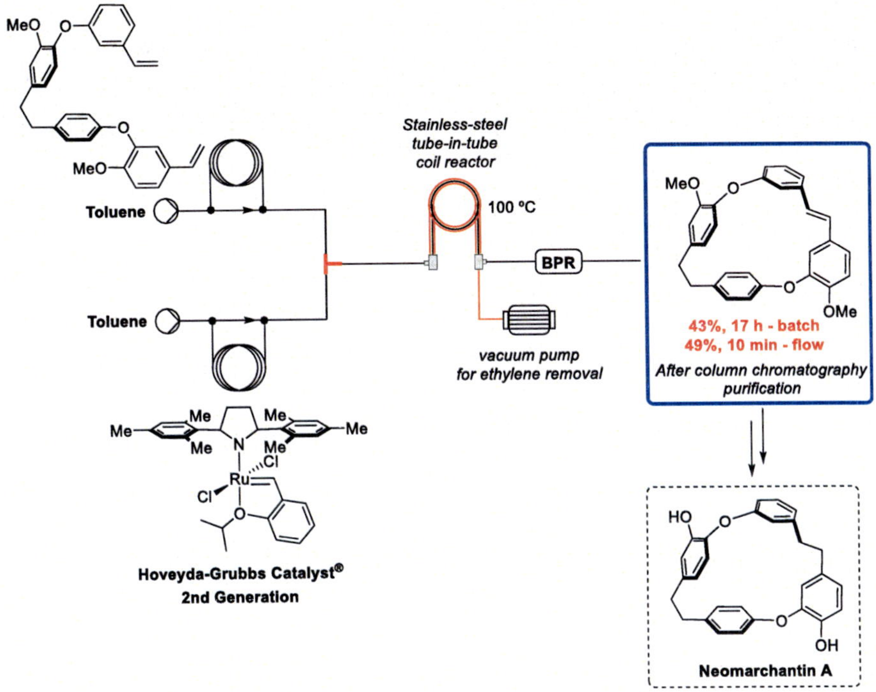

Fig. 10 Synthesis of neomarchantin A

with the desired gas, which will diffuse to the solution passing in the inner tube until saturation is reached.

This system also allows one to pull gasses out of the reaction mixture, and this was done by Collins and co-workers in the synthesis of Neomarchantin A, a macrocyclic bis-benzyl natural product (Fig. 10) [18]. The key step in this synthesis is a ring-closing metathesis (RCM) macrocyclization, which was performed under continuous flow conditions using a tube-in-tube device. The purpose of the tube-in-tube was to remove the ethylene formed from the RCM cyclization to drive the reaction towards the formation of the macrocycle. This was done by connecting a vacuum pump to the tube-in-tube device, so the ethylene formed was removed through the permeable inner tube. Unfortunately, the authors did not scale up this reaction in this paper and only demonstrated the metathesis protocol in up to 7 mg-scale. Due to the relevance of RCM in synthesis, the tube-in-tube/membrane reactors require further scale-up development [19].

As in the case of artemisinin, the natural product ingenol-3-mebutate is also an API (actinic keratosis, Picato™) and has been prepared in flow. This preparation was described in a patent registered by Alphora Research Inc. (2017) and describes the acylation of Ingenol with angelic anhydride in a continuous flow platform (Fig. 11) [20]. The big issue in this preparation is the chemoselectivity of the acylation reaction, which may take place formally at the C-3, C-5, or C-20 hydroxyl groups

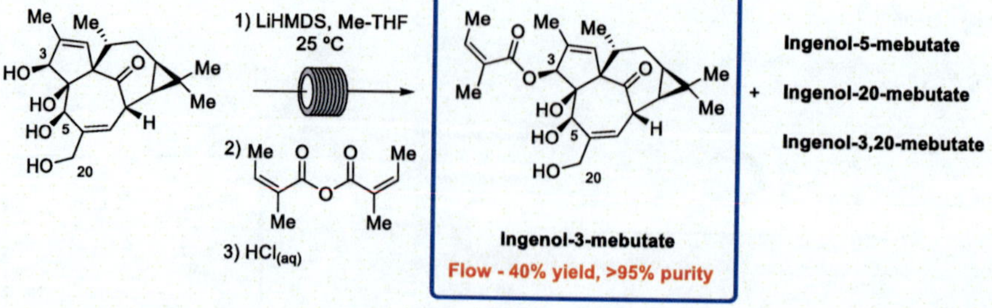

	HPLC (% a/a)				
	Ingenol	C-3	C-5	C-20	C-3,20
Batch	7.67	17.39	0.27	9.81	29.55
Flow	18.88	39.37	n.d	15.41	23.17

Fig. 11 Preparation of ingenol-3-mebutate registered by Alphora Research Inc.

or even at more than one. The initial experiments in batch were unable to provide the desired C-3 acylation as the major compound in the reaction mixture, favouring the formation of the product acylated at both C-3 and C-20. However, under continuous flow conditions, ingenol-3-mebutate was obtained as the major product and was isolated in 40% yield after purification by chromatography. The patent from Alphora Research Inc. did not describe the flow chem set-up in detail nor a robust scale-up. Again, by the relevance of this natural product/API, we have decided to describe it here.

To finish this first section of this chapter we present one recent example (2020) from the Pastre, Ley and Pilli groups working with a hybrid batch-flow synthesis of goniothalamin. They started the synthesis with cinnamaldehyde and allylmagnesium chloride (Fig. 12) [21]. In batch this is usually carried out at −78°C and the reaction takes 30 min to complete. Under continuous flow conditions, the reaction time was reduced to 0.4 min and can be run at room temperature, eliminating the necessity of cryogenic control. In the traditional synthesis, the secondary alcohol generated is purified by aqueous work-up and extraction with a separatory funnel. In the flow strategy, this alcohol was not isolated and met with an in-line stream of crotonic anhydride. The resulting mixture was feed into a coil reactor at 50°C with a residence time of less than 4 min. The crotonate ester was obtained in 96% yield from cinnamaldehyde in up to 20 g-scale. The final step, the RCM reaction, exhibited a better performance in batch than in flow, accounting for the choice to adopt a hybrid approach. The natural product goniothalamin was obtained in 75% overall yield after two telescoped reaction steps in flow and one batch procedure, in a total of 7.75 g.

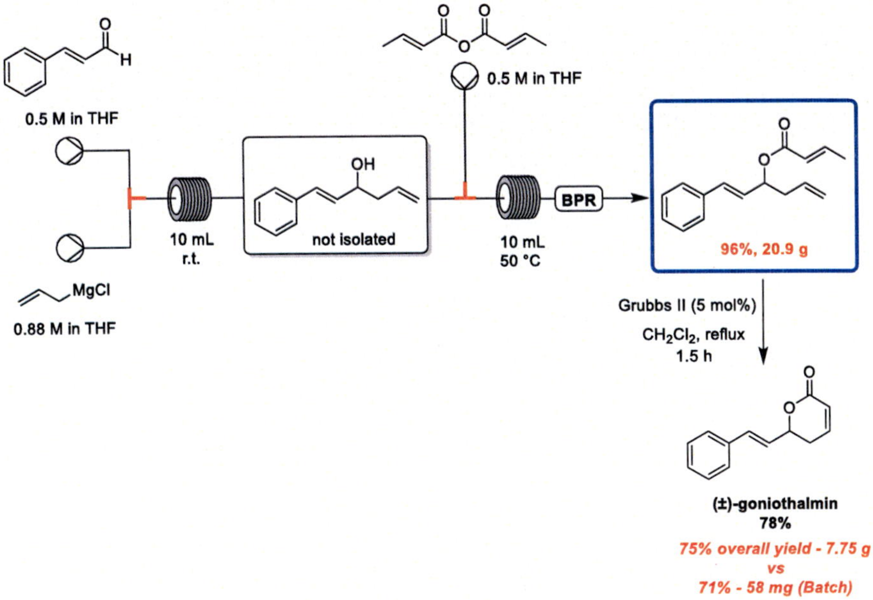

Fig. 12 Hybrid batch-flow total synthesis of goniothalamin

3 Good Manufacturing Practice (GMP) and Continuous Good Manufacturing Practice (cGMP)

The production of foods, cosmetics, medical devices, active pharmaceutical ingredients (APIs), medicines and much more consumer goods, requires the application of the Good Manufacturing Practices (GMP) regulations, which refer to the guidelines recommended by the agencies that control the authorization and licensing of their manufacture, as e.g. the American Food and Drug Administration (FDA). These regulations are based on rules that aim to ensure that a manufacturer supplies products with high quality, purity, safety, reproducibility, and efficiency [22–25].

One of the most important transformations that the industry has aimed at in the last few years is the implementation of continuous processes for manufacturing new medicines, thus justifying the investment of several companies that have constructed GMP continuous facilities [26, 27]. Nevertheless, although these efforts to integrate continuous processes in industry and the continuous support of contract manufacturing organizations (CMOs), up to now only a few APIs have been manufactured in advanced proof of concept plants (a preliminary version of GMP conditions) by the principal pharmaceutical companies. Continuous GMP (cGMP) conditions have been mostly applied to drug product formulation (DPF) or downstream processes. Among the Big Pharma players, we highlight GSK (two multistep continuous processes in their installations in Singapore), Eli Lilly (small-volume continuous manufacture (CM) concept in Kinsale, Ireland), consortium Novartis-MIT,

Pharmatech, Vertex, Pfizer, Johnson & Johnson and Amgen, with all of them making huge investments to develop their pharmaceutical ingredients. More recently, we have seen many industry-academia consortia resulting in the production from gram to metric ton quantities of pharmaceutical APIs. As an example of these consortia we can cite: Synthesis and Solid-State Pharmaceutical Centre (SSPC, Ireland), the Enabling Technology Consortium (members from the United States, European Union, and Asia), the Research Centre for Pharmaceutical Engineering (RCPE, Austria), and the Engineering Research Centre for Structured Organic Particulate Systems (C-SOPS, United States) [28].

Continuous manufacturing of APIs has proved to be valuable in many aspects and the big challenge is to perform it under cGMP conditions. The advantages of continuous API manufacturing range from increasing flexibility to produce smaller annual volumes of targeted therapies for smaller patient populations, increasing worker health and safety, improving product quality, higher pressure capabilities, a wider temperature operating window as well as reducing development costs for new medicines, decreasing cycle time, decreasing environmental impact, and decreasing costs of manufacture. Besides, it is already well known that making a combination between flow with chemo- or biocatalysis greatly improves the contributions with lower environmental impact and higher safety [26, 27, 29–31].

It is important to mention that cGMP syntheses are not clearly described as they have been under development, were discontinued or are part of the confidential and industrial development [32]. Additionally, many advanced studies that are in the proof of concept stage and represent a preliminary version of cGMP conditions are not clearly designed for cGMP. Based on the difficulty to precisely describe the status of each protocol we decided to highlight in this chapter the continuous manufacturing of APIs by the structural complexity, the relevance in the market, the robust reported gram to kg-scale production (with a few exceptions in mg-scale), merging both patents and journal publications in the field. We have selected examples involving only a few steps in flow (with additional steps in batch) up to end-to-end continuous production, trying to keep a chronological sequence from 2015 to mid-2020. The selected syntheses do not cover all the reported literature and are the author's choice based on an extensive literature search.

3.1 API Manufacturing Under Continuous Flow Conditions

To start this section, we describe vildagliptin which is an API developed by Novartis Pharma. This is an oral anti-hyperglycemic drug working as a dipeptidyl peptidase-4 (DPP-4) inhibitor used to treat type 2 diabetes mellitus (Fig. 13). In 2015, Pellegatti and co-workers developed a two-step flow process delivering the key intermediate of the Vildagliptin synthesis. The continuous flow protocol was aimed to mitigate the risks associated with the utilization of the Vilsmeier reagent on a large scale. Through this continuous flow process, the Vildagliptin intermediate was produced in 79% yield (12.7 g-scale), 99% purity and was scaled up delivering this advanced

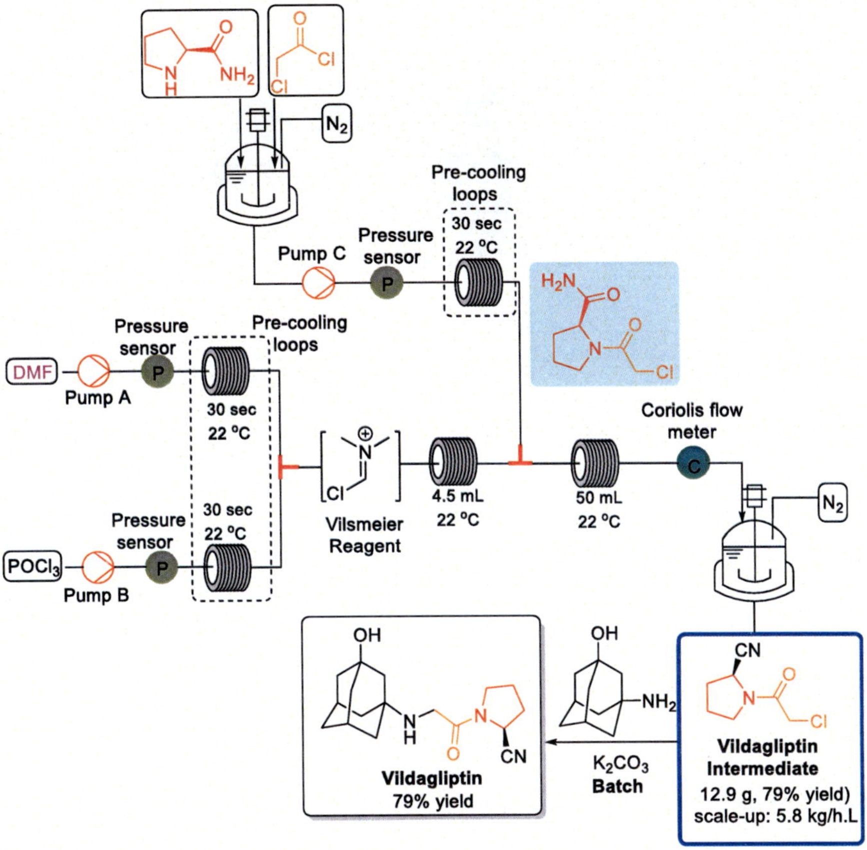

Fig. 13 Continuous flow synthesis of the vildagliptin intermediate

intermediate with a throughput of 5.8 kg/h L [33]. The last step of amination with an adamantyl derivative was performed in batch, also in 79% yield and leading to enantiopure vildagliptin.

Doravirine was approved by the FDA in 2018 to treat patients with HIV. However, the synthetic route for this compound was developed by Merck in 2015 using a continuous flow step for the aldol reaction to obtain the pyridone intermediate. Although relevant improvements were achieved, a robust scale-up was not exploited in this first investigation [34]. Subsequently, the same company designed and developed a pilot-plant for scale-up of the pyridone intermediate (Fig. 14). Researchers from Merck[25] had several problems to transpose this aldol reaction for the pilot-plant, and they used reactors with dual-tube heat exchangers to control the temperature and one Y-mixer which provided better mass transfer. After these adaptations, the pyridone intermediate was isolated in 68% yield and the pilot-plant was able to produce 178 kg in 24 h.

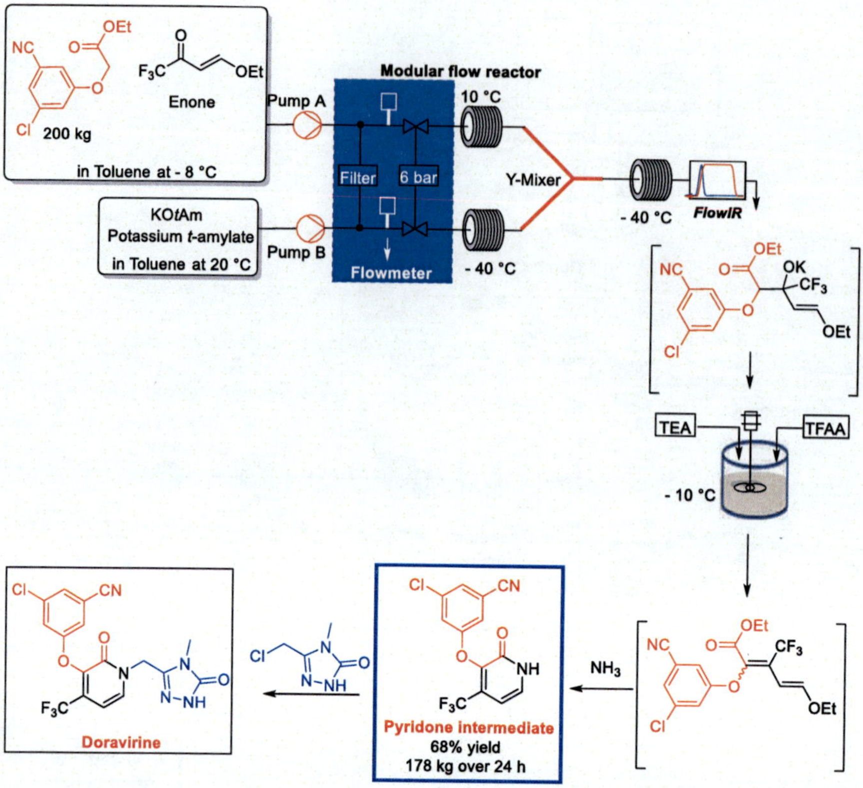

Fig. 14 Modular semi-continuous manufacturing of Doravirine intermediate

Vaborbactam is a β-lactamase inhibitor containing a cyclic boronic acid moiety (Vabomere API). It was approved in 2017 by the FDA and is used to treat complicated urinary tract infections. This compound had been manufactured by Rempex Pharmaceuticals in only batch processes [28, 35]. Nevertheless, the key Vaborbactam intermediate obtained by the Matteson homologation involves a problematic cryogenic step in batch mode, during the reaction of n-BuLi with CH_2Cl_2 and consequent formation of a borate complex after the Matteson homologation. In 2016 Rempex's researchers [36] applied similar conditions in a continuous process and obtained notable results with increased diastereoselectivity, yield, reproducibility, and productivity (Fig. 15). The work-up with $ZnCl_2$ was carried out in batch due to the better yield compared to flow conditions. The pilot-plant of Rempex was able to produce 880 kg/day of the Vaborbactam intermediate.

Crizotinib is an API classified as an anaplastic lymphoma kinase (ALK) and c-ros oncogene 1 (RSO1) inhibitor and was approved in 2011 by the FDA to treat patients with metastatic lung cancer (non-small cell lung cancer). Two patents describe the synthesis of this API or its intermediates. Nevertheless, in 2016 Asymchem Laboratories applied a two-step flow synthesis for the preparation of a Crizotinib

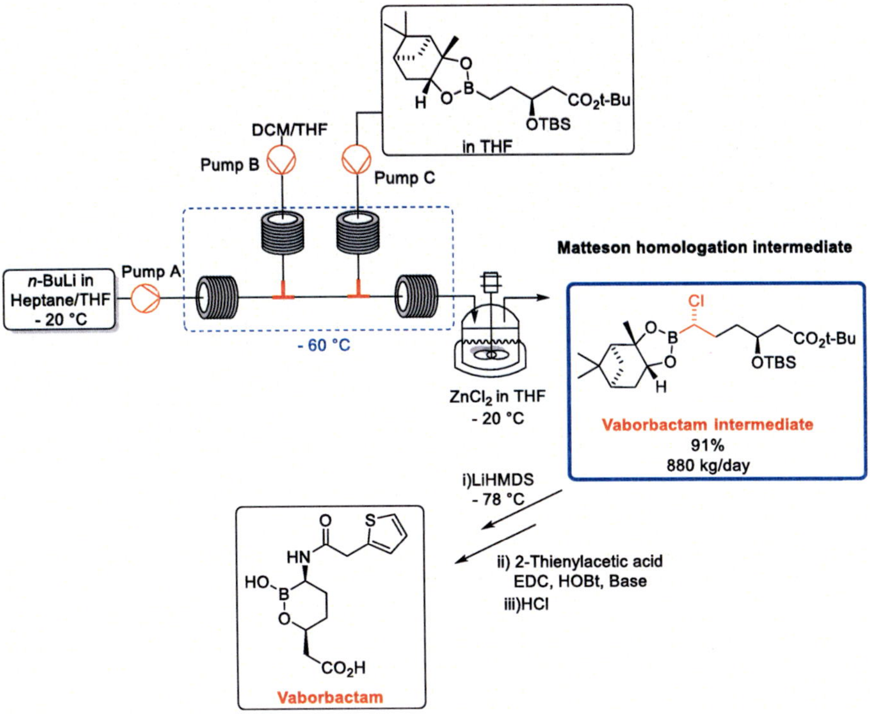

Fig. 15 Synthesis of Vaborbactam intermediate

intermediate (Fig. 16) [28, 37]. This method gave the pyrazole derivative in 90% yield, 93% purity, and in 400 g-scale, while the Crizotinib intermediate was isolated in 83% yield and 99% purity.

In 2017, the API Baricitinib was approved by the FDA for the treatment of rheumatoid arthritis. The batch process described by the Incyte company involves the preparation of three fragments, but only one of them has exhibited a problematic step. For the azetidinone intermediate, the synthetic route requires a high pressure of oxygen for efficient oxidation [28, 38]. In 2016, researchers from Lilly [39] applied a continuous flow approach to this synthetic step (Fig. 17), which provides greater safety and a more efficient oxidation process at high pressure (Novel Process Window concept – NPW). Another advantage of this synthetic route was the use of a simple set-up for preliminary scale-up. The oxidized product was isolated with 98% yield, 1.6 g-scale at 1.6 g/h.

Another API approved by the FDA in 2016 is Lifitegrast, which is used to treat keratoconjunctivitis sicca syndrome. SARcode Bioscience designed and developed the synthetic route for this drug, which involves the preparation of three main fragments [40, 41]. The synthesis of the carboxylated Lifitegrast intermediate involves a cryogenic reaction, which was considered a critical step in batch mode due to low yields. Owing to that, during 2016 and 2017, SARcode's researchers

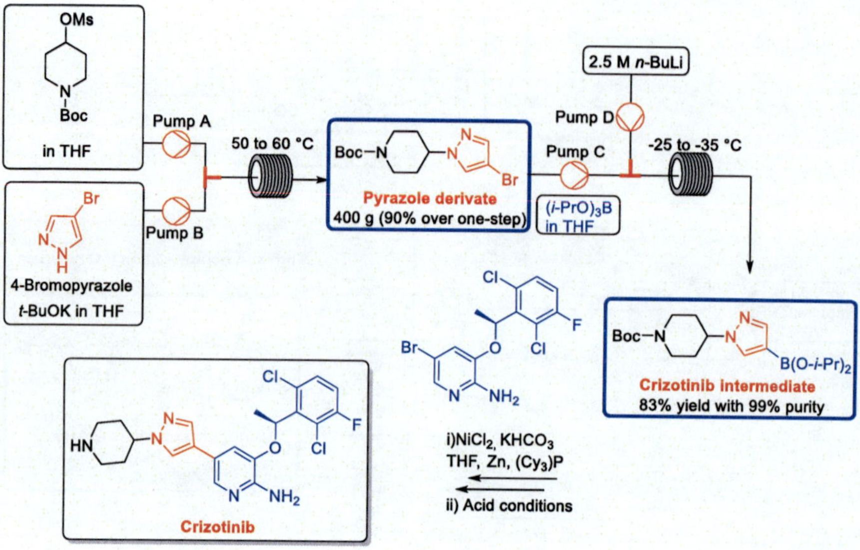

Fig. 16 Continuous scale-up synthesis of Crizotinib intermediate

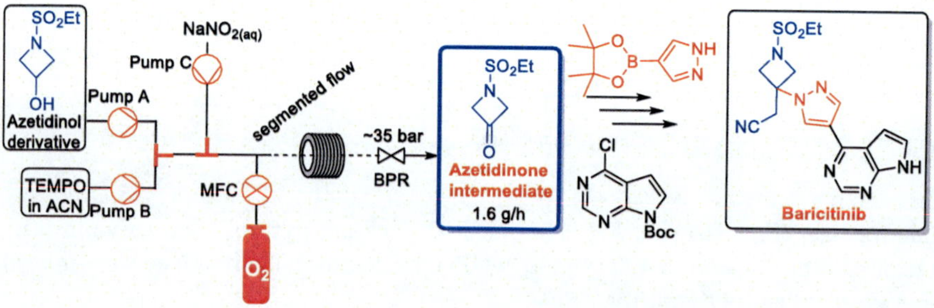

Fig. 17 One-step continuous flow oxidation to produce the azetidinone intermediate for the synthesis of Baricitinib

[28, 42] adapted the final step of the carboxylation reaction into a continuous flow process (Fig. 18). This mode of operation provided reproducibility with 88–91% isolated yields, 97–98% purity, and 4–5 kg-scales. The pilot-plant was able to produce a total of 22 kg of this fragment by the combination of continuous and batch processes.

In 2016, the MIT team headed by Jamison, Jensen, and Myerson published a new and disruptive technology, revealing a continuous flow multi-platform [43]. The authors developed a continuous manufacturing platform combining both synthesis and final drug product formulation into a single compact unit. They made a reconfiguration of their earlier platform for multiple syntheses of different compounds, with close-fitting integration of process streams for a reduced footprint. Innovations in the chemical reaction and purification step were presented, thus

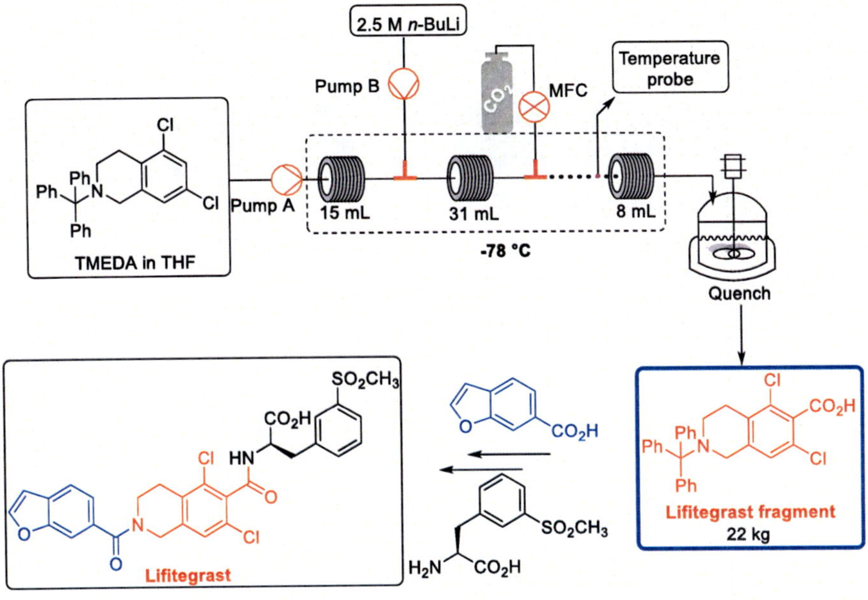

Fig. 18 Continuous flow carboxylation for manufacturing Lifitegrast fragment

giving a compact and reconfigurable system (*ca* 1/40 in the size) for crystallization and formulation of many drug products. To demonstrate the system versatilities four different pharmaceuticals were produced: diphenhydramine hydrochloride, lidocaine hydrochloride, diazepam, and fluoxetine hydrochloride all of them from their raw materials and in sufficient amounts to supply a large number (hundreds to thousands) of oral or topical doses per day.

The diphenhydramine hydrochloride API (also known as Benadryl) is an ethanolamine-based antihistamine which is used to treat the common cold, lessen symptoms of allergies, also acting as a mild sleeping aid. Typical batch processing requires five or more hours for this API to be obtained, but the continuous process transformation was complete in only 15 min and can be manufactured in its final dosage form. A real-time monitoring probe allowed the obtainment of the final dosage concentration (5 mL at 2.5 mg/mL) and the purity of the product is up to USP standards (Fig. 19). The system capacity is 4,500 doses per day [43].

Lidocaine hydrochloride is a well-known local anaesthetic and the class antiarrhythmic drug. The two-step synthesis was completed within 5 min in contrast with batch procedures under toluene reflux (60 min) or benzene reflux (4–5 h). Overall, complete conversion (99%) of the starting materials to the crude API was obtained in only 36 min giving the API in 88% yield and 97.7% purity (Fig. 20). In this case, this system is able to produce 810 doses (dosage strength = 20 mg/mL) of lidocaine hydrochloride per day [43].

Diazepam (Valium) is a central nervous system depressant [44]. When processed in the MIT machine, the complete conversion of the starting materials was

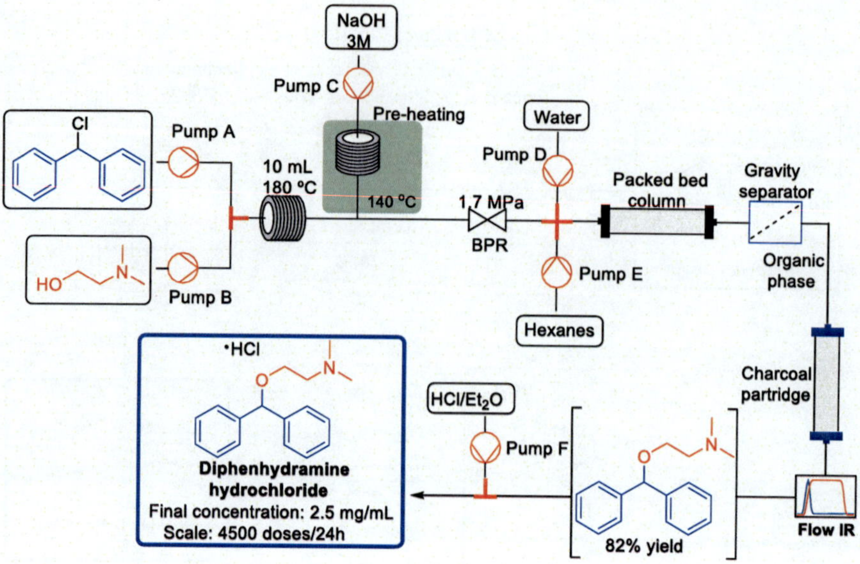

Fig. 19 Synthesis of diphenhydramine hydrochloride by the continuous manufacturing platform

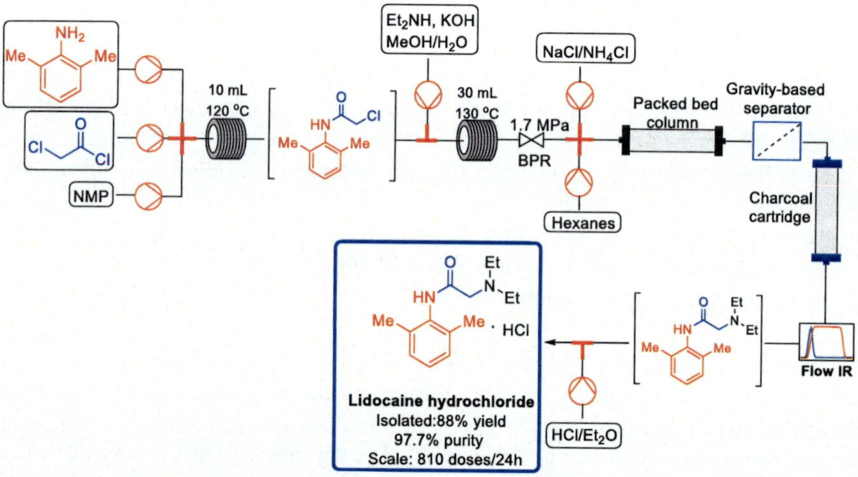

Fig. 20 Synthesis of Lidocaine hydrochloride by the continuous manufacturing platform

performed in only 13 min compared to 24 h of batch operation at room temperature. The dried diazepam crystals were obtained in 94% yield and purity that met with USP standards. At a dosage concentration of 1 mg/mL (one dose is 5 mL), this system is able to produce ~3,000 doses per day (Fig. 21) [43].

To demonstrate the versatility and capacity of this system to carry out a complex, fully integrated, and multistep synthesis, fluoxetine hydrochloride was prepared by

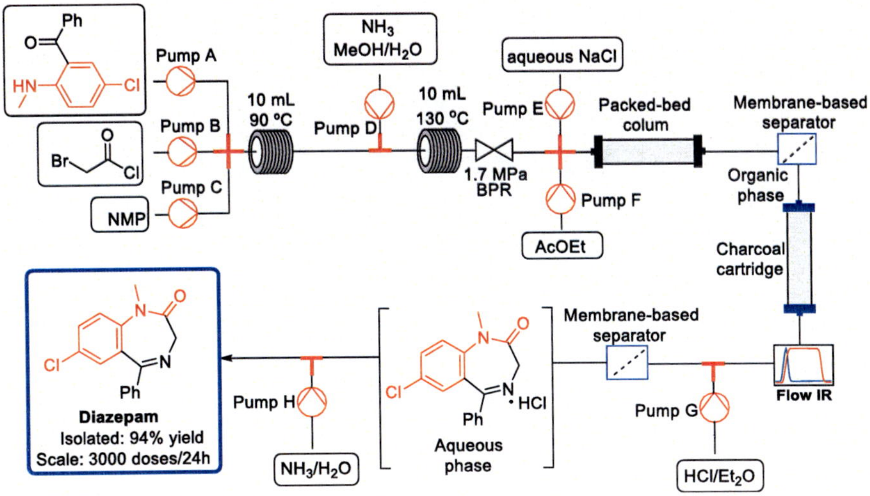

Fig. 21 Synthesis of diazepam by the continuous manufacturing platform

the MIT team. Fluoxetine (well-known as Prozac) is an antidepressant of the selective serotonin reuptake inhibitor (SSRI) class [45]. A series of individual reactions were previously carried out in both batch and flow, with purification and isolation of each intermediate. Using only continuous conditions, it is possible to perform the end-to-end synthesis of this API as a racemic mixture. The extraction and separation process yielded a solution of fluoxetine in *tert*-butyl methyl ether (TBME) with 43% yield allowing a potential production rate in up to 1,100 doses per day (one dose is 5 mL at 4 mg/mL) before the downstream processing. The fluoxetine hydrochloride crystals were provided in agreement with USP standards (Fig. 22) and the final redissolution in water yielded from 100 to 200 doses [43].

Brivaracetam is an anticonvulsant API, used for the treatment of partial-onset seizures in epileptic patients, and was approved in 2016 by the FDA. The company UCB Biopharma has developed the synthetic route for this compound in a three-step batch mode [46, 47]. However, the hydrogenation step to obtain this API is problematic due to the low diastereoisomeric ratio of the product brivaracetam (1:1 mixture of diastereomers), as well as the low productivity of the furanone formation (0.0013 kg/L/h). Therefore, UCB's researchers [48] discovered that the use of citric acid as an additive improves the diastereoselectivity. Hence, in 2017 [28, 49] they decided to apply a continuous process in this synthetic route combined with a batch process (Fig. 23). In contrast to the exclusive batch procedure, the combined continuous flow/batch process provides higher productivity of the furanone (21 kg/L/h) and improved yield (96 vs 88%) of the lactam derivative with three consecutive reactions at the hydrogenation step. The final hydrogenation process was carried out in a series of four reactors to increase conversion and diastereoselectivity after the improvement allowed by gas–liquid mass transfer.

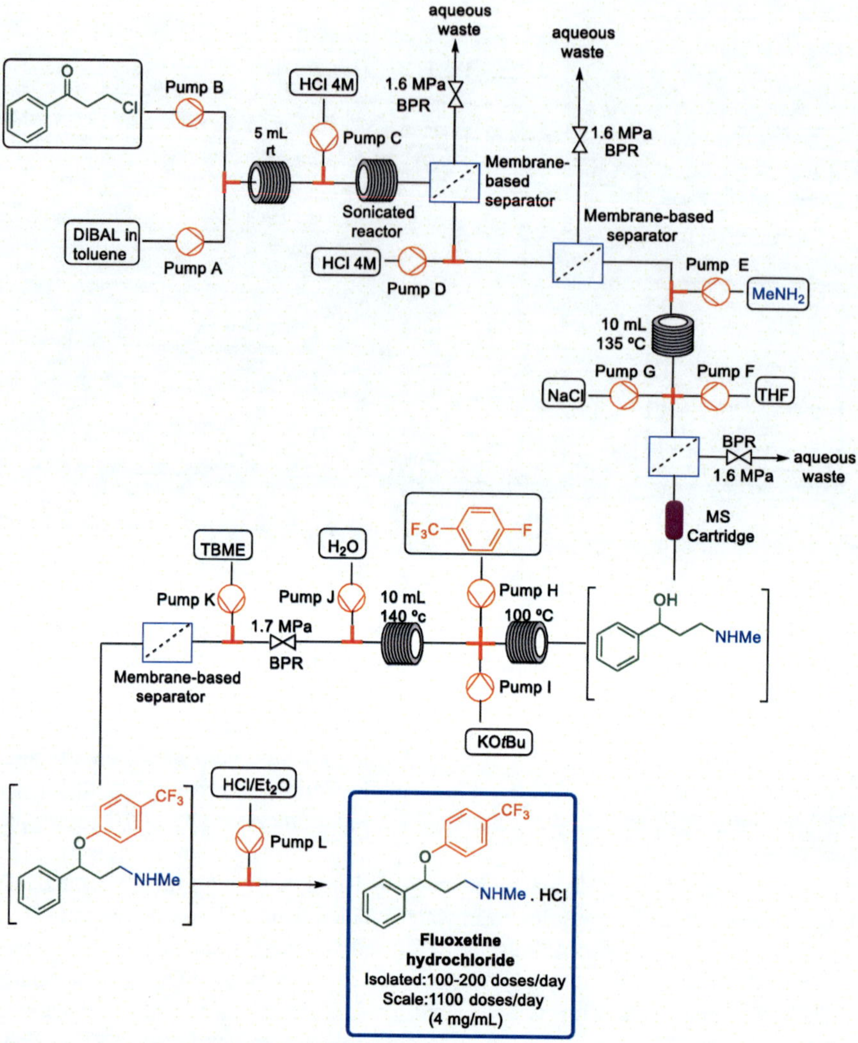

Fig. 22 End-to-end synthesis of Fluoxetine hydrochloride

In 2016, Noël and Hessel developed a flow protocol aimed at converting bulk alcohols into the corresponding chlorides, with subsequent reaction with amines to produce commercially available APIs. Depending on the starting material (alcohol) the system can offer different APIs such as cyclizine, cinnarizine, and buclizine, as follows [50].

Cyclizine is an antihistaminic used in the treatment of nausea, vomiting, and dizziness associated with motion sickness, vertigo, and consequences of anaesthesia and the use of opioids [51]. A continuous run on a 7 mmol/h scale was performed over 8 h in 94% yield and high purity (>99% GC–MS) (Fig. 24) [50].

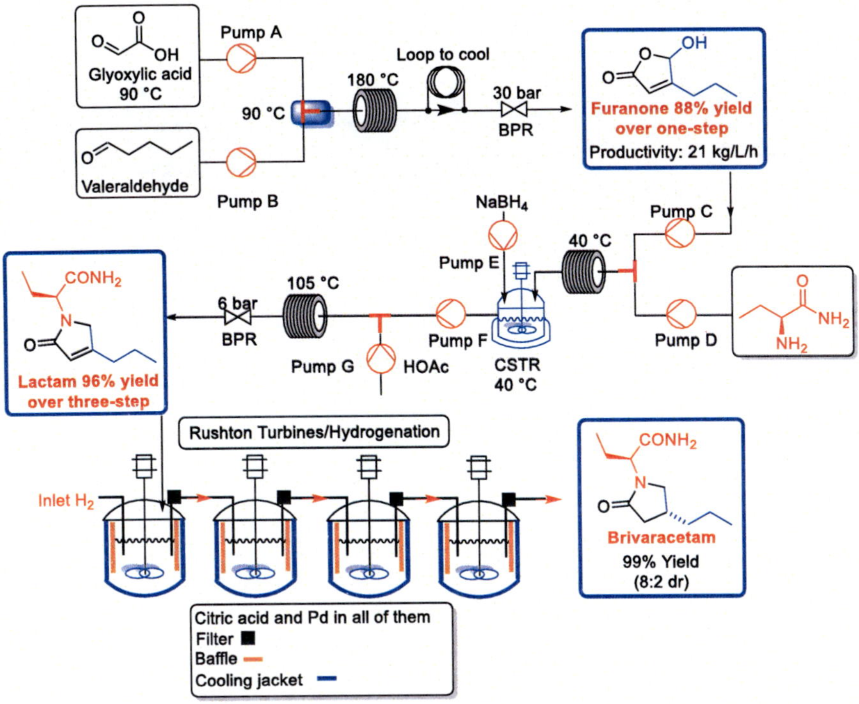

Fig. 23 Semi-continuous synthesis of Brivaracetam API

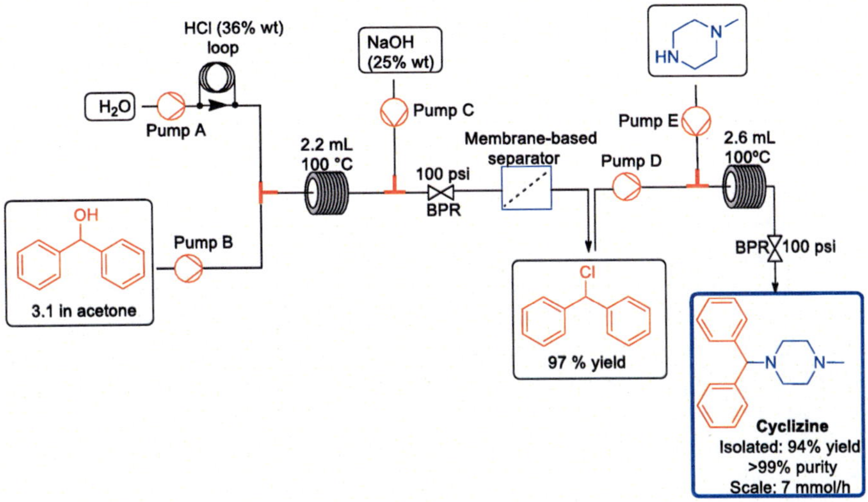

Fig. 24 Continuous flow synthesis of Cyclizine

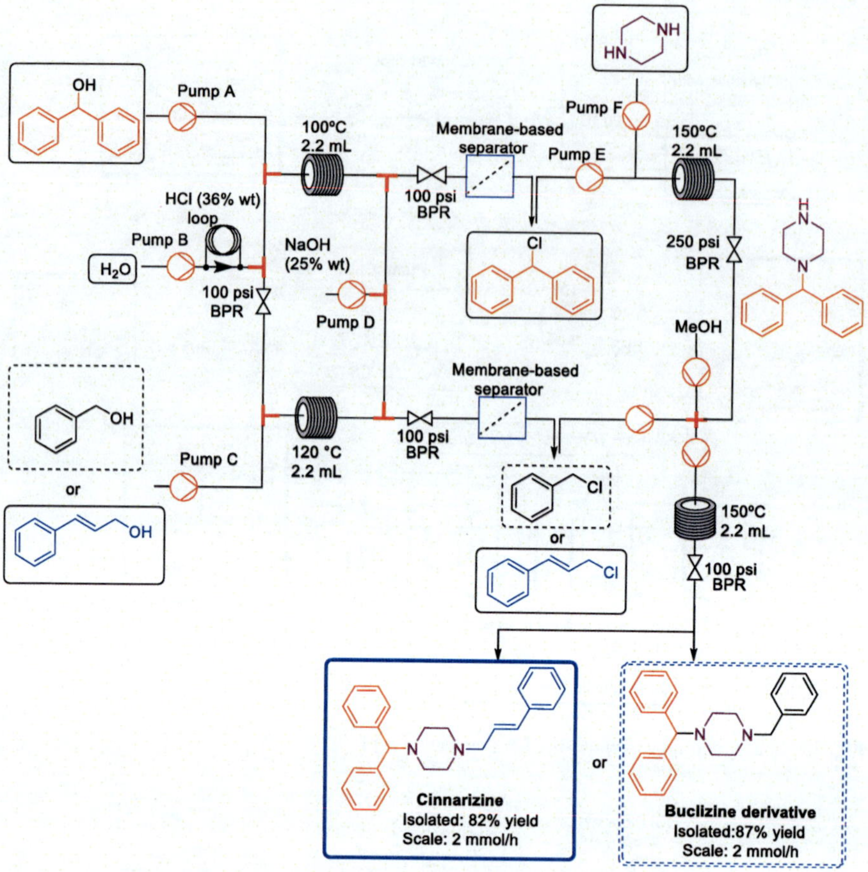

Fig. 25 Synthesis of Cinnarizine and Buclizine intermediate

Cinnarizine was first commercialized by Janssen Pharmaceuticals in 1955 and is used as an antihistaminic drug for the control of disorders in the sense of balance and motion sickness [52]. The Noël and Vessel's flow process allowed the obtainment of this API in mild conditions and high selectivity (Fig. 25). After the recrystallization step, cinnarizine was obtained in 82% isolated yield (scale-up: 2 mmol/h) [50].

Buclizine is an antihistamine and anticholinergic from the diphenylmethylpiperazine group [53]. Through this continuous process, it was possible to prepare a buclizine intermediate at 150°C with the full conversion of 1-(diphenylmethyl)piperazine, and in 89% yield based on diphenyl methanol (scale-up: 2 mmol/h) (Fig. 25). However, the greatest limitation of this method is the supply of hydrochloric acid. The use of HCl gas to supply hydrochloric acid could revolutionize this protocol, but unfortunately, it has not been developed yet [50].

AZD9291 acrylamide (generic name osimertinib) is also known as mereletinib (trade name Tagrisso). This medicine is an irreversible epidermal growth factor

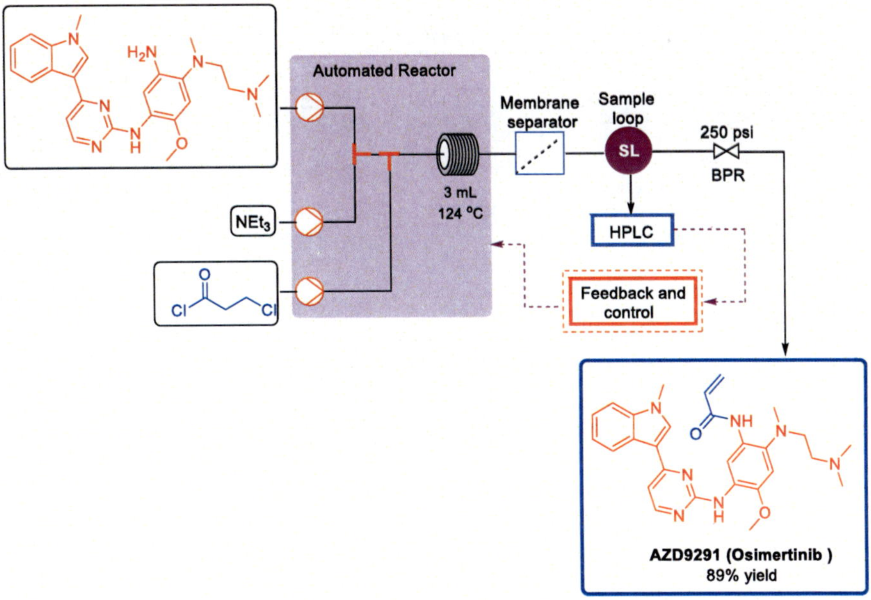

Fig. 26 Synthesis of AZD9291(Osimertinib) using a self-optimizing system

receptor (EGFR) kinase inhibitor and is used to treat non-small-cell lung carcinomas with specific mutations [54]. It was developed by AstraZeneca and approved as a cancer treatment in 2017 by both FDA and the European Commission. The API is commonly synthesized in batch, however, Bourne and col. presented a novel self-optimizing automated flow system for a 2-step telescoped synthesis of this API with an adaptive feedback control loop for the final bond-forming steps. The system presents very efficient and precise heating and cooling, reducing the time to reach the set conditions. The control is performed by an on-demand software and algorithms. Therefore, after the automated optimization (42 experiments, 10 g of processed raw material, 26 h), osimertinib was obtained in 89% yield giving ca. 240 mg per experiment (Fig. 26). The use of self-optimizing systems allows fast exploration and process optimization even in multistage reaction systems without intervention [55].

Elbasvir is a potent NS5A antagonist for the treatment of chronic hepatitis C. Administered in combination with grazoprevir, an HCV protease inhibitor, it has been clinically studied as a highly efficacious and well-tolerated oral regimen for the treatment of HCV infection, including patients with HIV co-infection. This API was approved by FDA in January 2016 and was developed by Merck. Knowles and col. introduced a novel visible light photoredox process for indoline oxidation involving an iridium photosensitizer and an environmentally-benign per-ester oxidant. Using a flow reactor, ca. 100 g of indoline substrate could be processed in 5 h, in 85% yield and 99.8% *ee* and was demonstrated on a lab-scale at a throughput of 0.04 mol/h. This provides one of the first well-succeeded demonstrations of

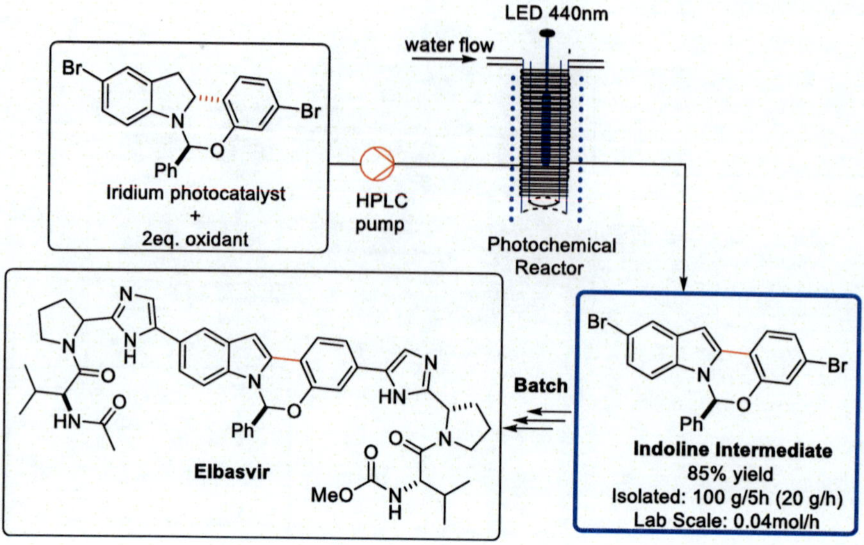

Fig. 27 Continuous photoredox process for the synthesis of Indoline intermediate

photoredox catalysis in the production of a pharmaceutical intermediate (Fig. 27) [56].

Menadione (vitamin K_3) is an intermediate in the chemical synthesis of vitamin K. Despite menadione supplements for direct consumption being banned by the FDA because of their potential toxicity for human use [57], this compound is a direct intermediate for the industrial production of Vitamin K and is considered a valuable API intermediate. In 2016, De Oliveira and McQuade reported a photocatalyzed protocol under continuous flow conditions using a simple home-made photoreactor and different porphyrin derivatives as photocatalysts. In the same paper, a scope of different naphthols is presented including the synthesis of the natural product juglone. After optimizations, vitamin K_3 was also obtained in 82% yield under continuous processing, against 20% yield obtained in batch. The simple set-up proved to be efficient and produces many oxidized products in gram-scale by using singlet oxygen, in only one reaction cycle, and short residence time (16.7 min) (Fig. 28) [58].

Ribociclib (which the succinate brand name is Kisqali) is a cyclin D1/CDK4 and CDK6 inhibitor and is used for the treatment of certain kinds of breast cancer. It is also being studied for the treatment of other drug-resistant cancers. Two different pharmaceutical companies have developed this API: Novartis and Astex Pharmaceuticals. In 2016, Novartis described an end-to-end two-step continuous flow process for Ribociclib. The whole process was performed in 90-min and delivered the desired product in high yield (92.5%) and purity (95%) with a throughput of 51.0 g/h (117 mol/h). After the integration of a continuous extraction dispositive, they demonstrated the feasibility to run in batch to the final API (with succinic acid) with a throughput of 0.85 g/min and 95% purity (Fig. 29) [59].

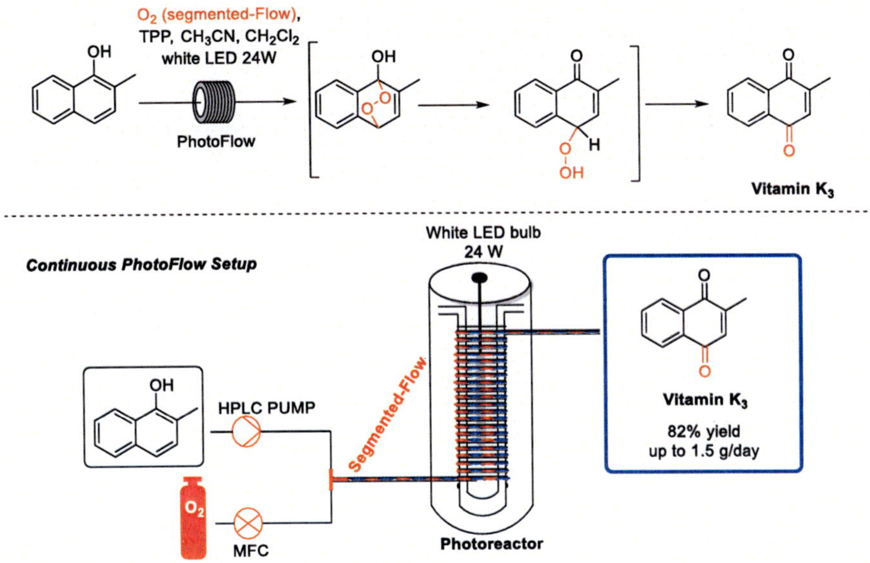

Fig. 28 Continuous flow synthesis of Vitamin K_3

Valacyclovir hydrochloride is considered to be a very similar drug to the antiviral acyclovir, which is used for the treatment of patients with *Herpes simplex* or *Herpes zoster* since 1995. This new API is classified as a prodrug and it has greater bioavailability than acyclovir. The synthetic route to obtain this compound involves a two-step protocol starting from acyclovir with an esterification reaction with Boc-L-valine and deprotection of the Boc-group. This route was described by Aurobindo Pharma first in batch mode. However, the product was obtained with low yield (30%) and 96.6% purity. Hence, in 2017 a continuous-flow approach [28, 60] was developed by the same company in the deprotection step (Fig. 30), which supplied an increased isolated yield (65%), 99% purity in a 31 g-scale.

Flucytosine is used to treat *Cryptococcal meningitis*, a fungal infection that is of particular concern to HIV patients with heavily compromised immune systems. This API was included in the 19th WHO Core List of Essential Medicines in 2015 and was approved by the FDA in 1971 (Ancoban, Valeant) for the treatment of fungal infections. The only generic version was introduced in the US market in 2011 by Sigmapharm company. In 2017, Sandford and col. described an inexpensive and readily scalable method for the synthesis of flucytosine as part of a long-term program developing the use of fluorine gas as a reagent for organic chemistry using both batch and continuous flow conditions. In batch this API could be synthesized in a moderate yield but leads to considerable waste, which contributes to higher production costs. Using continuous conditions, it was possible to prepare this API in a one-step synthesis, 100% conversion, and 63% yield. The scale-up was performed using a mesoscale Boostec® flow reactor, which allows the use of corrosive reagents and controlled exothermic processes (flucytosine reaction

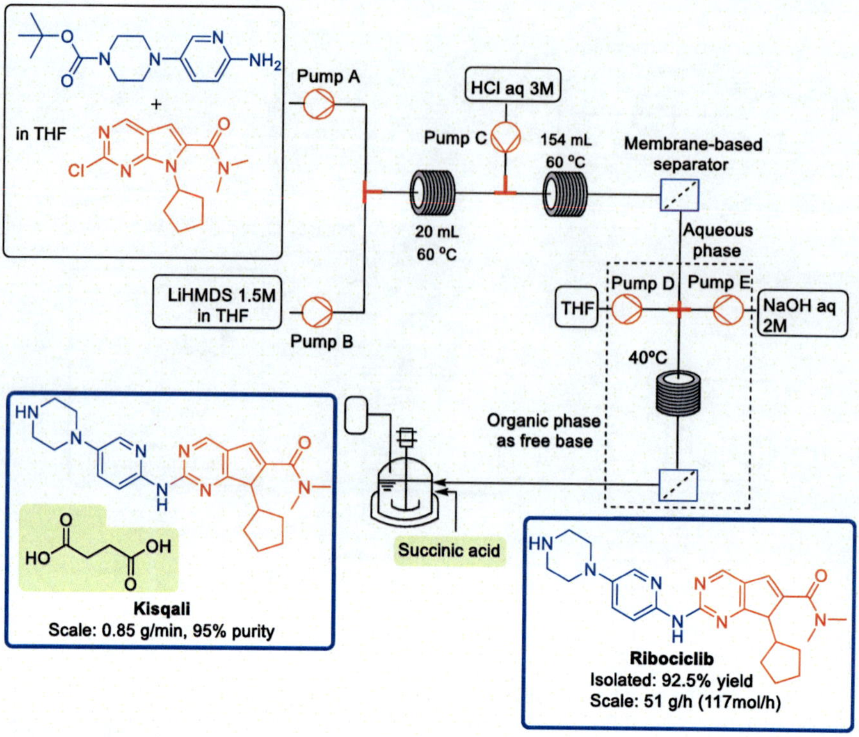

Fig. 29 Continuous flow synthesis of Ribociclib and its succinate Kisqali

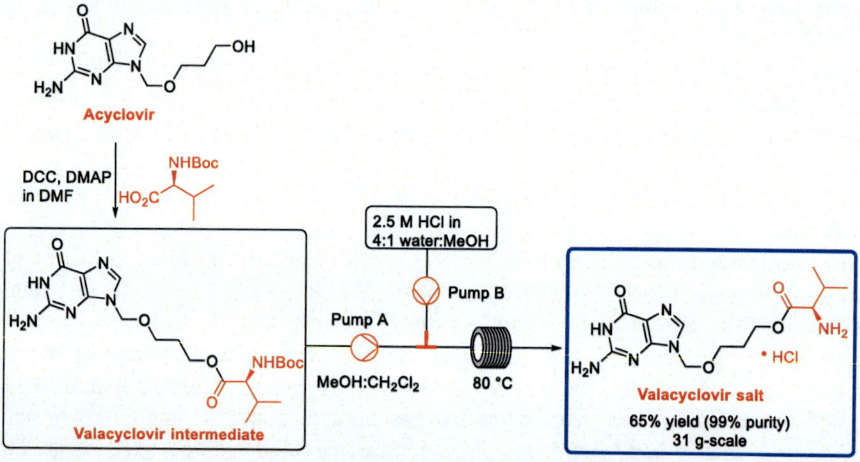

Fig. 30 Continuous flow synthesis of Valacyclovir salt

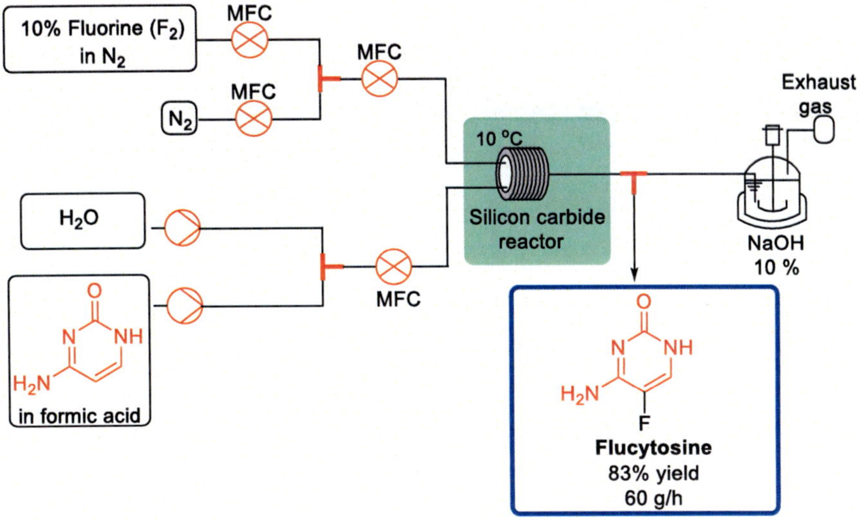

Fig. 31 One-step synthesis of flucytosine from cytosine using continuous flow

conditions). Flucytosine was produced in 58.0 g-scale, 83% yield, and 99.8% purity. Therefore, the one-step synthesis of flucytosine from cytosine using continuous flow direct fluorination was established and transferred to the pilot-scale (60 g production of pure flucytosine per hour per single reaction channel) (Fig. 31) [61].

Prexasertib is a checkpoint kinase inhibitor active against CHEK1 and CHEK2 with antineoplastic activity [62]. The Eli Lilly Company has developed a multistep continuous-flow process that produced 24 kg of this API (prexasertib monolactate monohydrate). The process is carried out by eight continuous unit operations conducted to produce the target at roughly 3 kg per day using small continuous reactors, extractors, evaporators, crystallizers, and filters in laboratory fume hoods. The continuous flow process was based in four reaction steps: Optimization of a hydrazine condensation reaction in flow (step 1) giving 26.4 kg of the first intermediate, total run time: 207 h and product throughput: 3.1 kg/day (Fig. 32). Then aromatic nucleophilic substitution with crystallizations (step 2) and deprotection with simultaneous gas and liquid handling (step 3) yielded 22.2 kg of the advanced intermediate, in 208 h total run, and a product throughput at 2.56 kg/h (Fig. 33). The formation of the final lactate salt (step 4) yielded 24 kg/290 h, and a final API throughput at 3 kg/day (Fig. 33). All these continuous unit and overall continuous processes afforded improved performance and safety relative to batch [63].

The MIT team and other institutions presented in 2018 a second-generation reconfigurable, continuous flow pharmaceutical manufacturing platform which was recently licensed by the company On-demand Pharmaceuticals (PoD machine). This new system features stand-alone upstream and downstream units for the synthesis, the purification, and formulation of APIs. Among many advantages, they highlighted improvements in manufacturing capabilities and reduction of the

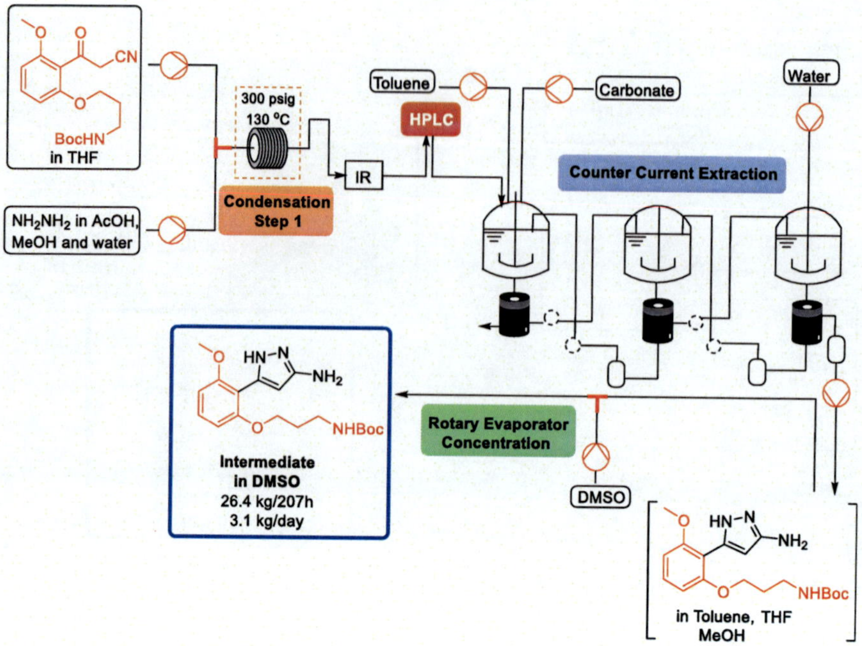

Fig. 32 Synthesis of the first Prexasertib intermediate in DMSO using a continuous process

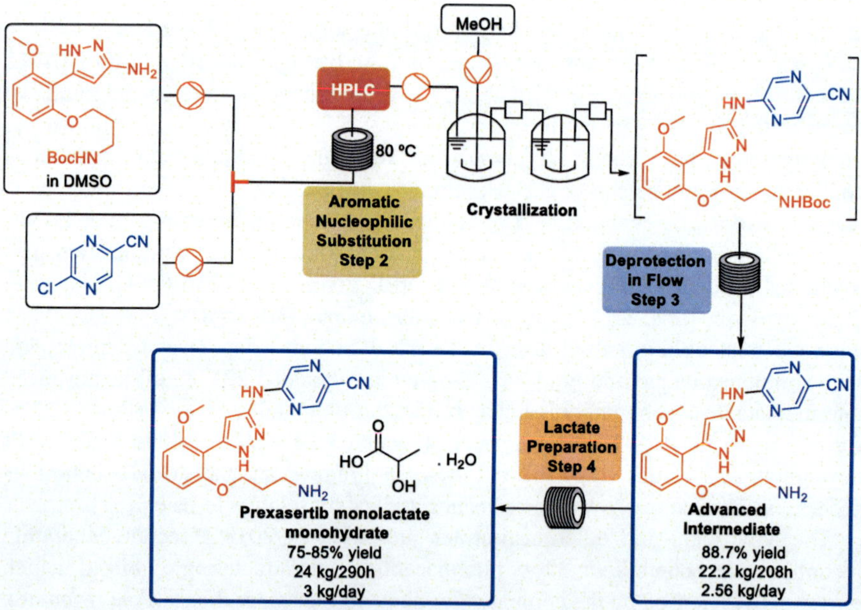

Fig. 33 Synthesis of Prexasertib monolactate monohydrate using a continuous processes

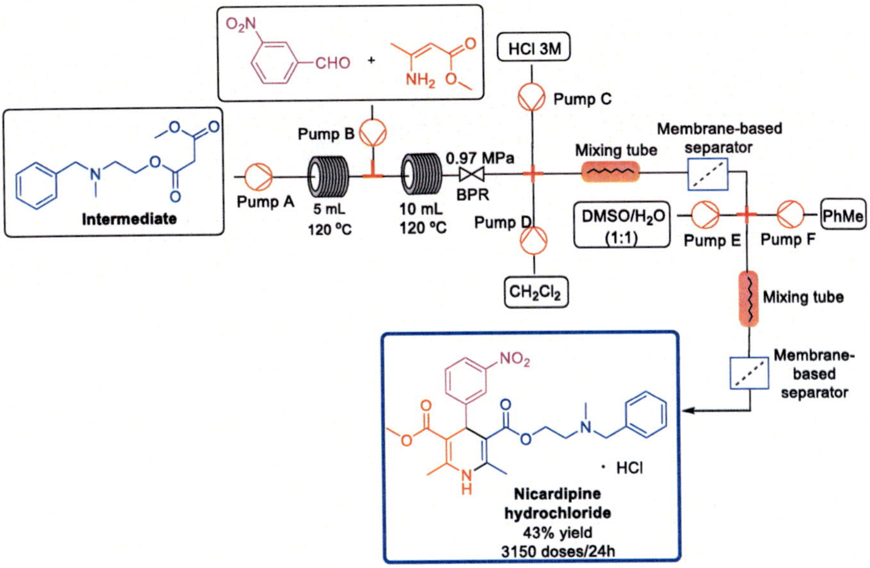

Fig. 34 Continuous flow process for the synthesis of Nicardipine hydrochloride

total volume of the system by more than 25% relative to the first-generation platform. This next-generation system includes continuous processing for crystallization, semi-continuous filtration, washing, dispensing, and drying. To showcase the capabilities of this platform a variety of APIs were prepared such as nicardipine hydrochloride, ciprofloxacin hydrochloride, neostigmine methylsulphate, and rufinamide, all of them in agreement with USP standards [64].

Nicardipine hydrochloride (Cardene) is a medication used to treat high blood pressure (antihypertensive drug), to control angina, and sometimes used to control Raynolds's phenomenon. This drug was approved by the FDA in December 1988 and is mostly manufactured by Baxter Healthcare Corporation USA [65]. During the Novartis consortium flow process, they have demonstrated the complete conversion of starting materials to the crude API, under elevated pressure and temperature, in only 80 min giving the nicardipine hydrochloride in 43% yield (3,150 doses per day) (Fig. 34) [64]. The dihydropyridine nucleus of the product is generated in an excellent multi-component reaction.

Ciprofloxacin is an antibiotic used to treat a significant number of bacterial infections like urinary tract infections and skin infections [66]. This API was patented in 1980 and is considered an essential drug by the World Health Organization's List of Essential Medicines. During the new continuous flow process, real-time monitoring using an SFU probe provided a final concentration of 1.5 mg/mL. HPLC analysis shows a 98.2% purity for this API, thus confirmed to USP standards. Overall, 2,700 doses of ciprofloxacin were produced (3.5 mg/mL solutions). After 6.8 h they demonstrated production of 9,600 doses per day (Fig. 35) [64].

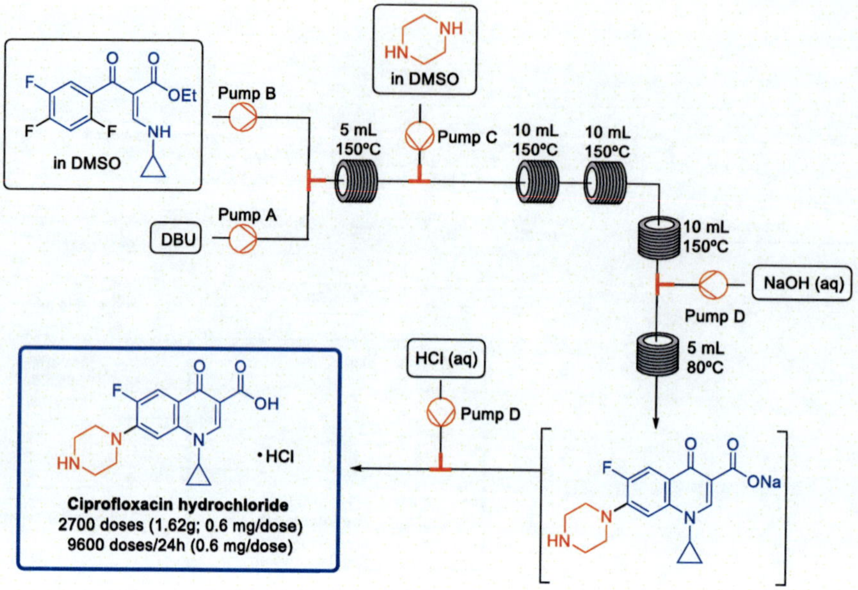

Fig. 35 Continuous flow process for the synthesis of Ciprofloxacin hydrochloride

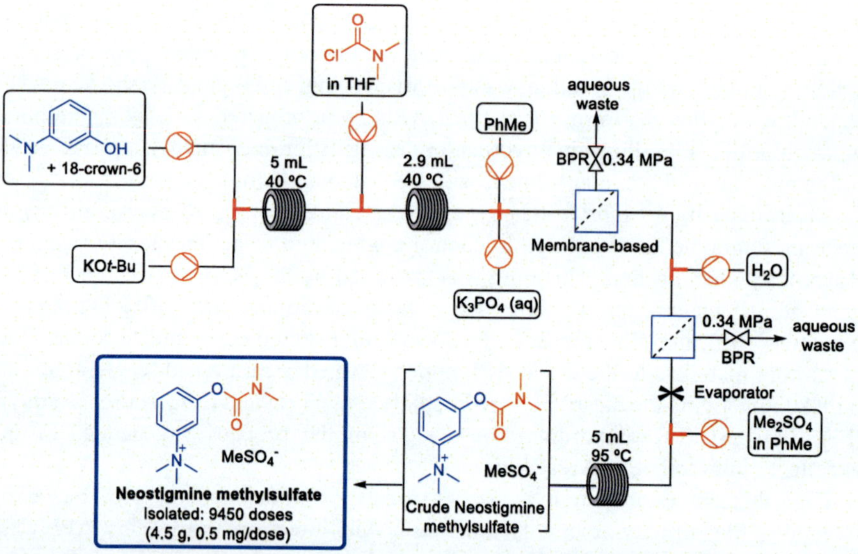

Fig. 36 Continuous flow process for the synthesis of Neostigmine methylsulphate

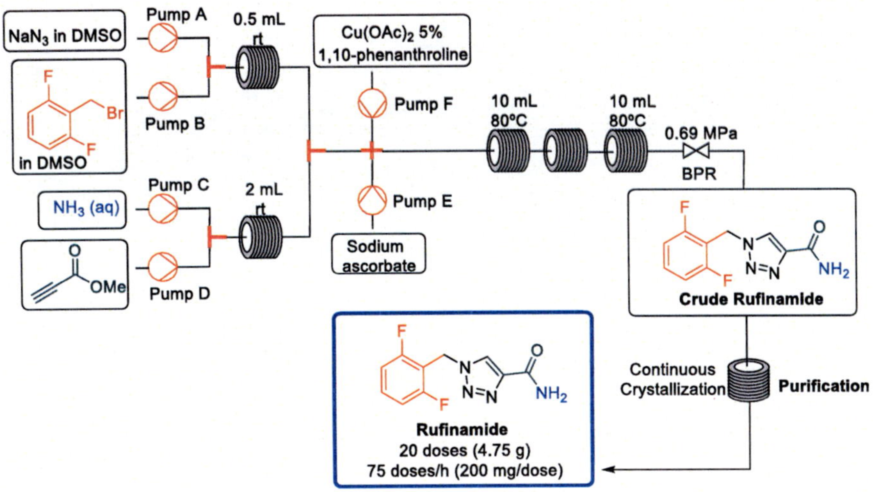

Fig. 37 Continuous flow process for the synthesis of Rufinamide

Neostigmine is a medication used to treat myasthenia gravis (rapid fatigue of the muscles), Ogilvie syndrome, and urinary retention without the presence of a blockage. Although a production rate cannot be accurately determined, 9,450 liquid doses (0.5 mg/dose) of Neostigmine were successfully produced (Fig. 36) [64].

Rufinamide is an anticonvulsant drug developed in 2004 by Novartis Pharma AG and currently manufactured by Eisai. In this second-generation continuous flow set-up, the consortium produced rufinamide crystals with 99.4% purity. In the downstream process, the redissolution in the formulation provided a final concentration of 40.4 mg/mL. In total, 20 doses of rufinamide, each dose corresponding to 200 mg (5 mL of 40 mg/mL suspension/dose) was manufactured (Fig. 37). After start-up and reaching steady-state (4.5 h), this corresponds to a production rate of 75 doses per day [64].

Eflornithine is also known as α-difluoromethylornithine (DFMO) and used for the treatment of sleeping sickness (the second stage of African trypanosomiasis) and hirsutism, and is considered to be an essential medicine by the WHO. The industrial synthetic route reports a 26% overall yield for this API. Different routes still utilize chlorodifluoromethane and report yields in the range of 37% to 40%. The use of fluoroform (an ozone-benign and nontoxic gas, but its release into the environment is forbidden under the Kyoto protocol) presents a viable cost-effective and more sustainable alternative. However, this compound should be transformed into less harmful substances before being released into the environment. Based on the previous works of fluoroform treatment, the Kappe group developed a scalable continuous difluoromethylation protocol using fluoroform with an attached telescoped high-T/p deprotection reaction and high throughput. They were interested in increasing the sustainability of the process by finding alternatives to the use of THF and $CHCl_3$ as solvents as they are classified as hazardous and problematic,

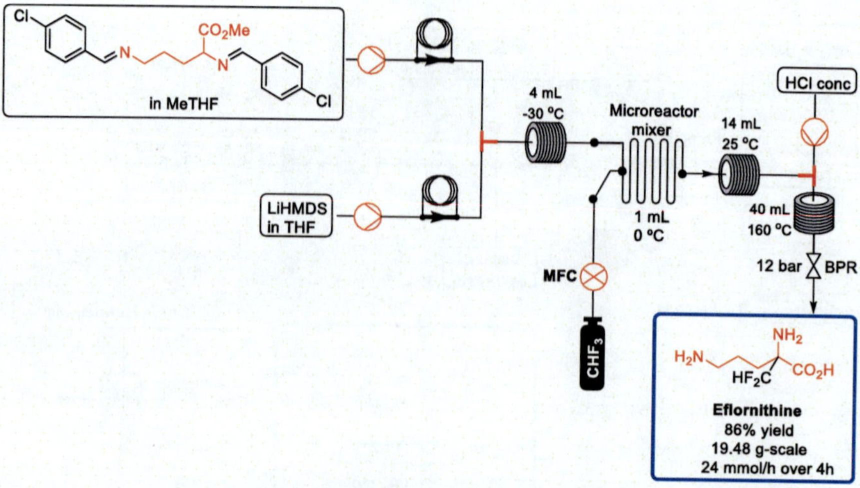

Fig. 38 Continuous flow synthesis of Eflornithine

respectively. Thus, the difluoromethylation and deprotection telescoped process delivered a throughput of 24 mmol/h. Eflornithine hydrochloride monohydrate was isolated in 86% yield, (significantly higher than previously reported yields, 37–40%) (Fig. 38). The overall processing time was decreased from 23 h in the case of the industrial process to 23.5 min under continuous flow, thus potentially minimizing manufacturing costs [67].

Hydroxychloroquine is an antimalarial drug developed for both treatment and prevention of the disease in response to the widespread malaria resistance to chloroquine, and is also an anti-inflammatory drug for the treatment of rheumatoid arthritis (RA) in patients with cardiovascular disease [68]. This API has been recently tested to treat the COVID-19 virus, but the results were not favourable [69]. A continuous processing of this API was recently developed by Gupton and co-workers (Medicines for All Institute) by optimizing already known continuous-flow methods for the synthesis of the two key intermediates from readily available starting materials and an oxime reduction step (Fig. 39). This efficient process has the potential to increase global access to this strategically important antimalarial drug and the authors are currently working to demonstrate that this fully integrated continuous-flow process can be scaled up to commercial operations [68].

Melitracen is also known by the brand name of Melixeran and has been used for the treatment of depression and anxiety. Kiil and col. redesigned a Grignard-based batch process for the preparation of Melitracen-HCl, to a continuous reactor system based on the API manufacturing strategy from H. Lundbeck A/S. The final flow set-up was operated for 5 h, thus producing melitracen HCl at 60 g/h, 85% yield, and 99% purity (Fig. 40) [70].

Verubecestat is a candidate drug for the treatment of patients with early symptoms of Alzheimer's disease. In 2016, Merck developed an efficient synthetic route for this API in batch, which involved the Mannich-type ketamine addition as the

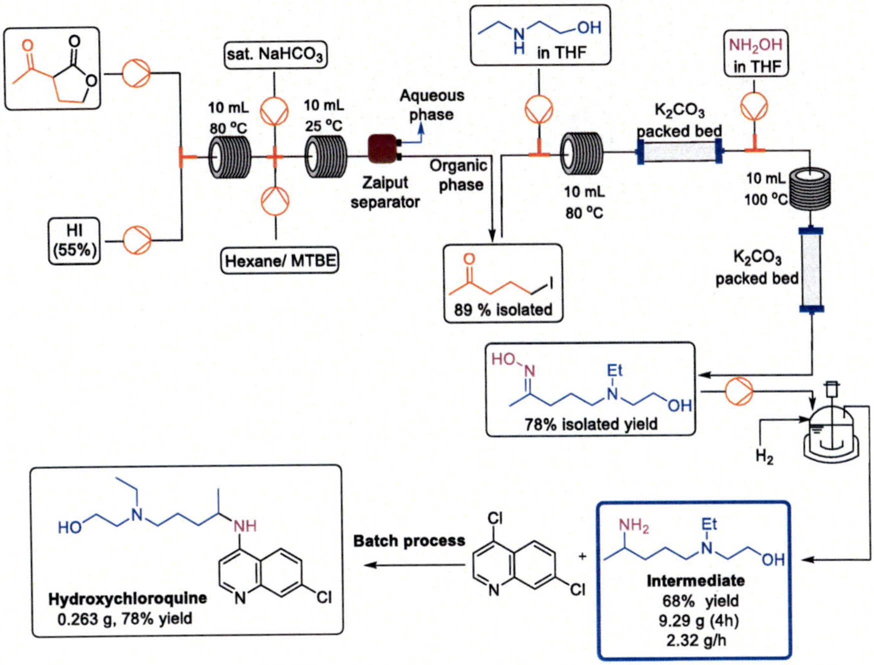

Fig. 39 Continuous-flow synthesis of hydroxychloroquine intermediate

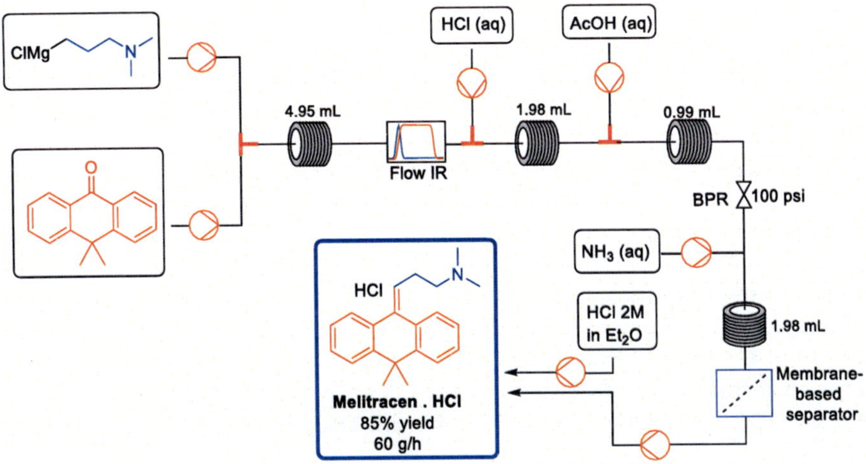

Fig. 40 Continuous-flow synthesis of Melitracen HCl

main and enantioselective step [71]. However, this reaction is extremely dependent on efficient mixing. In 2018 the same company [72] described an alternative continuous protocol with a static mixer in a flow set-up to induce fast mixing, and thus yield the verubecestat intermediate (Fig. 41). This adaptation provided an

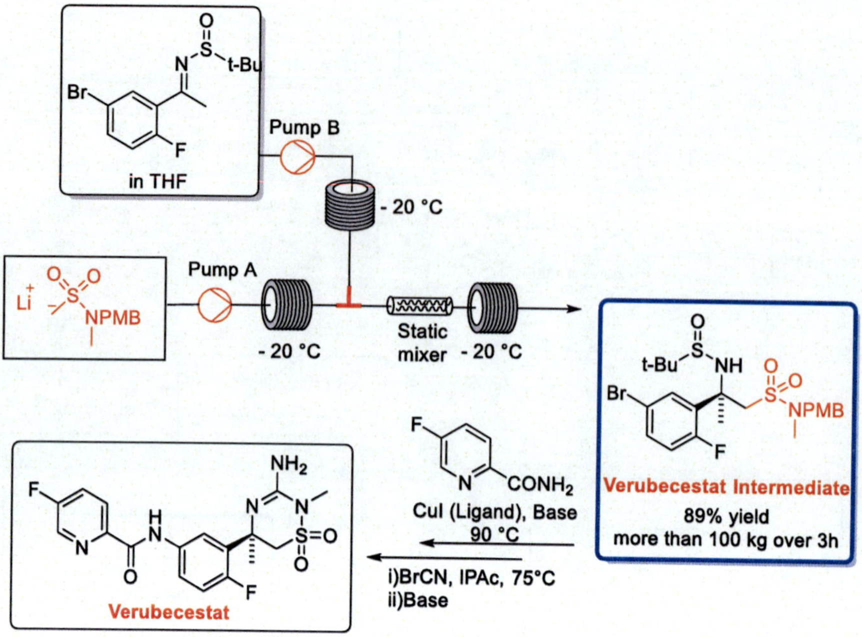

Fig. 41 Continuous flow synthesis of Verubecestat intermediate

improved yield and eliminated the complex cryogenic cooling necessary in batch. The pilot-plant produced the verubecestat intermediate in more than 100 kg, in 3 h and in 89% isolated yield.

Daclatasvir was considered by the FDA as one of the most effective and essential APIs for the health service. Since 2014, it is used in combination with other medications to treat patients with hepatitis C. The synthetic route to this medicine was developed by Bristol-Myers Squibb and was carried out in batch conditions but with long production times. In an attempt to solve this problem, in 2018, Singh and co-workers [73] developed an alternative approach that includes a continuous flow process in all steps. For the biphenyl synthesis, the researchers designed a novel micro electro-flow reactor that provided better yields (compared to batch) and eliminated long reaction times, despite the low scalability (98% yield and 0.82 g/day) (Fig. 42). Major improvements were achieved in other stages and provided the acylated intermediate in 98% yield and 6.3 g/day. The last Daclatasvir intermediate was then obtained in 99% yield at 672 g/day, and Daclatasvir itself finally produced in 84% yield and at 605 g/day.

In 2017, Copanlisib was approved by the FDA to treat patients with follicular lymphoma. The synthetic route for the preparation of this API involves a nitration reaction, which is extremely critical in batch because this reaction is exothermic and generally requires efficient temperature control with unstable nitrated compounds in some cases [74, 75]. Hence, in 2019, a patent was applied for by the Bayer company [76], which describes the continuous flow process as a strategic alternative for

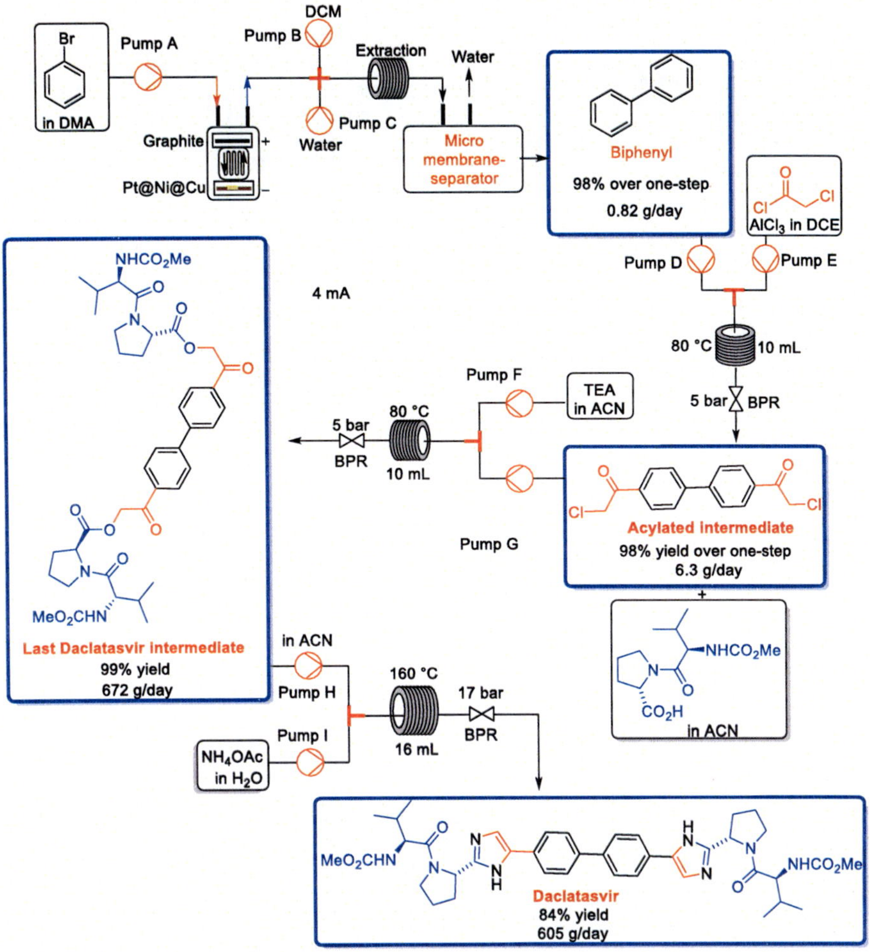

Fig. 42 Four-step continuous synthesis of Daclatasvir

manufacturing 2-nitro-vanillin (Fig. 43). This approach allowed improvements related to safety, selectivity, and productivity. 2-Nitro-vanillin was isolated in 62% yield and with a production of 0.95 kg over 15 h.

Adopting the same strategy (Fig. 44), in 2019 the Jiangsu Cale company [74, 77, 78] with the collaboration of other industries, applied a patent to synthesize the key nitrated intermediate for manufacturing Metronidazole, an antibiotic and antiprotozoal, which is approved by the FDA since 1960 for commercialization and is classified as essential for the health service. The synthetic route for the preparation of this intermediate in batch presents safety-issues, low reproducibility, and problems to scale-up. Thus, the continuous process provided better results relative to productivity and industrial scale. The Metronidazole intermediate was

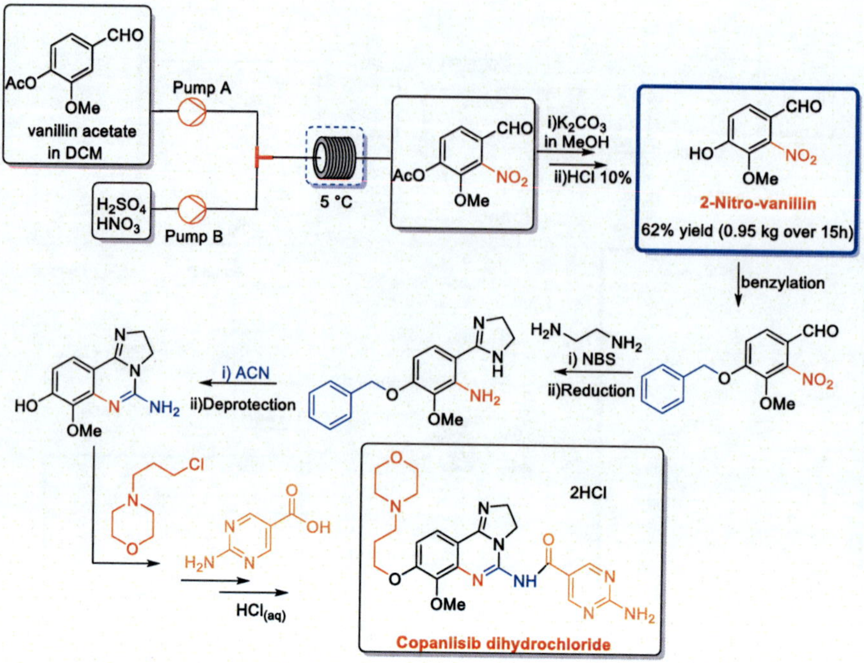

Fig. 43 Semi-continuous synthesis of 2-nitro-vanilllin in the production of Copanlisib

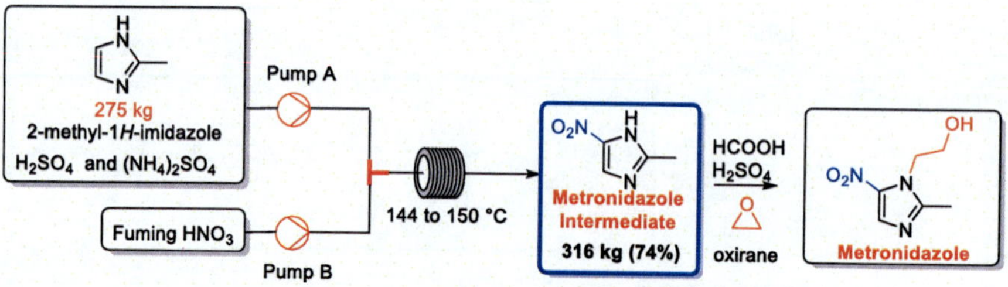

Fig. 44 One-step continuous nitration to synthesize the Metronidazole intermediate

isolated with 74% yield and the pilot-plant was able to produce 316 kg of the main Metronidazole intermediate.

As we have mentioned before, Rufinamide is an anticonvulsant API. There are several syntheses described for this medicine that involve multistep sequences, expensive reagents, and low yields and selectivity. The immediate precursor is the 1,2,3-triazole moiety which is obtained through a three-step protocol that requires high temperatures and pressures. In batch, these conditions are extremely critical because they need prolonged heating, which is a problem for scaling-up with a highly explosive azide intermediate. The Noël and Hessel groups [79] presented an excellent solution for this and developed a continuous approach which works under safe conditions at high-p, T, and provides improved yields. However, this procedure

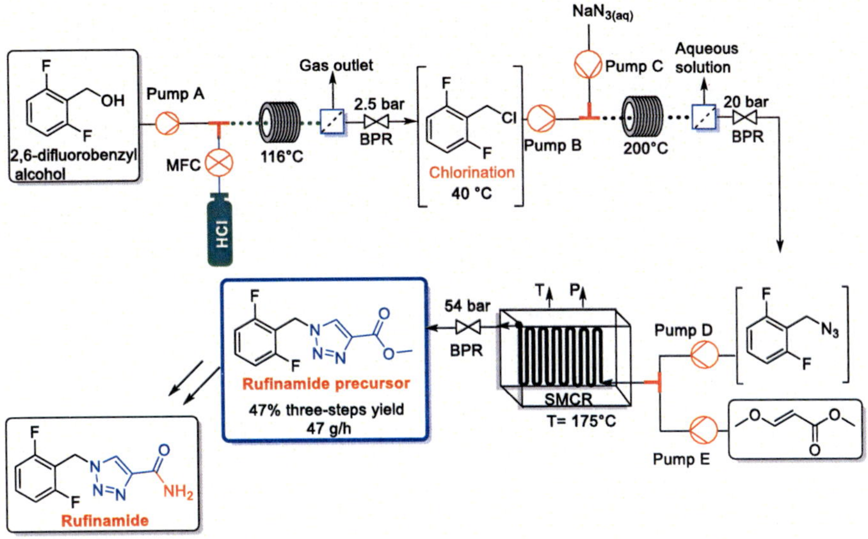

Fig. 45 Three-step continuous synthesis to the Rufinamide precursor

was developed in a lab-scale with 9 g/h of productivity. In 2019, Hessel [80] described similar conditions to obtain this intermediate, but on a large-scale using a mini-plant with a 3 m^2 flow set-up (Fig. 45). This scale-up provided higher safety, yields, and productivity in all steps, especially in the Huisgen cycloaddition reaction, which required the use of a stacked multichannel reactor (SMCR) developed by Kobe Steel Ltd. to promote high-pressure reactions (54 bar). The chlorinated product was isolated with 90% yield and 70.2 g/h productivity, while the azide was isolated with 99% yield and 78 g/h productivity. Thus, the direct Rufinamide precursor was obtained with 52% yield and 40.5 g/h productivity. After optimization of each step, they tested the mini-plant productivity for manufacturing this precursor over the three cascaded reactions, thus obtaining 47% overall yield and 47 g/h productivity.

In 2019, Gupton and co-workers [81] reported a protocol to obtain the previously mentioned Ciprofloxacin API in a four-step batch process. After optimizations of each step, they decided to transpose this protocol to continuous conditions and also develop a telescoped preparation (Fig. 46). The telescoped continuous process provided increased productivity, selectivity, and a lower reaction time relative to the batch process, giving this API in 83% yield and 4.7 g in 15 min. (15.8 g/h).

Fulvestrant is a selective estrogenic receptor used to treat metastatic breast cancer in women. Several batch methodologies describe the preparation of this API [82, 83], but in 2019 Kappe and co-workers [84] transposed a photochemical methodology for iodoperfluoroalkylation of alkenes and applied this procedure to synthesize a Fulvestrant fragment (Fig. 47). Additionally, deiodination with H_2 was continuously performed (H-Cube Pro reactor) and, after the alcohol protection with BzCl, the Fulvestrant intermediate was isolated in 73% yield and 123 g-scale. Unfortunately, the authors did not demonstrate the complete synthesis of Fulvestrant

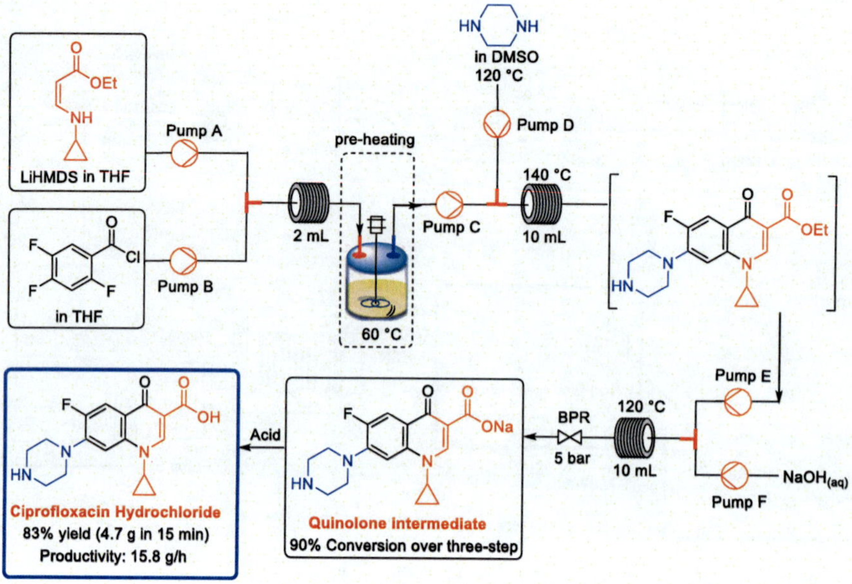

Fig. 46 The telescoped continuous process for ciprofloxacin hydrochloride

in flow, despite the major challenges being solved when they performed the synthesis of the fluorinated intermediate.

Nevirapine (trade name Viramune) is an antiretroviral medication used to treat and prevent HIV/AIDS, specifically HIV-1, and it is generally recommended for use with other antiretroviral medications. This anti-HIV was approved for medical use in the USA in 1996 and this treatment is on the WHO list of essential medicines. The continuous flow synthesis is initiated from two advanced starting materials, both prepared through relative low-cost strategies (Fig. 48) [85, 86]. Interestingly, the researchers calculated some parameters and obtained experimental data that allowed the correct design of the flow reactors. Cost-benefit methodologies were also developed aiming at joining the best results with lower costs, fulfilling the main principle of the Medicines for All Institute [85]. First, the primary amine starting material was deprotonated with NaH/diglyme, and after the transamidation reaction with the ester starting material, the intramolecular nucleophilic aromatic substitution was performed in a NaH packed bed column at 120–165°C to give Nevirapine after crystallization.

In 1992, (−)-Paroxetine was approved for the treatment of depression, panic disorder, and anxiety, being initially sold by GSK. The main moiety of this API is a chiral phenylpiperidine, with several batch methodologies having been described for the manufacture but involving many steps, chiral auxiliaries, and low productivity [87, 88]. In 2019, Kappe and co-workers [89] reported the catalytic enantioselective continuous preparation of the phenylpiperidine intermediate with one organocatalyzed and three-telescoped steps, employing several heterogeneous

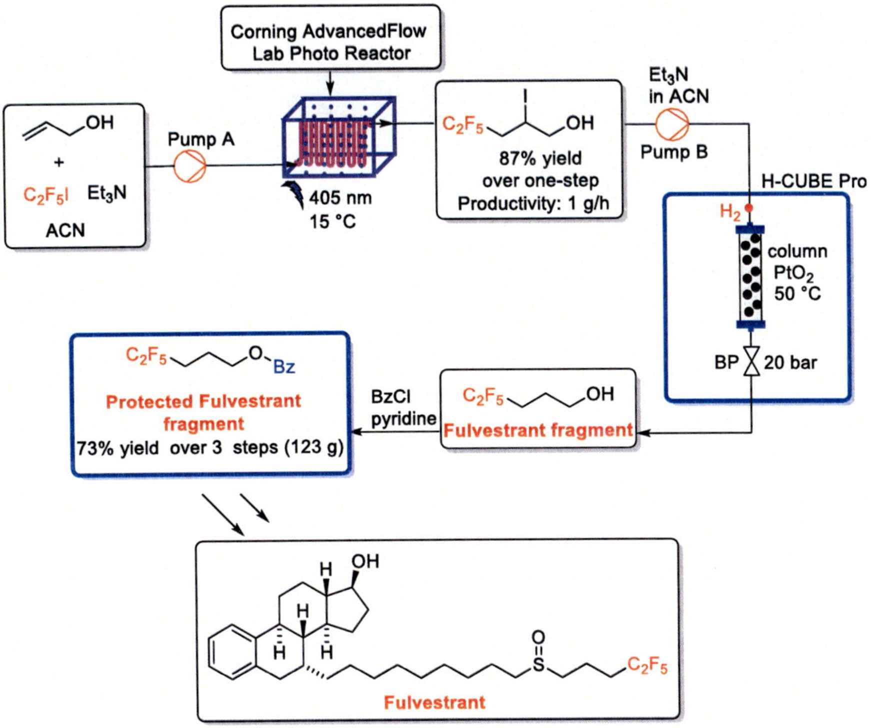

Fig. 47 Two-step telescoped continuous process to prepare Fulvestrant fragment

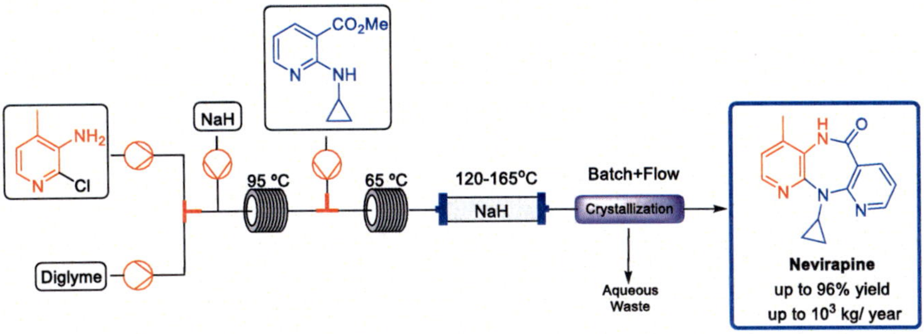

Fig. 48 Continuous flow synthesis of Nevirapine

catalysts in packed bed columns (Fig. 49). This approach provided significant improvements such as higher yields, better selectivity, and scale-up. The phenylpiperidine intermediate was isolated in 83% yield over the three-telescoped steps with 4.95 g/100 min and 96% *ee*.

BMS-919373 was developed by Bristol-Myers Squibb (BMS) and is classified as a candidate to treat patients with atrial fibrillation. In 2019, BMS's researchers [90]

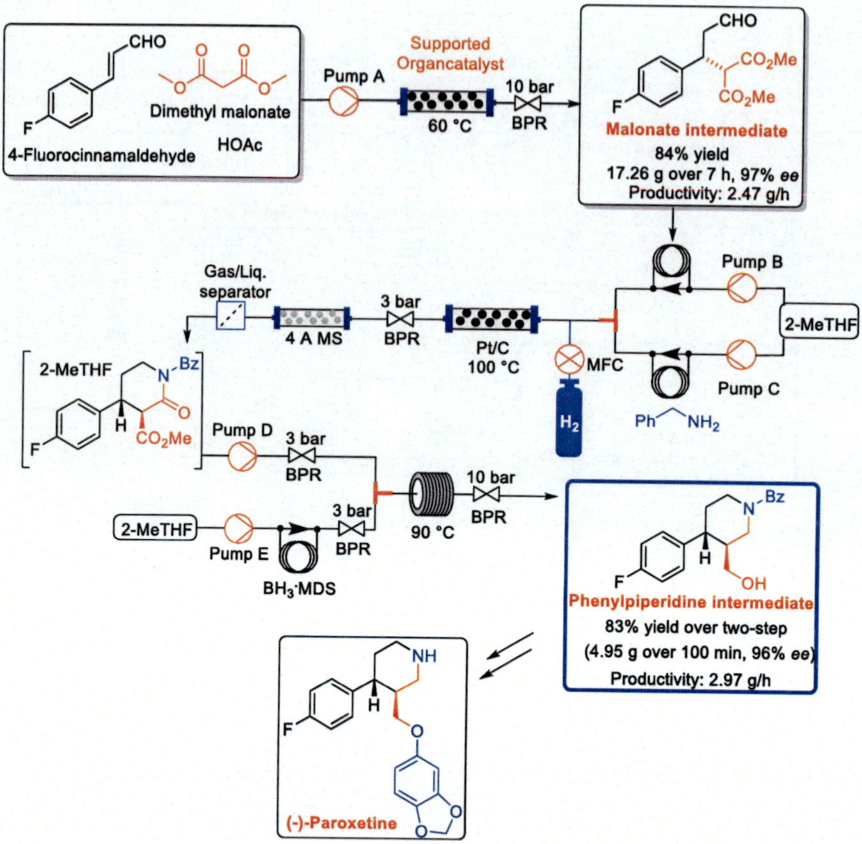

Fig. 49 Continuous synthesis of the phenylpiperidine intermediate

reported a synthetic approach in batch that allowed the scale-up of kilograms to support the clinical tests. However, the industrial batch-scale presented some failures such as limited conversion, low yield, and problems with the instability of the heteroaryl Grignard reagent. Therefore, in the same year, BMS's researchers [91] adopted a new continuous strategy with well-controlled cooling and a sequence of static mixers in the set-up to synthesize the key intermediate on a multikilogram scale (Fig. 50). The continuous approach was successfully implemented and the pilot-plant was able to produce 76 kg of the BMS-919373 intermediate in 8–10 days.

Lomustine is used as an anticancer chemotherapy medicine, being more applied for the treatment of brain tumours. Nowadays, this medicine is extremely expensive, and therefore, it is not available for most patients due to new regulatory challenges to synthesize it and handle with carcinogenic by-products (nitrosamines). Hence, in 2019, Thompson and co-workers [92] developed the telescoped continuous process for preparing this compound over only two-step for its manufacture (Fig. 51). The continuous synthetic approach was implemented with success and provided

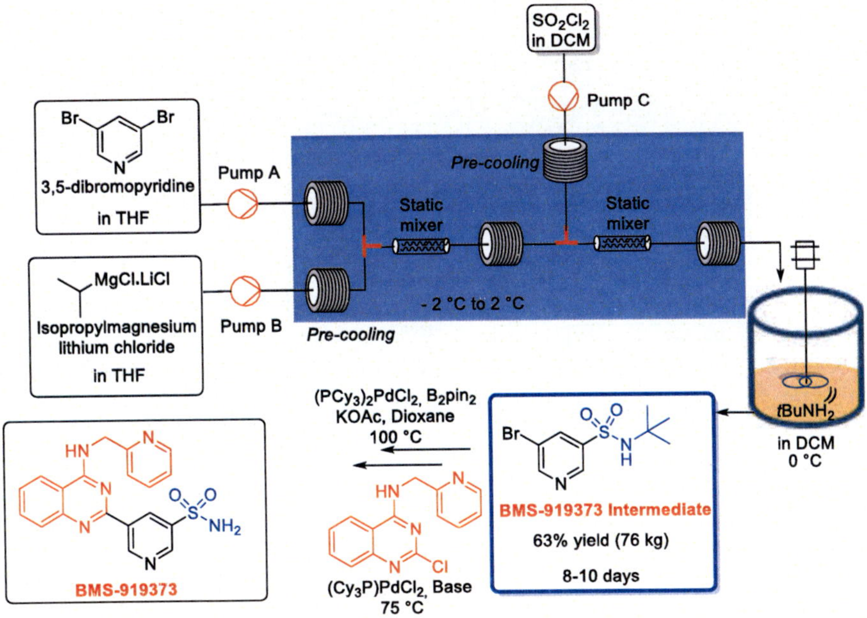

Fig. 50 Semi-continuous synthesis of BMS-919373 intermediate

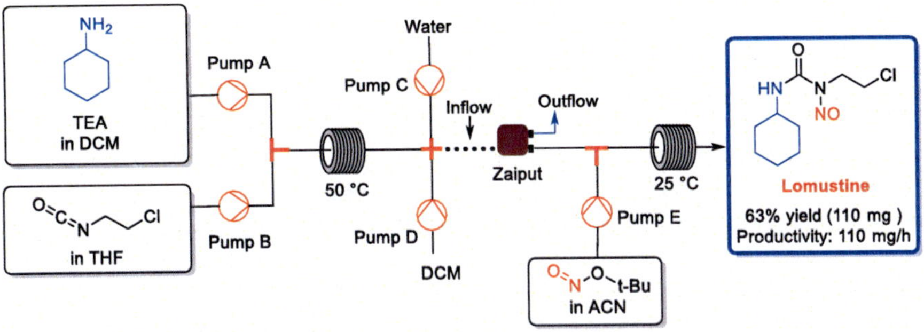

Fig. 51 Two-step telescoped continuous process to synthesize Lomustine

increased yield and safety. Lomustine was produced in 110 mg, 63% yield, and 110 mg/h. Despite the low scale, the continuous conditions allow a well-controlled selective nitrosation.

Capecitabine is known as a broad-spectrum anticancer 5-fluorouracil prodrug used clinically as the first-line treatment for metastatic colorectal cancer. It has also been used as an adjuvant in large-bowel colon cancer and advanced breast cancer chemotherapeutical treatments. In 2019, Miranda, Souza, and co-workers [93] reported the synthesis of Capecitabine starting from commercial 2′,3′-diacetoxy-5′-deoxy-5-fluorocytidine (Fig. 52). The synthesis comprises a late-stage carbamoylation and ester hydrolysis using the Schotten–Baumann reaction in a

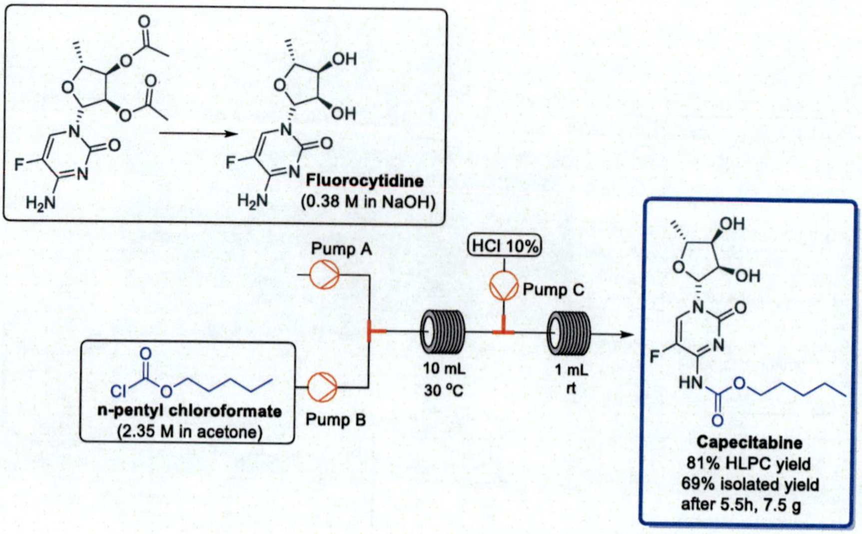

Fig. 52 Telescoped continuous process to synthesize Capecitabine

telescoped continuous flow process. Capecitabine was obtained under continuous flow conditions in 81% HPLC yield, and according to unpublished results from the authors, a 5.5 h continuous production gave this API in 7.5 g-scale and 69% isolated yield.

Emtricitabine (FTC) and Lamivudine (3TC) are similar nucleosides and are both classified as antiretrovirals already approved by the FDA for treating patients with HIV. These medicines have been considered essential by the FDA due to high demand. The Medicines for All Institute had reported a synthetic route in batch to obtain the key intermediate (dichloro-acetate) seeking for a lower cost for the manufacture. However, this strategy presented some problems relative to temperature control as well as mixing which provided by-products formation in greater amounts. Therefore, in 2020 the same group [94] described a continuous approach (Fig. 53) for the preparation of this intermediate and circumvented these problems. This new strategy provided better temperature control, mixing, selectivity, productivity, and yield. It was also possible to scale up these reactions. The dichloro-acetate intermediate was isolated in 98% yield, 260 g-scale in 141 g/h. The subsequent steps were not presented in this paper as they had previously been described by the same group.

Remdesivir is an antiviral API developed by Gilead Sciences and was initially positioned to treat hepatitis C. Nowadays, this API has been tested for treatment against SARS and MERS coronavirus, including the recent SARS-COV-2 (COVID-19). The batch route developed involves six-steps and includes a stereoselective cyanation reaction that requires well-controlled cryogenic conditions and extreme care because it may liberate large amounts of hydrogen cyanide. In 2020, Heumann and co-workers [95] reported this cyanation reaction in the batch process, however,

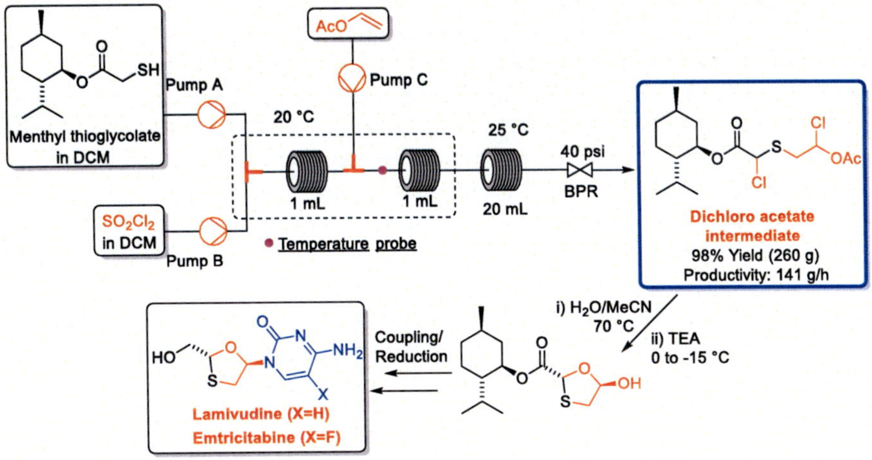

Fig. 53 Continuous flow synthesis of the dichloro-acetate intermediate to access Emtricitabine (FTC) and Lamivudine (3TC)

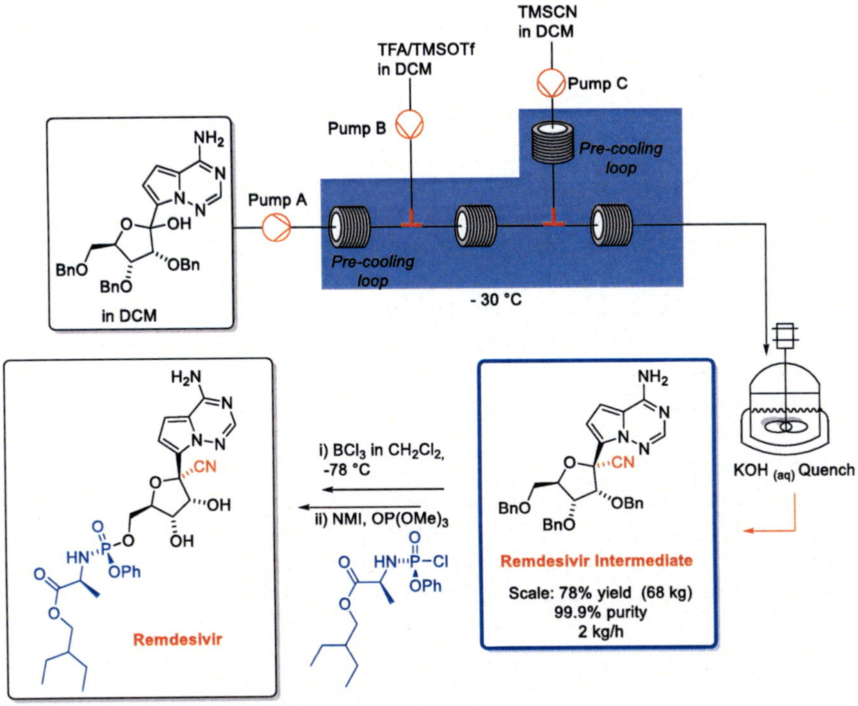

Fig. 54 Continuous flow cyanation scale-up for the manufacture of the Remdesivir intermediate

they faced several challenges relative to scale-up and safety. Therefore, the same researchers decided to transpose this methodology to a continuous process as an alternative to circumvent these problems (Fig. 54). The principal Remdesivir intermediate was isolated in 78% yield, in up to 68 kg, 99.9% purity over 4 days (2 kg/h).

Carfilzomib is an anticancer medicine approved by the FDA in 2011. This API is derived from the natural product epoxomicin and was initially developed by Proteolix's researchers. Few patents describe a synthetic route as well as purification of this medicine. Carfilzomib has an epoxyketone moiety, commonly synthesized from L-Boc-leucine and its production involves the morpholine amide formation as the first step, followed by enone formation with the addition of isopropenylmagnesium bromide and epoxidation in the last step. In 2020, the Amgen company [96, 97] developed a huge project for the preparation of this epoxyketone on industrial scale, employing a three-step telescoped continuous flow process to improve the commercial route (Fig. 55). This new strategy utilized continuous stirred tank reactors (CSTRs) and afforded an efficient asymmetric epoxidation protocol in kilogram-scale with improved diastereoselectivity. The (*R*)-enone intermediate was produced in 89% yield, 3.64 kg/26 h and the (*R*,*S*)-epoxyketone was subsequently prepared in batch in 79% yield, and 0.89 kg-scale in 9 h.

Finerenone is a novel selective nonsteroidal mineralocorticoid receptor antagonist that is in phase III clinical trials for the treatment of chronic kidney disease in people with type II diabetes. The Bayer company is a pioneer in manufacturing this API, which is described in the patent WO2008/104306 and involves the preparation of the product as a racemic mixture followed by chiral column chromatography to provide the desired (*S*)-Finerenone. To improve the desired (*S*)-Finerenone production, the same company decided to introduce a new redox recycling of the undesired enantiomer using a new flow electrochemical process (Fig. 56). Thus, the undesired (*R*)-enantiomer was first electrochemically oxidized to a mixture of atropisomers (15:85) which after heating gave a racemic mixture of the oxidized compounds. The reduction in the same system gave a racemic mixture of Finerenone again allowing the recovery of the correct (*S*)- enantiomer after new separations. Overall, the patent reports that up to 200 kg of Finerenone can be recycled for clinical trials [98, 99].

To finish our selection on API synthesis we highlight the recent work (2020) published by Pastre and co-workers on the synthesis of Lesinurad (Zurampic™, AstraZeneca) [100]. This API was approved in 2015 by the FDA for treating high blood uric acid levels associated with gout disease. Formally, this drug works inhibiting the urate anion exchange transporter 1 (URAT1), thus avoiding the formation of monosodium urate crystals in bone joints.

Although the authors have described different approaches for the synthesis of this API we highlight the most successful in Fig. 57. The isocyanate intermediate was continuously converted into the ester intermediate by the three-step protocol as previously optimized (88% overall yield). This ester intermediate was then brominated (81% yield) and hydrolysed (95% yield) successfully yielding Lesinurad in 68% overall yield (5 steps) and in up to 1.9 g-scale (2 h processing).

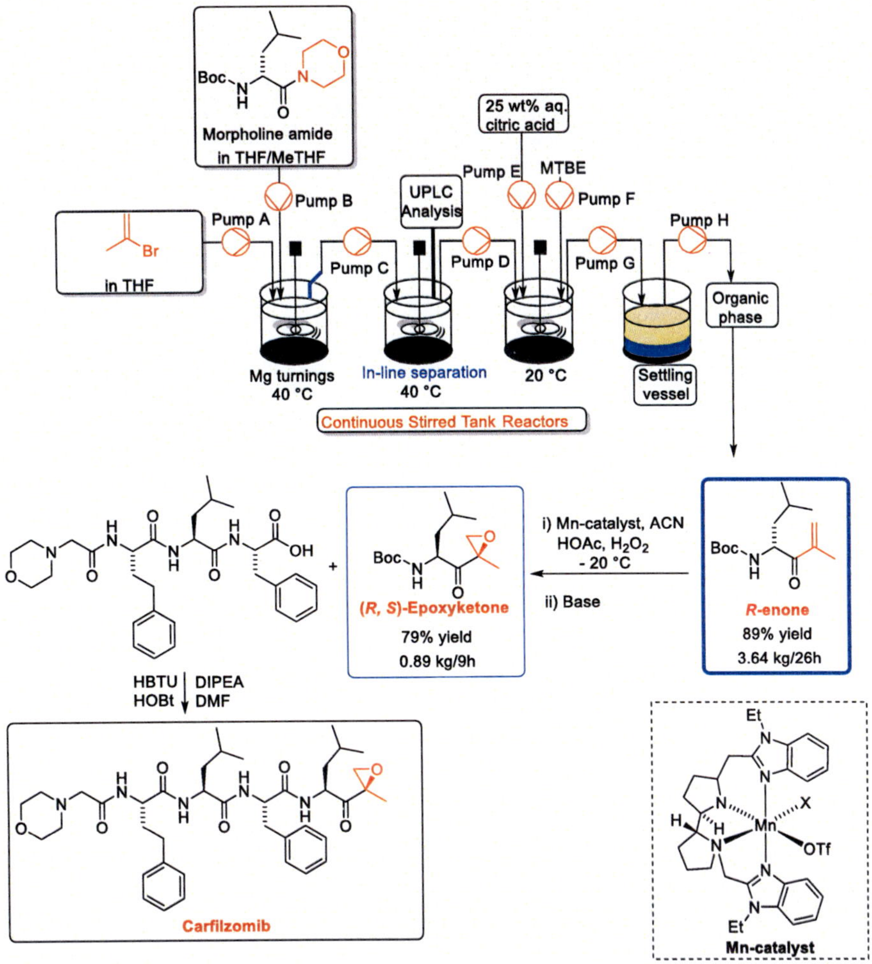

Fig. 55 Continuous flow process to synthesize both the *R*-enone and the related epoxyketone Carfilzomib intermediates

4 Conclusion and Perspectives

Continuous manufacturing has proven to be a quiet but very consistent revolution in the way of production of chemicals this century. Nowadays, many challenges are faced by industry to deliver better, safer, more sustainable, and cost-competitive products and our opinion is that continuous manufacturing has the greatest chance of success. Particularly, for the Pharma market, crucial questions must be considered such as cGMP conditions, costs, and payback on investments. However, it has been a tendency of this market that quick paybacks are not the main feature of this industry and more relevant questions are involved. From a technological point of

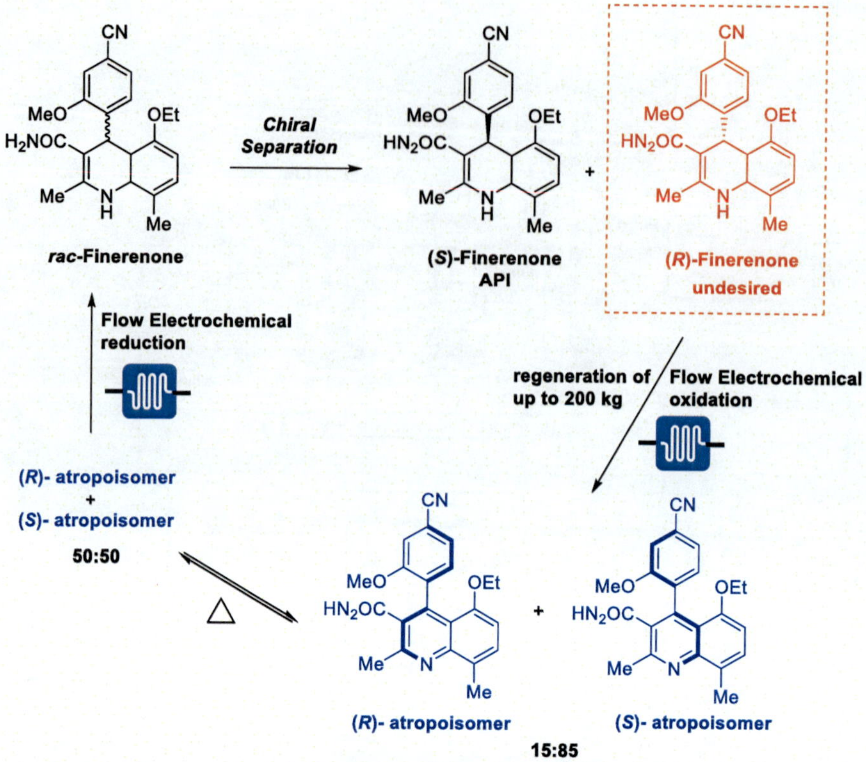

Fig. 56 Electrochemical recycling for the desired Finerenone enantiomer using continuous flow

view, continuous manufacturing is ready to face and solve many of the industrial challenges, and advanced solutions are now available. One important, urgent and critical point is how to educate the very next generation of chemists and chemical engineers aiming at becoming professionals for this market. Education in continuous flow chemistry is now an urgent demand for the very promising future of the chemical industry, and principally for the pharmaceutical and fine chemicals industries. The demand for fine chemicals and APIs is always increasing and traditional batch facilities should be improved or completely transformed by enabling technologies. We do not believe in a complete substitution of batch for flow as it does not always generate improvements, but combining batch and flow protocols can be a great solution.

In our opinion, the impact of continuous flow chemistry for the transformation and synthesis of complex organic molecules is unquestionable, especially as we show for bioactive natural products and APIs. This chapter demonstrates the tremendous gains being achieved by the major research groups and pharmaceutical companies. The obvious resistance to total change is due to the well-known big pharma industry conservatism, and the reluctance to re-invest in different and expensive equipment and physical space transformation.

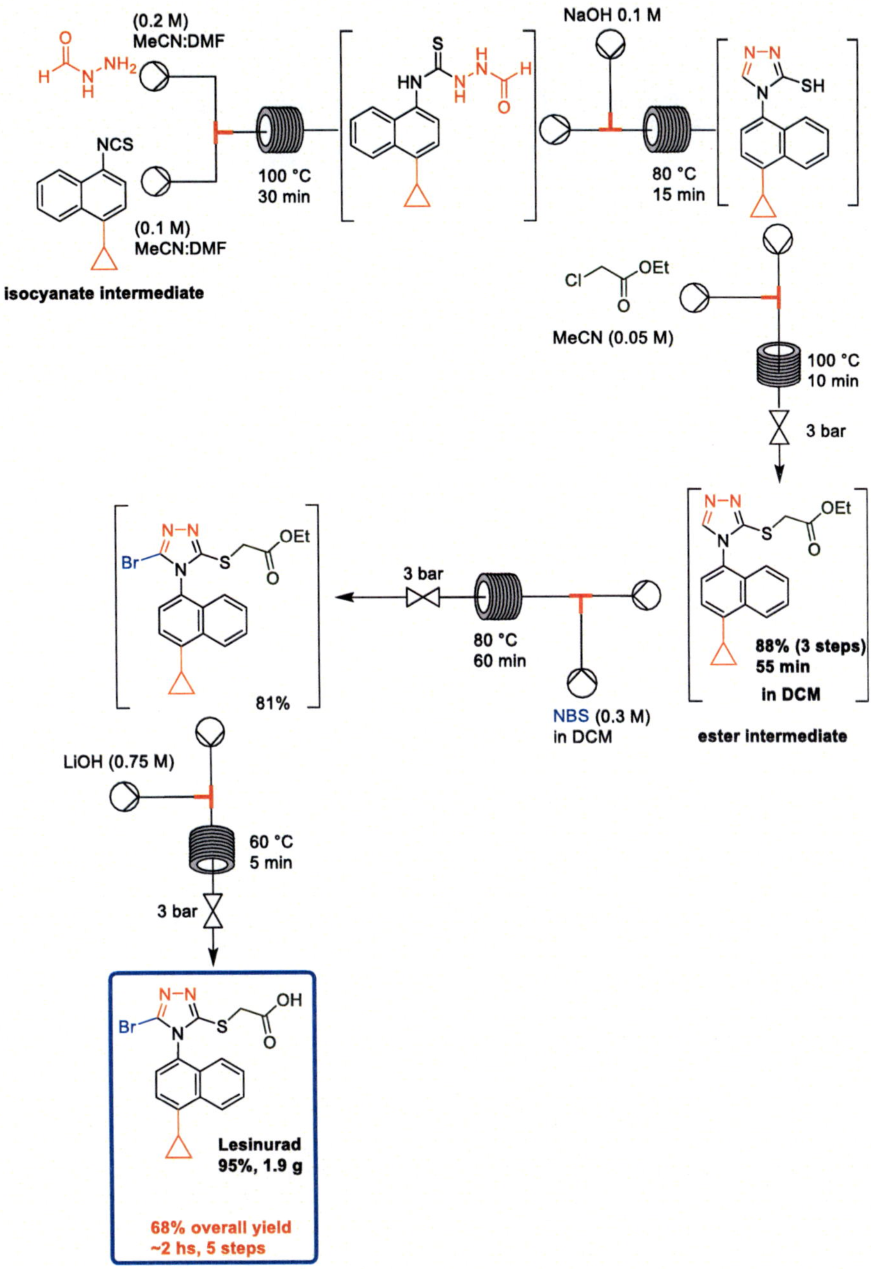

Fig. 57 Continuous flow synthesis of Lesinurad

However, the near future is very clear for the acceptance of continuous flow as being the new norm for the production of important chemicals.

Acknowledgments The authors would like to thank the São Paulo Research Foundation FAPESP (grant numbers: 2020/06874-6, 2019/27176-8 and 2013/07276-1), Fundação de Amparo à Pesquisa e Inovação do Estado de Santa Catarina – FAPESC (2020TR1451), the Conselho Nacional de Pesquisa – CNPq (Grant 407990/2018-6, K.T.O. research fellowship 303890/2019-3, A. A. Nunes de Souza and E.B.A Paez PhD fellowships) and the Coordenação de Aperfeiçoamento de Pessoal de Nível Superior – Brasil (CAPES) – Financial Code 001 and grant 88887.507339/2020-00 – 633/2020, for financial support.

Finally, K.T.O wishes to dedicate this chapter to his father, Sebastião de Oliveira, who passed away during the writing of this review – thanks dad!

Compliance with Ethical Standards

Funding: No funding was received for this chapter.

Informed Consent: No patients were studied in this chapter.

Ethical Approval: This chapter does not contain any studies with human participants or animals performed by any of the authors.

References

1. Lee SL, O'Connor TF, Yang X, Cruz CN, Chatterjee S, Madurawe RD, Moore CMV, Yu LX, Woodcock J (2015) J Pharm Innov 10:191–199
2. Fülöp Z, Szemesi P, Bana P, Éles J, Greiner I (2020) React Chem Eng 5:1527–1555
3. Ralph M, Ng S, Booker-Milburn KI (2016) Org Lett 18:968–971
4. Blackham EE, Booker-Milburn KI (2017) Angew Chem Int Ed 56:6613–6616
5. Yueh H, Gao Q, Porco Jr JA, Beeler AB (2017) Bioorg Med Chem 25:6197–6202
6. Kleinnijenhuis RA, Timmer BJJ, Lutteke G, Smits JMM, de Gelder R, van Maarseveen JH, Hiemstra H (2016) Chem A Eur J 22:1266–1269
7. Tanino K, Takahashi M, Tomata Y, Tokura H, Uehara T, Narabu T, Miyashita M (2011) Nat Chem 3:484–488
8. Fuse S, Mifune Y, Nakamura H, Tanaka H (2016) Nat Commun 7:13491
9. Lücke D, Dalton T, Ley SV, Wilson ZE (2016) Chem A Eur J 22:4206–4217
10. Carmona-Vargas CC, Alves LC, Brocksom TJ, de Oliveira KT (2017) React Chem Eng 2:366–374
11. Bédard AC, Longstreet AR, Britton J, Wang Y, Moriguchi H, Hicklin RW, Green WH, Jamison TF (2017) Bioorg Med Chem 25:6233–6241
12. Dai C, Snead DR, Zhang P, Jamison TF (2015) J Flow Chem 5:133–138
13. Lévesque F, Seeberger PH (2012) Angew Chem Int Ed 51:1706–1709
14. Triemer S, Gilmore K, Vu GT, Seeberger PH, Seidel-Morgenstern A (2018) Angew Chem Int Ed 57:5525–5528
15. Pieber B, Glasnov T, Kappe CO (2015) Chem A Eur J 21:4368–4376
16. Gilmore K, Kopetzki D, Lee JW, Horváth Z, McQuade DT, Seidel-Morgenstern A, Seeberger PH (2014) Chem Commun 50:12652–12655
17. Amara Z, Bellamy JFB, Horvath R, Miller SJ, Beeby A, Burgard A, Rossen K, Poliakoff M, George MW (2015) Nat Chem 7:489–495
18. Morin É, Raymond M, Dubart A, Collins SK (2017) Org Lett 19:2889–2892

19. Breen CP, Parrish C, Shangguan N, Majumdar S, Murnen H, Jamison TF, Bio MM (2020) Org Process Res Dev 24:2298–2303
20. Jordan RW, Dixon C, Gorin B (2017) US Patent 9,758,464 B2, 12 Sept 2017
21. Pastre JC, Murray PRD, Browne DL, Brancaglion GA, Galaverna RS, Pilli RA, Ley SV (2020) ACS Omega 5:18472–18483
22. World Health Organization (2019) Good manufacturing practices. https://www.who.int/biologicals/vaccines/good_manufacturing_practice/en/. Accessed 15 July 2020
23. ISPE Good Manufacturing Practice (GMP) Resources. https://ispe.org/initiatives/regulatory-resources/gmp. Accessed 15 Jul 2020
24. Inokon UM (2014) Approaches to GMP inspection. https://www.fda.gov/media/89231/download. Accessed 15 Jul 2020
25. FDA U.S. Food & Drug Administration (2018) Current good manufacturing practice (CGMP) regulations. https://www.fda.gov/drugs/pharmaceutical-quality-resources/current-goodmanufacturing-practice-cgmp-regulations. Accessed 15 Jul 2020
26. Martin LL, Peschke T, Venturoni F, Mostarda S (2020) Curr Opin Green Sustain Chem 25:100350
27. Burcham C, Florence A, Johnson M (2018) Annu Rev Chem Biomol Eng 9:253–281
28. Hughes DL (2018) Org Process Res Dev 22:13–20
29. Reizman BJ, Cole KP, Hess M, Burt JL, Maloney TD, Johnson MD, Laurila ME, Cope RF, Luciani CV, Buser JY, Campbell BM, Forst MB, Mitchell D, Braden TM, Lippelt CK, Boukerche M, Starkey DR, Miller RD, Chen J, Sun B, Kwok M, Zhang X, Tadayon S, Huang P (2019) Org Process Res Dev 23:870–881
30. Cole KP, Johnson MD (2017) Expert Rev Clin Pharmacol 11:5–13
31. Almaya A, de Belder L, Meyer R, Nagapudi K, Lin H-RH, Leavesley I, Jayanth J, Bajwa G, DiNunzio J, Tantuccio A, Blackwood D, Abebe A (2017) J Pharm Sci 106:930–943
32. Muldowney M (2019) Chim Oggi 37:48–50
33. Pellegatti L, Sedelmeier J (2015) Org Process Res Dev 19:551–554
34. Gauthier Jr DR, Sherry BD, Cao Y, Journet M, Humphrey G, Itoh T, Mangion I, Tschaen DM (2015) Org Lett 17:1353–1356
35. Hecker SJ, Reddy KR, Totrov M, Hirst GC, Lomovskaya O, Griffith DC, King P, Tsivkovski R, Sun D, Sabet M, Tarazi Z, Clifton MC, Atkins K, Raymond A, Potts KT, Abendroth J, Boyer SH, Loutit JS, Morgan EE, Durso S, Dudley MN (2015) J Med Chem 58:3682–3692
36. Felfer U, Stueckler C, Steinhofer S, Pelz A, Hanacek M, Pabst TH, Winkler G, Poechlauer P, Ritzen B, Goldbach M (2016) PCT Int Patent Application WO 2016/100043, 23 Jun 2016
37. Hong H, Gage J, Lu J, Li J, Shen L (2016) Chinese Patent Application CN 105906656, 31 Aug 2016
38. Rodgers JD, Shepard S (2012) US Patent 8,158,616 B2, 17 Apr 2012
39. Kobierski ME, Kopach ME, Martinelli JR, Varie DL, Wilson TM (2016) PCT Int Patent Application WO 2016/205487, 22 Dec 2016
40. Burnier J (2013) US Patent 8,378,105 B2, 19 Feb 2013
41. Zeller JR, Venkatraman S, Brot ECA, Iyer S, Hallet M (2015) US Patent 9,085,553 B2, 21 Jul 2015
42. Tweedle S, Venkatraman S, Zeller J (2017) US Patent 9,725,413 B2, 8 Aug 2017
43. Adamo A, Beingessner RL, Behnam M, Chen J, Jamison TF, Jensen KF, Monbaliu J-CM, Myerson AS, Revalor EM, Snead DR, Stelzer T, Weeranoppanant N, Wong SY, Zhang P (2016) Science 352:61–67
44. Calcaterra NE, Barrow JC (2014) ACS Chem Nerosci 5:253–260
45. Altamura AC, Moro AR, Percudani M (1994) Clin Pharmacokinet 26:201–214
46. Differding E, Kenda B, Lallemand B, Matagne A, Michel P, Pasau P, Talaga P (2005) US Patent 6,911,461 B2, 28 Jun 2005
47. Kenda B, Pasau P, Lallemand B (2013) US Patent 8,492,416 B2, 23 Jun 2013

48. Defrance T, Septavaux J, Nuel D (2017) PCT Int Patent Application WO 2017/076738 A1, 11 May 2017
49. Norrant E, Nuel D, Giordano L, Leclaire J, Septavaux J (2017) PCT Int Patent Application WO 2017/076737 A1, 11 May 2017
50. Borukhova S, Nöel T, Hessel V (2016) ChemSusChem 9:67–74
51. DrugBank (2005). https://www.drugbank.ca/drugs/DB01176. Accessed 4 Aug 2020
52. Lucertini M, Mirante N, Casagrande M, Trivelloni P, Lugli V (2007) Physiol Behav 91:180–190
53. Mostafa GA, Al-Badr AA (2011) Buclizine, comprehensive profile. profiles of drug substances, excipients and related methodology, vol 36. Elsevier, Amsterdam, pp 1–33
54. ChemoCare Osimertinib. http://chemocare.com/es/chemotherapy/drug-info/osimeritinib.aspx. Accessed 28 July 2020
55. Holmes N, Akien GR, Blacker AJ, Woodward RL, Meadows RE, Bourne RA (2016) React Chem Eng 1:366–371
56. Yayla HG, Peng F, Mangion IK, McLaughlin M, Campeau L-C, Davies IW, DiRocco DA, Knowles RR (2016) Chem Sci 7:2066–2073
57. Scott GK, Atsriku C, Kaminker P, Held J, Gibson B, Baldwin MA, Benz CC (2005) Mol Pharmacol 68:606–615
58. de Oliveira KT, Miller LZ, McQuade DT (2016) RSC Adv 6:12717–12725
59. Pellegatti L, Hafner A, Sedelmeier J (2016) Flow Chem 6:198–201
60. S. Jain, K. Ansari, S. Maddala, S. Meenakshisunderam, PCT Int. Patent Application WO 2017/149420 A1, 8 Sep 2017 (2017)
61. Harsanyi A, Conte A, Pichon L, Rabion A, Grenier S, Sandford G (2017) Org Process Res Dev 21:273–276
62. ClinicalTrials.gov (2016) https://clinicaltrials.gov/ct2/show/NCT02735980. Accessed 20 July 2020
63. K. P. Cole, J. McC. Groh, M. D. Johnson, C. L. Burcham, B. M. Campbell, W. D. Diseroad, M. R. Heller, J. R. Howell, N. J. Kallman, T. M. Koenig, S. A. May, R. D. Miller, D. Mitchell, D. P. Myers, S. S. Myers, J. L. Phillips, C. S. Polster, T. D. White, J. Cashman, D. Hurley, R. Moylan, P. Sheehan, R. D. Spencer, K. Desmond, P. Desmond, O. Gowran, Science 356:1144–1150 (2017)
64. Zhang P, Weeranoppanant N, Thomas DA, Tahara K, Stelzer T, Russell MG, O'Mahony M, Myerson AS, Lin H, Kelly LP, Jensen KF, Jamison TF, Dai C, Cui Y, Briggs N, Beingessner RL, Adamo A (2018) Chem A Eur J 24:2776–2784
65. MedLine Plus (2018) Nicardipine. https://medlineplus.gov/druginfo/meds/a695032.html. Accessed 23 Jul 2020
66. Hamilton RJ (2014) (6 th) Tarascon pharmacopoeia 2014 professional desk reference edition. In: Antimicrobials: quinolones. Jones & Bartlett Publishers, Burlington, pp 85–86
67. Köckinger M, Hone CA, Gutmann B, Hanselmann P, Bersier M, Torvisco A, Kappe CO (2018) Org Process Res Dev 22:1553–1563
68. Yu E, Mangunuru HPR, Telang NS, Kong CJ, Verghese J, Gilliland SE, Ahmad S, Dominey RN, Gupton BF (2018) Beilstein J Org Chem 14:583–592
69. Bajpai J, Pradhan A, Singh A, Kant S (2020) Indian J Tuberc. https://doi.org/10.1016/j.ijtb.2020.06.004
70. Pedersen MJ, Skovby T, Mealy MJ, Dam-Johansen K, Kiil S (2018) Org Process Res Dev 22:228–235
71. Thaisrivongs DA, Miller SP, Molinaro C, Chen Q, Song ZJ, Tan L, Chen L, Chen W, Lekhal A, Pulicare SK, Xu Y (2016) Org Lett 18:5780–5783
72. Thaisrivongs DA, Naber JR, Rogus NJ, Spencer G (2018) Org Process Res Dev 22:403–408
73. Mahajan B, Mujawar T, Ghosh S, Pabbaraja S, Singh AK (2019) Chem Commun 55:11852–11855
74. McMullen J, Marton CH, Sherry BD, Spencer G, Kukura J, Eyke NS (2018) Org Process Res Dev 22:1208–1213

75. Hentemann M, Wood J, Scott W, Michels M, Campbell A-M, Bullion A-M, Rowley RB, Redman A (2008) Int Patent Application WO2008/070150 A1, 12 June 2008
76. Peters J-G, Rubenbauer P, Götz D, Grossbach D, Mais F-J, Schirmer H, Stiehl J, Lovis K, Lender A, Seyfried M, Zweifel T, Marty M, Weingärtner G (2019) US Patent 10, 494,372 B2, 3 Dec 2019
77. Kraft MY, Kochergin PM, Tsyganova AM, Shlikhunova VS (1989) Pharm Chem J 23:861–863
78. Qi G, Qiu X, Shen J, Wu T, Xie W, Zhao G (2019) Chinese Patent Appl CN110156694A, 23 Aug 2019
79. Borukhova S, Noël T, Metten B, de Vos E, Hessel V (2013) ChemSusChem 6:2220–2225
80. Escribà-Gelonch M, Izeppi GAL, Kirschneck D, Hessel V (2019) ACS Sustain Chem Eng 7:17237–17251
81. Tosso NP, Desai BK, de Oliveira E, Wen J, Tomlin J, Gupton BF (2019) J Org Chem 84:3370–3376
82. Eve J, Brazier EJ, Hogan PJ, Leung CW, O'Kearney-McMullan A, Norton AK, Powell L, Robinson GE, Williams EG (2010) Org Process Res Dev 14:544–552
83. Caprioglio D, Fletcher SP (2015) Chem Commun 51:14866–14868
84. Rosso C, Williams JD, Filippini G, Prato M, Kappe CO (2019) Org Lett 21:5341–5345
85. Diab S, McQuade DT, Gupton BF, Gerogiorgis DI (2019) Org Process Res Dev 23:320–333
86. Verghese J, Kong CJ, Rivalti D, Yu EC, Krack R, Alcázar J, Manley JB, McQuade DT, Ahmad S, Belecki K, Gupton BF (2017) Green Chem 19:2986–2991
87. de Risi C, Fanton G, Pollini GP, Trapella C, Valente F, Zanirato V (2008) Tetrahedron Asymmetry 19:131–155
88. Porey A, Santra S, Guin J (2019) J Org Chem 84:5313–5327
89. Ötvös SB, Pericàs MA, Kappe CO (2019) Chem Sci 10:11141–11146
90. Wisniewski SR, Stevens JM, Yu M, Fraunhoffer KJ, Romero EO, Savage SA (2019) J Org Chem 84:4704–4714
91. Yu M, Strotman NA, Savage SA, Leung S, Ramirez A (2019) Org Process Res Dev 23:2088–2095
92. Jaman Z, Sobreira TJP, Mufti A, Ferreira CR, Cooks RG, Thompson DH (2019) Org Process Res Dev 23:334–341
93. Miranda LSM, Souza ROMA, Leão RAC, Carneiro PF, Pedraza SF, Carvalho OV, Souza SP, Neves RV (2019) Org Process Res Dev 23:2516–2520
94. De Souza JM, Berton M, Snead D, McQuade DT (2020) Org Process Res Dev 24:2271–2280
95. Vieira T, Stevens AC, Chtchemelinine A, Gao D, Badalov P, Heumann L (2020) Org Process Res Dev 24:2113–2121
96. Dornan PK, Anthoine T, Beaver MG, Cheng GC, Cohen DE, Cui S, Lake WE, Langille NF, Lucas SP, Patel J, Powazinik IV W, Roberts SW, Scardino C, Tucker JL, Spada S, Zeng A, Walker SD (2020) Org Process Res Dev 24:481–489
97. Beaver MG, Shi X, Riedel J, Patel P, Zeng A, Corbett MT, Robinson JA, Parsons AT, Cui S, Baucom K, Lovette MA, Içten E, Brown DB, Allian A, Flick TG, Chen W, Yang N, Walker SD (2020) Org Process Res Dev 24:490–499
98. Hughes DL (2020) Org Process Res Dev 24:1850–1860
99. Platzek J, Gottfried K, Assmann J, Lolli G (2017) PCT Int Patent Application WO/2017/032678, 2 Mar 2017
100. Damião MCFCB, Marçona H, Pastre JC (2020) React Chem Eng 5:865–872

Top Med Chem (2021) 38: 373–390
https://doi.org/10.1007/7355_2021_118

Published online: 7 August 2021

New Biomass Reagents for the Synthesis of Bioactive Compounds

Karen Fox, Rafael Luque, Luiz Antonio Soares Romeiro, and Maria Laura Bolognesi

Contents

Abstract In this chapter we discuss the possibility of using biomass as reagents for the synthesis of biologically active compounds. The selected examples only scratch the surface on the volume of research being carried out on the topics, but are significant of its strengths and challenges. As is delineated below, the data collected so far offers significant support for the continuation of an interdisciplinary research for the development of new pharmaceutically relevant products from biomass.

K. Fox
KelAda Pharmachem Ltd., Dublin, Ireland

Departamento de Quimica Organica, Universidad de Cordoba, Cordoba, Spain

R. Luque (✉)
Departamento de Quimica Organica, Universidad de Cordoba, Cordoba, Spain
e-mail: rafael.luque@uco.es

L. A. Soares Romeiro
Department of Pharmacy, Health Sciences Faculty, University of Brasília, Brasília, DF, Brazil

LADETER, Catholic University of Brasilia, Brasília, DF, Brazil
e-mail: luizromeiro@unb.br

M. L. Bolognesi (✉)
Department of Pharmacy and Biotechnology, Bologna, Italy
e-mail: marialaura.bolognesi@unibo.it

Keywords Biomass, Cashew nutshell liquid, Furfural, Sustainability, Waste valorization

1 Introduction

Waste is currently a startling problem that the planet is experiencing and will face in the future. Transitioning to a bio-based economy rather than relying heavily on fossil fuels [1] requires radical changes in our stance on waste and waste valorization. One solution to this ever-increasing problem is the sustainable transformation of biomass into chemicals and biofuels. Though this can be rather challenging, it is crucial to develop alternative sources to fossil fuels [2]. Promising ways of tackling this problem includes the valorization of waste products. Waste valorization is the process of recycling, reusing, and composting waste material to form new sources of energy, materials, and chemicals [3].

There are numerous sources of waste for use in valorization. One common source of waste for valorization is biomass processing [4]. Feedstock for biomass processing can include edible and non-edible crops, residues such as sugarcane or cut stems and algae (Fig. 1) [5].

Another common source of waste for use in valorization are waste streams and by-products produced from the industrial sector.

The valorization of this waste material can lead to novel or existing material products that can be processed into raw material as new sources of energy, fuels, and chemicals [6]. Although monomers (furfural, sugars) and fuels (e.g., H_2, CH_4, ethanol) tended to be the primary output of such processes in the past, more recently

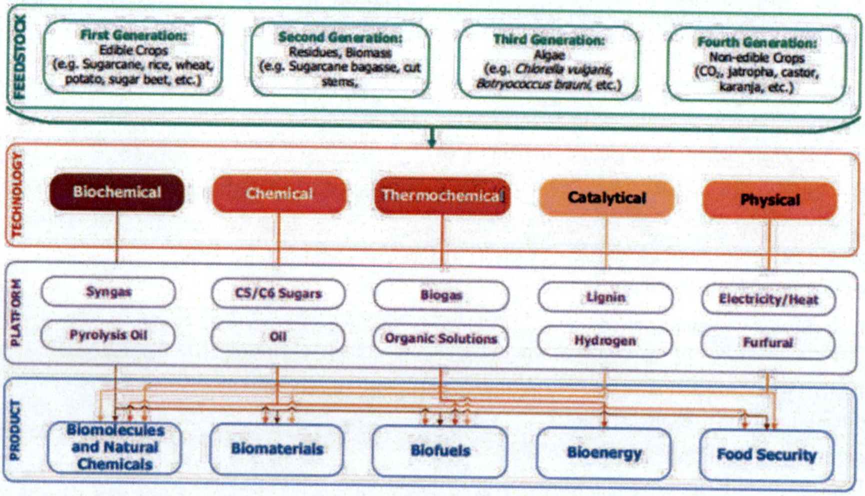

Fig. 1 Biomass sources and possible end targets [5]

attention has been devoted to the production of the so-called high-value chemicals [7]. These latter span from agrochemicals to fragrance chemicals and cosmetics, but also healthcare products, including drugs and food supplements [7]. In principle, these products should be more market-competitive and ensure the economic sustainability of the whole process.

Thus, an emerging area of waste valorization is the depolymerization and/or fermentation of biopolymers liberated from industrial waste streams which are easily transformed or used directly as valuable products [1]. In fact, through specific treatments followed by proper separation and purification procedures, pigments, pharmaceuticals, flavours, can be either extracted or produced further chemical manipulation [8]. In this context, this chapter will specifically explore the use of biomass and waste feedstocks as a starting material for the synthesis of new bioactive compounds with potential pharmaceutical applications.

2 D-mannose as a Biomass Reagent for the Synthesis of Bioactive Compounds

One example outlining the importance of biomass in the pursuit of bioactive molecules involves the synthesis of functionalized piperidine, pyrrolidine, and pyrrolotriazoles from D-mannose. Murphy et al. [9] developed a new strategy to convert biomass, mannose, a naturally occurring sugar, into high-value molecules, glycomimetics, for investigation in new drug discovery projects. This was carried out by subjecting D-mannose to two chemical reactions combined in a novel way (Fig. 2).

Firstly, methyl α-D-mannopyranoside was converted into the protected iodine intermediate **1** in three steps as previously described by Murphy et al. [10]. The dialkene **2** was synthesized via one-pot reductive fragmentation formulated by Davis and co-workers [11, 12] by utilizing *n*-butyllithium followed by in situ Wittig reaction. Removal of TES protecting furnished intermediate **3**. Reacting **3** with DIAD and DPPA gave desired azide **4**. Removal of isopropylidene using 2 M HCl gave **5** in a 72% yield. Removal of isopropylidene protecting group from **3** followed by addition of acetone under acidic conditions gave intermediate **6**. This intermediate was converted into azide **7** followed by acetonide removal to furnish **8**. With the various azidoheptadienes synthesized, the allylic azide rearrangement in tandem with Huisgen cycloaddition were investigated. Products **9–21** from the above tandem reactions with intermediates **4**, **5,** and **7** are shown in Fig. 3 below.

A variety of novel chiral building blocks from methyl α-D-mannopyranoside were converted into high-value *N*-heterocyclic products. Their biological activity has yet to be evaluated; however, their novel synthesis has the potential to generate higher value products for investigation in new drug discovery projects such as in the treatment of type-2 diabetes, Fabry and Gaucher's disease.

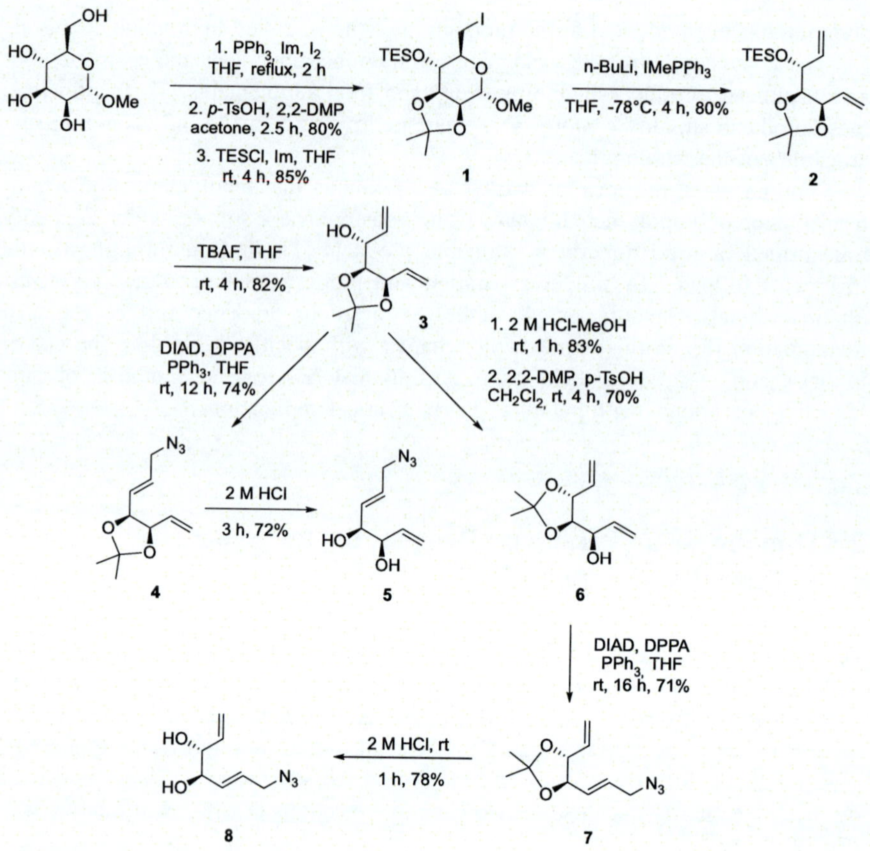

Fig. 2 Synthesis of azidoheptadienes [9]

3 Furfural as a Biomass Reagent for the Synthesis of Bioactive Compounds

Another such example of waste valorization in the pursuit of bioproducts for multidirectional use is outlined by Butin et al. [13]. Their research focused on the synthesis of the pharmacologically advantageous class of heterocycles indolo[3,2-*c*] quinolines and isocryptolepines from furfural. Furfural (Fig. 4) is a compound produced from the dehydration of sugars including xylose and arabinose obtained from agriculture and forestry waste [14]. Currently, 280,000 tonnes of furfural is produced from agricultural and forestry waste every year [15].

In the context of this book, it should be mentioned that the application of flow chemistry to bio-based furanoid platforms is gaining increasing attention. Although we were not able to find specific examples dealing with the use of flow chemistry for upgrading bio-based furanoids toward biologically active compounds, it should be

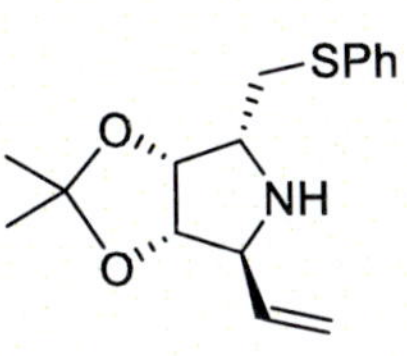

9

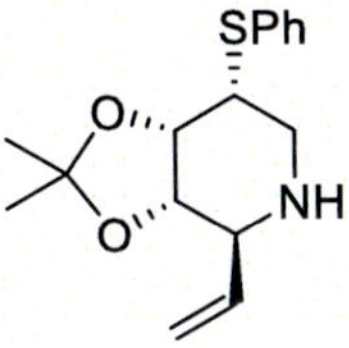

10

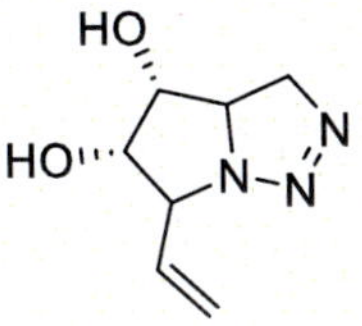

11

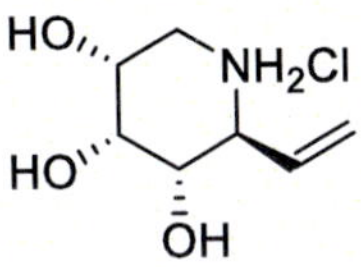

12

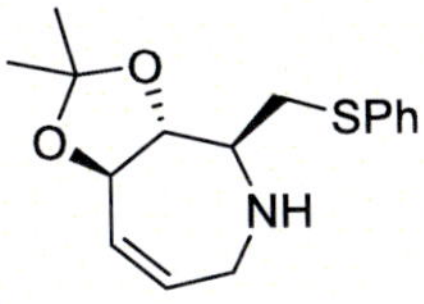

13

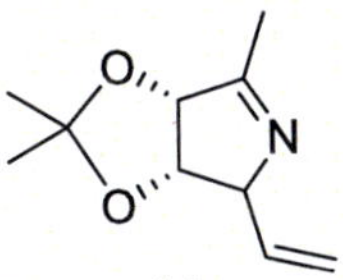

14

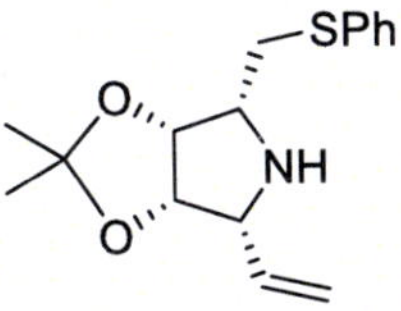

15

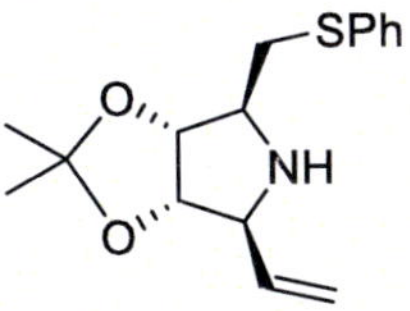

16

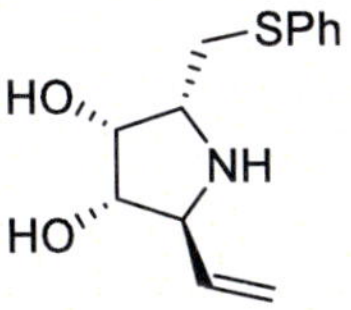

17

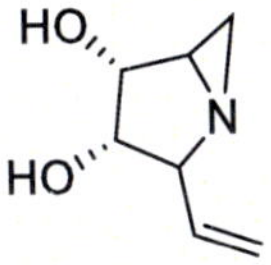

18

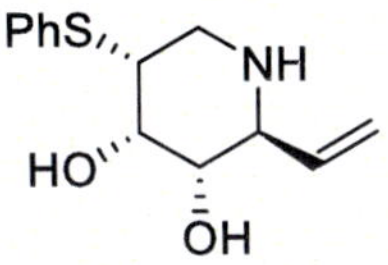

19

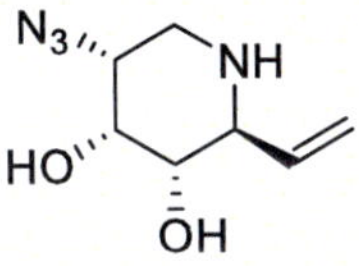

20

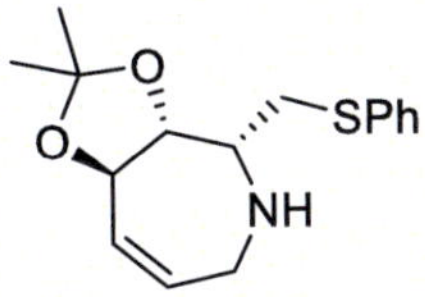

21

Fig. 3 Functionalized piperidine, pyrrolidine, and pyrrolotriazoles **9–21** [9]

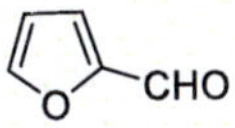

furfural

Fig. 4 Structure of furfural

noted that this is an emerging biorefinery concept [16]. Notably, chemists and chemical engineers are increasingly considering flow technologies in conjunction with the valorization of bio-based chemical platforms and this will likely be translated to the production of pharmaceutically relevant products in a near future.

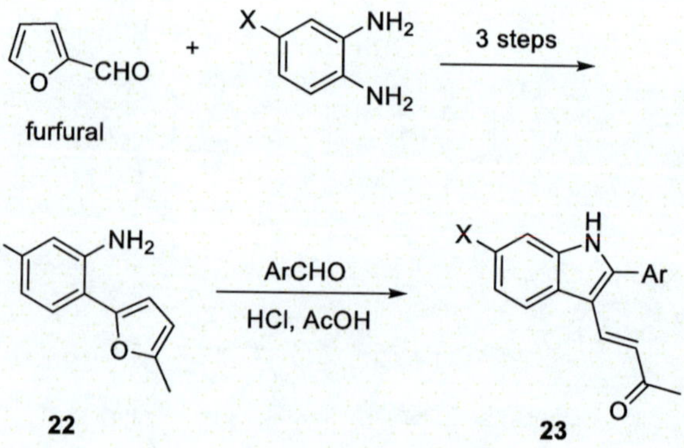

Fig. 5 Furan-to-indole recyclization

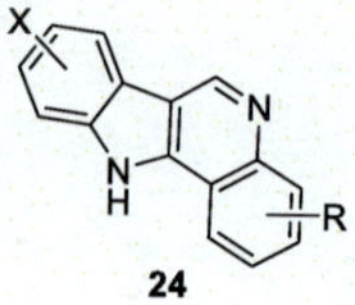

Fig. 6 Indolo[3,2-*c*]quinolines (**24**)

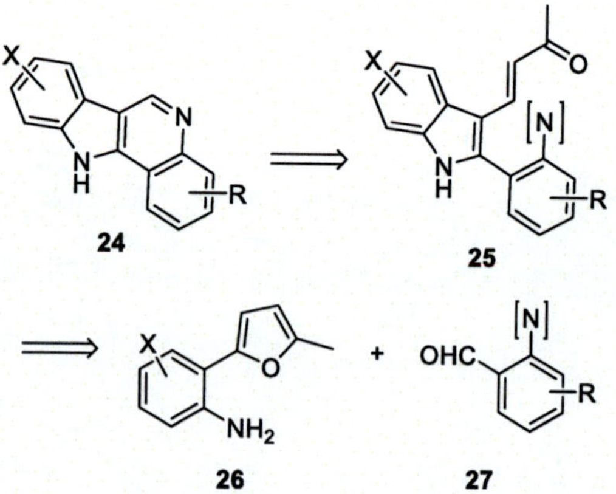

Fig. 7 Retrosynthesis of indolo[3,2-*c*]quinolines (**24**)

Butin et al. have had a long-standing interest in the development of batch techniques for furfural transformations [17–22]. Not long ago they reported a simple route for the formation of indoles of general structure **23** via furan-to-indole recyclization starting from 2-furylanilines **22** (Fig. 5).

Using this method, the synthesis of indolo[3,2-*c*]quinolines **24** (Fig. 6) was reported which has gained interest as pharmacologically attractive heterocyclic compounds. Indolo[3,2-*c*]quinolines are known for their antimalarial activity [23], antiproliferative properties [24] and their ability to form strong complexes with DNA [25].

Outlined in Fig. 7 is the retrosynthesis of indolo[3,2-*c*]quinolines (**24**). Butin and co-workers believed indoles, with latent amino group in the *ortho*-position of 2-aryl substituent, such as **25** could be suitable precursors of **24**. It was envisioned that,

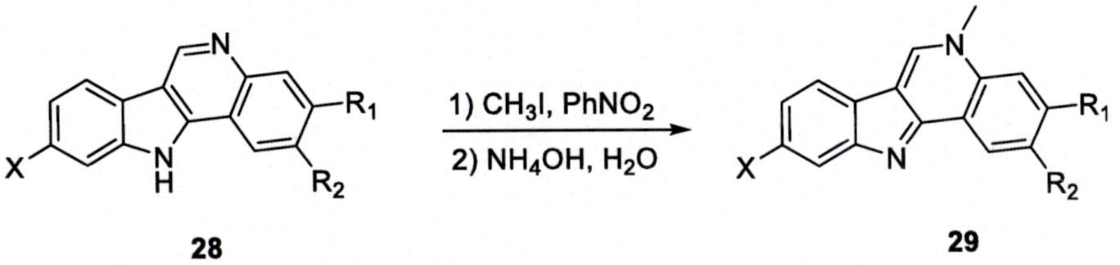

Fig. 8 Synthesis of isocryptolepine and its derivatives (**29**) from intermediate **28**

Fig. 9 General structure (**31**) of compounds inspired by cantharidin (**30**)

through Michael addition, the liberation of amine functionality would allow for easy pyridine ring formation. Sequentially, indoles such as **25** could be synthesized from a variety of aldehydes **27** and 2-furylanilines **26**.

Their rationale behind the selected starting 2-nitrobenzaldehydes was based on two main reasons: (1) there are a large range of substituted 2-nitrobenzaldehydes that can be readily synthesized and a huge number are commercially available; and (2) the reduction of nitroarenes to anilines is simply and environment-friendly. A number of 2-nitrobenzaldehydes **27** were reacted with 2-furylanilines **26** to afford the 2-(2-nitrophenyl)indoles. Although low yields were obtained for the formation of indoles **25**, from a preparative point of view they found this reaction to be highly convenient. The reductive cyclization of indoles **25** was examined. A number of various conditions were employed but to no avail. However, treatment of indoles **25** with Fe in AcOH afforded several indolo[3,2-*c*]quinolines, **24**. Furthermore, using this method, the facile synthesis of alkaloid isocryptolepine and its derivatives (**29**) by methylation of **28** was demonstrated (Fig. 8).

Alkaloid isocryptolepine has been found to demonstrate antiprotozoal activity [26–28]. It is believed the synthesis of isocryptolepine and its derivatives (**29**) would allow for further SAR studies to further understand its mode of action and to determine possible alternative therapeutic agents. All compounds synthesized are under investigation for antiplasmodic activity.

Another example of bioactive compounds derived from a bio-based furfural scaffold is represented by cantharidin (**30**) analogs whose general structure (**31**) is depicted in Fig. 9 [29].

Ananikov et al. first developed an easy "one-pot" protocol for the oxidation of 5-hydroxymethylfurfural (5-HMF, **32**) to 2,5-diformylfuran (2,5-DFF, **33**), by using a recyclable catalyst [Pip*(O)][BF_4] (4-acetamido-2,2,6,6-tetramethyl-1-

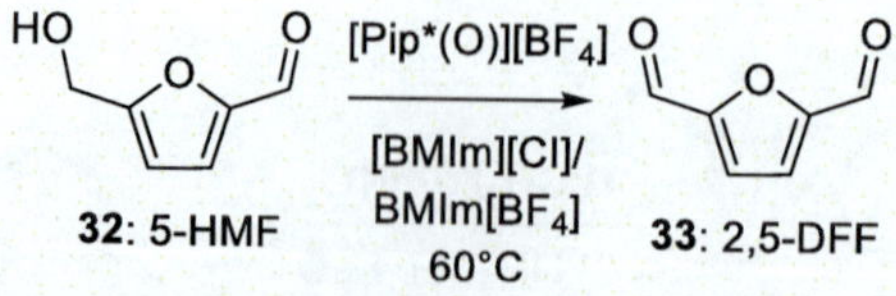

Fig. 10 "One-pot" synthesis of 2,5-diformylfuran (**33**: 2,5-DFF)

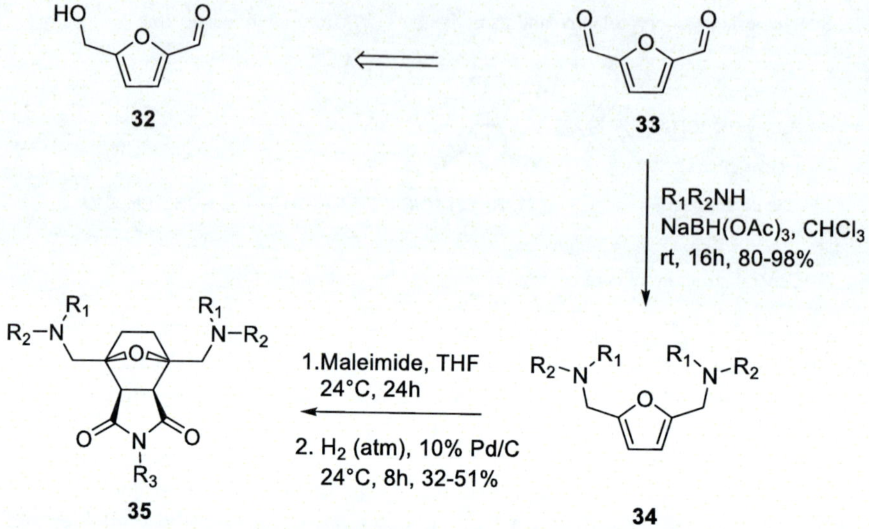

Fig. 11 Synthesis of bis(aminomethyl)furans (**34**) and norcantharimides (**35**)

oxopiperidinium tetrafluoroborate) in ionic liquid media in a 95% yield (Fig. 10) [30].

Then, with the specific aim of implementing the transition from a highly versatile bio-derived chemical platform, such as **32**, to pharmaceuticals, they exploited a reductive amination reaction on **33**. The synthesized bis(aminomethyl)furans (**34**) were further utilized as building blocks for the construction of new norcantharimide derivatives (**35**) carrying the structural features of cantharidin (**30** in Fig. 9), a naturally occurring biologically active compound (Fig. 11) [29].

This is the active component of Chinese blister beetles, which displays antitumor activity and induces apoptosis in many types of tumour cells. Cantharidin has been used as an anticancer agent in China for the treatment of hepatoma and oesophageal carcinoma for a long time [31]. However, its severe side effects limit its clinical application. Thus, the search of novel derivatives which might preserve the antitumor activity while showing lower cytotoxicity has been an active area of drug discovery research [31]. Using a one-pot procedure, which included carbonyl reduction, Diels–Alder reaction followed by hydrogenation of the double bond, several cantharidin analogues were obtained starting from 5-HMF (**32**) [32]. The cyclization process was diastereoselective and resulted in the formation of tricyclic products (**36** and **37**) with the *endo* configuration (Fig. 12). By exploiting this

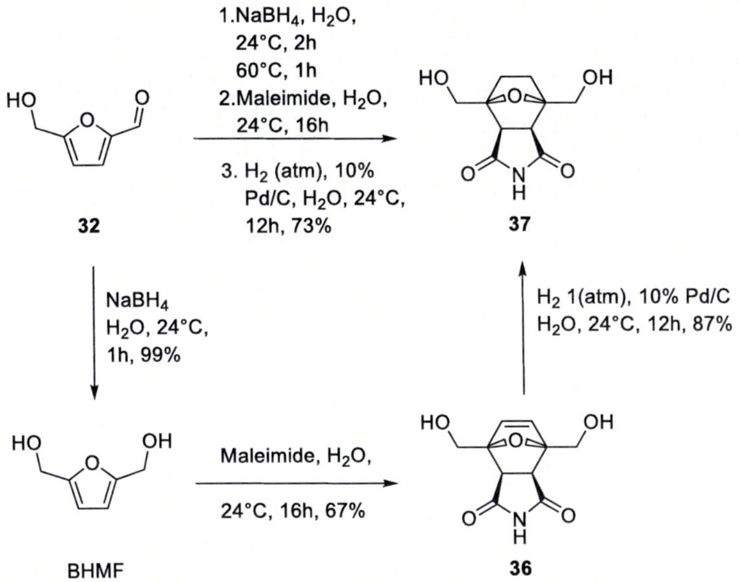

Fig. 12 One-pot and 3 steps synthesis of norcantharimides (**37**) from HMF (**32**)

protocol, with clear synthetic advantages with respect to the 3-step synthesis, several derivatives of general structures **31** were synthesized (Fig. 9). When tested against a hepatoma cell line, unfortunately they did not show an improved profile compared with the parent drug. However, the versatile synthetic methodology could be used for further medicinal chemistry activities around the cantharidin scaffold.

4 Cashew Nutshell Liquid (CNSL) as a Biomass Reagent for the Synthesis of Bioactive Compounds

Another enlightening perspective is to use food industry waste as starting material for the synthesis of novel biologically active compounds to be then further optimized into novel drugs. Historically, natural products from plants and animals were the source of virtually all pharmaceuticals. Prototypical examples are morphine isolated from poppy by Friedrich Serturner in 1806, or digoxin (digitalis), isolated by Claude-Adolphe Nativelle in 1869 from *Digitalis lanata* [33]. Even more recently, natural products have continued to enter clinical trials or to inspire the development of drug candidates, particularly in the anticancer and antimicrobial therapeutic areas [34]. In fact, the so-called semi-synthetic drugs are those therapeutic agents that are produced by natural sources via chemical reactions. Examples of semi-synthetic medicines in current pharmacopoeias are antibiotic penicillin and anticancer paclitaxel [35]. It might be envisaged that in the coming years, food waste will replace plants and animals as sources of new natural products [36], as well as new drugs.

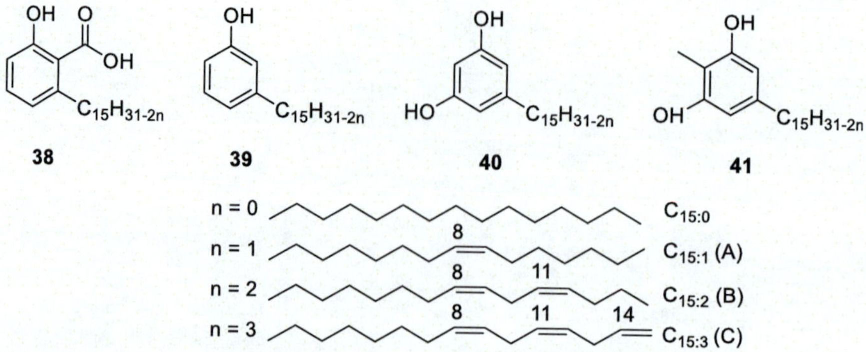

Fig. 13 CNSL phenolic lipids **38–41**

Being an inedible material, food waste displays evident environmental, financial, and ethical benefits over crops or animal grown for that purpose.

However, its use in a pharmaceutical context has been relatively under-explored.

Motivated by these considerations, we reasoned that obtaining new drugs from food wastes might fit the peculiar need of the so-called poverty-related diseases [37]. In fact, if access to medicines (including their availability and affordability) is a major global public health threat currently spanning all the therapeutic areas, it is a matter of special concern for those diseases disproportionately affecting low-income populations [37]. Recently, we explored the possibility of using cashew nutshell liquid (CNSL) as a sustainable, low-cost starting material for the development of new drugs against trypanosomiases [38]. CNSL, which is obtained as the by-product of cashew nut processing, has proven to be one of the most versatile food wastes for the production of monomers, polymers, and additives, but not drugs [39]. According to recent estimates, the global production of CNSL approaches one million tons annually [40]. CNSL mainly consists of phenolic lipids, i.e. anacardic acids (**38**) (71.7%), cardanols (**39**) (4.7%), cardols (18.7%) (**40**), and 2-methylcardols (2.7%) (**41**) (Fig. 13).

The pentadecyl alkyl side chain shows a variable degree of unsaturation, depending on the production method [41]. Although CNSL components have been reported to possess a wide range of biological activities [42], they are not potent enough for a realistic therapeutic application. To overcome this limitation, we argued to obtain semi-synthetic derivatives of CNSL [38]. In particular, by applying a medicinal chemistry strategy that foresees synergy and increased potency for hybrid drugs [43], we combined the chemical features of CNSL derivatives (**43–45**) with those of quinone **42**, endowed with anti-trypanosomal potential (Fig. 14) [44].

A small library was generated, with general structure **46** reported in Fig. 15. For the synthesis, we developed a protocol which was simple, accessible, and scalable. This, together with the fact that the starting material is cheap and easily available, makes the entire production process potentially highly affordable and local. Importantly, the fact that Africa is among the largest CNSL-producing countries provides

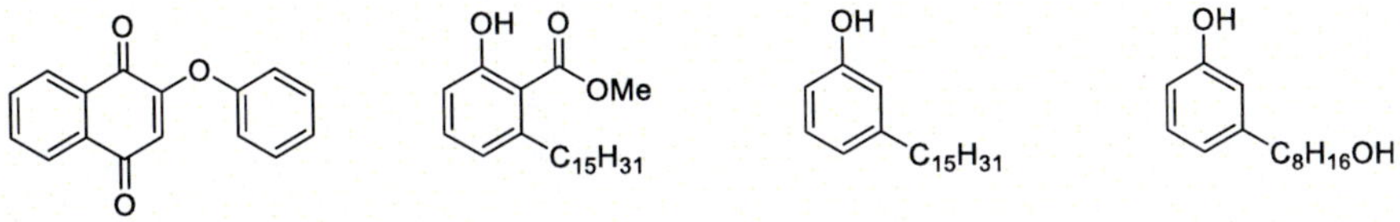

42 **43** **44** **45**

Fig. 14 Quinone (**42**) and CNSL derivatives (**43**–**45**)

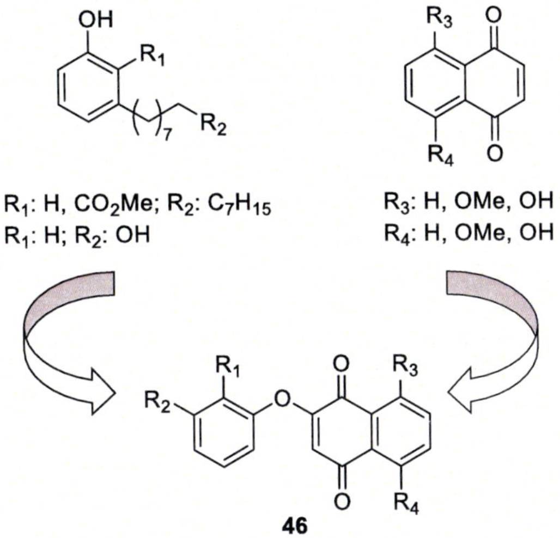

46

Fig. 15 Development of the CNSL-Quinone hybrid library of general structure **46**

unprecedented opportunities of engaging disease-endemic countries as crucial actors in drug discovery and development. When tested against *T. b. brucei*, the causative agent of African trypanosomiasis, the synthesized derivatives inhibited parasite growth, displaying rapid trypanocidal activity in the low micromolar range, and no discernible toxicity on human cell lines [38].

Along the same lines, we have explored the potential of CNSL for the production of drugs against Alzheimer's disease (AD). AD is the leading cause of dementia worldwide and represents a global health problem. In fact, if it was initially viewed as the epidemic of the developed world, in the next decades it will particularly impact low- and middle-income countries, in tandem with the ageing population [45]. Thus, if we want to guarantee universal coverage and equity of access to medications to the global AD population, drugs must be affordable and accessible to avoid that a high price might be a barrier to treatment [45]. On these premises, we were motivated to develop bio-based HDAC inhibitors for AD, inspired by drug vorinostat [46]. Vorinostat (or SAHA, **47**), first approved for the treatment of lymphoma, has shown efficacy in animal models and is currently investigated in clinical trials for AD [47]. Thus, it has high potential for the development of new AD drugs. Particularly, we were intrigued by the fact that the C8-acids **48** and **49** (Fig. 16), easily obtainable from CNSL, might be a suitable starting point for the synthesis of **50** and **51** SAHA-derived hydroxamates.

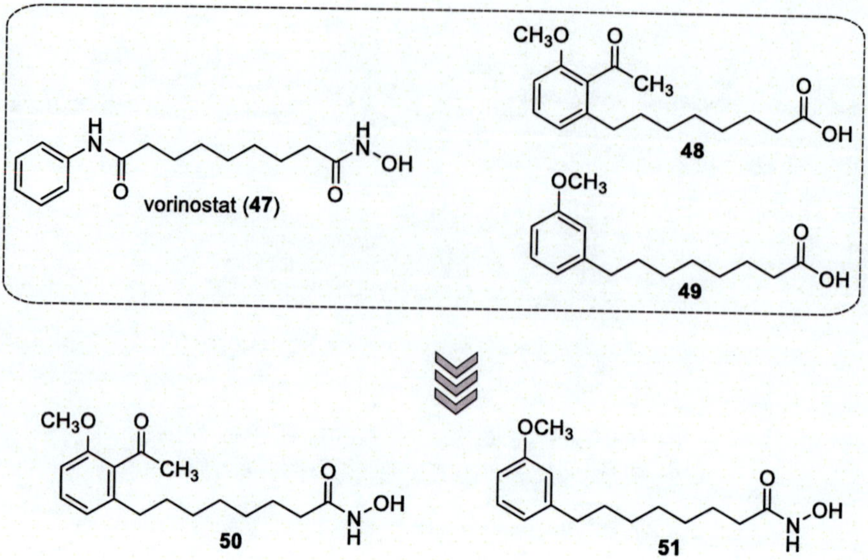

Fig. 16 SAHA (**47**) and CNSL-based hydroxamates **50** and **51**

To obtain key intermediates methyl 8-(3-methoxyphenyl)octanoate (**52**) and methyl 8-(3-methoxy-2-carbomethoxyphenyl)octanoate (**53**), a synthetic sequence was developed (Fig. 17) [48].

Thus, individual reactions of **38** and **39** with iodomethane in acetone afforded the corresponding methyl *O*-methoxyanacardate **54** *O*-methoxycardanol **55** mixtures in high yields. Next, treatment with ozone air at −70°C, followed by reduction with sodium borohydride provided the corresponding alcohols **56** (80%) or **57** (50%). Two oxidation steps on **56** or **57**, first with PCC and then with Jones Reagent, led to acids **48** and **49**. Finally, **48** and **49** were converted to the respective methyl esters **52** and **53** by treatment with iodomethane in acetone and then to the target hydroxamates **50** and **51** by aminolysis [46]. Despite **50** and **51** were effectively obtained from CNSL in 5 synthetic steps, the developed protocol is not compatible with sustainability criteria. Definitively, oxidation reaction that eliminates the use of chromium compounds and greener alternatives to methyl iodide as alkylating agent are needed. In spite of this evident shortcoming, **50** and **51** displayed a HDAC inhibitory profile similar to vorinostat, together with a more promising safety for **51**. Moreover, both compounds, and particularly **51** were able to effectively modulate glial cell-induced inflammation and to revert the pro-inflammatory phenotype, thus emerging as sustainable hit compounds for the treatment of AD.

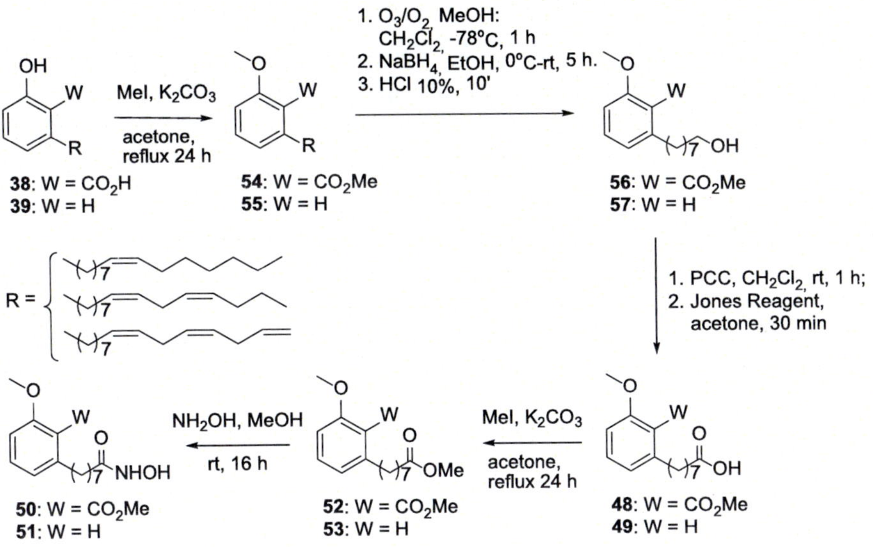

Fig. 17 Synthesis of hydroxamates form CNSL

5 *Penicillium antarcticum* as a Biomass Reagent for the Synthesis of Bioactive Compounds

Another example of new biomass reagents in the quest for bioactive molecules has been reported by Murphy et al. [49] dealing with the identification of itaconate derivatives synthesized by *Penicillium antarcticum* (a marine derived fungus) and their potential as inhibitors of human mesenchymal stem cell (MSC) chondrogenic differentiation.

Investigations into the factors that influence MSC differentiation have been comprehensively studied; however, full pharmacological control of MSC fate has not been fully defined. It is widely believed that small molecules (SMs) could provide a great starting point for the pursuit of compounds for the selective regeneration of MSC differentiation. SMs have several advantages when compared to the use of growth factors in the treatment of MSC differentiation. SMs are considerably more stable, they can directly trigger intracellular targets, are non-immunogenic, and are easier to produce at scale [50–52]. They believed high throughput screening of large numbers of isolated SMs from natural sources could detect novel lead compounds for the development of new drug targets for the modulation of differentiation of human MSCs.

For this study, Murphy and co-workers developed a new HTS platform using automated miniaturized assays. This screening platform proved to be cost effective and tested a plethora of metabolite libraries for regenerative medicine studies. They examined and optimized that the metabolite libraries could be analysed using assay miniaturization to fit 96-well plates.

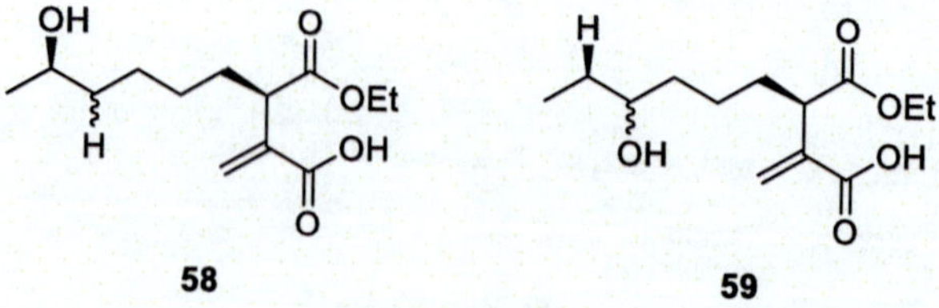

Fig. 18 Active fungal metabolites **58** and **59**

Firstly, they began by optimizing both osteogenic differentiation and high throughput chondrogenic assays. Osteogenic differentiation assays evaluated the ability of MSCs to differentiate into osteoblasts [53] by measuring intracellular alkaline phosphatase (ALP) activity and calcium mineralization. Assays for screening for chondrogenic differentiation focused on MSCs ability to differentiate into osteoblasts [54].

Once the conditions for optimum human MSC differentiation were established, 12 marine derived fungal metabolites were analysed. Of these, compounds **58** and **59** (Fig. 18) showed the greatest potential as possible targets in the quest for modulation of human MSCs differentiation.

Currently, these newly detected bioactive metabolites are under investigation to fully elucidate their participation in the modulation of differentiation of human MSCs and as possible new drug candidates, for as of yet, untreated diseases.

6 Conclusions

The main purpose of this chapter is the appreciation of the promise of an emerging research field, i.e., the production of bioactive compounds based on renewable carbon sources. This is pushing back the frontiers of waste valorization towards more ambitious and articulated targets and potential impact on human health. However, the reported real-world examples have highlighted that to find new drugs starting from biomass is not an easy task. It combines the hurdles of two very challenging endeavours such as drug discovery and waste valorization. Admittedly, in many cases the developed compounds did not show an improved potential compared to the parent drugs. The drugs are unique chemicals that must possess finely-optimized features to reach, recognize, and modulate their biological targets. On the other hand, the overall production process must be both sustainable and competitive in terms of cost with traditional pharmaceutical manufacturing. This requires specific expertise in the development of green industrial processes. Thus, it is imperative that experts from the two communities come together to create the critical mass for the development of sustainable technologies for the actual production of new, effective drugs based on food waste utilization. In this respect, advancement may come from flow technologies. Thanks to its inherent properties, scalability, less hazard, and potential downstream cost reduction, flow chemistry offers promise. On the other hand, it might be more complex to implement in the sense that the bio-based chemical substrates are usually either sensitive compounds, viscous or poorly soluble materials difficult to process under flow conditions. Thus, further refinements are needed before flow technologies are fully integrated in the production process of drugs and drug candidates.

To ensure healthy lives and promote well-being for all at all ages is the no. 3 United Nation's Sustainable Development Goal (SDG 3). This along with the sustainable development goals of food security, environmental protection, and energy efficiency makes the development of waste valorization towards the production of new medicines worthy of further pursuit.

Compliance with Ethical Standards

Funding: MLB was supported by a grant of Visiting Professor (PVE-CNPq #401864/2013-8) from the program "Ciência sem Fronteiras", Brazil National Council for Scientific and Technological Development (Grant CNPq 401864/2013-8). LASR is a fellow holder for Productivity in Technological Development and Innovative Extension – DT (CNPq 308486/2020-0).

Ethical Approval: This manuscript is a review of previously published accounts, as such, no animal or human studies were performed.

Informed Consent: No patients were studied in this chapter.

References

1. Koutinas AA, Vlysidis A, Pleissner D, Kopsahelis N, Garcia IL, Kookos IK, Papanikolaou S, Kwan TH, Lin CSK (2014) Chem Soc Rev 43:2587–2627
2. Namlis K-G, Komilis D (2019) Waste Manag 89:190–200
3. Arancon RAD, Lin CSK, Chan KM, Kwan TH, Luque R (2013) Energy Sci Eng 1:53–71
4. Klass DL (1998) Biomass for renewable energy, fuels, and chemicals. Academic Press, San Diego
5. Cardona-Alzate CA, Serna-Loaiza S, Ortiz-Sanchez M, Sustain J (2020) Dev Energy Water Environ 8:88–117
6. Kabongo JD (2013) Idowu SO, Capaldi N, Zu L, Gupta AD (eds) Encyclopedia of corporate social responsibility. Springer, Berlin, pp 2701–2706
7. Mascal M (2019) ACS Sustain Chem Eng 7:5588–5601
8. Fava F, Totaro G, Diels L, Reis M, Duarte J, Carioca OB, Poggi-Varaldo HM, Ferreira BS (2015) New Biotechnol 32:100–108
9. Chadda R, McArdle P, Murphy PV (2017) Synthesis 49:2138–2152
10. Zhou Y, Murphy PV (2008) Org Lett 10:3777–3780
11. Brock EA, Davies SG, Lee JA, Roberts PM, Thomson JE (2011) Org Lett 13:1594–1597
12. Palmer AM, Jäger V (2001) Eur. J. Org. Chem.:1293–1308
13. Uchuskin MG, Pilipenko AS, Serdyuk OV, Trushkov IV, Butin AV (2012) Org Biomol Chem 10:7262–7265
14. Mamman AS, Lee JM, Kim YC, Hwang IT, Park NJ, Hwang YK, Chang JS, Hwang JS (2008) Biofuels Bioprod Bioref 2:438–454
15. Zeitsch KJ (2000) The chemistry and technology of furfural and its many by-products. Elsevier, Amsterdam
16. Gérardy R, Debecker DP, Estager J, Luis P, Monbaliu JM (2020) Chem Rev 120:7219–7347
17. Pilipenko AS, Mel'chin VV, Trushkov IV, Cheshkov DA, Butin AV (2012) Tetrahedron 68:619–627

18. Nevolina TA, Shcherbinin VA, Serdyuk OV, Butin AV (2011) Synthesis:3547–3551
19. Butin AV, Tsiunchik FA, Kostyukova ON, Uchuskin MG, Trushkov IV (2011) Synthesis:2629–2638
20. Butin AV, Nevolina TA, Shcherbinin VA, Trushkov IV, Cheshkov DA, Krapivin GD (2010) Org Biomol Chem 8:3316–3327
21. Butin AV, Dmitriev AS, Uchuskin MG, Abaev VT, Trushkov IV (2008) Synth Commun 38:1569–1578
22. Dmitriev AS, Abaev VT, Bender W, Butin AV (2007) Tetrahedron 63:9437–9447
23. Go M-L, Ngiam T-L, Tan AL-C, Kuaha K, Wilairat P (1998) Eur J Pharm Sci 6:19–26
24. Filak LK, Mühlgassner G, Bacher F, Roller A, Galanski M, Jakupec MA, Keppler BK, Arion VB (2011) Organometallics 30:273–283
25. Ibrahim E-S, Montgomerie AM, Sneddon AH, Proctor GR, Green B (1988) Eur J Med Chem 23:183–188
26. Whittell LR, Batty KT, Wong RP, Bolitho EM, Fox SA, Davis TM, Murray PE (2011) Bioorg Med Chem 19:7519–7525
27. Van Baelen G, Hostyn S, Dhooghe L, Tapolcsányi P, Mátyus P, Lemière G, Dommisse R, Kaiser M, Brun R, Cos P (2009) Bioorg Med Chem 17:7209–7217
28. Van Miert S, Hostyn S, Maes BU, Cimanga K, Brun R, Kaiser M, Mátyus P, Dommisse R, Lemière G, Vlietinck A (2005) J Nat Prod 68:674–677
29. Galkin KI, Kucherov FA, Markov ON, Egorova KS, Posvyatenko AV, Ananikov VP (2017) Molecules 22:2210
30. Kashparova VP, Khokhlova EA, Galkin KI, Chernyshev VM, Ananikov VP (2015) Russ Chem Bull 64:1069–1073
31. Liu D, Chen Z (2009) Anti Cancer Agents Med Chem 9:392–396
32. Kucherov F, Galkin K, Gordeev E, Ananikov V (2017) Green Chem 19:4858–4864
33. Calixto JB (2019) An Acad Bras Cienc 91:e20190105
34. Harvey AL, Edrada-Ebel R, Quinn RJ (2015) Nat Rev Drug Discov 14:111–129
35. Mathur S, Hoskins C (2017) Biomed Rep 6:612–614
36. Kühlborn J, Groß J, Opatz T (2020) Nat Prod Rep 37:380–424
37. Bolognesi ML (2019) Chibale K (ed) Annual reports in medicinal chemistry, vol 52. Academic Press, San Diego, pp 153–176
38. Cerone M, Uliassi E, Prati F, Ebiloma GU, Lemgruber L, Bergamini C, Watson DG, de Ferreira TAM, Roth Cardoso GSH, Soares Romeiro LA, de Koning HP, Bolognesi ML (2019) ChemMedChem 14:621–635
39. Lochab B, Shukla S, Varma IK (2014) RSC Adv 4:21712–21752
40. Campaner P, D'Amico D, Longo L, Stifani C, Tarzia A (2009) J Appl Pol Sci 114:3585–3591
41. Voirin C, Caillol S, Sadavarte NV, Tawade BV, Boutevin B, Wadgaonkar PP (2014) Polym Chem 5:3142–3162
42. Hamad FB, Mubofu EB (2015) Int J Mol Sci 16:8569–8590
43. Viegas-Junior C, Danuello A, da Silva Bolzani V, Barreiro EJ, Fraga CA (2007) Curr Med Chem 14:1829–1852
44. Pieretti S, Haanstra JR, Mazet M, Perozzo R, Bergamini C, Prati F, Fato R, Lenaz G, Capranico G, Brun R, Bakker BM, Michels PA, Scapozza L, Bolognesi ML, Cavalli A (2013) PLoS Negl Trop Dis 7:e2012
45. Prince M (2017) Lancet 390(10113):e51–e53
46. Soares Romeiro LA, da Costa Nunes JL, de Oliveira Miranda C, Roth Cardoso GSH, de Oliveira AS, Gandini A, Kobrlova T, Soukup O, Rossi M, Senger J, Jung M, Gervasoni S, Vistoli G, Petralla S, Massenzio F, Monti B, Bolognesi ML (2019) ACS Med Chem Lett 10:671–676
47. Fischer A, Sananbenesi F, Mungenast A, Tsai LH (2010) Trends Pharmacol Sci 31:605–617
48. Lemes LFN, de Andrade Ramos G, de Oliveira AS, da Silva FMR, de Castro Couto G, da Silva Boni M, Guimarães MJR, Souza INO, Bartolini M, Andrisano V (2016) Eur J Med Chem 108:687–700

49. Marchese P, Mahajan N, O'Connell E, Fearnhead H, Tuohy M, Krawczyk J, Thomas OP, Barry F, Murphy MJ (2020) Mar Drugs 18:192
50. Blaich G, Janssen B, Roth G, Salfeld J (2010) Pharmaceutical sciences encyclopedia: drug discovery, development, and manufacturing, pp 1–35
51. Pyeritz RE, Engl N (2008) J Med 358:2829–2831
52. Egusa H, Saeki M, Doi M, Fukuyasu S, Matsumoto T, Kamisaki Y, Yatani H (2010) J Oral Biosci 52:107–118
53. Krause U, Seckinger A, Gregory CA (2011) Mesenchymal stem cell assays and applications. Springer, New York, pp 215–230
54. Li J, Dong S (2016) Stem Cells Int. Article ID 2470351. https://doi.org/10.1155/2016/2470351

Top Med Chem (2021) 38: 391–420
https://doi.org/10.1007/7355_2021_114

Published online: 9 April 2021

Flow Chemistry Supporting Access to Drugs in Developing Countries

Cloudius R. Sagandira and Paul Watts

Contents

Abstract While there are a variety of companies that formulate medicines in developing countries, there are a few active pharmaceutical ingredients (APIs) manufacturers with the consequence that these need to be imported from overseas; generics principally from India and China. Consequently, the medications are inaccessible to most patients due to high cost, unguaranteed supply chain and quality, as most patients fall in the low-middle income bracket in developing economies resulting in avoidable loss of life and unnecessary huge health burden.

C. R. Sagandira and P. Watts (✉)
Department of Chemistry, Nelson Mandela University, Port Elizabeth, South Africa
e-mail: s211176125@mandela.ac.za; Paul.Watts@mandela.ac.za

Herein, we assess how developing economies can take advantage of the limited existing batch manufacturing infrastructure to set up state-of-the-art continuous flow manufacturing infrastructure to enable local pharmaceutical manufacturing as well as support access to drug discovery. This potentially provides paradigm shift in developing countries' pharmaceutical industry that could lead to better drug discovery access as well as enabling local drug manufacturing with the consequence of improving access to medicines.

Keywords Active pharmaceutical ingredients, Developing countries, Drug discovery, Flow Chemistry

1 Introduction

1.1 *Disease Burden and Pharmaceutical Landscape in Developing Countries*

Developing countries have significantly different disease profiles to developed countries due to both poverty and geography. Communicable diseases such HIV/AIDS, tuberculosis and malaria are the leading causes of death in developing countries whereas developed countries are hugely affected by non-communicable diseases such as cancer and cardiovascular diseases [1–4]. According to recent United Nations Industrial Development Organisation (UNIDO) and World Health Organisation (WHO) data, 30% of world's population has no access to life saving medicines with the number going as high as 50% in developing economies such as Africa [5–9]. Unfortunately, adequate treatment for these serious communicable diseases is beyond the capacity of most developing countries with the consequence of huge health burden, which impedes economic development [10]. Due to the significant divergence between mortality and disease profiles in developed and developing economies, drug development efforts benefit developed countries more than developing countries resulting in a '10/90 gap' [11]. While there are a variety of companies that formulate medicines in developing countries, there are a few active pharmaceutical ingredients (APIs) manufacturers with the consequence that these need to be imported from overseas; generics principally from India and China [12]. Consequently, the medications are inaccessible to most patients due to high cost, unguaranteed supply chain and quality, as most patients fall in the low-middle income bracket in developing economies. Although these countries hugely benefit from donor medication and tiered pricing [10], serious drug shortages are widely experienced resulting in avoidable loss of life and unnecessary health burden. Furthermore, due to over reliance on pharmaceutical importation, developing countries are left exposed and vulnerable in the face of pandemics which disrupt supply chain such as the novel coronavirus (COVID-19) owing to world-wide lockdowns being implemented to manage the pandemic. As a result of the COVID-19 pandemic, the US Food and Drug Administration (FDA) has the same concern that the medical product supply chain will undoubtedly be affected, including potential disruptions to supply or shortages of critical drugs [13]. This situation can be alleviated with local pharmaceutical production, which has been inadequate in developing economies.

In recent years, regional bodies such as the African Union (AU), intergovernmental organisations such as United Nations (UN) and international donors such as the Bill and Melinda Gates Foundation have intensified efforts to improve medicines accessibility in developing economies [8, 14, 15]. In 2018, AU in partnership with African Development Bank and the African Import and Export Bank discussed practical steps towards fund creation to enable local pharmaceutical manufacturing [16]. Additionally, AU held the 7th Strategic Stakeholders Retreat on Industry under the theme: "Implementation Review of the Plan of Action for the Accelerated Industrial Development of Africa" [17]. These efforts are in line with its aspiration to guarantee high standard of living, quality of life and well-being of the African citizens as envisaged by Africa's Agenda 2063 [14]. The Bill and Melinda Gates Foundation through the Medicine for All initiative seeks cheaper and more efficient ways of manufacturing drugs. In Brazil, the Medicine Without Borders project has similar goals. The South African government funded Nelson Mandela University to investigate efficient and cost-effective local drug manufacturing of generic drugs using flow chemistry technology with the grand vision of improving accessibility of important drugs.

1.2 *Flow Chemistry in the Pharmaceutical Industry*

The economic pressures on the pharmaceutical industry to expeditiously provide cost-effective and high-quality therapeutics have driven the adoption of new enabling technologies. Flow chemistry is increasingly playing an important role in the pharmaceutical industry. It is delivering significant benefits in the area of drug discovery and development by overcoming most limitations associated with the traditional batch process by availing new chemical processing windows. Well-documented flow process benefits such as labour, cost and energy efficiency have seen numerous syntheses of commercial APIs being redesigned around flow chemistry [18–22]. These benefits are seen throughout all areas of pharmaceutical pipeline, from library synthesis in drug discovery, rapid reaction optimization in development to manufacturing.

Although this technology is being implemented in developed economies such as the USA, Europe, India and China due to its associated benefits, its adoption in established pharmaceutical industry is still slow due to its disruptive nature. However, developing countries can take advantage of the limited existing batch manufacturing infrastructure to set up state-of-the-art continuous flow manufacturing infrastructure to enable local pharmaceutical manufacturing, particularly of generics. This will help vulnerable populations to access affordable quality medications, reducing the level of dependence on imports and donations thus improving the health and the economy simultaneously. This chapter discusses how flow chemistry is being applied in the synthesis of APIs in developing economies with the vision of enabling local drug manufacturing. Although there is an appreciable amount of drug discovery efforts in developing economies, it is ironic that nothing much is being done towards drug manufacturing to complement the on-going drug discovery. Consequently, if there is any drug discovery breakthrough, one has to outsource the manufacturing to developed economies due to lack of local capacity. In an effort to close this gap, developing economies have been taking advantage of emerging and enabling technologies such as flow chemistry to build local drug manufacturing capacity and these are discussed herein.

2 Continuous Flow Synthesis of APIs in Developing Countries

2.1 Lamivudine

Lamivudine **1** (Fig. 1) is an antiretroviral drug originally marketed by GSK, used for the treatment of HIV/AIDS and hepatitis B [6, 12, 23, 24]. It is potent for both HIV-1 and HIV-2, and is commonly used in combination with other antiretroviral drugs such as zidovudine and abacavir [6, 23]. However, this important drug is still not easily accessible in developing countries [6, 23, 24]. As part of the on-going research towards the development of cost- effective routes for important medicines and capacity building for local manufacturing in Africa using flow chemistry technology with the grand vision improving drug accessibility, the Watts group in South Africa demonstrated a three-stage continuous flow synthesis of lamivudine [24, 25]. In the first stage, L-Menthyl glyoxalate **2** was treated with 1,4-dithiane-2,5-diol **3** in a 2 ml glass reactor at 90°C under 10 psi back pressure using acetonitrile as a solvent (Fig. 2). The eluent was subsequently treated with acetic anhydride/pyridine in acetonitrile at ambient temperature in a 1.9 ml (0.2 ml static mixer +1.7 ml LTF reactor volume) reactor affording acetylated 1,3-oxathiolane **4** (95% conversion) in 9.7 min residence time. The desired isomer of **4a** was afforded in 45% isolated yield after an off-line recrystallisation procedure with 1% triethylamine/hexane.

The second stage involved an *N*-glycosidation reaction. Acetylated 1,3-oxathiolane **4a** was initially treated with pyridinium triflate in acetonitrile in 0.2 ml mixer at ambient temperature (Fig. 3). Silylated cytosine **5** was subsequently introduced into the reaction stream using a T-mixer and the resultant mixture was

Fig. 1 Lamivudine structure

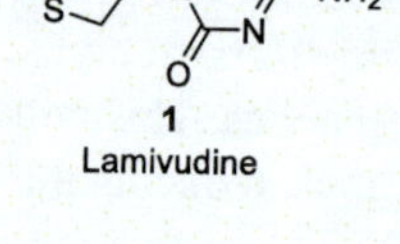

1

Lamivudine

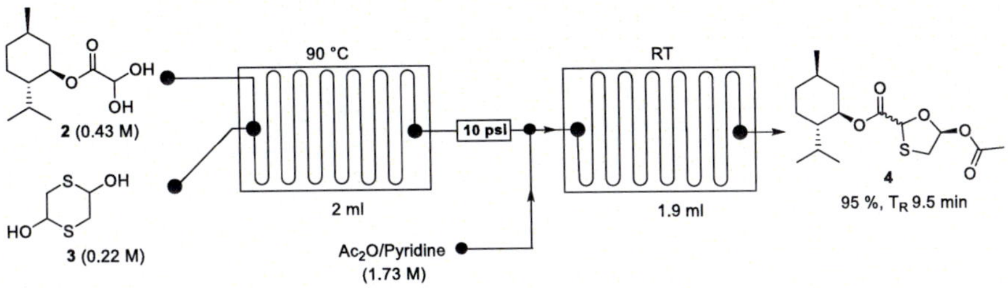

Fig. 2 Continuous flow synthesis of 1,3-oxathiolane **4**

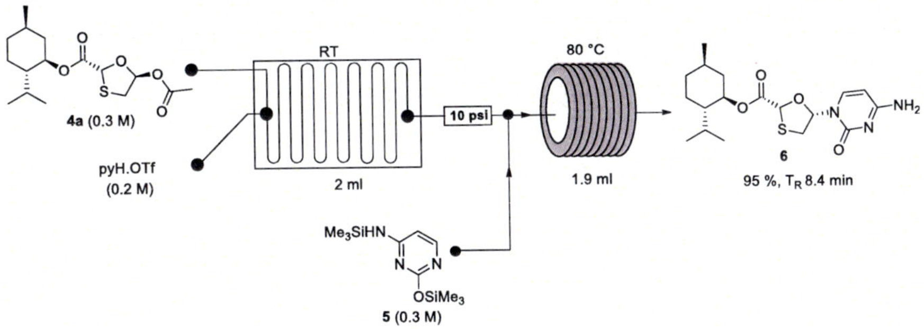

Fig. 3 Continuous flow glycosylation

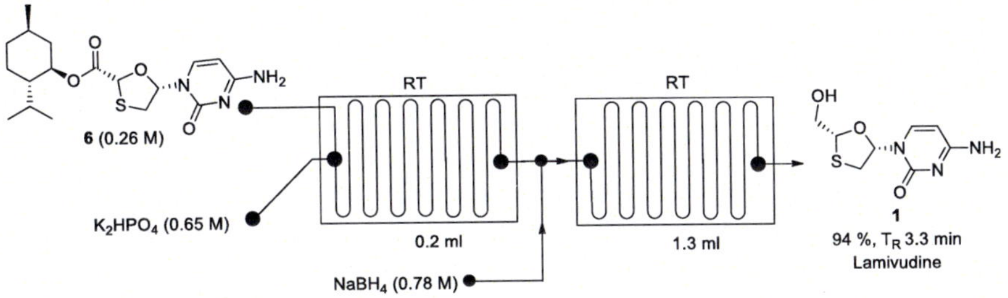

Fig. 4 Continuous flow reduction of intermediate **6** with $NaBH_4$

heated at 80°C in a 3.8 ml PTFE tubing reactor to afford **6** (95% isolated yield) in 8.4 min residence time.

In the final stage, intermediate **6** was treated with aqueous K_2HPO_4 in a 0.2 ml reactor at ambient temperature to remove the menthyl group (Fig. 4). The resulting product was subsequently reduced with $NaBH_4$ in a 1.3 ml reactor (0.2 ml static mixer +1.1 ml LTF reactor) at ambient temperature to afford lamivudine **1** in 94% isolated yield (99.8% ee) with 3.3 min residence time. This is an effective and scalable 21.4 min total residence time continuous flow lamivudine synthesis procedure with 41% overall yield which is comparable to literature [24, 26]. Beyond process efficiency over batch in terms of reaction time and yield, the authors demonstrated reaction telescoping thereby avoiding isolation and purification in some instances. Long term the L-menthol could be recycled in the process.

2.2 *Tenofovir*

Tenofovir disoproxil fumarate **7** and tenofovir alafenamide fumarate **8** (Fig. 5) are nucleotide prodrugs for the treatment of HIV/AIDS and hepatitis B [26, 27]. They are usually used in combination with emtricitabine and efavirenz. The de Souza group [27] in Brazil reported a chemo-enzymatic continuous flow process for (*R*)-propylene carbonate, an important intermediate in the synthesis of tenofovir prodrugs (Fig. 6). They started from glycerol carbonate **9** derived from glycerol, a very cheap and renewable raw material. Glycerol carbonate **9** was pumped through a $NaAlO_2$ packed-bed reactor held at 90°C to afford (±) glycidol **10** in 66% yield in 7 min residence time. (±)- Gycidol **10** was then pumped through a tube-in-tube reactor saturated with H_2 for further hydrogenolysis catalysed by a 10% Pd/C packed-bed column reactor fitted with a 40 psi back-pressure regulator to afford 1,2-propanediol **11** in greater than 99% conversion at room temperature with 17 min residence time. 1,2-Propanediol **11** protection was accomplished in the presence of triphenylmethyl chloride **12**, DABCO and TEA in a 3 ml coil reactor under 6 bar back-pressure regulator at 30°C to afford trityl ether **13** in 63% conversion with 15 min residence time. The use of flow remarkably reduced the reaction from 12 h in batch to 15 min in flow. Kinetic resolution of trityl ether **13** was achieved in flow by pumping it through a Novozym 435 immobilised column reactor in the presence of vinyl acetate as acyl donor at 30°C to afford intermediate (*R*)-**15** in 47% conversion with excellent enantiomeric ratios (>170) in 7 min residence time. An impressive space-time yield of 98 g h^{-1} was achieved under the above conditions. Dual removal of the acetate and trityl groups of intermediate (*R*)-**15** was performed in a $NaHSO_4{\cdot}SiO_2$ packed-bed reactor held at 30°C to afford (*R*)-1,2-propanediol **16** in 95% conversion with 30 min residence time. The final step involved the treatment of (*R*)-1,2-propanediol **16** with DMC and DBU in a 3 ml coil reactor with a 6 bar

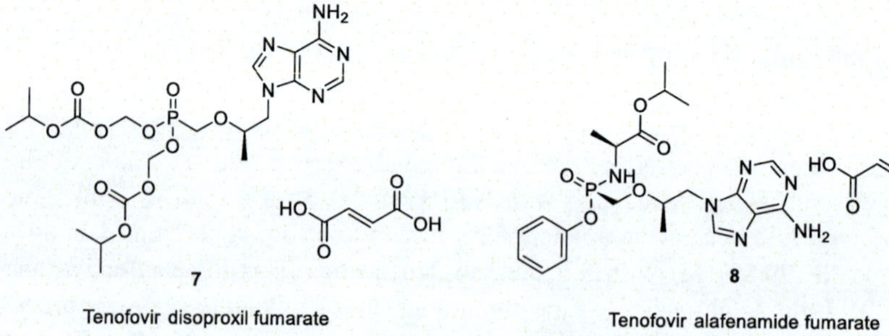

Fig. 5 Structures of tenofovir

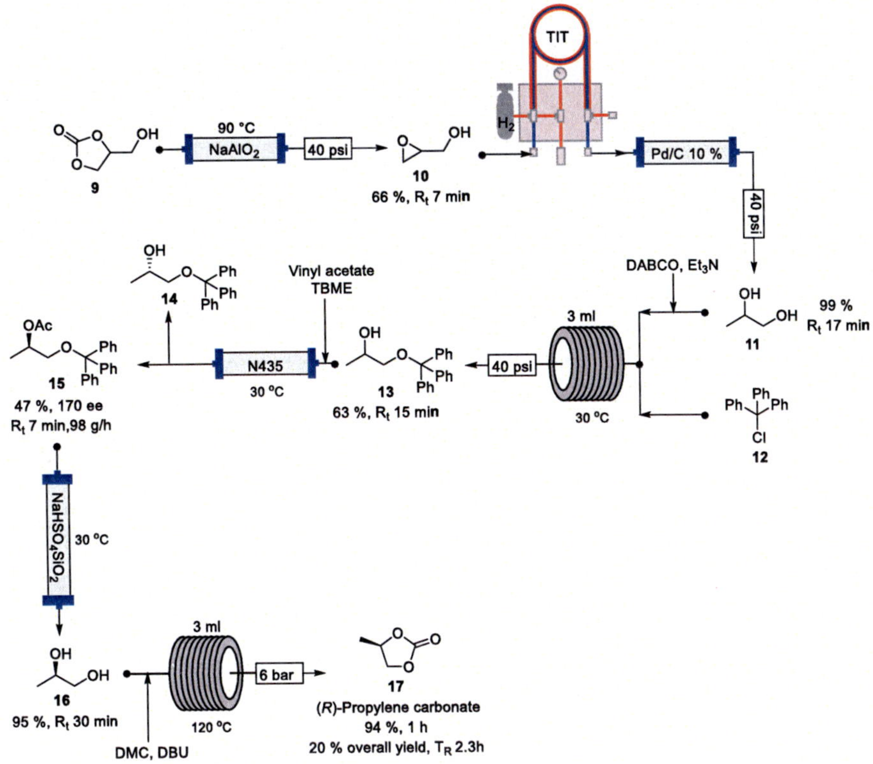

Fig. 6 Continuous flow synthesis of (R)-propylene carbonate **17**

back-pressure reactor at 120°C to afford the desired (*R*)-propylene carbonate **17** in 94% isolated yield with 1 h residence time. This seven-step continuous flow strategy afforded the desired important intermediate for tenofovir synthesis, (*R*)-propylene carbonate in 20% overall yield with 2.3 h total residence time. The study excellently showcases the application of flow chemistry in enzymatic kinetic resolution of racemic mixtures affording pure products in excellent conversion and selectivity. Furthermore, flow chemistry allowed for dual deprotection thus shortening the synthetic route.

2.3 *Efavirenz*

Efavirenz **18** (Fig. 7) is a non-nucleoside reverse transcriptase inhibitor (NNRTI) discovered at Merck laboratories [26, 28]. It is one of the preferred agents used in combination therapy for first-line treatment of HIV/AIDS. Watts group investigated the synthesis of a efavirenz key intermediate, *tert*-butyl-4-chloro-2-(2,2,2-trifluoroacetyl)phenyl carbamate **21** in flow [25, 28]. Unlike the Seeberger group

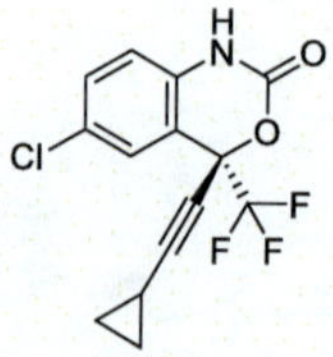

Fig. 7 Structure of efavirenz

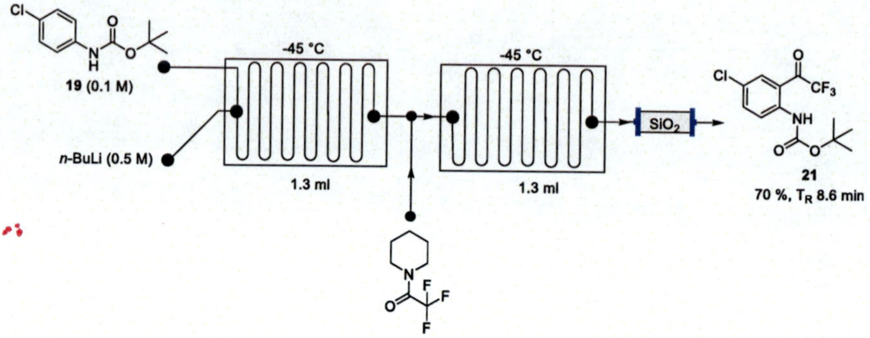

Fig. 8 Continuous flow synthesis of efavirenz key intermediate **21**

[29] procedure which utilised 1,4-dichlorobenzene as the starting material, Watts and co-workers treated *N*-Boc-4-chloroaniline **19** with *n*-BuLi in a 1.3 ml reactor (0.2 ml static mixer + 1.1 ml LTF reactor) at −45°C (Fig. 8). The dianion eluent was subsequently trifluoroacetylated with piperidine trifluoroacetic acid **20** in a 1.3 ml reactor (0.2 ml static mixer +1.1 ml LTF reactor) at −45°C and quenched inline via a silica packed column. The key intermediate **21** was afforded in a significantly higher yield (70%) in 8.6 min residence time compared to batch (28%). The authors avoided intermediate isolation and purification by reaction telescoping. The success of this reaction sets a good platform for total synthesis of efavirenz to be investigated [30].

2.4 *Darunavir*

Miranda and co-workers [31] reported the continuous flow synthesis of non-peptidal bis-tetrahydrofuran moiety of Darunavir **22** (Figs. 9, 10, and 11). Darunavir is a highly active HIV protease inhibitor, which is usually used when standard treatment fails. The authors started by performing ozonolysis of alkene (±)-**23** in an Ice Cube™ flow reactor with an ozone module at −30°C affording product (±)-**24** in

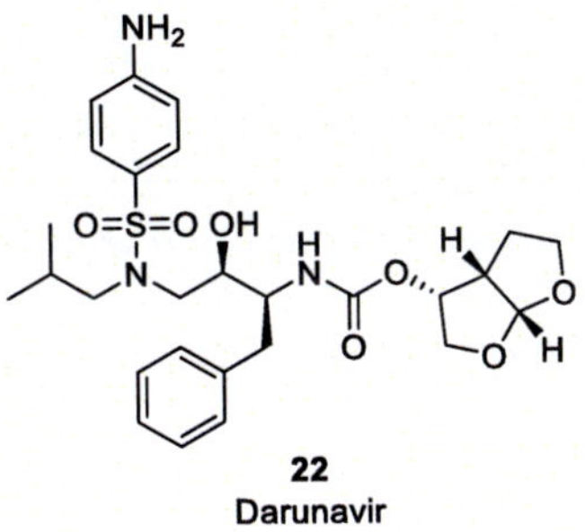

Fig. 9 Structure of darunavir

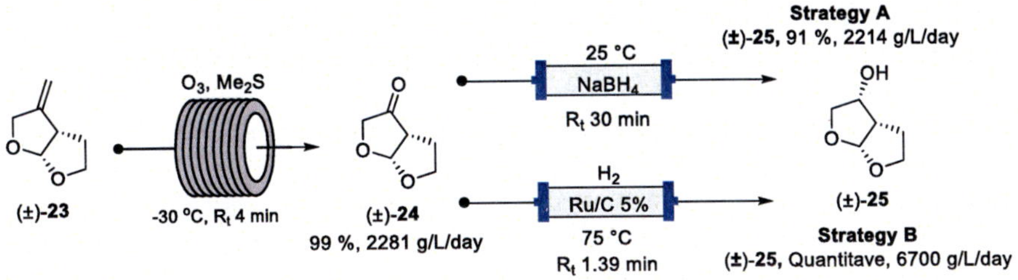

Fig. 10 Continuous flow ozonolysis and carbonyl reduction

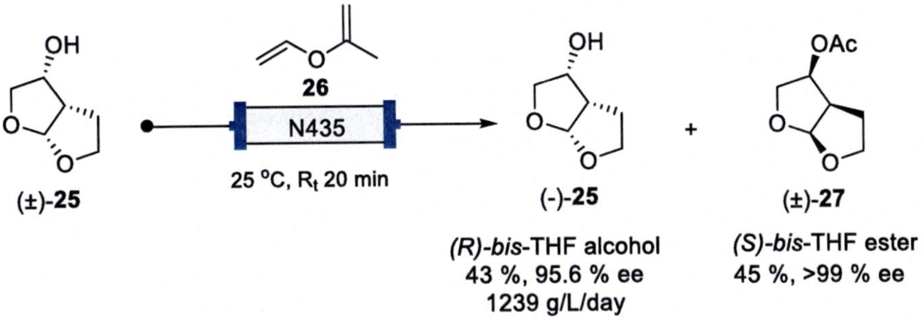

Fig. 11 Continuous flow enzymatic resolution

99% yield with 4 min residence time and space time yield of 2,281 g/l/day, a significant improvement compared to batch (3 h, 152 g/l/day). The next step involved carbonyl group reduction of (±)-**24**, which was accomplished using two strategies (Fig. 10). In the first strategy, (±)-**24** was reduced in an immobilised borohydride packed column reactor at 25°C to afford the racemic alcohol (±)-**25** in 91% isolated yield with 30 min residence time and space time yield of 2,214 g/l/day. A much higher space time yield of 6,700 g/l/day for (±)-**25** could be achieved by using the second strategy: hydrogenation at 75°C using a H-Cube Pro™ fitted with Ru/C 5% packed column reactor.

The final step involved enzymic kinetic resolution of (±)-**25** in the presence of vinyl acetate in a lipase-N435 packed column reactor at 25°C to afford the desired alcohol (−)-**25** in 46% yield (96% ee) with 20 min residence time and 1,239 g/l/day thus efficient than batch (41% conversion, 92% ee, 2 h) (Fig. 11). Most importantly, the authors proved that the use of flow chemistry represents an important synthetic improvement towards the bicyclic core of Darunavir owing to reduced reaction times, excellent enzymatic kinetic resolution and reduced workup manipulations.

2.5 *Isoniazid*

Watts group [25, 32] recently filed a patent on a two-step continuous flow procedure for the preparation of isoniazid **28** (Figs. 12 and 13). Isoniazid **28** (Fig. 12) is an important first-line antitubercular drug. Tuberculosis (TB) is the ninth leading cause of mortality globally, with developing economies being the most affected. In 2016, Africa accounted for 25% of all the global TB deaths. Furthermore, it is the leading killer of HIV-positive people, thus further exacerbating the already unbearable HIV/AIDS burden in Africa. Their approach started with hydrolysis of nitrile 4-cyanopyridine **29** in MeOH/H_2O with aqueous NaOH at 95°C in a 1,7 ml reactor. Subsequently, the eluent stream was treated with hydrazine hydrate at 105°C in a 2.4 ml reactor fitted with a 35 psi back-pressure regulator to afford the desired product isoniazid **28** in 96% overall conversion with 21 min total residence time over the two steps. The study demonstrated how total synthesis of drugs can be done in a single flow by telescoping in continuous flow systems.

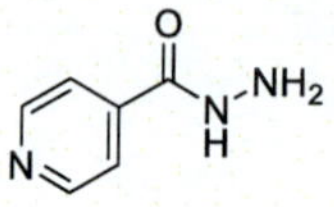

28
Isoniazid

Fig. 12 Structure of isoniazid

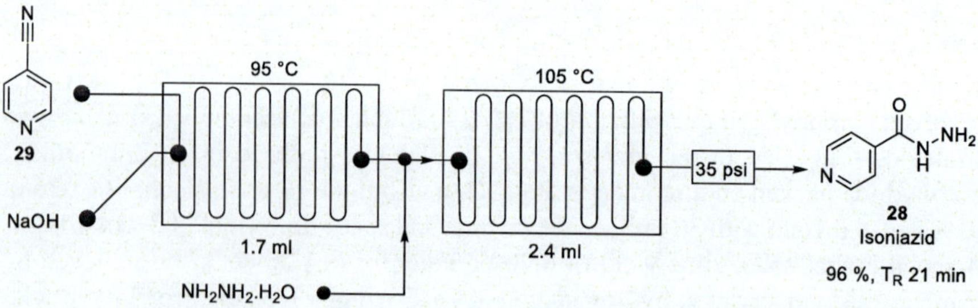

Fig. 13 Telescoped synthesis of isoniazid in flow

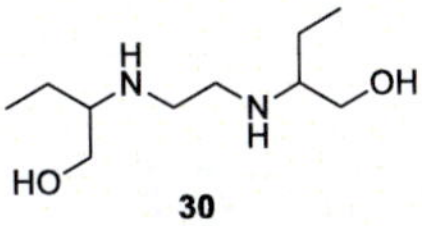

30
Ethambutol

Fig. 14 Structure of ethambutol

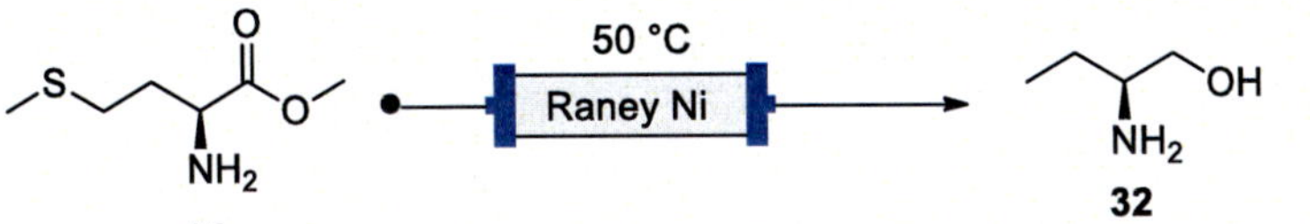

31

32
97 %, Rt 12 min

Fig. 15 Continuous flow desulphurisation of L-Methionine methyl ester **31**

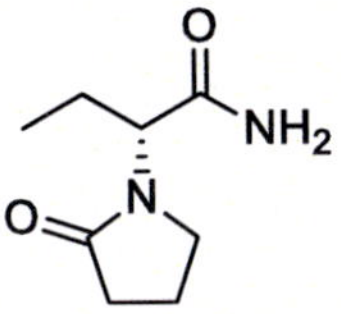

33
Levetiracetam

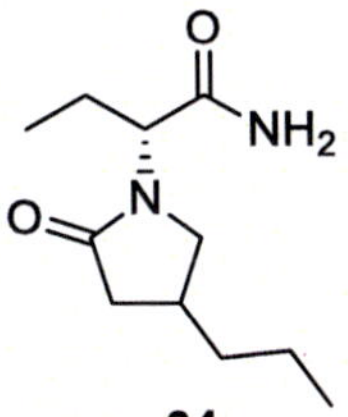

34
Brivaracetam

Fig. 16 Structures of levetiracetam and brivaracetam

2.6 *Ethambutol*

De Souza and co-workers [33] demonstrated a continuous flow strategy towards the synthesis of *(S)*-2-aminobutan-1-ol **32**, ethambutol **30** key intermediate (Figs. 14 and 15). Ethambutol **30** (Fig. 14) is a potent drug for treating tuberculosis. *(S)*-2-Aminobutan-1-ol **32** (97% yield) synthesis was achieved by desulphurization and ester reduction of L-methionine methyl ester **31** in MeOH/H_2O using a Raney Ni packed column at 50°C with 12 min residence time. This became the first Raney Ni H_2 free carboxylic ester reduction in the literature. Moreover, a cascade desulphurization/reduction was possible in flow accompanied by excellent yield.

2.7 *Levetiracetam and Brivaracetam*

Levetiracetam **33** and brivaracetam **34** (Fig. 16) are both antiepileptic drugs, which share the same key intermediate (*S*)-2-aminobutanamide **36**. De Souza group demonstrated three different strategies towards key intermediate in continuous flow **36**

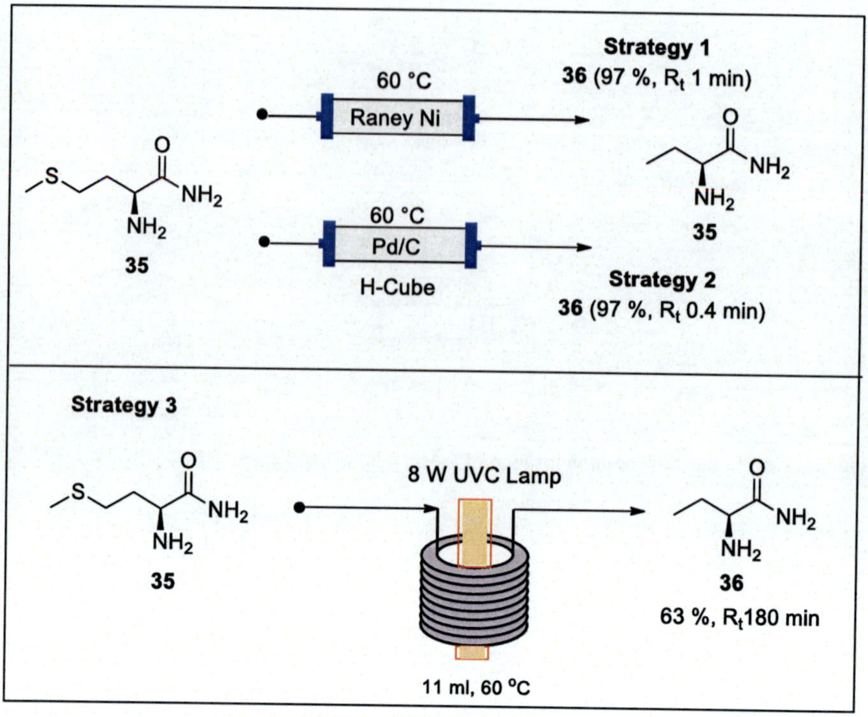

Fig. 17 Continuous flow desulphurisation of L-methionamide **35**

(Fig. 17) [33]. The first strategy involved desulphurization L-methionamide **35** in MeOH using a Raney Ni packed column at 60°C to afford (*S*)-2-aminobutanamide **36** in 97% conversion with 1 min residence time. The second strategy demonstrated desulphurization of L-methionamide **35** at 60°C in H-Cube Pro™ fitted with Pd/C packed column reactor in the presence of H_2 under 40 bar back-pressure regulator affording the desired (*S*)-2-aminobutanamide **36** in 97% conversion with 0.4 min residence time. This H_2-Pd/C desulphurization procedure has productivity due to lower substrate/catalyst ratio compared to the Raney Nickel process. However, the process suffers from reduced Pd/C cartridge half-life due to sulphur poisoning. In the third strategy, photo-desulphurization was performed in 11 ml PFA coil reactor using a 8 W-UVC lamp at 60°C in 89% conversion and 71% selectivity of the desired (*S*)-2-aminobutanamide **36** (effectively 63% yield) with 180 min residence time. Unlike in batch, flow chemistry accurately allowed for the control of light needed.

2.8 *Capecitabine*

Capecitabine **37** (Fig. 18) marketed as Xeloda by Roche is an anticancer prodrug of 5-fluorouracil used for the treatment of colorectal, breast and gastric cancer. Its

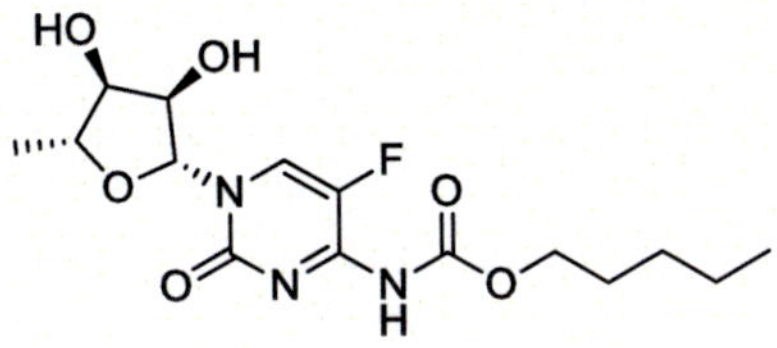

37
Capecitabine

Fig. 18 Structure of capecitabine

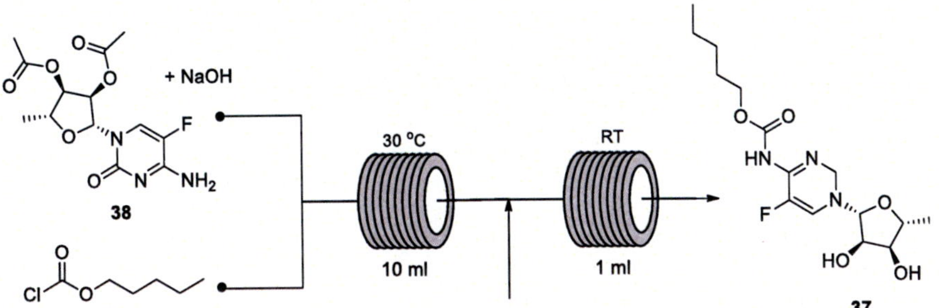

Fig. 19 Continuous flow 2′,3′-diacetoxy-5′-deoxy-5-fluorocytidine **38** with n-pentyl chloroformate **39**

synthesis involves late-stage carbamoylation and ester hydrolysis [34]. Miranda and co-workers [34] in Brazil demonstrated the sequential Schotten-Baumann carbamoylation and acetylation in a continuous flow system towards capecitabine **37** (Fig. 19). The process involved the reaction of 2′,3′-diacetoxy-5′-deoxy-5-fluorocytidine **38** with *n*-pentyl chloroformate **39** in the presence of NaOH in a 10 ml PFA coil reactor at 30°C. The eluent flow stream was subsequently quenched in line with HCl in a 1 ml PFA coil reactor at room temperature to afford capecitabine **37** in 81% yield with 30 min total residence time. The continuous flow strategy was time economical compared to the batch process (5 h, 82% yield). Flow chemistry avoided the use of hazardous solvents, low-temperature intermediate workup and isolation making the process less laborious, greener and more time economical.

2.9 *Daclatasvir*

De Souza group in collaboration with Kappe group [35] demonstrated a telescoped two-step procedure towards 1*H*-4-substituted imidazoles which are key building

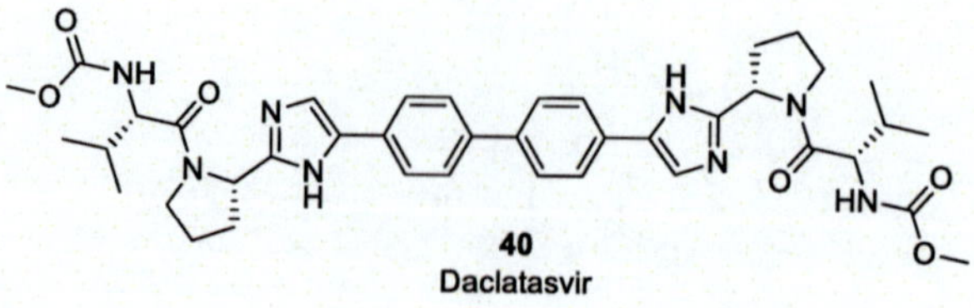

Fig. 20 Structure of daclatasvir

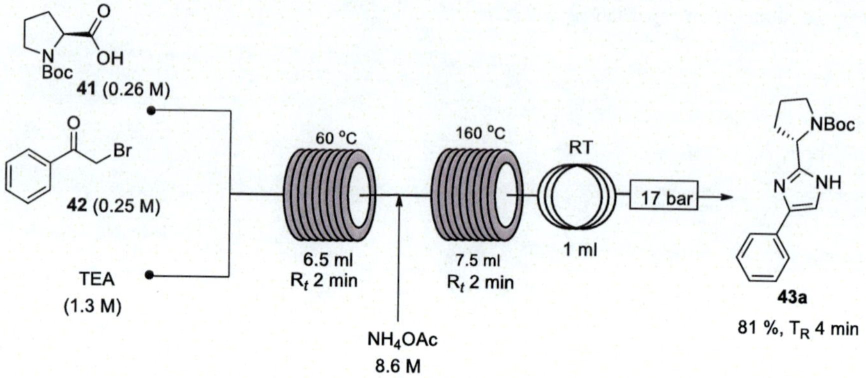

Fig. 21 Two-step flow synthesis of imidazole **43a**

blocks in the synthesis of NS5A inhibitors such as daclatasvir **40** (Figs. 20 and 21). Daclatasvir **40** (Fig. 20) is a potent drug for the treatment of hepatitis C. NS5A inhibition drugs usually contain a 4-phenyl imidazole moiety derived from L-proline and 1*H*-4-substituted imidazoles. Their model telescoped strategy started with reacting Boc-L-proline **41** with α-bromoacetophenone **42** in the presence of TEA in a 6.5 ml PFA coil reactor at 60°C for 2 min. The eluent stream was subsequently treated with NH_4OAc in a 7.5 ml stainless steel coil reactor at 160°C for 2 min then cooled in a heat exchanger at room temperature. The system was fitted with 17 bar back-pressure regulator. The desired imidazole **43a** was afforded in 81% yield with 4 min total residence time. With a flow model strategy for imidazoles in hand, the authors demonstrated the synthesis of various imidazoles **43a-m** in 39–94% yield with 2–5 min total residence time (Fig. 22).

More specifically on daclatasvir **40**, the authors applied the developed continuous flow strategy in the synthesis of the symmetrical core unit of daclatasvir **40**, biphenyl bisimidazole **46** (Fig. 23). Having made the acyloxy-derivative **44** in batch by stirring dibromo intermediate, *N*- Boc-L-proline **41** and TEA for 10 min at room temperature, the resulting solution was then treated with NH_4OAc **45** in a 7.5 ml stainless steel coil reactor at 160°C for 3 min then cooled in a heat exchanger at room temperature affording the desired biphenyl bisimidazole **46** in 71% yield. Although the yield is comparable to literature, the reaction time in flow is significantly lower than the reported batch processes. Beyond reaction telescoping, de Souza

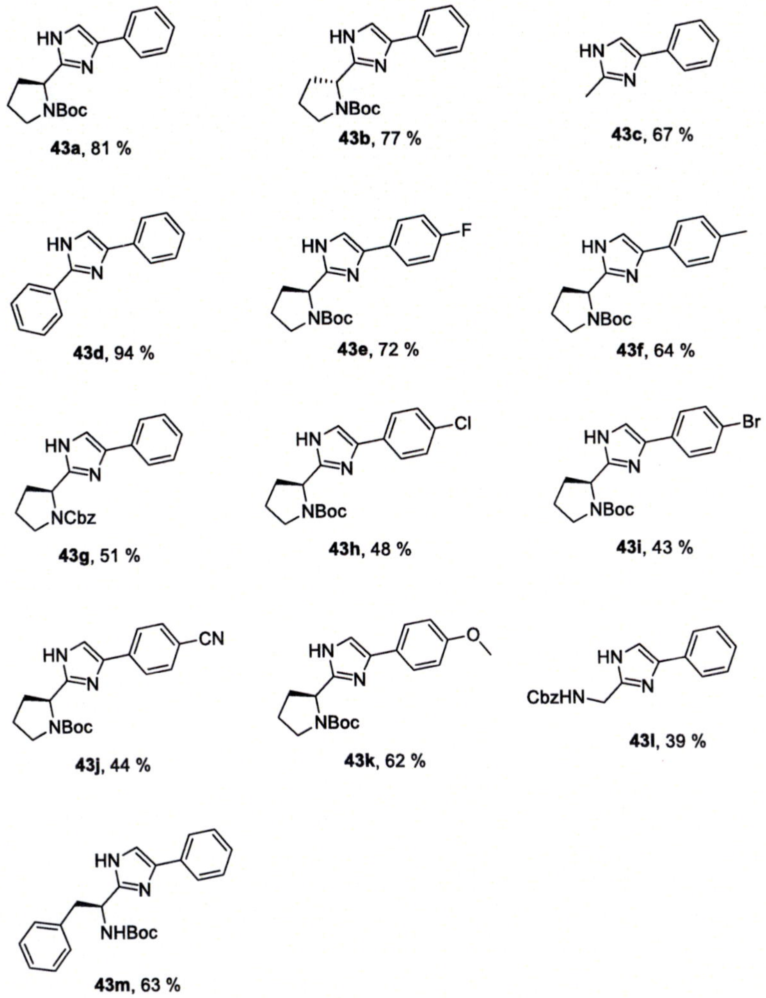

Fig. 22 Continuous flow synthesis of various imidazoles

demonstrated reaction superheating above solvent boiling point in flow owing to system pressurization, which is exclusive to continuous flow resulting reduced reaction times.

2.10 *Clozapine*

Riley group [25, 36] in South Africa demonstrated a batch-flow hybrid synthesis of an atypical antipsychotic agent used in the treatment of schizophrenia, clozapine **47**

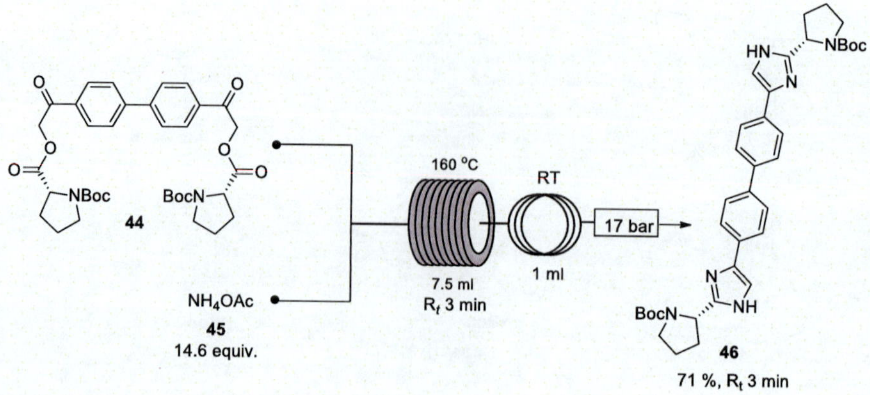

Fig. 23 Continuous flow synthesis of bisimidazole **46**

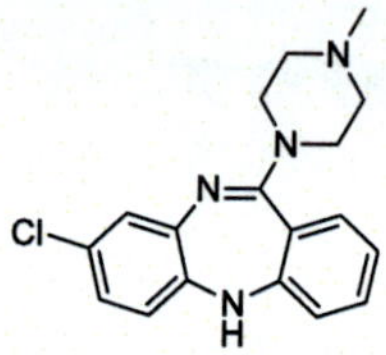

Fig. 24 Structure of clozapine

marketed by Novartis (Figs. 24 and 25). Their work highlighted the integration of downstream processes such as on-the-fly purification by trituration, filtration, drying and solvent swap in a four-stage continuous flow synthesis of clozapine **47**. The first stage involved copper mediated Ullman/Goldberg coupling between anthranilic acid **48** and 2-bromo-4-chloronitrobenzene **49** in isopropanol in the presence of potassium carbonate under reflux for 8–12 h in batch with an integrated inline workup and purification to afford the desired diaryl product **50** in 80% yield. In the second stage, a pure stock solution of diaryl **50** was pumped through a 2 ml mixing chip at 25°C under novel reductive conditions (sodium dithionite/aqueous ammonia) in 92 s residence time. Isolated yield of 79% for product **51** as a standalone step after on-the-fly trituration precipitated and vacuum dried which was subsequently re-solubilised in anhydrous THF prior to passage into the next stage. The third stage involved intramolecular amide coupling. Pumping **51** through a polymer supported EDC afforded product **52** in 76% isolated yield as a standalone step with 3.5 h residence time. Without any purification, product **52** was pumped into a solvent swapper to enable the use of toluene in last stage which needs to be heated in excess of 100°C in batch. The final step involved a $TiCl_4$ mediated chlorination of **52** followed by immediate treatment with *N*-methyl piperazine **53** to displace the

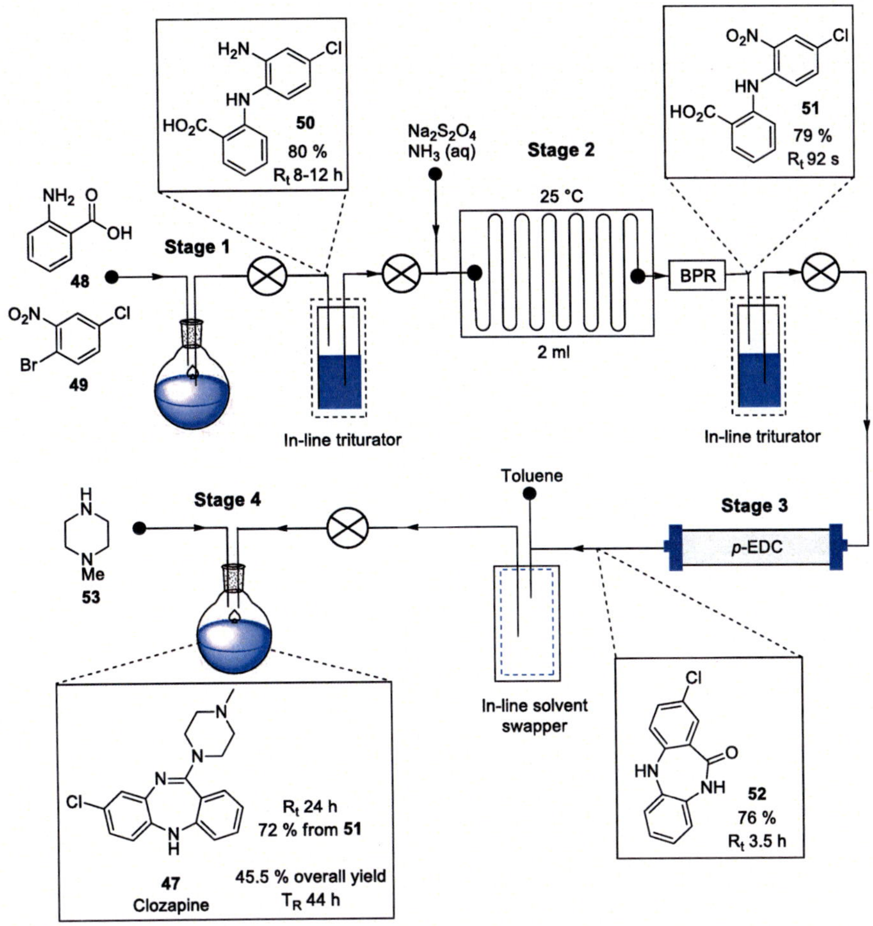

Fig. 25 Batch-flow hybrid synthesis of clozapine **47**

chloride and heated at 110°C for 24 h. Clozapine **47** was afforded in 72% over two stages (3 and 4).

The telescoped hybrid process on 1 g scale afforded clozapine **47** with an overall yield of 45.5% in 44 h, which is better than a purely batch process (27%, 132 h reaction time) [36]. This remarkable improvement can be attributed to the use of flow chemistry technology in the second and third stage of the process further reaffirming the power of the technology. This is an excellent example of batch-flow integration, in line work-up, purification and solvent swap in flow.

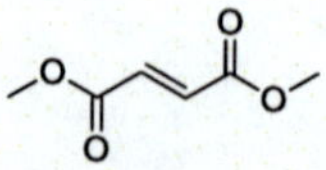

Fig. 26 Structure of dimethyl fumarate

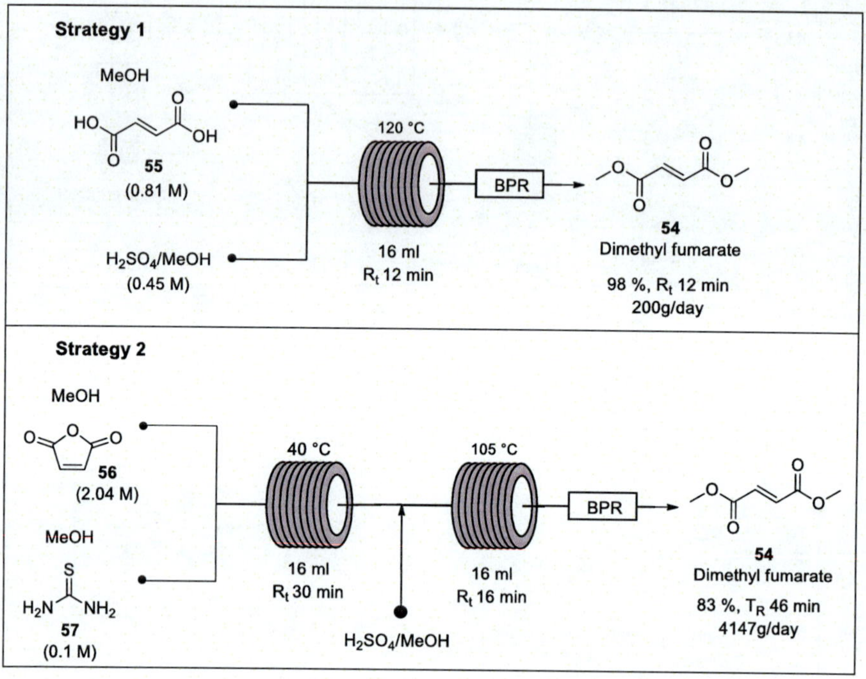

Fig. 27 Continuous flow cascade reactions for dimethyl fumarate **54**

2.11 *Dimethyl Fumarate*

Most recently, de Souza and co-workers [37] reported two continuous flow synthesis strategies of dimethyl fumarate **54** (Figs. 26 and 27), a prodrug used in the treatment of psoriasis and multiple sclerosis. The prices of treating multiple sclerosis are high reaching 75,000 USD/annum for 87 g dimethyl fumarate (862 USD/g) [37] thus inaccessible to most patients in developing economies. The authors demonstrated two strategies towards dimethyl fumarate starting from either fumaric acid or maleic anhydride. In the first strategy, fumaric acid **55** in methanol was treated with methanolic H_2SO_4 at 120°C in a 16 ml PFA coil reactor fitted with a back-pressure regulator to afford dimethyl fumarate **54** in 98% yield with 12 min residence time and space time yield of 200 g/day. The second strategy involved the reaction of

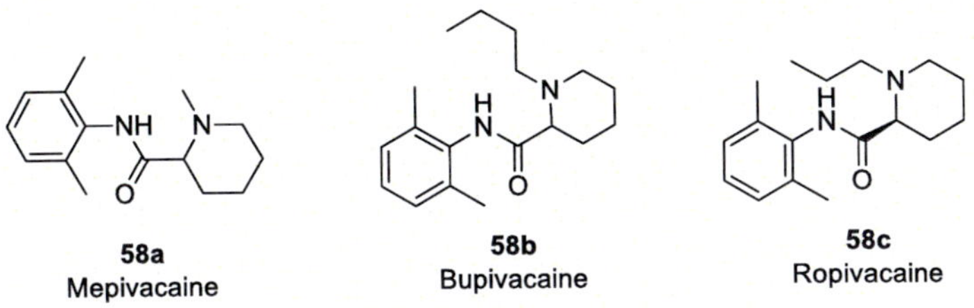

Fig. 28 Structure of mepivacaine and its analogues

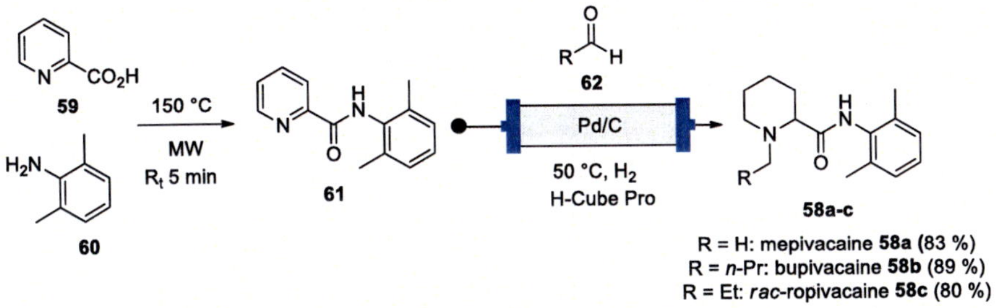

Fig. 29 Continuous flow synthesis of mepivacaine its analogues

maleic anhydride **56** in methanol with thiourea **57** in methanol in a 16 ml PFA coil reactor at 40°C for 30 min. The eluent stream was subsequently treated with methanolic H_2SO_4 at 105°C for 16 min in a 16 ml PFA coil reactor fitted with a back-pressure regulator to afford dimethyl fumarate **54** in 83% with a good space time yield of 4,147 g/day. Accurate heating and system pressurisation was achievable in flow.

2.12 *Mepivacaine and Its Analogues*

Mepivacaine **58a**, bupivacaine **58b** and ropivacaine **58c** are potent local anaesthetics (Fig. 28). Suveges and co-workers [38] in Brazil in collaboration with Kappe group developed a continuous flow tandem hydrogenation/reductive amination strategy towards racemic amide anaesthetics mepivacaine **58a**, bupivacaine **58b** and ropivacaine **58c** (Fig. 29). In the first step, picolinic acid **59** was reacted with α-2,6 xylidine **60** in the presence of PCl_3 in a microwave reactor at 150°C for 5 min to afford 2′,6′-picolinoxylidide **61** in 95% isolated yield. 2′,6′-picolinoxylidide **61** in the presence of aldehyde was subsequently subjected to hydrogenation/reductive amination in a H-Cube Pro™ flow hydrogenator using 10% Pd/C cartridge set at 50 bar of H_2 and 50°C (55°C for bupivacaine **58b**) to afford mepivacaine **58a**, bupivacaine **58b** and ropivacaine **58c** in 83%, 89% and 80% yield, respectively.

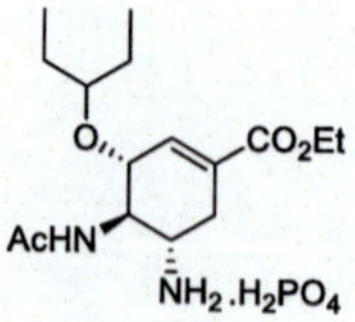

Fig. 30 Structure of oseltamivir phosphate

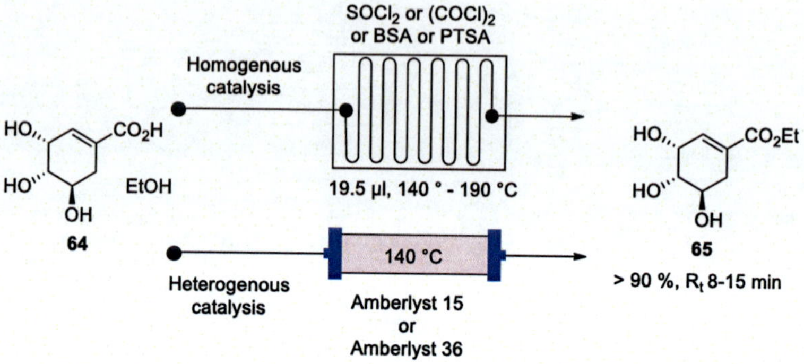

Fig. 31 Continuous flow strategies for shikimic acid esterification

2.13 Oseltamivir Phosphate

Sagandira demonstrated total flow synthesis of an anti-influenza drug, oseltamivir phosphate **63** marketed by Roche (Fig. 30) [25, 39–41]. According to the WHO, this is one of the most important drugs to guard against an influenza pandemic [42]. Their procedure started with shikimic acid **64** esterification with ethanol to afford ethyl shikimic acid **65** in the presence of various esterifying agents. The authors developed eight continuous flow esterification strategies affording ethyl shikimate in >90°C in less than 10 min residence time (Fig. 31) [41].

Ethyl shikimate **65** was then treated with mesyl chloride in the presence of TEA at room temperature in a 0.8 ml PTFE continuous flow coil reactor under sonication to afford the desired mesyl shikimate **66** in 92% isolated yield of *O*-mesylate with 12 s residence time (Fig. 32). Mesyl shikimate **66** was treated with aqueous NaN_3 at 0°C in a 19.5 µl glass reactor to selectively afford the desired azide **67** in 91% isolated yield in 30 s residence time. Other azidating agents such as DPPA, TMSA and TBAA gave comparable results [40]. Azide **67** aziridination with either $(MeO)_3P$ or $(EtO)_3P$ in a 19.5 µl glass reactor at 190°C afforded the desired aziridine **68** in 94% isolated yield with 12 s residence time as a standalone step. Aziridine **68** ring opening was achieved in a 19.5 µl reactor in the presence of 3-pentanol and boron

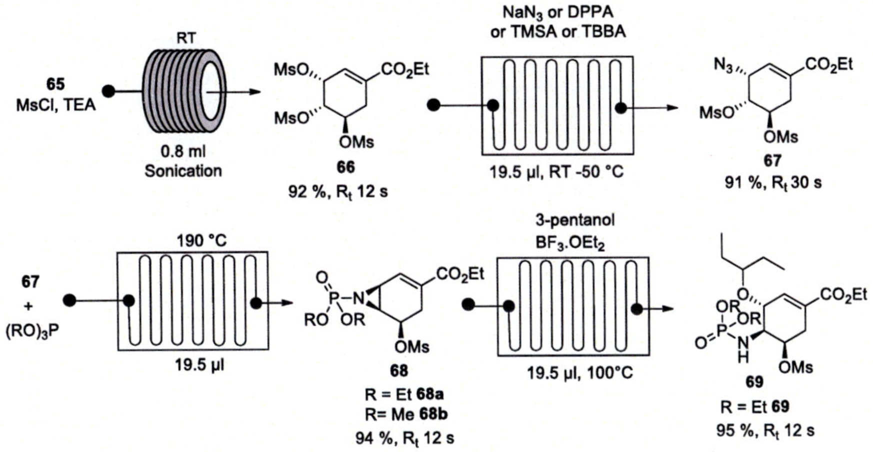

Fig. 32 Continuous flow synthesis towards oseltamivir phosphate [39]

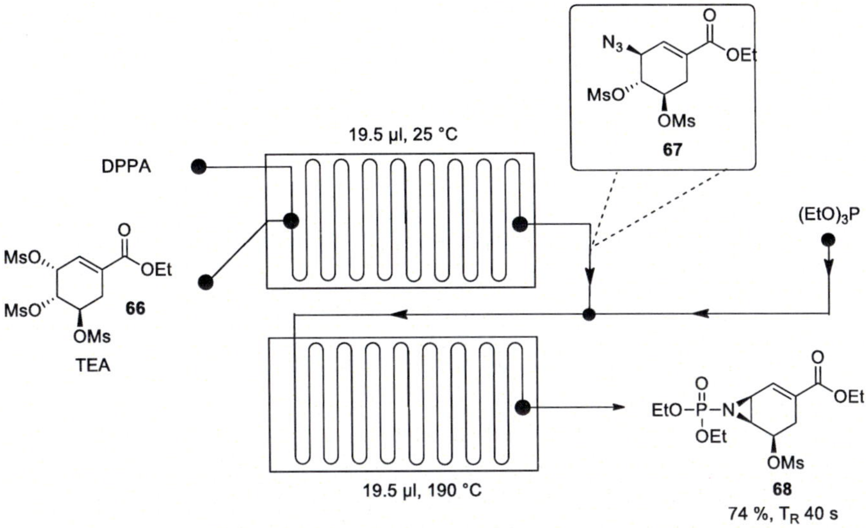

Fig. 33 Two-step flow synthesis of aziridine **68** towards oseltamivir phosphate [39]

trifluoride etherate at 100°C, 12 s residence time to afford 3-pentyl ether **69** in 95% isolated yield (Fig. 32) [39].

Due to azide handling safety concerns, the authors further demonstrated multistep synthesis of aziridine **68** from mesyl shikimate **66** via in situ azide **67** formation using DPPA subsequently followed by aziridination with $(EtO)_3P$ to afford aziridine **68** in 74% yield in 40 s residence time over the two steps (Fig. 33) [39].

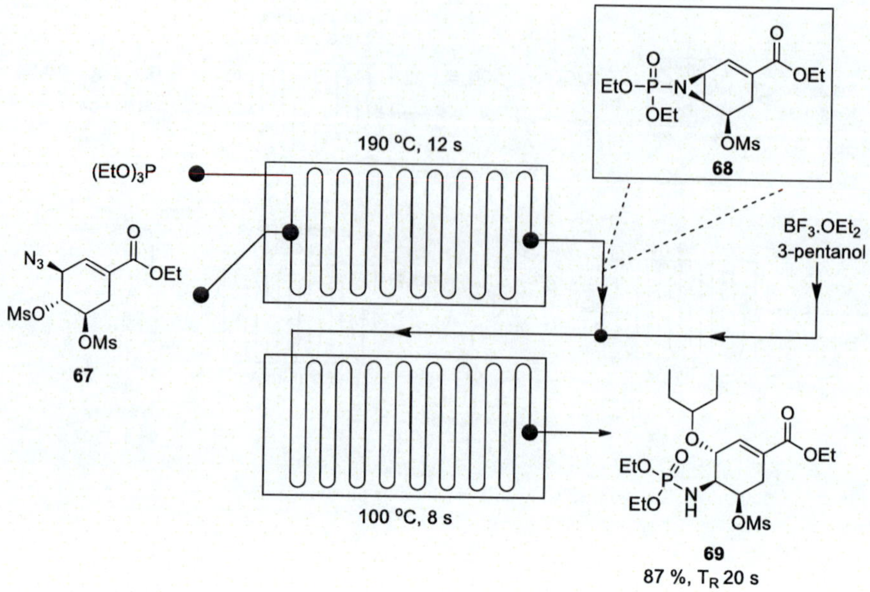

Fig. 34 Two-step flow synthesis of 3-pentyl ether **69** towards oseltamivir phosphate [39]

Telescoped synthesis of 3-pentyl ether **69** from azide **67** via in situ aziridine **68** formation gave 87% yield with 20 s residence time over the two steps (Fig. 34).

Acetamide **71** was formed via a tandem of reactions; N-P bond cleavage of 3-pentyl ether **69** using H_2SO_4 in a 19.5 μl glass reactor at 170°C forming intermediate **70** in situ, subsequently followed by acetylation with Ac_2O in a 19.5 μl glass reactor at room temperature to afford acetamide **71** in 93% isolated yield in 30 s total residence time (Fig. 35). Acetamide **71** azidation was accomplished using NaN_3 in a 19.5 μl glass reactor at 190°C and 45 s residence affording azide **72** 89% isolated yield [40]. Azide **72** reduction was performed at room temperature in a 0.8 ml PTFE continuous flow coil reactor under sonication using $NaBH_4$ in the presence of catalytic $CoCl_2$ affording 93% isolated yield of oseltamivir **73** in 5 s residence time. In the final step, oseltamivir phosphate **63** was afforded in 97% isolated yield by treating oseltamivir **73** with H_3PO_4 in a 0.8 ml PTFE continuous flow coil reactor under sonication at 50°C and 60 s residence time (Fig. 35) [39].

This process elegantly handled the hazardous azide chemistry involved in this procedure. The authors demonstrated various strategies towards oseltamivir phosphate **63** affording 45–58% overall yield with 11.5–18.5 min total residence time starting from shikimic acid and about 3.5 min starting from ethyl shikimate [39, 41]. The overall yield of the process is literature comparable, however, processing time is significantly shorter than all the reported procedures, which are mostly greater than 30 h [43, 44]. Without doubt, this presents a safe, efficient and scalable procedure for the synthesis of the drug.

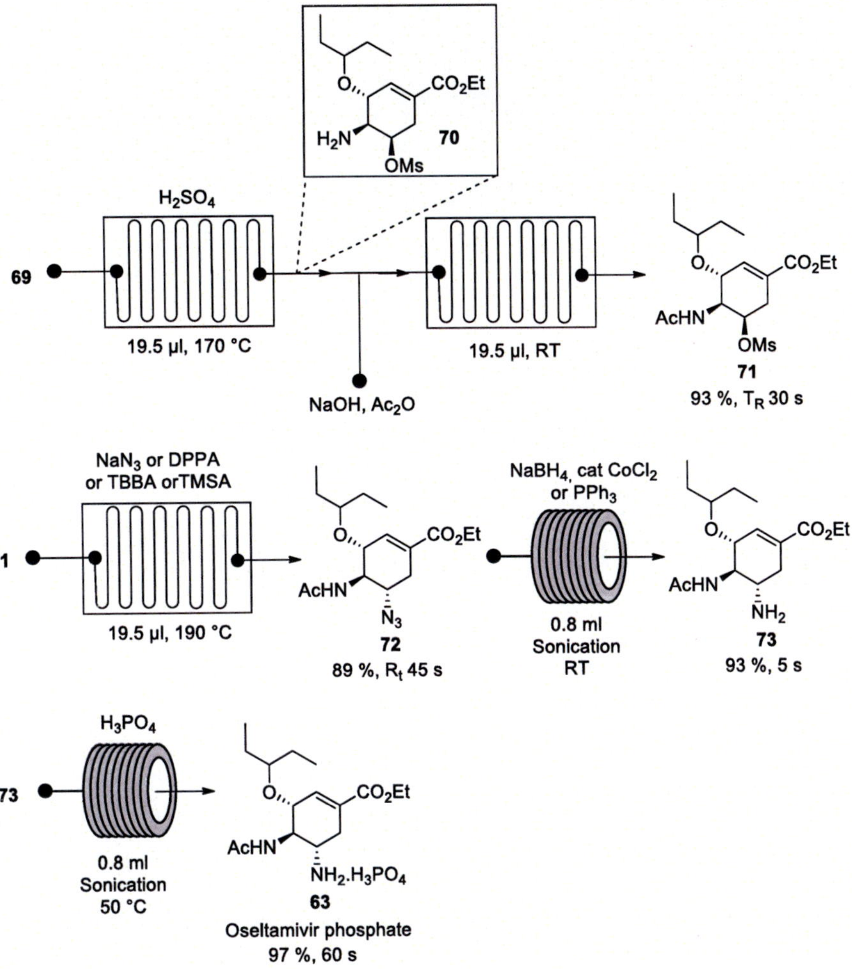

Fig. 35 Continuous flow synthesis towards oseltamivir phosphate [39]

2.14 Nevirapine

Nevirapine **74** was the first commercially available non-nucleoside transcriptase inhibitor (NNRTI) for HIV treatment and it is still a widely prescribed HIV antiviral agent (Fig. 36). 2-Chloro-3-amino-4-picoline (CAPIC) **79** is an important building block for the preparation of nevirapine. Gupton and co-workers [45] demonstrated a continuous flow synthesis procedure of the 2-bromo-4-methylnicotinonitrile **79a**, a CAPIC **79** derivative (Fig. 37). The authors started with Knoevenagel condensation of malononitrile **75** and acetone **76** in the presence of basic Al_2O_3-packed column reactor at 95°C and 0.9 min residence time affording isopropylidenemalononitrile **77** in situ. A column reactor packed with molecular sieves was used to absorb water at

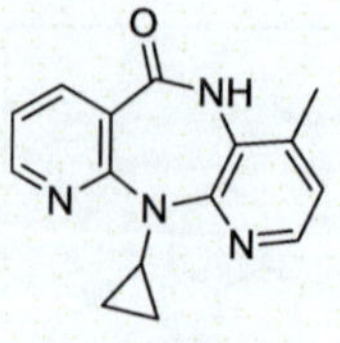

Fig. 36 Structure of nevirapine

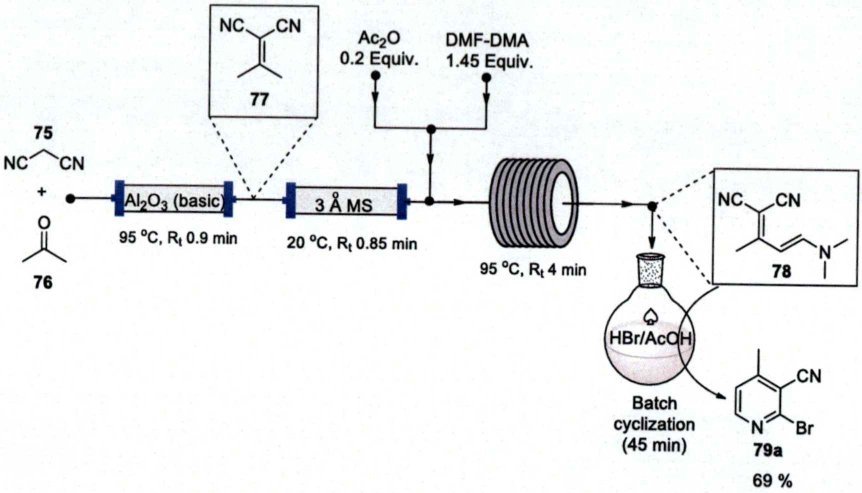

Fig. 37 Continuous flow synthesis of **79a**

20°C and 0.85 min residence time prior to the addition of DMF-DMA in a coil reactor held at 95°C with 4 min residence time to afford enamine **78**. Enamine **78** cyclization in the presence of HBr was performed in batch to afford 2-bromo-4-methylnicotinonitrile **79a** in 69% yield. Previously, the authors had demonstrated enamine **78** cyclization in the presence of HCl afford CAPIC **79** in 81 yield [46].

Gupton and co-workers [47] later reported continuous flow synthesis of nevirapine starting from CAPIC **79** (Fig. 38). CAPIC **79** was treated with NaH in a thin-film reactor at 95°C. The resulting salt was subsequently treated with **80** in a continuous stirred tank to afford intermediate **81** in situ. Intermediate **81** was pumped through a NaH-packed column reactor held at 165°C to afford nevirapine **74** in 92% yield with excellent Volume-Time Output (VTO) of $7.01 \times 10^{-3}\ m^3\ h\ kg^{-1}$. The continuous flow procedure resulted in overall yield improvement from 63% to 92% while reducing the (Process Mass Intensity) PMI value from 46 to 11.

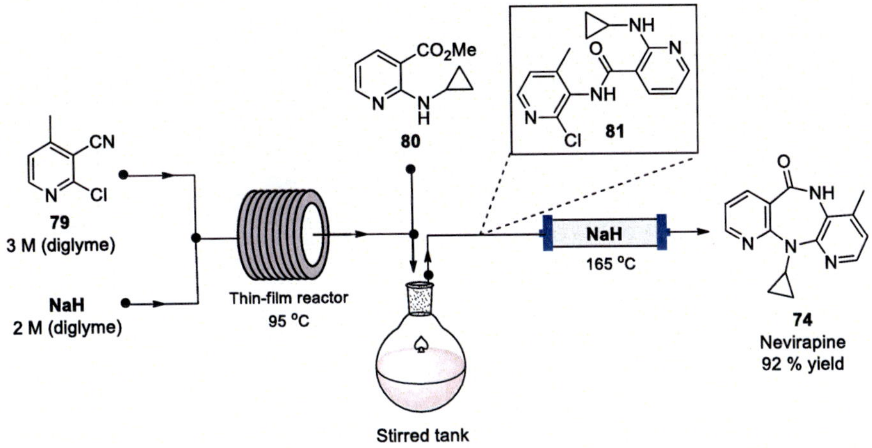

Fig. 38 Continuous flow synthesis of nevirapine **74**

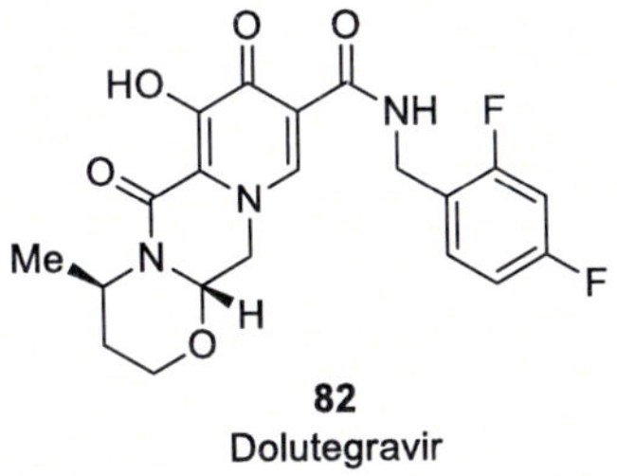

Fig. 39 Structure of Dolutegravir

2.15 *Dolutegravir*

Dolutegravir (DTG) **82** is an integrase inhibitor co-developed by GlaxoSmithKline and Shinogi that was approved by the FDA in 2013 for the treatment of HIV (Fig. 39). Gupton and co-workers [48] demonstrated a 7-step flow total synthesis of Dolutegravir in three separate flow operations (Fig. 40). The first operation is a 3-step telescoped synthesis of pyridone **87** starting from methyl 4-methoxyacetoacetate **83** and DMF-DMA **84** in 56% overall yield with 73.3 min total residence and 3.4 g h^{-1} throughput (Fig. 40, Flow operation 1). The second operation is a 3-step telescoped synthesis of DTG-OMe **90** starting from pure pyridone **87** and difluorobenzylamine **88** in 48% overall yield (7:1 dr) with 190 min total residence time (Fig. 40, Flow operation 2). The major diastereomer was separated by silica gel chromatography affording analytically pure DTG-OMe **90** and was subsequently demethylated in the presence of LiBr in the last flow operation affording Dolutegravir **82** in 89% yield with 31 min residence time from DTG-OMe **90** (Fig. 40, Flow operation 3); 24% overall yield with 4.5 h total

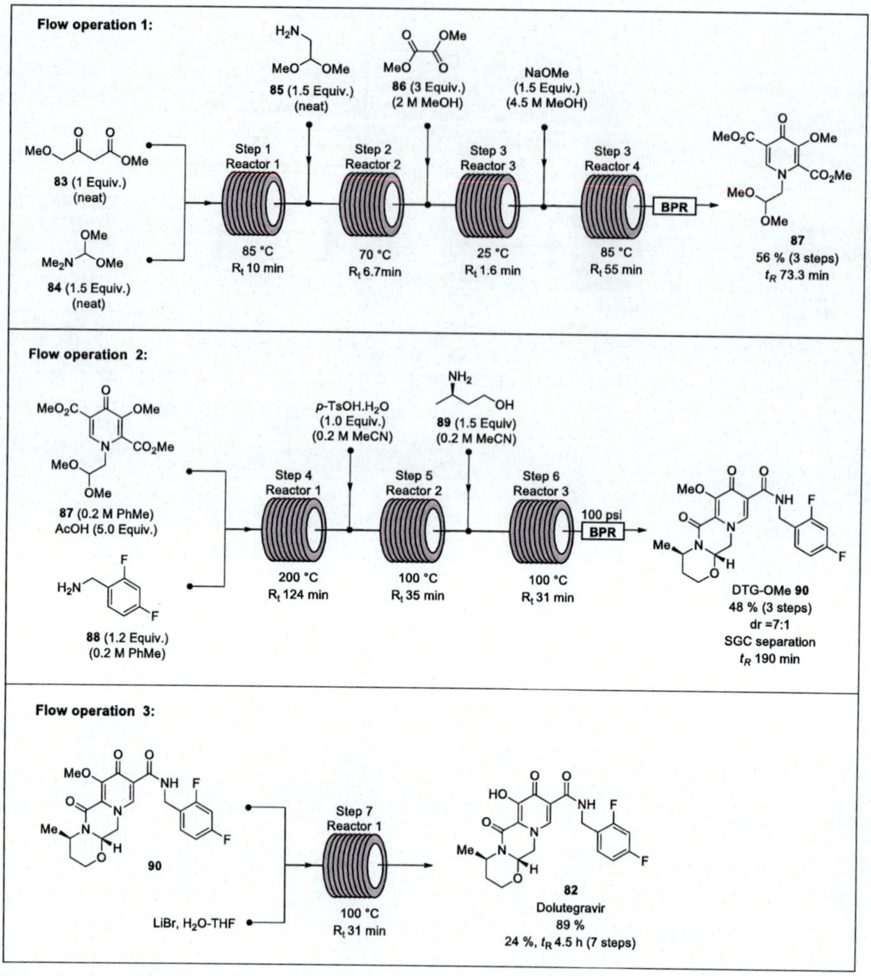

Fig. 40 Continuous flow synthesis of Dolutegravir **82**

residence time over the 7 steps (Fig. 40). The authors demonstrated rapid synthesis, telescoping of multiple steps to avoid intermediates isolation and high throughput synthesis in continuous flow [48]. Furthermore, the synthetic procedure could be adapted to other two important HIV antiviral agents, Cabotegravir and Bictegravir by switching the benzylamine and amino alcohol used in the synthesis.

3 Conclusion and Outlook

Flow chemistry has evolved into a powerful and enabling tool in the pharmaceutical industry owing to its well-documented advantages over the conventional batch process. Although this technology is being implemented in developed economies, its adoption in established pharmaceutical industry is still slow due to its disruptive nature. However, developing economies can take advantage of the limited existing batch manufacturing infrastructure to set up state-of-the-art continuous flow manufacturing infrastructure to enable local pharmaceutical manufacturing as well as support access to drug discovery. By exclusively reviewing the use of flow chemistry in developing economies in the past 5 years, this review exhibited and reaffirmed the advantages of flow chemistry such as reduction in cost, energy requirements and safety concerns in APIs synthesis, which are invaluable in the pharmaceutical industry. This potentially provides paradigm shift in developing countries' pharmaceutical industry that could lead better drug discovery access as well as enabling local drug manufacturing with the consequence of improving access to medicines.

Compliance with Ethical Standards

Funding: We thank the National Research Foundation and Nelson Mandela University for their financial support (Grant number – 85103).

Informed Consent: No patients were studied in this chapter.

Ethical Approval: This Chapter does not contain any studies with human participants or animals performed by any of the authors.

References

1. Department for International Development, UK Govenment Policy Paper, Increasing people's access to essential medicines in developing countries: a framework for good practice in the pharmaceutical industry, https://apps.who.int/medicinedocs/en/m/abstract/Js18384en/. Accessed 8 Apr 2020
2. Lewinberg A (2015) The pharmaceutical industry in the developing world: generic and branded. http://individual.utoronto.ca/adamlewinberg/Access/Box5.htm
3. Tannoury M, Attieh Z (2017) Curr Ther Res 86:19–22
4. Boutayeb A (2010) Preedy VR, Watson RR (eds) Handbook of disease burdens and quality of life measures. Springer, New York, pp 531–546
5. Bate R (2008) Local pharmaceutical production in developing countries. Campaign for fighting diseases, https://www.unido.org/sites/default/files/2016-01/Local_Pharmaceutical_Production_web_0.pdf. Accessed 8 Apr 2020
6. Riley DL, Strydom I, Chikwamba R, Panayides JL (2019) React Chem Eng 4:457–489
7. UNIDO (2018) Pharmaceutical production in developing countries: a pathway to health and industrialization, (UNIDO). https://www.unido.org/our-focus-advancing-economic-competitiveness-investing-technology-and-innovation-competitiveness-business-environment-and-upgrading/pharmaceutical-production-developing-countries. Accessed 8 Apr 2020

8. WHO (2011) Investment in pharmaceutical production in the least developed countries. A Guide for Policy Makers and Investment Promotion Agencies (United Nations Conference on Trade and Development), https://apps.who.int/medicinedocs/en/m/abstract/Js18660en/. Accessed 8 April 2020
9. WHO (2018) WHO methods and data sources for global burden of disease estimates 2000-2016 (WHO), https://www.who.int/healthinfo/global_burden_disease/estimates/en/index1.html. Accessed 8 Apr 2020
10. Stevens H, Huys I (2017) Front Med 4:1–6
11. Luchetti M (2014) BJMP 7(4):a731
12. de Souza ROMA, Watts P (2017) J Flow Chem 7:146–150
13. Hahn SM (2020) Coronavirus (COVID-19) supply chain update. (U.S. FDA) https://www.fda.gov/news-events/press-announcements/coronavirus-covid-19-supply-chain-update, Accessed 8 Apr 2020
14. Mohammed AJ (2016) Agenda 2063: the Africa we want. (AU), https://au.int/en/agenda2063/overview. Accessed 8
15. AU (2012) Pharmaceutical manufacturing plan for Africa: business plan. (WHO). https://apps.who.int/medicinedocs/en/m/abstract/Js20186en/. Accessed 8 Apr 2020
16. AU, Fund for African Pharmaceutical Development Meeting, Cairo, Egypt. 7–9 August 2018, https://au.int/en/newsevents/20180807/fund-african-pharmaceutical-development-fap-d-meeting-7-9-august-2018-cairo. Accessed 8 Apr 2020
17. AU, Strategic Stakeholders Retreat on Industry, Nairobi, Kenya. 19, February 2018. https://au.int/sites/default/files/pressreleases/33849-pr-pr_028_-_7th_strategic_stakeholders_retreat_on_industry.pdf. Accessed 8 Apr 2020
18. Baraldi PT, Hessel V (2012) Green Process Synth 1:149–167
19. Hughes DL (2018) Org Process Res Dev 22:13–20
20. Akwi FM, Watts P (2018) Chem Commun 54:13894–13928
21. Fitzpatrick DE, Ley SV (2016) React Chem Eng 1:629–635
22. Fanelli F, Parisi G, Degennaro L, Luisi R (2017) Beilstein J Org Chem 13:520–542
23. Mandala D, Watts P (2017) ChemistrySelect 2:1102–1105
24. Mandala D, Chada S, Watts P (2017) Org Biomol Chem 15:3444–3454
25. Sagandira CR, Moyo M, Watts P (2020) Arkivoc 2020:29–34
26. Mandala D, Thompson WA, Watts P (2016) Tetrahedron 72:3389–3420
27. Suveges NS, Rodriguez AA, Diederichs CC, de Souza SP, Leão RAC, Miranda LSM, Horta BAC, Pedraza SF, de Carvalho OV, Pais KC, Terra JHC, de Souza ROMA (2018) Eur J Org Chem 23:2931–2938
28. Chada S, Mandala D, Watts P (2017) J Flow Chem 7:37–40
29. Correia CA, Gilmore K, McQuade DT, Seeberger PH (2015) Angew Chem Int Ed 54:4945–4948
30. Chada S, Watts P. US 2020/0062722 A1, 27 Feb 2020
31. Leão RAC, De Lopes RO, De Bezerra MAM, Muniz MN, Casanova BB, Gnoatto SCB, Gosmann G, Kocsis L, De Souza ROMA, De Miranda LSM (2015) J Flow Chem 5:216–219
32. Mangwiro R, Watts P. A continuous flow synthesis method for the manufacture of isoniazid. UK Patent No. 1914685.1
33. Aguiar RM, Leão RAC, Mata A, Cantillo D, Kappe CO, Miranda LSM, De Souza ROMA (2019) Org Biomol Chem 17:1552–1557
34. Miranda LSDM, De Souza ROMA, Leaõ RAC, Carneiro PF, Pedraza SF, De Carvalho OV, De Souza SP, Neves RV (2019) Org Process Res Dev 23:2516–2520
35. Carneiro PF, Gutmann B, De Souza ROMA, Kappe CO (2015) ACS Sustain Chem Eng 3:3445–3453
36. Neyt NC, Riley DL (2018) React Chem Eng 3:17–24
37. Lima MT, Finelli FG, De Oliveira AVB, Kartnaller V, Cajaiba JF, Leaõ RAC, De Souza ROMA (2020) RSC Adv 10:2490–2494
38. Suveges NS, de Souza ROMA, Gutmann B, Kappe CO (2017) Eur J Org Chem:6511–6517

39. Sagandira CR, Watts P (2020) Synlett 31(19):1925–1929
40. Sagandira CR, Watts P (2019) Beilstein J Org Chem 15:2577–2589
41. Sagandira CR, Watts P (2019) J Flow Chem 9:79–87
42. WHO (2018) New guidance on clinical management of influenza infections. https://www.who.int/influenza/resources/documents/clinical_management_2012/en/. Accessed 8 Apr 2020
43. Magano J (2011) Tetrahedron 67:7875–7899
44. Magano J (2009) Chem Rev 109:4398–4438
45. Longstreet AR, Opalka SM, Campbell BS, Gupton BF, McQuade DT (2013) Beilstein J Org Chem 9:2570–2578
46. Longstreet AR, Campbell BS, Gupton BF, McQuade DT (2013) Org Lett 15:5298–5301
47. Verghese J, Kong CJ, Rivalti D, Yu EC, Krack R, Alcázar J, Manley JB, McQuade DT, Ahmad S, Belecki K, Gupton BF (2017) Green Chem 19:2986–2991
48. Ziegler RE, Desai BK, Jee JA, Gupton BF, Roper TD, Jamison TF (2018) Angew Chem Int Ed Engl 57(24):7181–7185

Top Med Chem (2021) 38: 421–480
https://doi.org/10.1007/7355_2021_135

Published online: 10 September 2021

Drug Discovery Automation and Library Synthesis in Flow

Paul Richardson and Irini Abdiaj

Contents

Abstract The spiraling costs, competitive nature, and stringent timelines associated with Drug Discovery fuel investigations into new technologies that can potentially alleviate the pressures associated with these factors. Whereas advantages of the implementation of Flow Chemistry in the Development phase of a campaign appear

---

P. Richardson (✉)
Pfizer Medicine Design, Pfizer, La Jolla, CA, USA
e-mail: Paul.F.Richardson@pfizer.com

I. Abdiaj
Janssen Research and Development, Toledo, Spain
e-mail: iabdiaj@its.jnj.com

obvious specifically toward the large-scale synthesis of the molecule of interest, in early Discovery it is often harder to justify the time to investigate/develop and validate enabling technologies particularly given the fact that there may be no near-term tangible return on this investment. The current chapter takes a detailed look at several case studies on innovative flow-based technologies developed to expedite the Drug Discovery process and evaluates the overall advantages/disadvantages of each approach as well as their overall sustainability in terms of potential uptake within the industry.

Keywords Closed-loop discovery, Drug discovery, Flow chemistry, High throughput experimentation, Parallel medicinal chemistry, Photoredox catalysis, Radial synthesis, SAR optimization, Segmented-flow, Suzuki-Miyaura coupling

Abbreviations

$[Ir\{dF(CF_3)ppy\}_2(dtbbpy)]PF_6$	[4,4′-Bis(1,1-dimethylethyl)-2,2′-bipyridine-N1, N1′]bis[3,5-difluoro-2-[5-(trifluoromethyl)-2-pyridinyl-N]phenyl-C]Iridium(III) hexafluorophosphate
AI	Artificial intelligence
Amphos	Di-*tert*-butyl(4-dimethylaminophenyl)phosphine
API	Active pharmaceutical ingredient
BCR-Abl	Breakpoint cluster region Abelson tyrosine kinase
Bpin	Pinacolato boron
BPR	Back pressure regulation
CataCXium-A	Di(1-adamantyl)-*n*-butylphosphine
CML	Chronic myeloid leukemia
CSS	Central switching station (reactor and inline analysis)
CV	Collection vessels
DABCO	1,4-Diazabicyclo[2.2.2]octane
DAD	Diode Array detector
DBU	1,8-Diazabicyclo[5.4.0]undec-7-ene
$(dF(CF_3)ppy)$	2-(2,4-difluorophenyl)-5-(trifluoromethyl) pyridine
DMF	*N,N*-dimethylformamide
DMSO	Dimethyl sulfoxide
DoE	Design of experiments
DPP4	Dipeptidyl peptidase-4
dppf	1,1′-Bis(diphenylphosphino)ferrocene
dpy	Dipyridine
dtbbpy	4,4′-Di-*tert*-butyl-2,2′-dipyridyl
dtbpf	Bis(di-*tert*-butylphosphino)ferrocene
ELSD	Evaporative light scattering detector
ESI	Electrospray ionization

GC-MS	Gas chromatography-mass spectrometry
1,2-dimethoxyethane	Glycol ether
H-Gly-Pro-AMC	(*S*)-*N*-(2-Aminoacetyl)-1-(4-methyl-2-oxo-2*H*-chromen-7-yl)pyrrolidine-2-carboxamide
HTE	High-throughput experimentation
I.D.	Inner diameter
IC_{50}	Half maximal inhibitory concentration
MFC	Mass flow controller
MISER	Multiple injections in a single experimental run
ML	Machine learning
MS	Mass spectrometer
MW	Molecular weight
NMP	*N*-methylpyrrolidone
PFA	Perfluoroalkoxyalkane
PMC	Parallel medicinal chemistry
PTFE	Polytetrafluoroethylene
QC	Quality control
RDS	Reagent delivery system
RSPU	Reaction segment preparation unit
SAR	Structure-activity relationships
SFC	Supercritical fluid chromatography
SM	Standby module
S-Phos	2-Dicyclohexylphosphino-2′,6′-dimethoxybiphenyl
TPSA	Topological polar surface area
UIB	Universal Interface box
UPLC	Ultra high performance liquid chromatography
Xantphos	9,9-Dimethyl-9*H*-xanthene-4,5-diyl)bis(diphenylphosphane
Xphos	2-Dicyclohexylphosphino-2′,4′,6′-triisopropylbiphenyl

1 Introduction

There are numerous recent reviews highlighting the advantages of flow chemistry with several specifically focused on the potential benefits to be realized within drug discovery [1–6]. While core benefits of the technology are well recognized (for example, automation, speed, reproducibility, enhanced safety, smaller footprint), it can sometimes be challenging to see how these can be an immediate advantage to a conventional medicinal chemist [7]. Whereas, as we will see, there are multiple impressive flow chemistry-based technology platforms in current use to facilitate medicinal chemistry programs (for example, library synthesis [8], reaction screening), it is important to bear in mind that the development (conceive/build/test) of these requires a relatively significant investment in terms of cost/resource and time

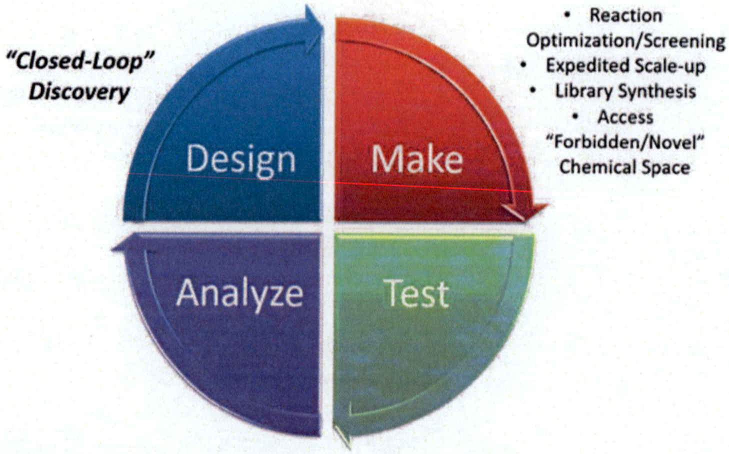

Fig. 1 Conventional "Design, Make, Test, Analyze" cycle highlighting tasks that flow chemistry approaches can potentially impact

with the realization that the payback for this will be provided through their continued utility (naturally with minor tweaks in terms of enhancements) typically by dedicated specialist users over time on multiple medicinal chemistry programs [9–11]. However, in the regular classical discovery research laboratory, it appears that except for several specific applications (for example, flow-based hydrogenation), the uptake of flow chemistry on a routine day-to-day basis has been somewhat limited [12].

At the outset, it is instructive to consider the reasons behind this trend with several possible causes. The first to consider is somewhat counterintuitive particularly considering that it is often cited as one of the main drivers for incorporation of flow within early-stage medicinal chemistry programs, and that is increased speed and efficiency. As has been well documented, in a conventional setting, drug discovery programs operate with a "design-make-test-analyze" based cycle, with the speed at which a team can move through each iterative sequence often being defined as a quantitative measure of project progress (Fig. 1) [13]. Whereas clearly, the situation is not as straightforward as this, one can distill out the baseline fact that the "faster one can synthesize-relevant drug-like compounds" is likely to determine a go/no-go decision for a specific project. With this statement, it is important to also highlight the importance of a team reaching a "no-go" decision. While clearly not the optimal outcome, given the established high costs associated with drug discovery in general, being able to rapidly reach a judicious decision regarding a program long-term feasibility for success prior to it progressing to development (or the clinic) represents an opportunity to dedicate resources (time/cost) elsewhere.

Playing devil's advocate as to why implementation of flow might not lead to an increase in terms of compound synthesis speed and subsequent cycle time leads to an examination of the evolving role of the conventional medicinal chemist within drug discovery [14]. The first consideration is around the mindset of how chemistry is traditionally done with the desired analogues typically being processed in a singleton (one by one discretely) fashion in a batch-wise manner using conventional glassware. Scale also plays a role here with initial compound amounts required for first tier assays only being ~10–20 mg. Given this, and the fact that many medicinal chemists are unfamiliar with the techniques of flow chemistry, the path of least resistance (and probably the most expeditious approach) for making a new analogue will be through a traditional synthetic batch-type approach. Furthermore, the commercially available flow chemistry equipment is typically not amenable to working on relatively small scales (10–50 mg) without a degree of expertise/familiarity.

Despite this, there is no doubt that uptake of flow chemistry will continue to increase within medicinal chemistry laboratories as scientists emerging from academia now have a greater appreciation of the concepts and experimental skills required in this space. Over the past 10–15 years, this uptake has been evidenced particularly in the case of flow hydrogenation wherein the use of the H-Cube is now widely adopted to safely execute this chemistry in a discovery-type setting. The current chapter will focus on several technologies with the potential to "move the needle" more significantly regarding the "Design-Make-Test-Analyze" cycle. The high-throughput and automated nature of continuous approaches are implemented to provide "industrial-scale" solutions to discovery chemistry problems. For example, we will consider chemistry/screening optimization approaches that evaluate 1000s of reaction conditions on nanomole scale. This methodology is further expanded to specifically examine photoredox chemistry (access "novel" chemical space) using a droplet-based approach. Furthermore, the development of a novel-modular-based reaction platform for the multistep synthesis of complex biologically active compounds in parallel is described. Finally, the "total package" discovery/compound optimization loop developed by Cyclofluidics is teased out. It should be noted that there are many other similar endeavors in the literature describing optimization/library/synthesis platforms. The decision made here to discuss a number of them in a level of detail was carried out with the intention of addressing the key question as to whether or not they "either have or will provide a suitable return on the resource-investment" required to develop the technology.

2 Nanomole Reaction-Screening and Micromole-Scale Synthesis in Flow

2.1 Introduction

Reaction optimization, through a sequential investigation of the various parameters involved in the process, has long been recognized as a critical endeavor undertaken in process chemistry for the development of a compound. In this environment, the route to access a compound of interest is likely to be well established thus enabling a focused approach on the key problematic bond-forming transformations. In this setting as well, utilization of DoE (Design of Experiments) approaches evaluating the continuous variables (temperature, time, concentration, stoichiometry, etc.) of a reaction is often implemented to define synergistic interactions between these variables, the optimal operating conditions, and also critically the overall "robustness" of the process, namely how the limits in tolerance that can be afforded for changes in each variable before the reaction will "fail" [15]. Further advantages of reaction screening in the development setting are that material availability for testing is not an issue and markers of both the product and usually potential impurities/by-products from the process under evaluation will exist thus expediting the identification of suitable analytical methods for reaction monitoring. One common misconception though within development is that there is ample time to allow extended reaction optimizations to be conducted through sequential rounds of screening. This is typically not the case as significant time pressures exist with compounds of interest at this point being on a critical path in terms of timelines with various supplies required for defined toxicity, regulatory, and clinical studies.

The argument for evaluating and optimizing chemistries within discovery has both pros and cons. From one perspective, a breakthrough in identifying an innovative approach to a lead compound can be adopted and streamlined from an early stage and can in principle expedite the future delivery of batches of material. Furthermore, value can also be derived from a negative outcome from a reaction screen as evaluation of a broad parameter space for a speculative bond-forming reaction leading to no positive results can eliminate the need to further explore the approach if the molecule of interest should progress into development. However, this latter statement also encompasses the crux of the whole debate around at what point does one embark on reaction screening/optimization campaigns?

In the discovery space, molecule attrition is a critical consideration with most compounds evaluated not progressing. A second factor linked to this is not only the number of molecules being examined but also the goal for the route being developed to access them. Whereas in development, the focus is usually on the best route to make a specific single molecule, in discovery the desired methodology is likely to focus on a series of molecules with different approaches being examined simultaneously to vary different vectors from a central scaffold. An extension of this is that the chemistry being evaluated may be intended for the construction of chemical libraries, and with this approach in mind, a wide substrate scope takes precedent over optimal yields being obtained for a single congener within a chemical series. This places a further onus on effective analytics with the requirement of searching for multiple products in a screen readout. One area in which high-throughput experimentation (HTE) has enjoyed an exponential growth is late-stage functionalization in which chemistries are evaluated for their ability to modulate the properties of lead molecules through the addition of small fragments (methyl and fluorine proving particularly popular) [16]. An established application of this is through metabolism studies in which "hotspots" are identified through subjecting a compound to chemical or biocatalytic oxidative conditions [17]. A note of caution though within this latter application is that a further complexity is introduced from the analytical standpoint as while identifying suitable conditions that enable the "desired transformation" (oxidation), often multiple sites exist within the molecule of interest as to where this transformation can occur, and to rigorously characterize the product, often a scale-up, isolation, and structural elucidation is carried out thus exponentially increasing the workload required.

The key question as ever here is what time and resources are available, and how technology can be exploited to ensure a seamless workflow that minimizes bottlenecks within the screening process. Over the past decade, reaction screening as an enabling tool has assumed a pivotal role within many research organizations both within academia and industry with a wealth of resources documenting best practices available to facilitate further uptake taking into account potential differing levels of capital investment available to achieve this. Over time, the evolution of chemical technologies and the design of the accompanying equipment/tools have allowed more reactions to be examined per day whilst driving the possible reaction scale-down. This is an important consideration within discovery as one would like to evaluate as much chemical space (discrete variables such as reagents being more important as one often wants the reaction to work with the view that it can be optimized later if needed or an alternative approach devised to make a compound) as possible whilst using the least amount of starting material. Chemical space for a reaction is broader as aspects of cost of a specific reagent or the ability to use for example a proprietary ligand are of less impact within an early-stage discovery environment. The ability to screen on milligram scale was a breakthrough enabling extension into the discovery space though often no markers for the desired product exist and rapid analytical methods are essential to allow rapid turnaround of a specific screen. Over time, implementing a screening paradigm within a research group has a further benefit, which is only now starting to be realized with the utilization of the vast amounts of reaction data that is accumulated over time and typically curated in a standard fashion. As with all service-based technologies of value, the risk is that submissions will rapidly exceed the capacity limits (time/resource/equipment) and so with large amounts of data in hand, it can be envisioned that this can be used to provide a database of "best standard" conditions for a specific transformation (to test before submission for screening) and/or incorporated into a "training set" for a model that can be built to predict potential conditions for a novel transformation.

The best way to highlight several themes emerging from reaction screening can be through consideration of the following literature-based example. Herein, we will see both how the desired extent of chemical space to be covered can rapidly expand whilst highlighting the benefits of reaction screening in identifying the synergistic interactions of various parameters, which would more than likely be missed in a "one-by-one" sequential experimental optimization mode. The work of Doyle and MacMillan explores an alternative paradigm for cross-coupling through the combination of photoredox- and nickel-catalysis [18]. This method is particularly attractive to the medicinal chemistry community as it not only utilizes readily available and stable carboxylic acids as coupling partners but is also well-suited for the formation of sp^3-sp^2-bonds thus increasing the "three-dimensional" shape of the targets of interest [19]. The model system initially disclosed featured the coupling between *N*-Boc proline (the use of amino-acids as substrates is a further benefit of this methodology) with iodotoluene with the combination of $Ir[df(CF_3)ppy]_2(dtbbpy)PF_6$ and NiC_{12}.glyme (glycol ether), dtbbpy as the catalyst system. With DMF as solvent, and Cs_2CO_3 as the base, under the irradiation of white light achieved using a 26-W compact fluorescent bulb, the desired product was obtained in 78% yield. One important aspect of this work was the ability to utilize a bench-stable Ni(II) salt to mediate the reaction with comparable efficiency to a Ni(0) source with the reduction to the active species taking place through a photocatalytic process. The use of such a species facilitates the screening of alternative ligands for this reaction. While the original report provides several examples of heterocyclic halides being effective coupling partners, there are no reports of 5-membered rings being exemplified, and as a model reaction, the coupling of the protected 4-iodopyrazole **1** was evaluated through conventional reaction screening and serves as an example to illustrate how rapidly the scope of the chemical space can expand in such an optimization campaign (Fig. 2).

Two series of screening experiments were conducted in tandem to identify the best base or ligand for the coupling process. With the amidine-derivative held constant, a series of 86 bases were evaluated, while 76 ligands were then examined with DBU employed as the base. Herein, a total of 162 experiments have been carried out with the results highlighting the pitfalls associated with an approach focused on optimizing one parameter at a time. Although each set of reactions demonstrated a local optimum, these experiments fail to consider the importance of the interactions between the discrete variables thus emphasizing the exponential value of multiparameter optimization-based approaches in identifying the best possible conditions for a transformation. Also, of note here is that the potential effect of solvent (arguably the most important variable for a reaction) has been essentially ignored. To carry out the full matrix of base/ligand combinations in five solvent systems would entail 32,680 experiments being run. Within a conventional screening paradigm, this inevitably would be impractical based on both material- (particularly in a discovery setting) and resource-requirements leading to investigations into alternative approaches to high-throughput reaction optimization focusing on leveraging automation as well as miniaturization of scale.

As noted in the earliest phase of a medicinal chemistry program, availability of materials is often a limiting factor with a focus on the utilization of advanced intermediates for the synthesis of key analogues (often through sub-optimal chemistries) prioritized over route-scouting or HTE screening. Interestingly when balancing the resource perspectives involved, the onus placed on expert purification in accessing 1–2 analogues using poor chemistries is often overlooked against the efforts put into reaction optimization screening which potentially long-term could facilitate access to multiple compounds for consideration in an expeditious manner. The other pivotal factor to consider for screening within discovery is that the workflows must be efficient yielding results in a matter of days as prolonged optimization campaigns lead to longer overall cycle times thus diminishing the value of the endeavor. So, with limited material, and a desire to increase the coverage of chemical space, it is necessary to decrease the scale of experimentation. The advent of advanced liquid handling technologies and importantly the ability to

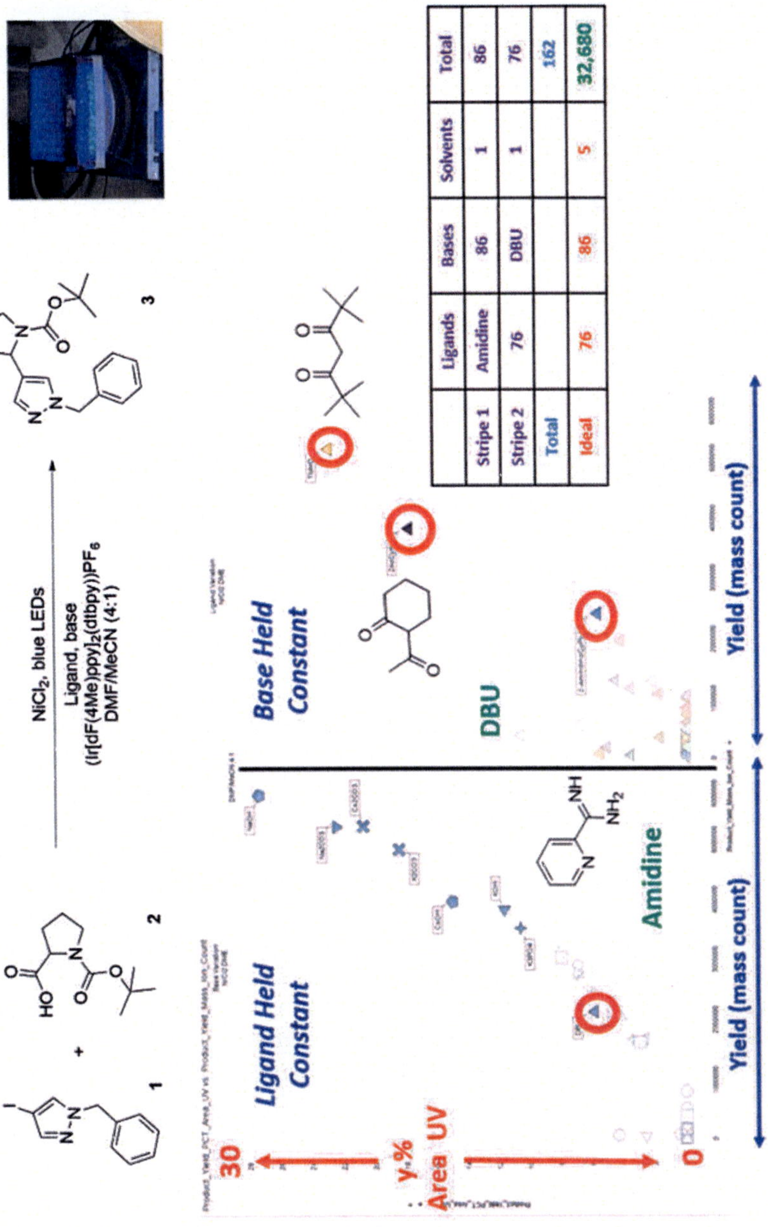

	Ligands	Bases	Solvents	Total
Stripe 1	Amidine	86	1	86
Stripe 2	76	DBU	1	76
Total				162
Ideal	76	86	5	32,680

Fig. 2 Discrete ligand- and base-screening for a Ni-catalyzed photoredox coupling

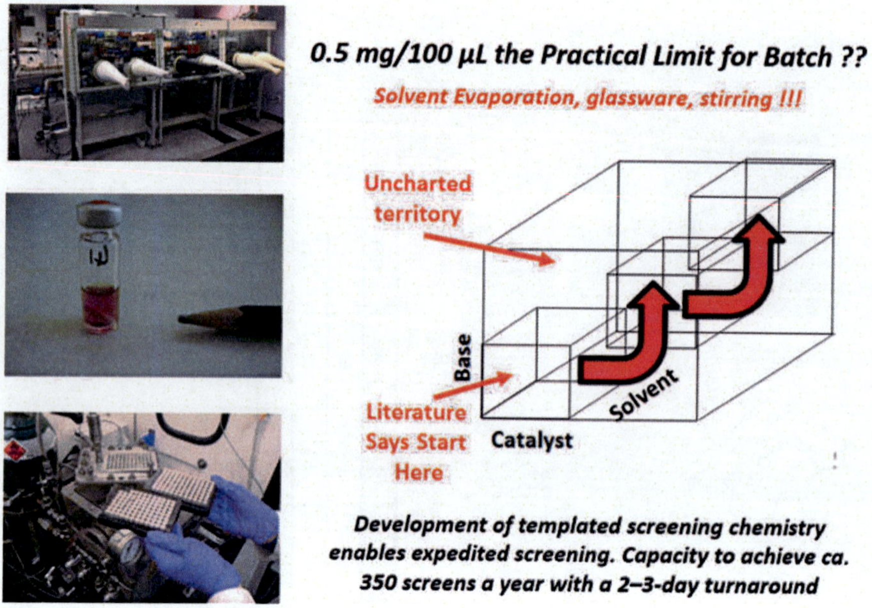

Fig. 3 Practical limitations of batch HTE

rigorously exclude air and moisture in glove-box environments have enabled robust reaction screens to be carried out using >1 mg quantities of material at 50–100 μL volume per reaction (Fig. 3). It has therefore become routine to run screens of hundreds of reactions, which in combination with high-throughput analytics such as UPLC/MS achieve the desired rapid turnaround from design all the way through, to results in 2–3 days, thus drastically altering the cost/benefit analysis of prospective reaction screening. Further efficiencies can be gained in this space through the development of templated screens for common chemistries (Suzuki coupling, amide-bond formation) with extensions of this approach leading to the preparation of pre-prepared reagent screening plates for distribution to the wider synthetic chemistry group as needed though in the latter case, considerations should be made in terms of the necessary analytics to process the outcomes (equipment, level of expertise required) [20]. Overall, the approach to screening described herein expands the potential structural diversity of investigated compounds and will save both time and resource downstream in the development phase if efficient chemistries discovered/optimized through HTE are already in place.

At this juncture, it appears that we have reached the practical limit for reaction screening in batch, as there are engineering challenges associated with achieving further reductions in scale. Many of these are substantial and potentially insurmountable such as the ability to accurately bring together extremely small charges of materials that are often heterogeneous, effective agitation of reaction mixtures, prevention of loss of volatile solvents, suitable vials/plates for the discrete reactions, and the incorporation of general analytical approaches to assay reaction outcomes.

2.2 Nanomole-HTE in Batch

Scientists at Merck recognized the limitations of the conventional batch reaction screening and proposed an approach to enable reaction optimization campaigns to be conducted at nanomole scale through exploitation of equipment and technologies commonly used in the development and execution of biological assays [21]. The initial report of this work demonstrated optimization of a Pd-mediated Buchwald-Hartwig coupling using drug-like molecules evaluating 1,536 reactions in 2.5 h using as little as 0.02 mg per reaction with 20 nL volumes.

The decision to focus on C-N bond formation was based not only on the widespread use of this reaction within medicinal chemistry but also on the high failure rate of such reactions when applied particularly in the late stage of syntheses with highly functionalized polar medicinally relevant substrates (for example **4**, **5**). The failure of such compounds specifically within broadly-used catalytic manifolds notably can when applied in parallel-medicinal chemistry campaigns (so-called libraries) serve to enrich screening sets in molecules (based on the substrates that work) with poorly suited physicochemical properties to become successful drug candidates.

The challenge regarding the volatility of the solvents was proposed to be overcome by running the reactions at ambient temperature and to achieve this, DMSO (similar solvents like DMF and NMP were also considered) was selected as the solvent owing to its low volatility, and its ability to effectively solubilize a wide range of substrates. In addition, many biological assays are conducted using DMSO-stock solutions thus providing a potential link between the chemistry and biology platforms. However, DMSO is not a conventional solvent for Pd-catalyzed C-N coupling owing to both its incompatibilities with the strong bases typically used in these reactions as well as issues with solvent-metal coordination preventing the formation of the required Pd-ligand complexes. To address these challenges, it was proposed to use the more recently developed highly hindered electron-rich ligands (e.g., **7**) to effectively protect the Pd from DMSO coordination while employing non-nucleophilic organic superbases (e.g., **8**), which while both soluble and compatible with DMSO should also be basic enough to promote the C-N coupling at ambient temperature. It is important to note at this juncture that by choosing DMSO as the reaction medium for this endeavor that one is effectively ignoring the opportunity to screen alternatives and potentially identify better solvents for the reaction.

Proof of concept for the proposed reaction conditions with a focus on assaying the performance and screening for the best organic superbase/ligand was achieved in a conventional HTE-setting utilizing a 96-well array with glass micro-vials. The approach was then translated to nanoscale with reactions run on a 1 μL-scale in a 1,536-well plate without stirring. To achieve this and facilitate mixing, reagent dosing was achieved using a TTP LabTech Mosquito High Throughput Screening nanoliter liquid-handling robot inside a glove box. To prepare the discrete reactions, the robot combined the various components from different wells of the source plate in a single pipette tip and then dosed as a single reaction drop into the plate thus ensuring proper mixing and negating the need for stirring. The reported approach is material-sparing, tolerates a degree of heterogeneity, and is performed identically to the same reactions performed in glass micro-vials.

The initial series focused on an array of 8 × 12 electrophiles/nucleophiles resulting in successful product formation in ~50% of cases (Fig. 4). A degree of tolerance of both polar functional groups and heterocycles was observed. For 32 electrophile/nucleophile substrate combinations that failed to yield product under the original conditions, a nanomole reaction screen was carried out using eight organic superbases and six ligands for a total of 1,536 reactions. With catalyst loadings increased to 20 mol% while substrate concentration was reduced to 0.05 M, a complete set of 48 reactions for a specific reactant combination utilized less than 1 mg of each substrate (50 nmol, ~ 0.02 mg per reaction). Reactions were run for 2 h, but clearly, now the bottleneck becomes reaction analysis.

To expedite the analyses with the goal of completing the whole screening cycle of 1,536 reactions within 24 h, MISER (multiple injections in a single experimental run) LC-MS was utilized based on its ease of data acquisition and analysis. With a row of 48 reactions all leading to the same product, multiple injections were carried out in discrete isocratic runs with mass detection set for the desired molecular ion. With a run time of 22 s per sample, the whole plate could be analyzed in ~9 h. To gain further efficiences in terms of the reaction analysis, it was noted that different combinations of reagents (ie. different rows in the plate) will lead to products that have different molecular weights thus allowing combination and simultaneous monitoring of up to four rows of reactions allowing a commensurate four-fold reduction in analysis time (~2.5 h).

While this substrate-focused approach is impressive in enabling the rapid identification of suitable conditions for a range of complex Pd-catalyzed C-O and C-N cross-coupling reactions in nanomole scale, there are several limitations to this plate-based approach. Specifically, there are restrictions placed on several reaction variables most notably solvent and temperature while the analytics is an offline method and presents only low-resolution data. In addition, this approach only offers limited possibilities for direct translation to compound production and isolation with separate experiments being required to access tangible amounts of material for structural characterization and advancing through a project-testing cascade.

2.3 *Nanomole HTE in Flow: Instrument Development*

The advantages of screening chemistries in flow parallel those of flow chemistry in general though the variables involved present a dichotomy in terms of their ease of realization on a conventional flow reaction screening platform. For example, when considering continuous variables, the use of flow allows the accurate control and ease of variation of reaction conditions such as temperature and residence time whilst facilitating mixing and heat transfer. The parameter space for the study of the reactions performed in this manner is also significantly extended allowing the use

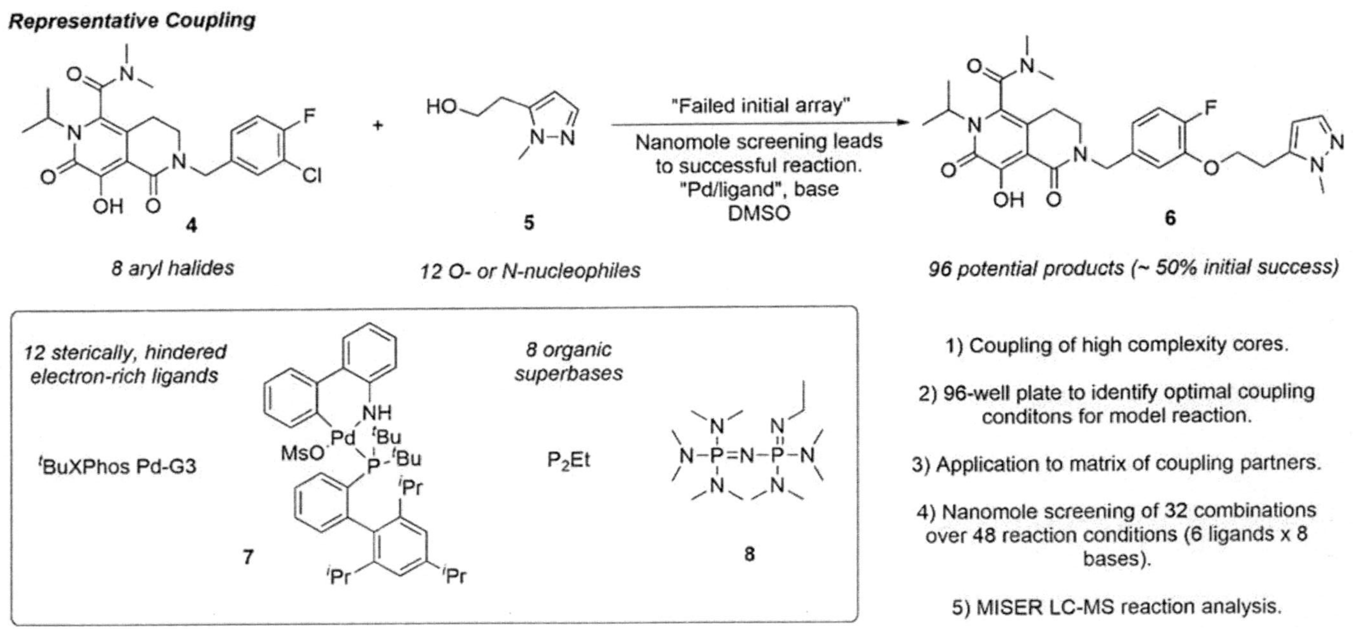

Fig. 4 Overview of Merck nanomole batch-HTE workflow

of higher temperatures and pressures (350°C and 200 bar), thus leading to accelerated reaction rates. This is particularly notable for heterogeneous reactions, nevertheless care must be exercised to avoid clogging of the system. By-product formation is reduced as the mixture is moved out of the reaction zone soon after the desired product is formed. Solvent and reagent use can also be minimized while flow is broadly acknowledged as a safer approach to perform hazardous chemistries. In addition, flow systems are also widely promoted for photochemistry, an area in which reaction screening is rapidly expanding given the advances in this field, in that mixtures can be acutely and uniformly irradiated in a controlled manner for a set time period owing to the high surface to volume ratio of micro-reactors. However, while the advantages of using flow systems for the optimization of continuous variables of a reaction are readily achieved, the analogous variation/screening of discrete variables presents greater challenges [22]. Reports on studies on optimizations of reaction solvents in flow are rare and involve the preparation of multiple stock solutions with the reactions then being evaluated in a "segmented-flow" manner [23].

A solution to this was provided by workers from Pfizer who sought to develop a fully automated system for HTE screening using flow chemistry technology [24]. The goals of the outset of this endeavor included integration of an in-line LC-MS system to allow real-time reaction monitoring and to maximize throughput, as well as being able to utilize diverse solvents (regardless of their volatility) with common reagent stocks. The system was intended to be material-sparing (0.05 mg of material per reaction) intending to achieve a 10-fold reduction in material needs compared to a conventional plate-based HTE, while realizing a 10-fold increase in productivity over the batch process with the goal to evaluate ~1,500 reactions in a 24 h period. Finally, this system was envisioned to allow direct scale-up of compound syntheses/isolation either through repeated injections or through the translation of the identified conditions to commercially available flow-reactor systems.

The key to achieving the over-arching goal was to recognize the possible synergy between continuous and segmented flow systems [25]. In the latter, the discrete "reactions" are separated ("spaced") within the flow stream typically using either an immiscible fluorous solvent or inert gas bubbles. Thus, each reaction can be prepared and analyzed separately as it emerges from the flow reactor, which is typically a heated coil. If the "spacer" element were to be removed, then the reaction "segment" would simply diffuse into the surrounding "carrier" solvent. The concept behind the flow HTE-system deliberately exploits this phenomenon allowing the segment to diffuse into the carrier solvent of the flow system. With sufficient dilution, then this carrier solvent can then be considered to be the "reaction" solvent (Fig. 5). The benefits of this are that only one concentrated stock solution of each of the various reaction components (reagents, ligands, catalysts, etc.) needs to be made up in any suitable inert solvent, while screening of the reaction solvent is achieved simply through switching the carrier solvent.

The system configuration is shown in Fig. 6 and given the potential air and moisture sensitivity of the catalyst/ligand systems it was assembled in a glove-box environment. A further advantage of this is the long-term stability of the stock solutions under these conditions. Basically, the reaction segments are prepared

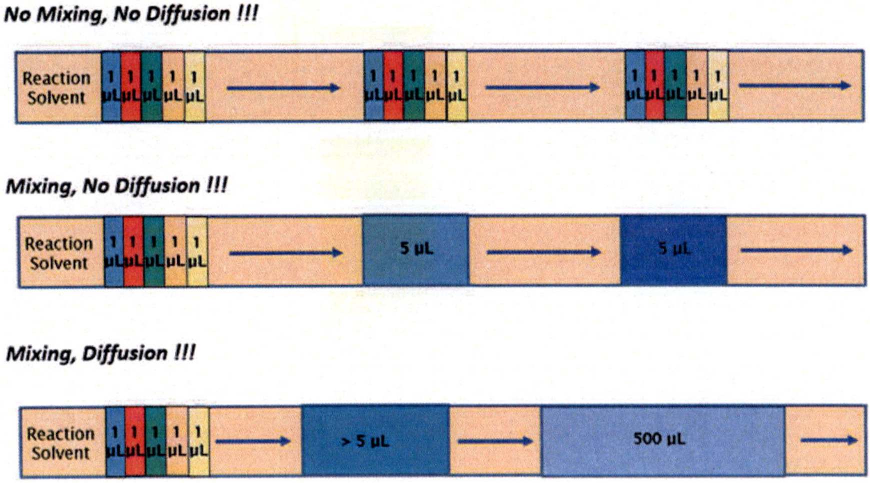

Fig. 5 Concepts of flow-screening system: Combining "continuous" and "segmented" flow

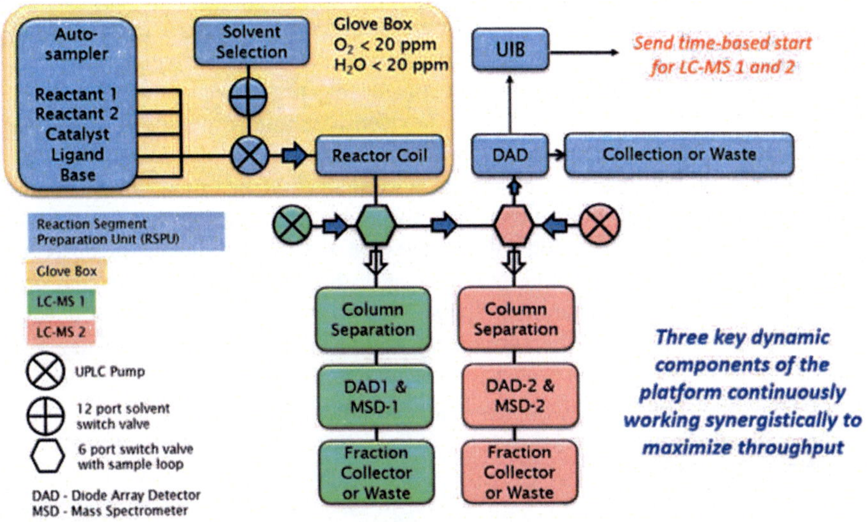

Fig. 6 Configuration of nano-HTE flow-screening system

using a modified HPLC with a well-plate sampler (the so-called RSPU – reaction segment preparation unit). A user-defined program guides the accurate aspiration and injection of microliters (0.5–100 µL volumes) from up to 192 source vials. This is determined by the configuration of the autosampler, which was initially set up to accommodate 2 × 96-well plates, however both 24- and 384-well plates are also compatible with the current instrument. Furthermore, the source vials contain a micro-stir bar enabling the stock solutions to be agitated and warmed throughout

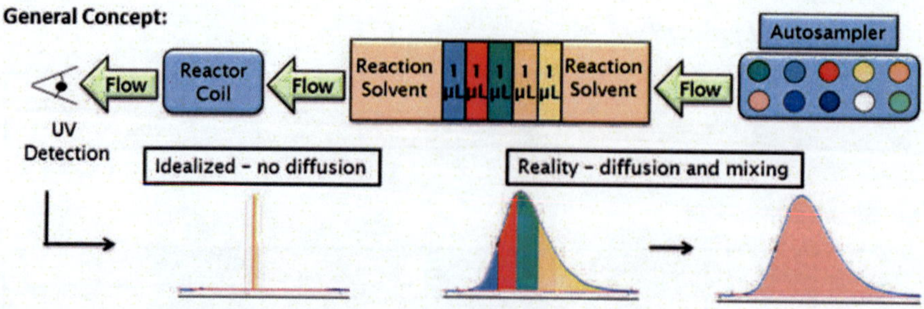

Fig. 7 General concept of flow-screening system

the process while the crimp seal cap prevents evaporation. For a model reaction to validate the system, a Suzuki-Miyaura coupling was chosen thus requiring five components to make up a reaction segment, specifically the aryl halide, aryl boronic acid, catalyst, ligand, and the base. One thing that is crucial for the success of this approach is the timings needed to have all aspects of the system working in a synchronized fashion to ensure continuous operation. With the autosampler, a reaction segment is prepared in 45 s and then injected into the flowing solvent stream. The flow rate for this solvent stream (driven by a quaternary pump) will determine the residence time within the reactor while the solvent is predetermined by a programmed method using the 12-port solvent selection valve.

Residence time in the flow reactor is a key consideration within the design of the experiments as this will play a pivotal role in determining the overall reaction throughput of the system. For the model reaction, a residence time of one minute was selected. While one could contend that this is not sufficient for most reactions to reach completion, an assumption was made that the initial conversion observed within this residence time would be reflective of the potential overall success of a reaction. For the model system as shown, this was demonstrated to be a valid hypothesis proven through the subsequent scaling of both positive and negative control results from the 1,536 flow-screening results. The staggered timings between the preparation of the segments by the RSPU and the residence time enable multiple reactions to flow through the reactor coil at the same time without risk of diffusing into each other (Fig. 7).

On entering the solvent stream (with subsequent diffusion), the reaction then proceeds through a Hastelloy reactor coil (0.5 mm i.d., 710 μL internal volume). This coil was originally designed for the Accendo Conjure flow reactor and sits above a stirrer hotplate thus enabling the reaction to be heated with the temperature monitored by a probe. While for the proof-of-concept study, these reactor coils were used, it is important to note that the modular nature of the system with a range of flow-based reactors can easily be adapted for use in a plug-and-play manner. After validation, this was demonstrated through the extension of the system's use to photoredox-mediated transformations using a chip-type reactor.

In order to ensure maximal time efficiency and allow potential future incorporation of a real-time feedback loop, the reaction samples must be analyzed as soon as

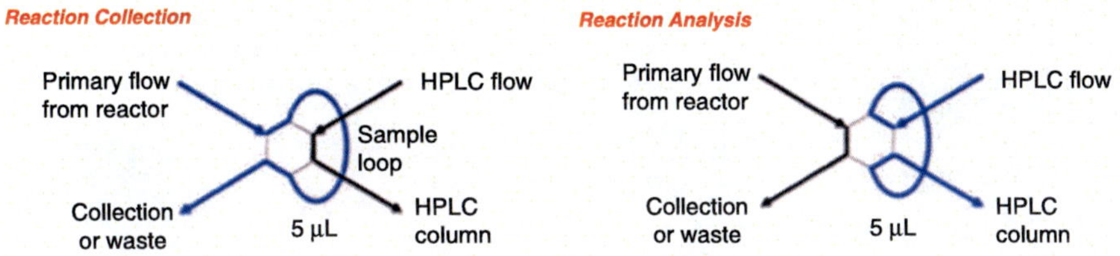

Fig. 8 Six-point valve controls the emerging flow stream from the reactor

they emerge from the reactor. To achieve this, two Agilent 1200 UPLC/MS systems were positioned post-reactor outside the glove box so that whilst one was analyzing a reaction segment, the other was waiting to be triggered to analyze the next emerging segment. As such, when the system is in operation, the three dynamic components (two UPLC/MS instruments and the RSPU) are continuously working in tandem for maximum throughput. To control the analysis, on emerging from the reactor, a segment encounters a six-point switching valve (Fig. 8), which directs it to the vacant LC/MS system for detailed analysis (using a 0.8-min run method). The excess segment is directed through a diode array to enable visualization of the segment with excess then either being sent to a fraction collector or diverted to waste depending on the instrument's mode of operation.

The overall system is run by three computers with the first two used to control the UPLC/MS systems while the third coordinates the RSPU for segment preparation injection, queueing, and overall run control. Sequence tables are loaded to the Agilent ChemStation software to guide the software for run control and analysis. Outside the glove box between the two UPLC/MS systems is a central stack that serves to split and direct the flow to the waiting LC/MS. A detailed flow schematic of the operation of the system is shown in Fig. 9.

2.4 *HTE in Flow: Instrument Validation*

Prior to executing a reaction screen using the system in flow, key experiments needed to be conducted to ensure homogeneous mixing of the reaction components within a specific segment as well as to assess the extent of diffusion of the segment into the surrounding carrier solvent, which represents the cornerstone of this technology. As noted a pharmaceutically relevant Suzuki-Miyaura reaction between 6-bromoquinoline **9** and an indazole boronic acid **10** was chosen to validate the platform with a discrete example selected for mixing/diffusion studies using Pd $(OAc)_2$ with PPh_3 as the ligand, aqueous NaOH as the base with methanol as the solvent (Fig. 10). Assessment of homogeneity in a "real" system presents several challenges, namely that each of the reaction components may respond differently in the LC-MS (thus complicating quantification) while analysis of the output trace will

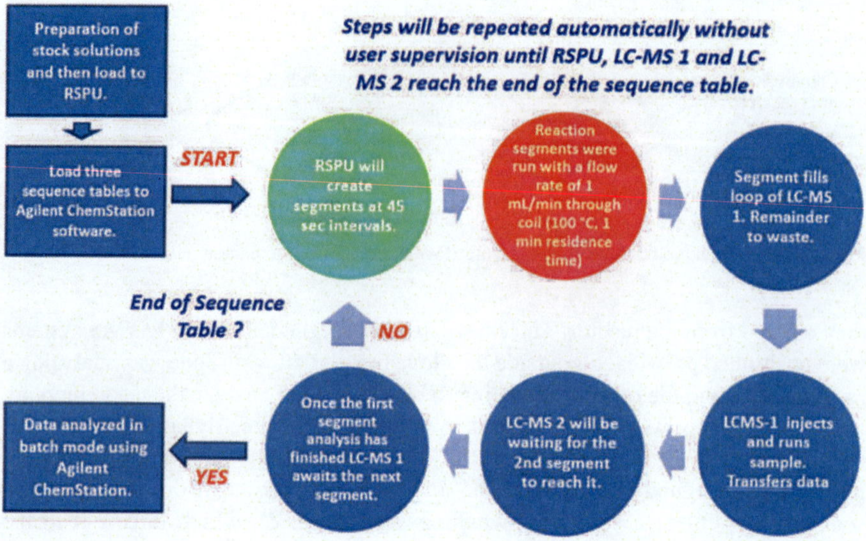

Fig. 9 Operational flow schematic for the nano-HTE flow system

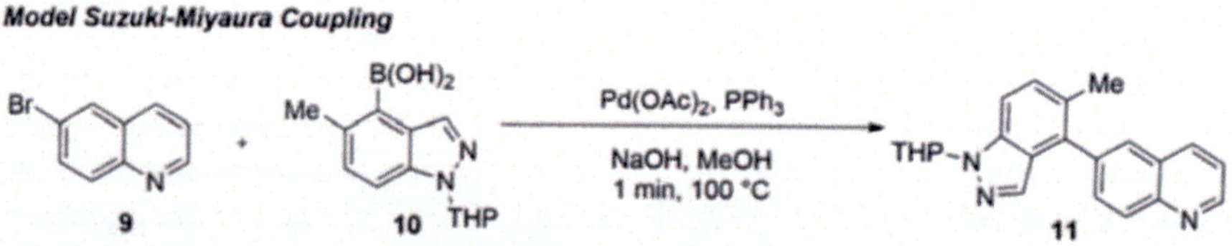

Fig. 10 Suzuki-Miyaura coupling for system validation

further be complicated by the extent of conversion to product **11** or unproductive side reactions of the starting materials.

An elegant solution to overcome this was developed through the preparation of stock solutions of the individual reaction components each with different inert internal standards, which can then be independently monitored in the LC/MS output. Judicious selection of these internal standards was required to ensure that they do not interfere with the reaction, that they also all respond in an almost identical manner in the chromatographic output, and finally that they are compatible with solubilization of their assigned reaction component. For this purpose, UV active non-polar solvents were used for the readily soluble components with 6-bromoquinoline **9** being dissolved in 1,3-diethylbenzene, $Pd(OAc)_2$ in 1,3,5-triethylbenzene and the ligand in toluene. The boronic acid **10** proved more challenging to dissolve in similar media so herein DMF was used with 4,4′-di-*tert*-butylbiphenyl added as an internal standard while the aqueous base was not monitored owing to the difficulty in identifying a suitable aqueous soluble internal standard. The stock solutions were prepared with concentrations such that equal volume injections of each effectively dictated the stoichiometry of the reaction. For the control reaction, 1 μL of each (5 μL total) equated to a ratio of 1: 1: 2.5: 0.125: 0.0625 for aryl halide **9**/boronic acid **10**/base/ligand/Pd. The LC/MS trace for the reaction run at 100 °C with a residence time of 1 min through the Hastelloy coil demonstrates the relative ratio of the internal standards confirming the effectiveness of the mixing and the stoichiometry (Fig. 11). Consistent use of these internal standards throughout a screening run enables an in situ QC check of any segment to ensure that all the reaction components are present and that the system is operating accurately.

Demonstration of diffusion required the development of an additional series of experiments with again the utilization of the inert internal standards critical to the quantification and proof of concept. With this experiment, reaction segments of increasing volume were created using 1, 2, 4, 8, and 16 μL of each component thus leading to overall segments with total volumes of 5, 10, 20, 40, and 80 μL. It should be noted that this is the practical limit given that the autosampler uses a 100 μL syringe. These experiments not only confirm that efficient mixing and diffusion into bulk solvent is occurring over a range of reaction volumes but also highlight the scalability of the technology which is a key consideration if direct extension to compound preparation is to be considered. Segment preparation in the conventional manner consists of sequential aspiration of the components in series before injection into the flowing solvent stream. To determine mixing/diffusion, the output was collected in a 96-well plate in 40 μL fractions, with each of these being subsequently analyzed offline by LC-MS using a standard method. The results are graphically represented in Fig. 12. As can be seen in all cases, the segments diffused showing a bell-shaped distribution across a significantly increased volume with, for example, the smallest 5 μL (1 μL each component) injection now being spread across ~8 fractions (320 μL). In addition, the ratio of each of the internal standards is approximately equivalent throughout the segment demonstrating homogeneous diffusion/consistent stoichiometries into the surrounding carrier solvent across a range of injection volumes. This also suggests that larger volumes can be injected for direct scale-up of screening results to produce meaningful quantities of materials for biological assays. The homogeneous diffusion of the segment into the carrier solvent also implies that solvent screening for optimization will be possible in a continuous flow manner while using the same reactant/reagent stock solutions.

2.5 *HTE in Flow: Suzuki-Miyaura Coupling*

Switching now to a screening paradigm, a comprehensive assessment was made in terms of the reaction under consideration not only in terms of the ligand/base but also in the nature of the coupling partners involved. This latter point is of interest because it has a direct impact on the logistics of the reaction not only from a reactivity standpoint but also from an economics and ease of commercial accessibility perspective. For the study, it was planned to evaluate a series of four electrophiles **9a–d** based on the quinoline as well as a series of three nucleophiles **10a–c** based on the indazole [26]. In addition, the reverse combination was considered using the bromo-

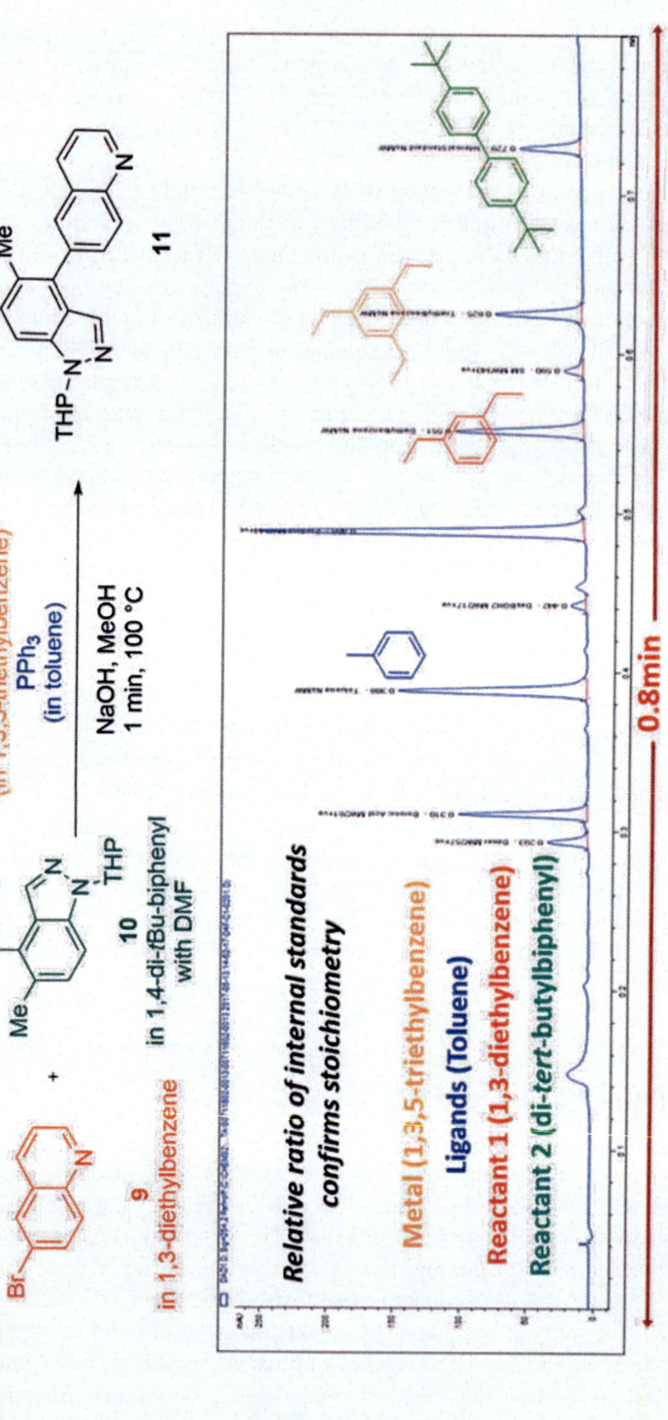

Fig. 11 Use of internal standards to monitor system performance and ensure mixing

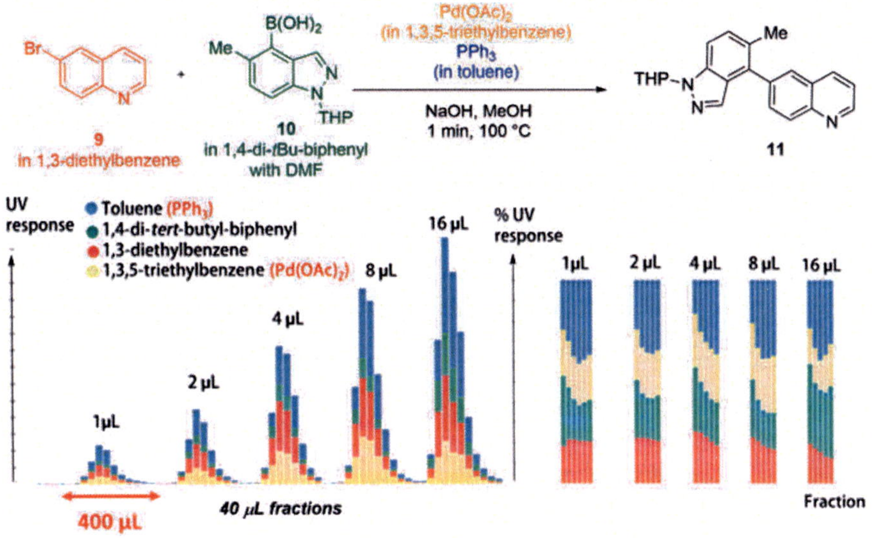

Fig. 12 Collection of multiple fractions demonstrating diffusion

indazole derivative **10d** and three boron-based quinoline nucleophiles **9e–g**. Herein, we get some insight into the ease of accessibility of the various coupling partners with more options being readily available from the quinoline fragment **9**. One point to make herein concerns the potential complexity of the analytics notably when considering the the automated readout/processing of the data is that although one is utilizing different coupling partners, the identity of the desired product remains the same throughout, thus simplifying the back-end analysis as will be demonstrated. While $Pd(OAc)_2$ was maintained as the sole "Pd" source, 11 ligands (plus one blank), 7 bases (again with a blank), and four solvents completed the matrix for a total of 5,760 reactions (Fig. 13).

While there are numerous ligand systems disclosed for Pd-mediated couplings (and the ability to expand a wider chemical space is offered by this system), the selection of the subset of these for the proof-of-concept study was dictated by several criteria. These had demonstrated success in previous batch Suzuki-Miyaura screens, an internal principal component analysis of the ligand property space, coverage of the main ligand classes from the literature, and importantly commercial availability and cost (Fig. 14) [27]. A range of both inorganic and organic bases previously demonstrated to be successful for the transformation were employed while the solvents were selected to display a range of polarities/dipole moments and the presence/absence of hydrogen-bond donors. It has been long established that Suzuki-Miyaura reactions benefit from the presence of water as a co-solvent not only to aid in solubilization of the inorganic base but also to promote the formation of boronate complexes involved in the transmetallation and as such the pump was set to provide a 9:1 solvent/water ratio.

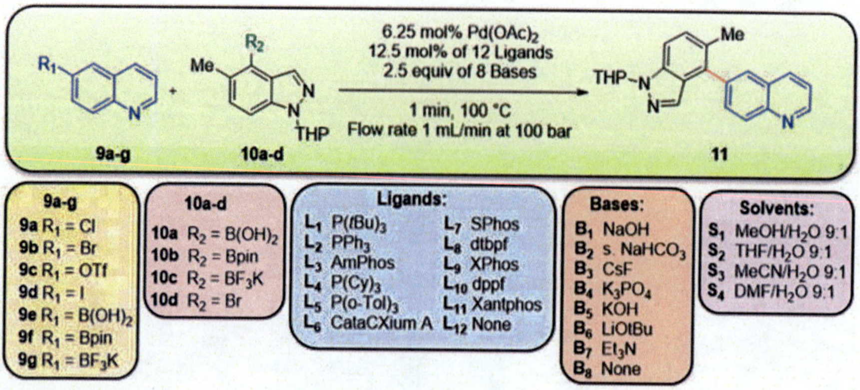

Fig. 13 Full matrix of 5,760 reactions under evaluation

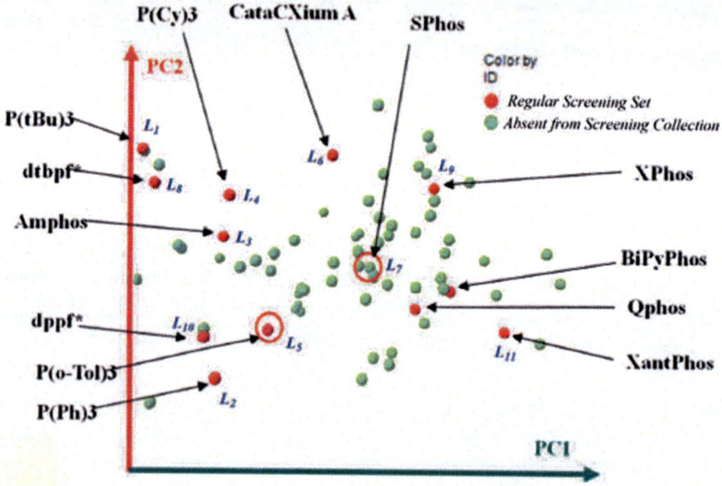

Fig. 14 Principal component analysis of screening ligand property space

Throughout the discussion so far, the key assumption has been that dilution of the reaction segment into the bulk carrier solvent will occur to a sufficient degree that the latter can be considered to dominate any reactivity effects seen in the transformation under consideration. Despite this appearing to be a rational supposition, there remain valid concerns that the solvents utilized for the preparation of the stock solutions may play some role in influencing the observed reaction outcomes, and it is important to probe and eliminate this possibility prior to running a full matrix of experiments. To achieve this aim, an initial run of experiments was carried out evaluating 6-bromoquinoline **9b** and the indazole boronic acid **10a** across the screening space (12 ligand/8 base/4 solvents) for a total of 384 experiments. Given the common nature of the solvents utilized here to make the stock solutions, we can infer that if there are changes in reactivity on moving across the solvents, then these are directly attributable to the carrier solvent mediating the reaction based on a ~100:1 dilution factor. The results are presented in Fig. 15.

As can be seen, the solvent has a striking effect on the outcome of the reaction with the best results being obtained in MeOH while both THF and DMF are poor selections for this reaction. The latter is of particular note as DMF has been utilized for the preparation of the stock solution of the indazole **10a**, and this factor appears to have a negligible effect on either the numerous successful reactions in MeOH or even the complete lack of reactivity seen in THF. For the screen, each reaction used 1 μL of each stock solution equating to 0.4 μmol of material. For the bromoquinoline **9b**, this is 83 μg of material or 124 mg for 1,500 reactions, which can be completed in 24 h. Note the segments flowed at 1 mL per minute with 100 bar of pressure. As noted, the short reaction time is necessary to expedite throughput with the theory that relative initial conversions would allow for judicious selection of conditions for further investigation and scale-up. Once a run is initiated, the system operates in a fully automated fashion, and in contrast to 1,500 reactions in 24 h, a typical batch screen is estimated to evaluate 192 reactions in the same time frame including the subsequent off-line analyses.

Following this study, the complete matrix of 5,760 experiments was run for 3–4 days of continuous operation. To expedite data analysis, the Agilent ChemStation software was used to identify key peaks within the LC-MS spectra in real-time followed by off-line refinement of the data utilizing the iChemExplorer software before export into Spotfire® for visualization. The latter presents a versatile platform for the manipulation of the data into a range of flexible formats to easily identify key reactivity trends. Given the abundance of data from the whole experiment, it was found to be more facile to treat the data in batches of 1,500 reactions with the processing of each of these datasets taking approximately one hour. The complete set of data for the 5,760 reactions is presented in heatmap form in Fig. 16. To orient the viewer, the color represents the extent of conversion to desired product ranging across a spectrum from blue (0%) to red (100%). The x-axis is divided into four main sections representing the reaction solvents, with further subdivisions on each trellis for the 12 ligands evaluated. The y-axis is somewhat more complex with again initially being broken into four sections to reflect the nature of the indazole component **10**, either nucleophile or electrophile. For each of the three nucleophilic indazole components evaluated, the results are further divided based on the nature of the quinoline electrophile **9a–d** while for the indazole bromide **10d** three boron-based quinolines **9e–g** were evaluated. Each section is further broken down to show the reactions with the 8 base systems that were studied.

Despite representing an attractive depiction of a large amount of information, is there really anything that can be learned from this high-level visualization of the data? Despite the relative complexity, one can observe stark differences in reactivity as a function of solvent, which is critical as it supports the hypothesis that diffusion into carrier solvent as a tool to carry out continuous flow-screening (Fig. 17). It can also be seen that 4-chloroquinoline not unexpectedly is the worst electrophile.

A better visualization of this can be provided by a box plot comparing the four quinoline-based electrophiles in their reactions with the three indazole-based

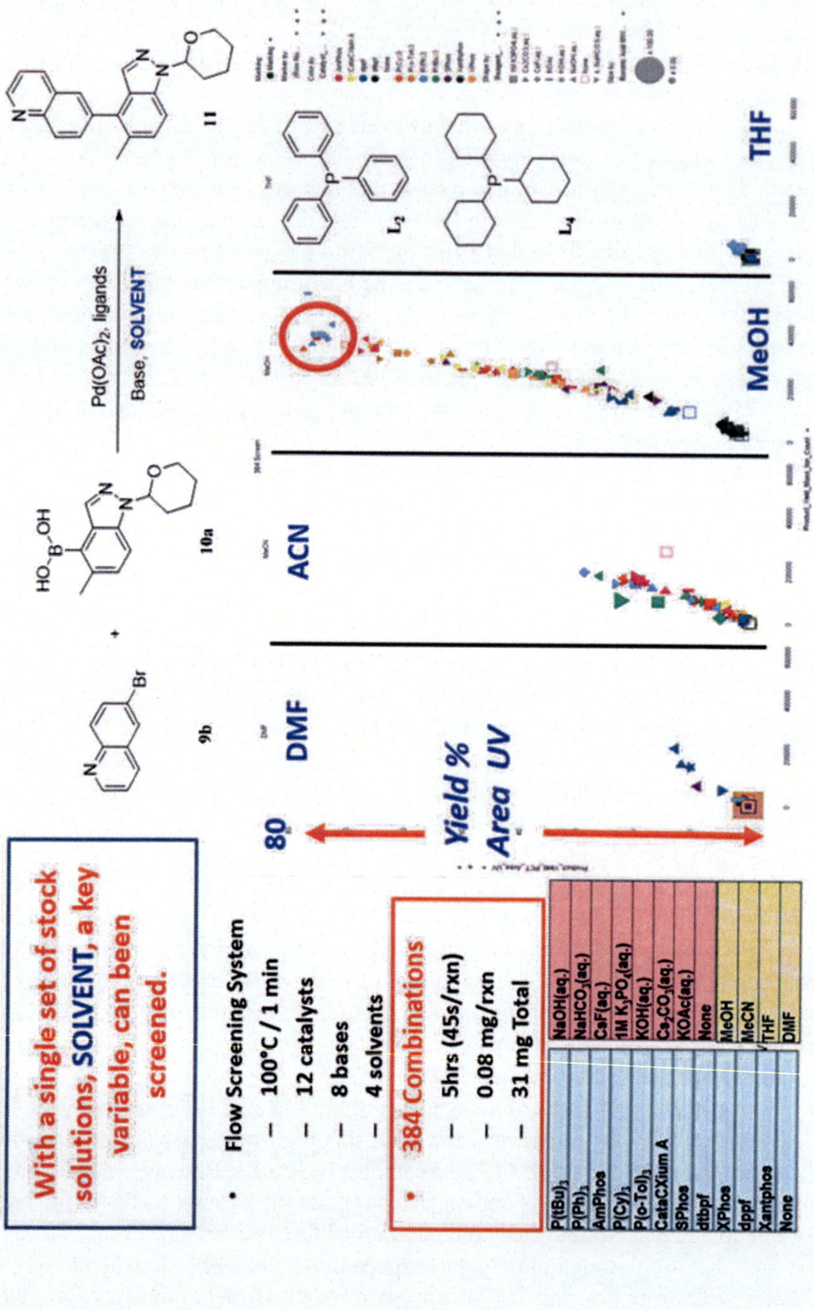

Fig. 15 Does solvent make a difference?

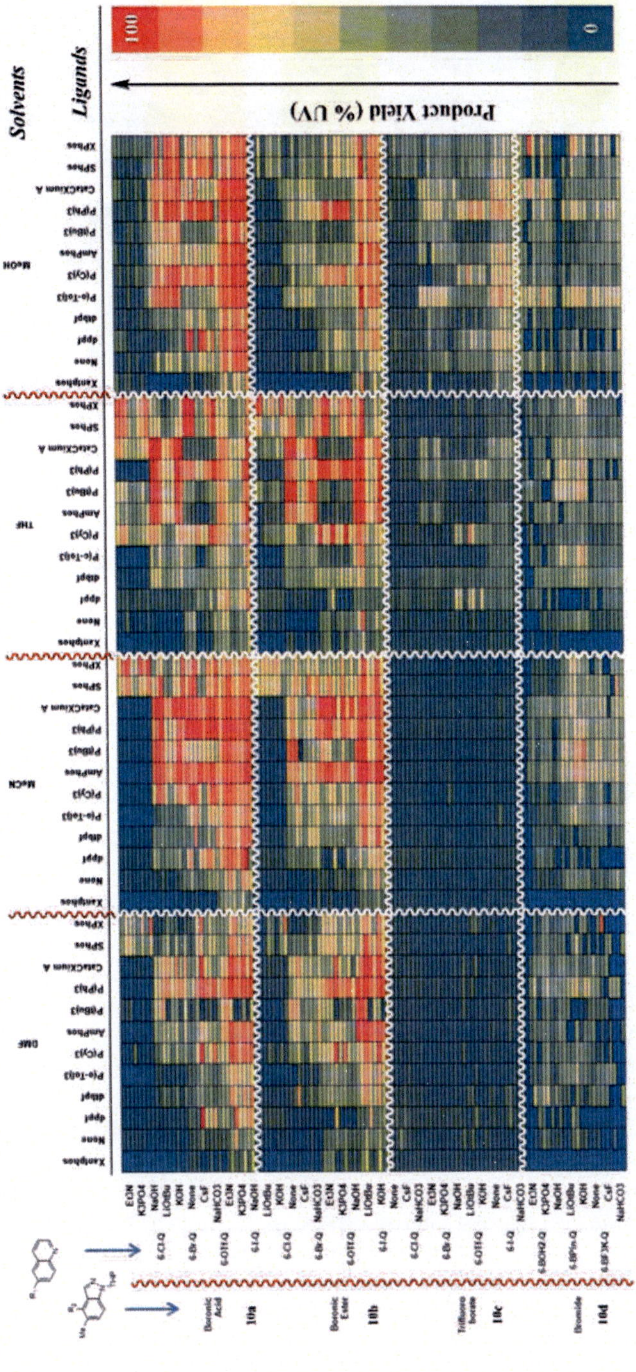

Fig. 16 Heatmap depicting outcomes from the full matrix of 5,760 reactions

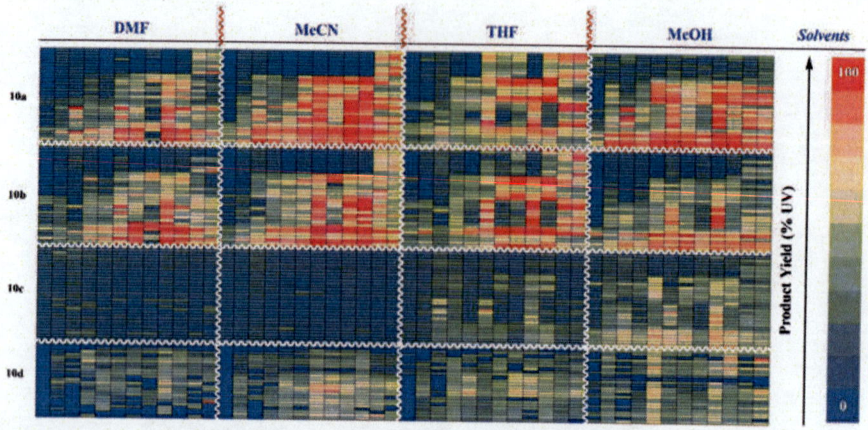

Fig. 17 Analysis of 5,760 reactions highlighting variations in solvent reactivities

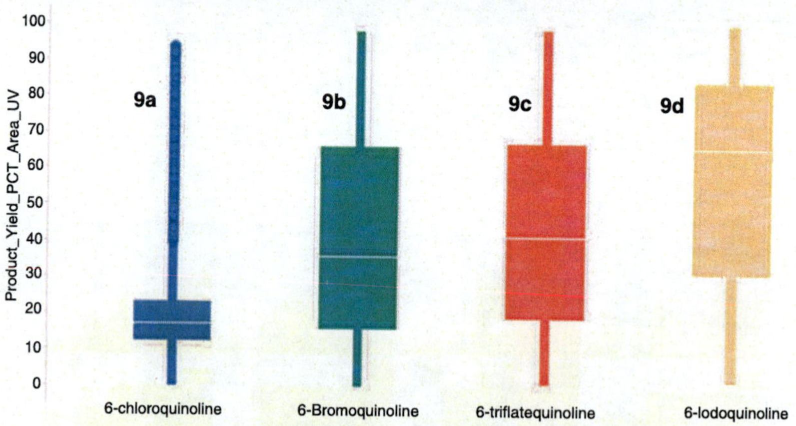

Fig. 18 Comparison of quinoline-based electrophiles **9a–d**

nucleophiles (boronic acid **10a**, BPin **10b**, BF_3K **10c** derivatives) (Fig. 18). While 6-chloroquinoline (**9a**) is clearly the worst substrate, a learning herein is that there are conditions in which this does show good reactivity specifically when ligands such as XPhos and SPhos are utilized.

Further learnings on analysis of the data provide us with the observation that higher levels of conversion occur when the indazole component is utilized as the nucleophilic partner rather than the electrophile **10d** (Fig. 19). Comparison of the boron-sources indicate that the indazole BF_3K derivative **10c** is far inferior to both the corresponding boronic acid **10a** and BPin ester **10b**. A possible rationale for this observation might be that the one-minute residence time is not sufficient for in situ hydrolysis of the trifluoroborate to the boronic acid to enable the reaction to proceed [28]. A closer examination demonstrates that when the BF_3K derivative **10c** has

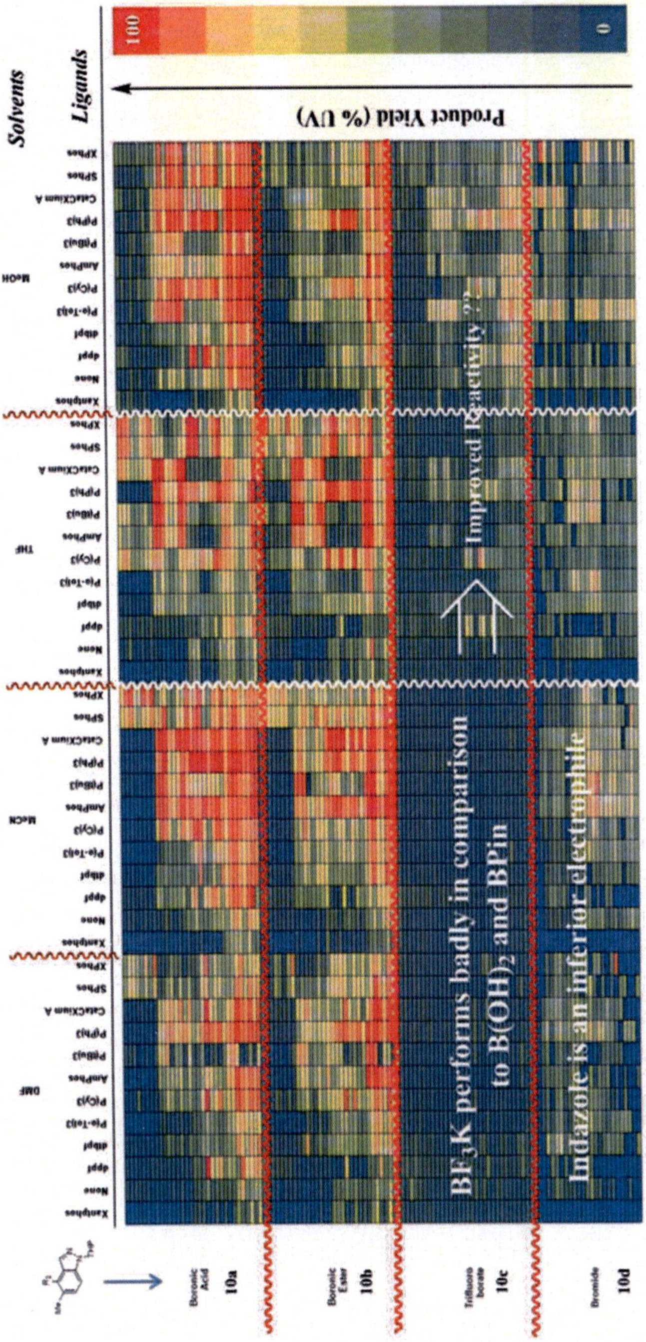

Fig. 19 Indazole-based **10a–d** reactivity trends

shown some level of reactivity, the use of MeOH (and to a lesser degree THF) as the solvent is key thus reinforcing the use of continuous flow-screening to identify reactivity trends through variation of the bulk carrier solvent.

Typically, within screening campaigns, the full range of coupling partners is not available, and thus a more focused set of the data is studied. In these cases, of most interest is generally the identity of the ligand/catalyst system that performs best in mediating the reaction. Looking at a subset of the data (1,536 reactions, one day instrument time) capturing the coupling of the four electrophiles with the indazole boronic acid **10a** alone allows a series of reactivity trends to be drawn regarding the ligand selection. As noted, the 6-chloroquinoline **9a** is the least prolific electrophile requiring specific ligands to be utilized to observe reactivity, and the data clearly shows that MeOH is a superior solvent for the transformation when using the 6-bromoquinoline **9b**. In addition, from a ligand perspective, it can also be determined that Xantphos performs poorly in the transformation while PPh_3 gives high conversions to the desired product for all electrophiles (except the aforementioned chloro-derivative) particularly in MeOH or MeCN (Fig. 20).

One of the challenges with the availability and analysis of large datasets particularly in the reaction screening/optimization space is deriving results that provide tangible value in a timely fashion amenable to a range of chemistry functions ranging from medicinal to process chemistry. For the former, one endeavor critical to project progression at an early stage is the use of parallel medicinal chemistry approaches in making series of related compounds using a common set of conditions for a specific transformation. An example of how this information can be derived from the dataset considered herein is through filtering the conditions and clustering those that provide >85% conversion for all the quinoline electrophiles **9a–d** screened. The data demonstrates that for 384 conditions evaluated for each electrophile, only three will work well independent of the substrate chosen and all utilize either X-Phos or S-Phos as the ligand with MeCN as the solvent (Fig. 21). The related nature of the conditions identified for each of the electrophiles provides further validation as to the reproducible nature of the flow-screening method for reaction optimization. Again, it should be noted here that this task is facilitated by all the reactions leading to the same reaction product thus simplifying the analytics, though the screening platform is highly amenable to screening different pairs of reaction partners (providing discrete products) to enable a scope evaluation in tandem with condition screening for a specific transformation with the analytics being developed to manipulate the data in a split-batch fashion.

While a degree of investment is required to develop the nanomole flow-screening capability, it is interesting to take a rough look at the economics of this endeavor when compared to traditional batch-screening. The intent herein is not to provide a rigorous analysis though instead to do a "back of the envelope" calculation based on gram-scale pricing of the reagents involved derived from a commercial supplier's catalog (Sigma-Aldrich). Figure 22 shows the results with each trellis representing a specific electrophile with the x-axis showing the % yield of a transformation (by UV) while the y-axis depicts the cost/reaction. The amount of data enables high-level conclusions to be drawn such as the ~10-fold increase in cost of utilizing the

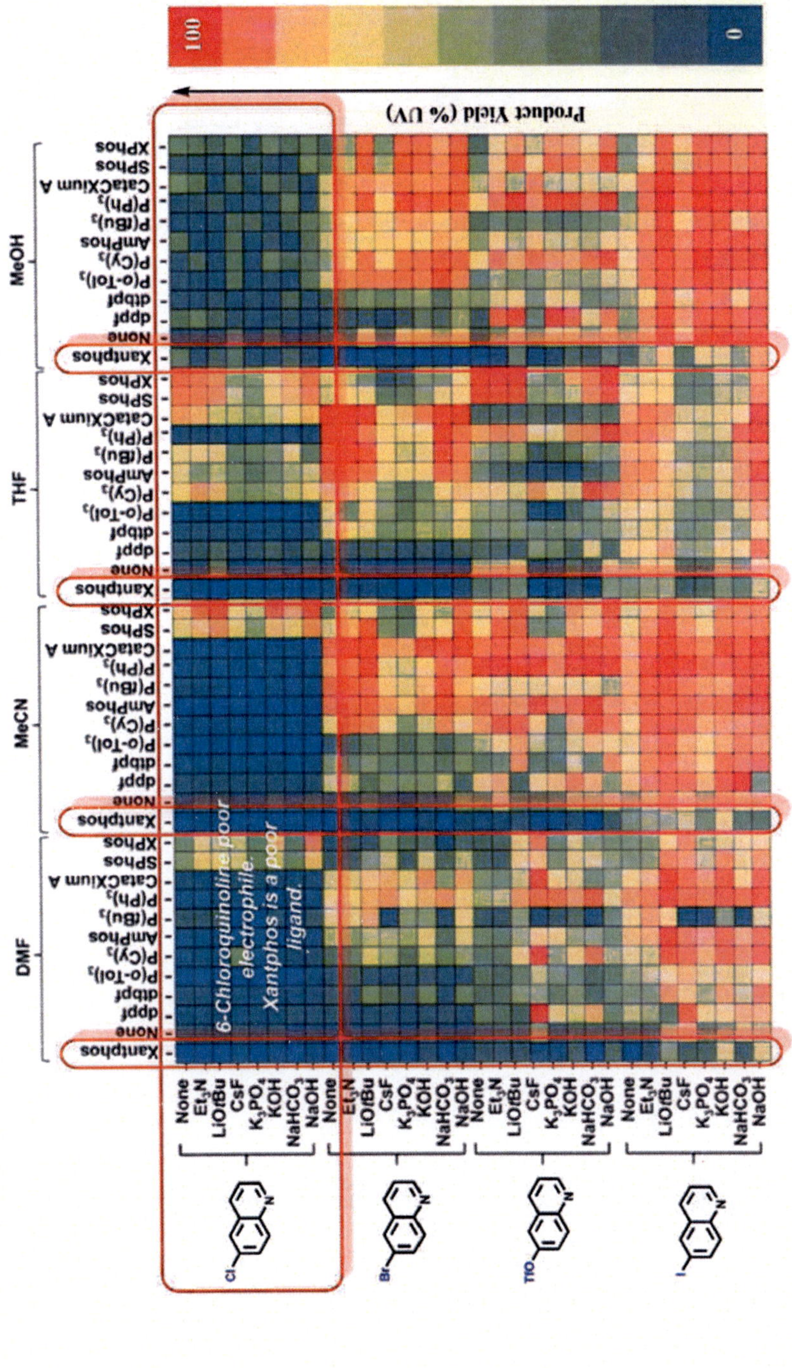

Fig. 20 Electrophile-/ligand-based reactivity trends

Reaction Condition	Solvent	Base	Ligand	6-Cl-Q (9a)	6-Br-Q (9b)	6-OTf-Q (9c)	6-I-Q (9d)
A	MeCN	LiOtBu	SPhos	85	92	91	93
B	MeCN	NaOH	SPhos	86	93	91	88
C	MeCN	NaOH	XPhos	93	90	92	86

Fig. 21 Robust coupling conditions for all four quinoline-based electrophiles **9a–d**

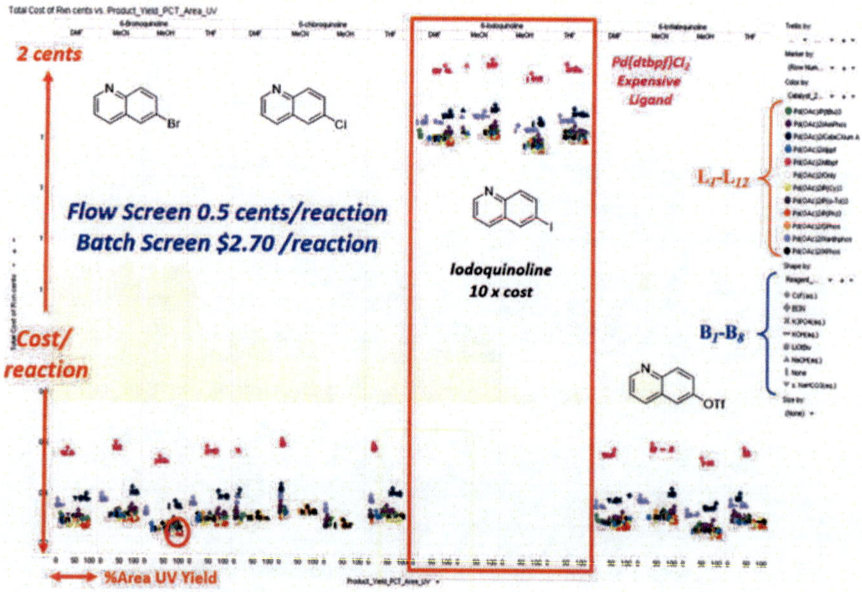

Fig. 22 Crude "cost-analysis" of flow-screening reactions

iodoquinoline **9d** as the electrophilic partner in the reaction as well as highlighting the Pd(dtpf)Cl_2 as an expensive ligand option. Overall, flow-screening was estimated to equate to 0.5 cents/reaction compared to $2.70 for a corresponding batch campaign though again it must be emphasized that this is by no means a rigorous treatment of the costs involved.

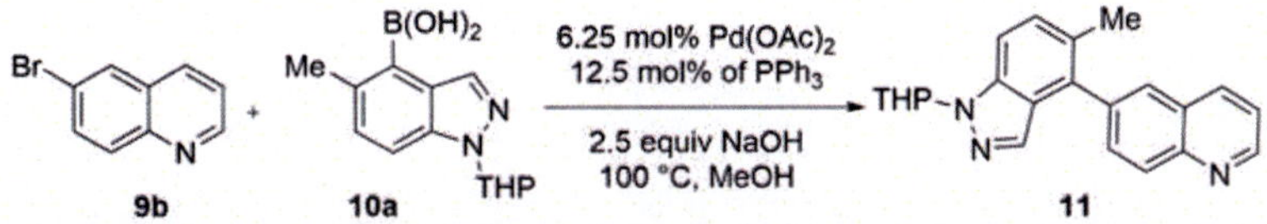

Entry	Method	Time (min)	Conc.	Scale	Yield
A	Screening flow (100 segments)	75	0.8 mM	0.32 mmol	59%
B	Vapourtec flow	10	0.05 M	0.87 mmol	42%
C	Batch	10	0.1 M	0.15 mmol	79%

Fig. 23 Scale-up of screening results

One of the goals at the outset of the development of the flow-screening capability was to demonstrate translation of the results to preparative scale, both through the injection of multiple segments on the platform and through reproduction in both batch and using the commercial flow instrumentation (Fig. 23). Selecting the coupling of the bromoquinoline **9b** with the indazole boronic acid **10a** as the model reaction, 100 consecutive segments were injected through the system (8 μL per reaction component, 3.2 μmol per segment) over 75 min (reflecting as noted previously that multiple reaction segments can process through the reactor simultaneously) to provide potentially a maximum output of 110 mg of material. The combined fractions were collected off-line, evaporated, and purified by chromatography to provide 65 mg (59% yield) of material (Entry A). Translation of the same conditions to a Vapourtec R series instrument was to some degree complicated by a two-pump setup being operated for multiple reaction inputs though 42% yield of material was still obtained after purification running the reaction on a 0.87 mmol scale (Entry B). Translation to batch was also demonstrated with the reaction being run in a more concentrated fashion (0.1 M) on 0.15 mmol scale with a 79% yield being achieved after 10 min under otherwise identical conditions to those identified on the screening platform (Entry C).

One assumption that was made at the outset is that the initial conversions observed after 1 min of reaction time in flow would reflect the eventual reaction outcome, and to probe the veracity of this as well as further validate the results obtained from the platform, a further series of positive and perhaps, more importantly, negative control reactions were carried out in batch (Fig. 24). For the former, a reaction of both the iodo-**9d** and bromoquinoline **9b** with the BPin ester **10b** and boronic acid **10a**, respectively, was scaled using conditions differing in the nature of the ligand, base, and solvent. In both cases, high yields (>90%) observed in the flow-screen gave isolated yields of 67 and 93% after 10 min of reaction time in batch thus demonstrating their validity (Entries A and B). For negative controls, a coupling of the 6-chloroquinoline (**9a**) and **10b**, which provided <2% in the screen was carried out and led to no conversion in batch after reaction overnight. Although this may represent somewhat of a loaded example in choosing the 6-chloroquinoline (**9a**), which has been acknowledged to be the worst electrophile, a further case was

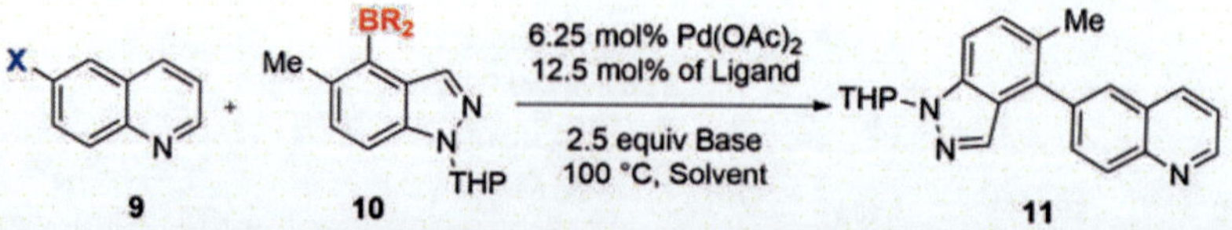

Entry	X	BR_2	Solvent	Ligand	Base	Screen (%)	Yield (%)	Time
A	I (**9d**)	Bpin (**10b**)	DMF	Amphos	NaOH	93	93	10 min
B	Br (**9b**)	$B(OH)_2$ (**10a**)	THF	CataCXium-A	$NaHCO_3$	96	67	10 min
C	Cl (**9a**)	Bpin (**10b**)	MeCN	P(*o*-Tol)$_3$	NaOH	2	0	18 h
D	Br (**9b**)	Bpin (**10b**)	MeOH	Xantphos	$NaHCO_3$	2.3	< 10	18 h

Fig. 24 Positive- and negative-controls for screening data

selected utilizing the 6-bromoquinoline (**9b**) with **10b** and herein <10% conversion was observed after analysis at both 4 and 18 h supporting the validity and reproducibility of reaction outcomes identified through flow-screening (Entries C and D).

The proposed implementation of this methodology in identifying broadly applicable reaction conditions for parallel library medicinal chemistry has already been highlighted, but an additional application in this space is to rapidly identify effective conditions across a broad parameter space for monomers that may have historically represented poor coupling partners in such campaigns. From this perspective, bromo-oxindole **12** was selected as a model electrophilic substrate given that it had failed to undergo the desired cross-coupling in any of the 13 libraries in which it has been selected as a monomer (Fig. 25). The continued selection though of its motif reflects its desirable properties from a medicinal chemistry perspective. Nevertheless, its lack of reactivity highlights a "gap" alluded previously that compounds with drug-like physicochemical attributes often are the poorest participants in cross-coupling chemistry.

A flow-screen was carried out evaluating 12 ligands, 8 bases, and 6 solvent systems for a total of 576 reactions taking ~7 h to complete (Fig. 26). Each reaction was run on 0.4 μmol of material equating to a total of 50 mg of the bromo-oxindole **12** for the whole campaign. The heatmap is in stark contrast to the previously studied cross-coupling between the quinoline and indazole partners in that only a few conditions provided moderate conversion (45–65%) to the desired product after the one-minute reaction time. As a negative control, the legacy conditions used in the previous library campaigns ($Pd(OAc)_2$, $P(Cy)_3$, $NaHCO_3$, THF/H_2O) failed completely, while the screen identified CataCXium A as a uniquely effective catalyst in either aqueous THF or possibly aqueous MeOH. A range of bases showed potential but Et_3N appeared to be optimal. Scaling these conditions with a 90-min reaction time led to a 79% isolated yield of the desired product thus validating this screening approach.

This flow-based screening platform represents a significant advance in the ability to screen reactions, solvents, and conditions in a material-sparing manner. Although

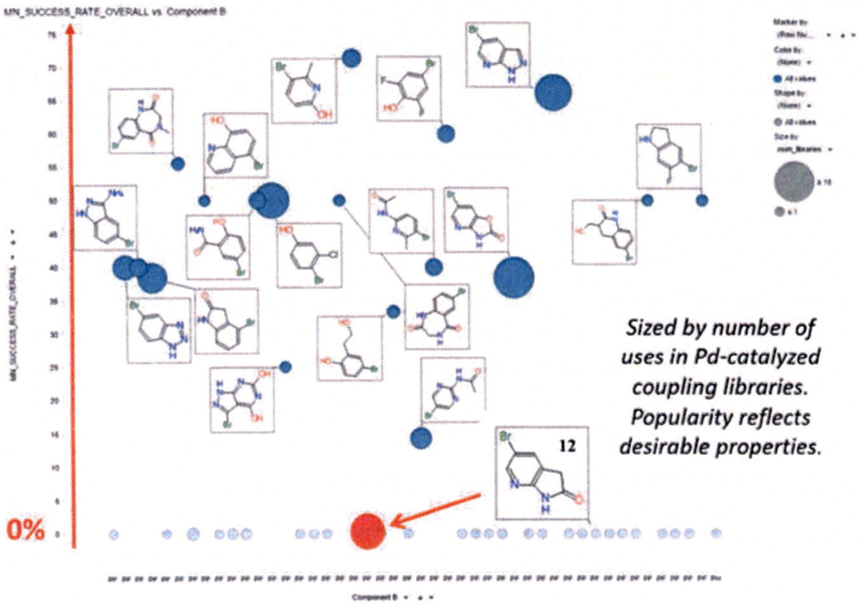

Fig. 25 Identification of a "poorly-performing" reaction monomer

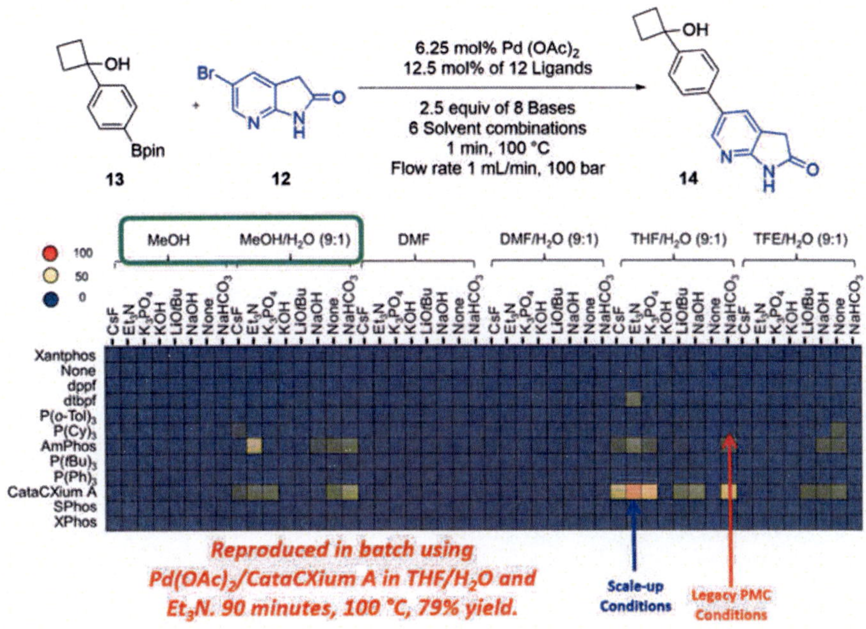

Fig. 26 Identification/scale-up of coupling conditions for **12**

direct comparison of batch- and flow-screening approaches are challenging owing to limited information regarding the various setups, the methodology developed allowing the analysis of 1,500 reactions in 24 h with high-resolution reaction data available benchmarks this favorably against plate-based methods. While heterogeneous reactions still present a challenge herein, the flow-screening presents many advantages consistent with the technology in avoiding solvent evaporation, improved mixing, and uniform and precise temperature control.

3 HTE in Flow for Photochemical Reaction Discovery

3.1 *Introduction*

The implementation of continuous flow technology is critical toward enhancing the application of photochemical reactions as a high-throughput tool for reaction discovery [29]. Micro reaction technology and adjunct concepts from continuous flow chemistry provide an important background to drive photochemical reactions most efficiently and sustainably [30–32]. Furthermore, translating small scale-up photoredox reactions (typically performed in batch) to a continuous flow process is important for industrial scale-up as this represents the most effective means to provide consistent and efficient irradiation to the reaction mixture. Nevertheless, this process can be both time and resource intensive.

Droplet microfluidics can be an enabling platform for reaction discovery by promoting the HTS of drug compound libraries and the optimization of flow specific parameters. Segmentation of samples with an immiscible phase enables the simultaneous handling of numerous samples over extended periods of time [33–35]. From a material consumption standpoint, microfluidics screens are typically performed at nanoliter to femtoliter scale, which translates to a reduction of starting material usage by three to eight orders of magnitude relative to a traditional multi-well plate-based screen. The design of a flow-based screening platform that interfaces with pre-plated compound libraries would allow for straightforward integration into existing pharmaceutical HTE workflows and infrastructures. The combination of droplet microfluidics, MS, and photoredox catalysis can be a breakthrough for accelerating pharmaceutical discovery and development, with a concerted emphasis on time and material efficiency.

3.2 *System Setup and Validation*

Stephenson *et al.* have presented a platform where they combine an Automated Droplet Generator with a screening plate of picomole scale reactions and a high-throughput ESI-MS analysis [36]. This setup could enable the high-throughput handling of pharmaceutical libraries of 384 or 1,536-well plates (Fig. 27). An automated droplet generator was used to generate reaction droplets at around

10 nL in volume continuously in a rapid manner. The reactor was made from PFA tubing connected to a syringe pump and downstream to a robotic arm which moves the tubing across, up, and down the plate to generate droplets. Each plate contains the reaction mixture on the bottom phase with perfluorodecalin on the top. The robotic arm will move the tubing up and down between the phases to generate the reaction droplets, which are segmented by Partial Flow Dilution. The size of droplets can be controlled based on the speed that the tubing is moved. These reaction droplets contain picomolar amounts of reagents. Following the droplet generation, irradiation and ESI-MS analysis of each droplet is performed using a sheath sprayer which enables inline dilution of samples as they are introduced into the ESI source. In contrast with UPLC methods (which can screen up to 1,000 reactions per day), the use of ESI-MS enhances furthermore the throughput by being able to screen up to 3,000 droplets/h.

This setup was validated using the radical trifluoromethylation, developed by the same group, as late-stage functionalization of four substrates (**15–18** – 3 of which were pharmaceutical compounds provided by Pfizer), which were either approved therapeutics or drug candidates (Fig. 28). Sequential reaction droplets of 4 nL were segmented from PFD 8 nL and irradiated for 10 min. The droplets then flowed through the sheath sprayer for inline dilution before the ESI-MS analysis with a frequency of 0.3 samples/s. All reactions provided a good *m/z* response of the desired product validated through a calibration curve carried out on different droplet size reactions. A direct comparison between in-droplet and non-droplet batch reactions clearly showed the yield improvement of the reaction while carried out in the flow system.

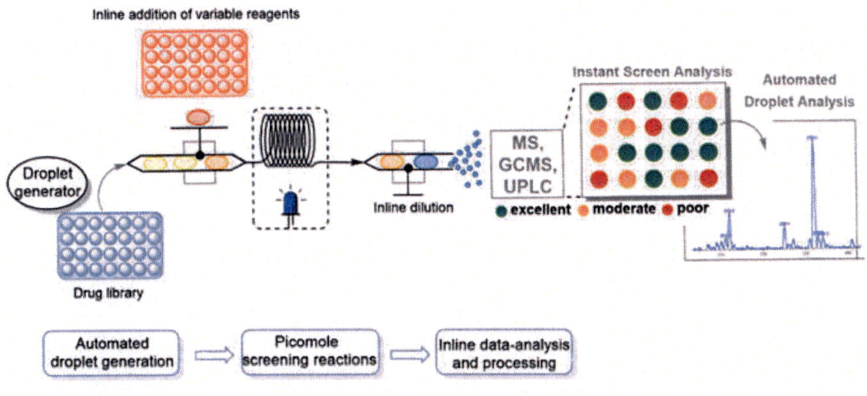

Fig. 27 Picomolar reaction screening microfluidic platform setup

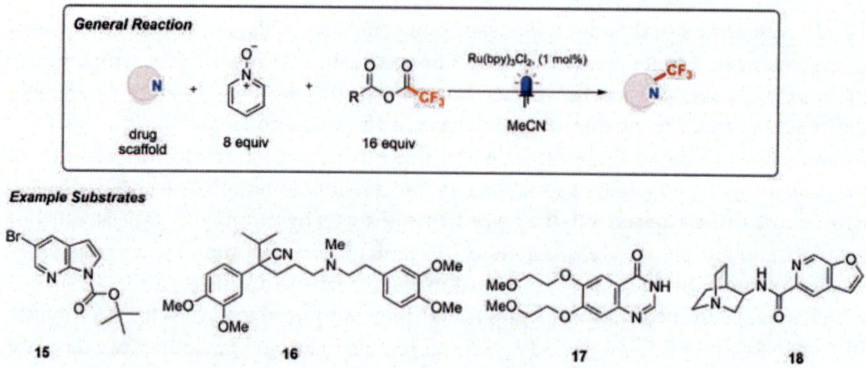

Fig. 28 Validation of methodology through trifluoromethylation reaction

3.3 *In Droplet HTE Reaction Discovery*

The real benefit of this proof-of-concept experiment is the use of the platform as a tool for in-droplet reaction discovery. For this purpose, a visible-light-driven alkene aminoarylation reaction was selected for validation (Fig. 29). To perform residence time optimization, an oscillatory system was created by alternating two syringe pumps between withdrawal and infusion modes while maintaining a constant flow rate of 200 nL/min. The setup could potentially accommodate more than 100 samples per incubation plate.

With the optimized setup, a set of 10 sulfonylacetamides combined with 10 alkenes were screened for the synthesis of 100 possible distinct alkene aminoarylation products from the Smiles-Truce rearrangement reaction. The droplet reactions were performed in 500 pmol scale and they were irradiated using a Cree LED array photoreactor for 30 min at a flow rate of 200 nL/min for a total of 100–200 droplets per incubation period. Afterward, ESI-MS analysis was performed at a throughput of 0.3 samples/s. From this screening, 37 hit conditions were identified which presented a significant product profile in the MS analysis. The library generation elucidated reactivity trends and structure-reactivity relationship of the reported alkene aminoarylation methodology. It was noted that electron-deficient sulfonylacetamides generally are more reactive and compatible with a broader scope of alkenes. In accordance with the batch-reported reactions, *trans*-anethole was one of the main by-products observed during evaluation of the reaction scope. The knowledge generated has significantly expanded the scope of the Smiles-Truce rearrangement reaction and gives important insights regarding future mechanistic studies.

Nine of the droplet reactions were selected for scale-up and yield validation. Initially, 0.01 mmol scale reactions were performed in the same reactor setup and run continuously at a flow rate of 400 nL/min for a residence time of 15 min. Upon irradiation, purification was performed using mass-triggered HPLC. Seven out of nine reaction products (**19–23**, **25**) were successfully isolated and confirmed the hit

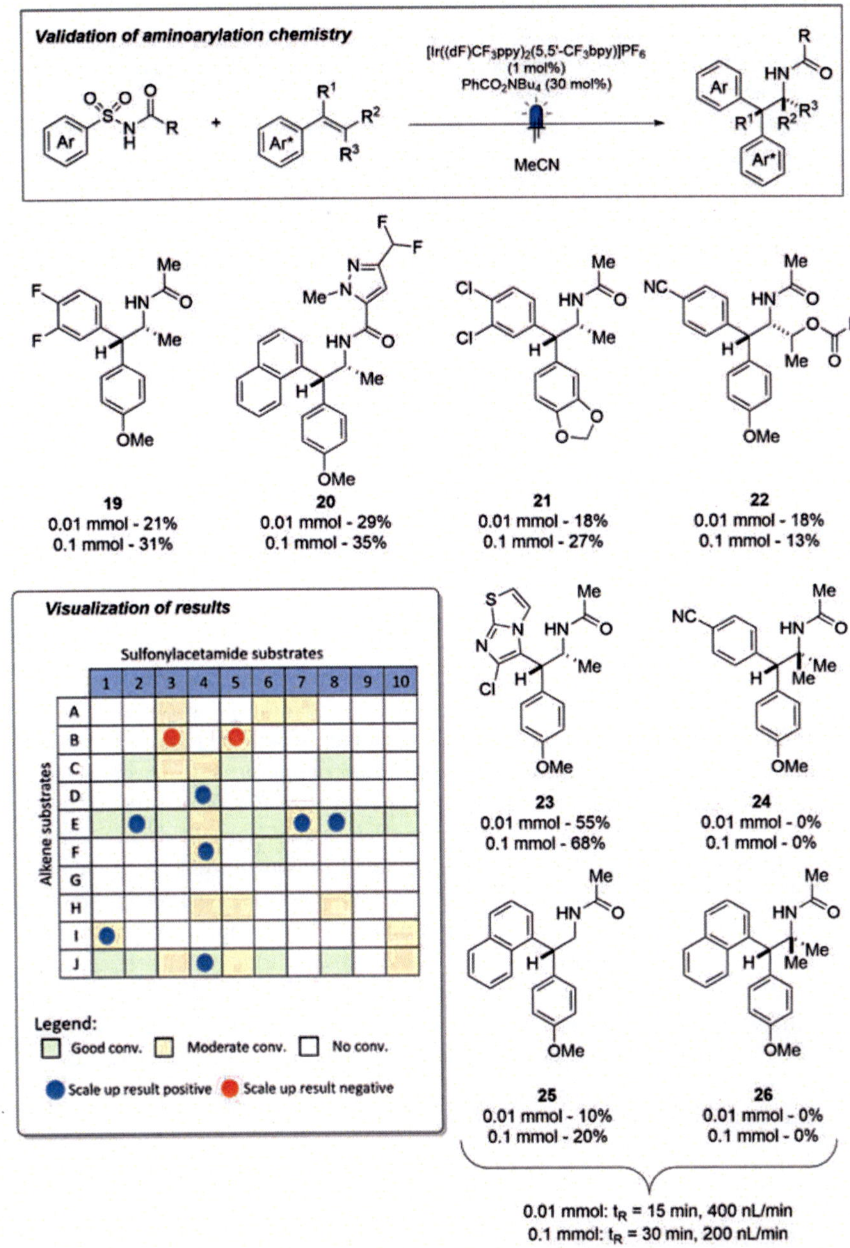

Fig. 29 In-droplet HTE reaction discovery depicting the reaction screening heatmap with the substrates for scale-up

result obtained in the microfluidic system. Control reactions were performed in the two reactions which were not able to yield the desired product (**24**, **26**). The false-positive result was due to a possible by-product formed with the same *m/z* as the expected product.

To demonstrate the transferability of the droplet screen results in microscale flow, a further scale-up in 0.1 mmol scale was set up to generate milligram amounts for drug discovery applications. In this case, a new setup was built using a 100 μL volume PFA reactor (0.03″ internal diameter) flowing with a flow rate of 3 μl/min for 30 min residence time. The isolated yields were reproducible as in the 0.01 mmol scale confirming the droplet reactions MS analysis results.

4 Automated Radial Synthesis and Parallel Functionalization of Rufinamide in Flow

4.1 Introduction

In recent years, flow chemistry techniques have been widely applied in process development for the synthesis of different APIs [37–40]. Nevertheless, its application in early phases of drug discovery combined with automation and PMC has been limited to very few examples. Applying flow chemistry in the early phases of drug discovery, simply put, means running the process continuously, as on an assembly line, thereby creating new chemical entities sequentially, which are thereafter ready and available for biological testing [41]. Most of the continuous processes for scaling up flow chemistry are carried out using a linear approach. For library synthesis and broader exploration of new chemistry, though, a radial approach can be more convenient. Seeberger *et al.* have described a radial synthesizer that can synthesize 18 different compounds of two derivative libraries without manual reconfiguration [42]. The term radial consists of a series of individually accessible reactors arranged around a central switching station. Single transformations or multistep sequences are performed as sequential but not simultaneous series of continuous operations. By decoupling the subsequent steps, the reactors can be re-used under different conditions, and the residence time is independent of the previous steps. Therefore, making the approach more versatile and efficient in order to obtain maximum returns, with minimum equipment.

4.2 Instrument Design

The radial synthesizer is composed of four sections (Fig. 30):

1. *Solvent and reagent delivery system (RDS)*

   Stores up to 12 solvents, liquid reagents, homogeneous solutions, or synthesized intermediates in pressurized vessels. A syringe pump, accessing the desired

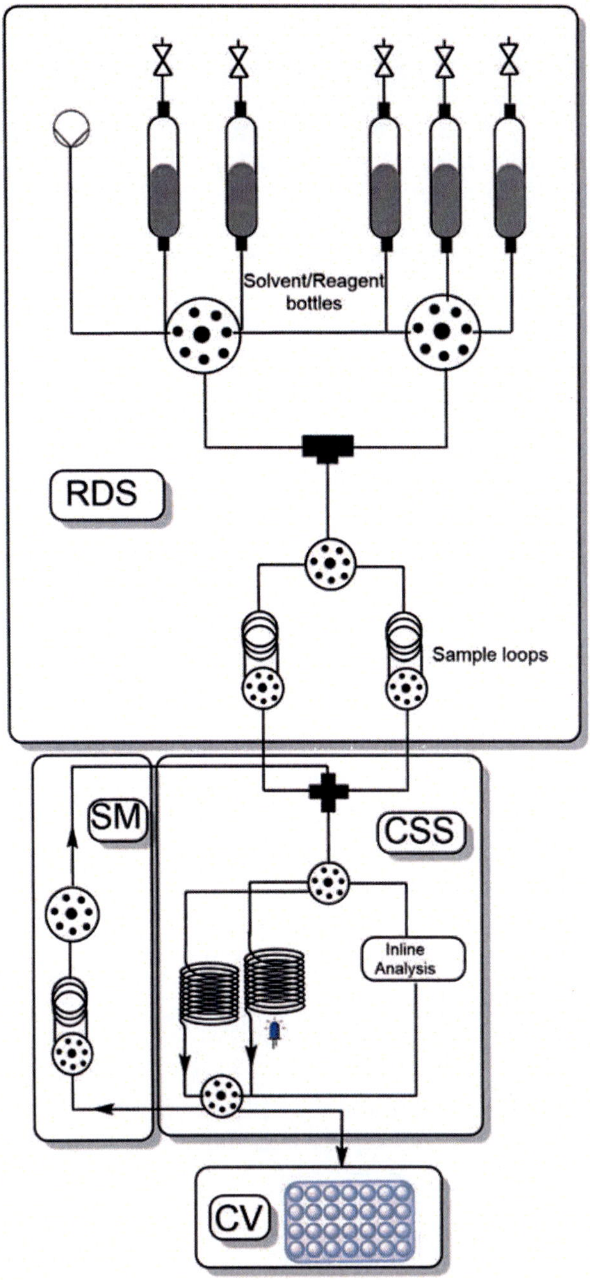

Fig. 30 The radial synthesizer

reagents via a 16-way valve, loads, 2 sample loops of 0.5 mL. The solutions can be diluted before entering the loop via the solvent delivery section. This enables reaction optimization for both concentration and solvents without the need for preparing new solutions. The solutions are then pushed from the sample loops by N_2 gas to a 4-way connector, which mixes the two solutions and directs them to the CSS, also known as the heart of the system.

2. *Central switching station (reactor and inline analysis) (CSS)*
   The CSS diverts the stream coming from RDS to the desired module, reactors, or inline analysis modules. When the residence time is completed, the solution exits the module and returns to the core. It then passes through a Flow IR module, before being sent to the collection vessel or the storage vessel in the RDS in case of intermediates. Alternatively, if they are going to be used immediately for a subsequent reaction step, they are loaded into a 0.5 mL loop within the standby module.
3. *The standby module (SM)*
   The standby module is composed of a series of bottles/containers used to store intermediates of reactions while waiting to be used in subsequent steps.
4. *Collection vessels (CVs)*
   The collection vessels represent the endpoint of the system. They are used to collect the final reaction fractions after the inline analysis has been performed.

There are essentially six possible "pathways" the solution can take in the Radial Synthesizer (not including options with regard to the choice of different reactor modules). The reagents can enter only from the reagents delivery system (single-step reaction/1st steps) or from both reagents delivery system and standby module (cyclic steps reactions) and the intermediate/product can be sent by the forwarding valve to the reagents delivery system (RDS), to the standby module (SM), or a final collection vessel (CV).

1. **R-R Pathway**: the reagents enter the CSS only from the RDS and the intermediate is stored in the RDS.
2. **R-S Pathway**: the reagents enter the CSS only from the RDS and the intermediate is stored in the SM.
3. **R-C Pathway**: the reagents enter the CSS only from the RDS and the intermediate is stored in the CVs.
4. **S-R Pathway**: the reagents enter the CSS from both RDS and SM and the intermediate is stored in the RDS.
5. **S-S Pathway**: the reagents enter the CSS from both RDS and SM and the intermediate is stored in the SM.
6. **S-C Pathway**: the reagents enter the CSS from both RDS and SM and the intermediate is stored in the CVs.

The entire system is pressurized with N_2 delivered from three Mass flow controller and the pressure is regulated by a digital BPR. The flow within and the addition of the reagents into the closed system are made possible by the controlled venting of N_2 at different points within the system. The flow rates of the solutions through the system are controlled by two MFCs in the RDS and three MFCs between RDS. The precise range of residence time depends on temperature (150°C max) and Pressure (9 bar max). All the modules are controlled from a software written in LabVIEW. The software allows the end-users to enter information related to reagent type, residence time, flow rate, system pressure, reaction steps, etc. habilitating the instrument to work fully autonomous. All data are processed and saved in a database for analysis or reproduction.

4.3 *Synthesis of Rufinamide and Derivatives in Flow*

The synthesis of rufinamide (**28**) in flow for scale-up has been previously described by Jamison et al. [43] They report a convergent route featuring copper tubing reactor-catalyzed cycloaddition reaction. The same route can be transferred to the radial synthesizer to synthesize rufinamide in a convergent or linear pathway with small re-optimizations and using CuI as catalyst in order to have a PTFE tube reactor as a more versatile reactor. For the convergent route, independent optimizations (one for the synthesis of the azide and the other one for the synthesis of amide) were run before the final optimization of the cycloaddition reaction. The three reactions were optimized in the synthesizer by screening solvents, stoichiometries, concentration, temperature, catalyst, and residence time. The conversion was monitored using inline Flow IR and the yield was measured offline by analysis of the collected samples. To optimize the last step, the intermediates of the first and the second step were stored in the RDS and SM and used for conditions screening directly from there. For the linear process, the most tricking step was the first one. During optimization, it was found that the intermediate of this step (6) was insoluble in the dilutions used for the convergent synthesis so inline dilution from 1.5 to 1 M was necessary to enable the continuous three-step synthesis. Both processes are concluded with purification by crystallization. The advantage of implementing two different synthetic pathways to a target molecule on the same instrument consists in the direct comparison for both synthetic routes. For the aforementioned the formation of the insoluble intermediate, which required further dilution in the linear route, was a key challenge for the purification of the product. In this regard, the convergent route seems to be the most efficient, providing both higher isolated yield (70% compared to 45%) and ^{1}H-NMR yield (88% compared to 83%).

Furthermore, the practicality of having both routes optimized in one system allows easy derivatization of the principal core by combining the functional groups used for the synthesis. There are two principal hotspots for functionalization in rufinamide (**28**): the arene and the amine. The arene core of the rufinamide is introduced initially in both the convergent and the linear pathways. With the convergent route, variation of this group would require two new potential re-optimizations, whereas three steps of the linear process could be affected. Likewise, the amine is introduced in the last step of the linear pathways so possibly one re-optimization is needed, whereas two re-optimizations might be required if the derivatives are obtained by the convergent route (Fig. 31).

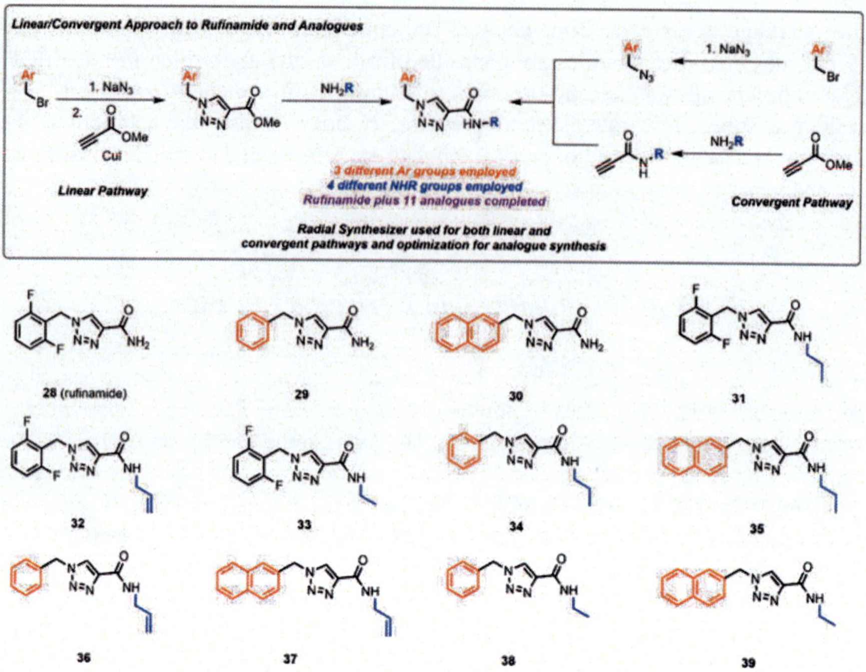

Fig. 31 Library synthesis of rufinamide (**28**) and derivatives **29–39**

With this in mind three different sets of derivatives were prepared using the pathways with less re-optimization for the synthesis of each class. In total, 18 derivatives of rufinamide were prepared following the same synthetic process in a sequential manner using a single reactor.

Two more benzyl bromides and three more amines were loaded into the delivery system RDS. The benzyl bromide derivatives class was diverted using the convergent route. The reactivity of the naphthyl and phenyl derivatives was similar to original benzyl group. Inline dilution 1:3 of naphthyl-methyl bromide and sodium azide was necessary due to solubility. Derivatives **29** (78% yield) and **30** (77% yield) were synthesized in two subsequent runs with a water/acetonitrile flush between synthesis.

The amine derivative class was synthesized using the linear route. As the amine diversification was the last step of the route, only conditions for step three were screened. Alkylamines showed lower reactivity than ammonia so the temperature was increased to 100°C and residence time to 20 min to complete the conversion. Allylamines were even less reactive and required stop flow to complete conversion in one hour at 100°C. Using the optimized conditions, derivatives **31**, **32**, and **33** were obtained in 86%, 58%, and 74% yield, respectively, with only wash cycles between each synthesis.

Finally, the third library of derivatives was prepared by combining the two benzyl moieties with different amines. In this case, the linear pathway was selected to run

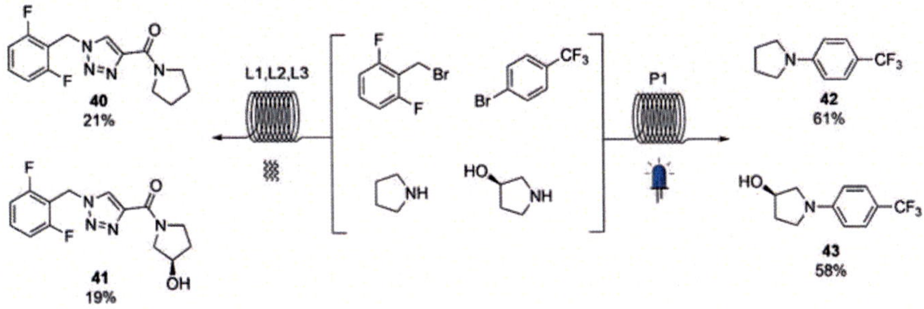

Fig. 32 Synthesis of rufinamide (**40**, **41**) and pyrrolidines **42**, **43** derivatives using a single loading of the RDS and both reaction modules (thermal and photochemical) of the radial synthesizer with no reconfiguration between syntheses; L1 (Reaction conditions: Aryl bromide + NaN_3); L2 (Reaction conditions: Methyl propiolate + CuI); L3 (Reaction conditions: Amines); P1 (Reaction conditions: $(Ir[(dF(CF_3)ppy)]_2(dtbppy))PF_6$; $NiBr_2$)

the library due to the solubility issues of the benzyl derivatives. No further optimization was required for the introduction of the amine. Six new derivatives (**34–39**) of rufinamide were prepared in 75–83% yields with washing cycles in between the synthesis of each derivative (Fig. 31).

To validate and confirm the flexibility of the radial synthesizer, a photochemical module was added to the platform. This allowed for additional diverse chemical processes to be performed on the same system with the same reagents and without reconfiguration (Fig. 32). The dual nickel/photocatalyzed cross-coupling reaction is a powerful methodology to build C-C and C-N bonds. Chemistry was initially validated with standard examples reported by MacMillan et al. [44] The coupling of the amine with the benzene was performed as a two-step process (generation of the intermediate and storage, use of the stored intermediate to react in the reactor, and collection in the collection vessel) using the photochemical reactor and the flow NMR module as an in-line analysis tool. Solutions of different aryl bromides (4-trifluoromethylbromide), cyclic amines (pyrrolidine, pyrrolidin-3-ol and piperidin-4-ol), $NiBr_2.3H_2O$, $(Ir[dF(CF_3)ppy]_2(dtbpy))PF_6$ (Ir cat) and 1,4-diazabicyclo[2.2.2]octane (DABCO) in dimethylacetamide (DMA) were loaded in the reagents pool of the radial synthesizer. Different combinations of reactants, as well as different ratios of $NiBr_2$, Ir cat, and DABCO, were screened under different residence times and light voltage. With the optimized conditions in hand, pyrrolidines **42** and **43** were obtained in moderate yields (Fig. 32).

Finally, the RDS was loaded with solutions for both rufinamide derivatives and metallaphotoredox catalysis. The same amine solutions were used sequentially with two different reactors for the synthesis of two new derivatives from one side using the thermal reactor and the synthesis of metallaphotoredox compounds using the photochemical reactor. No reconfiguration or additional modification to the reactor was required to perform these 4 consecutive synthesis processes.

5 Integrated Design-Make-Test Cycle Platform

5.1 Introduction

While iterative cycles of synthesis followed by in vitro screening represent the normal paradigm in drug discovery, it was recognized that exponential efficiency gains could be realized if these two activities could be combined on a single microfluidics-based platform. Furthermore, a "hit-to-lead" campaign could be effectively automated if such an approach were to integrate an algorithm based on activity data as it is generated informing the platform of what analogue to make next [45, 46]. Whilst this sounds like a relatively simple proposition, undertaking the task to develop such a platform for discovery represents a significant challenge. Cyclofluidics Ltd. undertook this endeavor, and the details were recently described in a publication from David Parry [47]. The "closed-loop" platform featuring design, synthesis, and assay of new molecules with immediate feedback of results informing the next compound to be made was initially conceived in 2007 based on work from GSK with the new company being started in 2008 (Fig. 33).

While the assembly of a multidisciplinary team is essential to realize this goal, from a chemistry perspective, several questions need to be addressed prior to

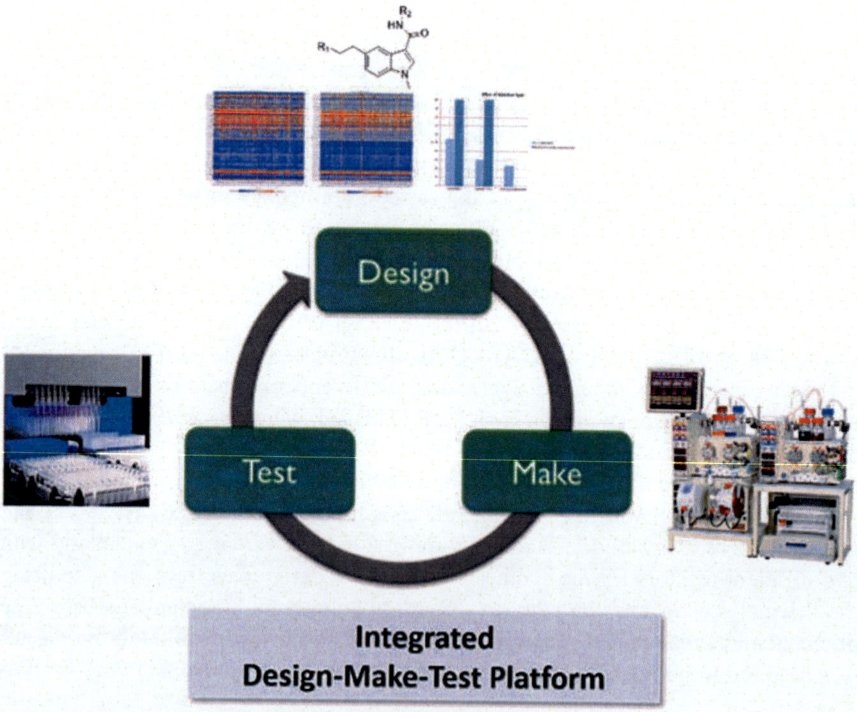

Fig. 33 Integrated design-make-test cycle platform

development. With a closed-loop system, making the compounds in flow was the logical choice though decisions needed to be made regarding the delivery of reagents, methods of synthesis, as well as how to achieve in-line purification and subsequent dilution to deliver a sample in a suitable form and of sufficient purity to be directly introduced into a biological assay. From a resource perspective, it made sense to utilize commercial flow-based Vapourtec apparatus, while the scale of syntheses was planned to be executed on microfluidic scale (<1 mm diameter tubing and μL volumes) enabling a material-sparing approach for the reagents being utilized. The initial prototype platform carried out the compound syntheses on a "glass-chip" while reagents were delivered from a carousel capable of storing/dispensing 24 samples.

To analyze and purify the reactions from the flow platform, an HPLC system was integrated with quantification achieved using ELSD. The desired product was isolated by sampling the middle of the corresponding HPLC peak followed by a subsequent dilution to provide a suitable sample for an in-line biological assay. For the initial platform, feasibility studies were initiated to carry out the relevant biological assays in flow initially on a custom glass-chip featuring ~80 μm channels though this was replaced at an early stage with 75 μm ID capillary tubing owing to both cost and ease of access/replacement. Several further challenges had to be surmounted in the eventual successful realization of this goal including ensuring solubility as well as utilization of a nano-HPLC pump to accurately deliver reagents/solvents at low flow rates and high pressures over a defined gradient to allow granular IC_{50} delineation between test compounds.

While the initial platform allowed proof of concept of the integrated "make-test" scenario, several significant drawbacks were identified with the prototype that would need to be addressed for a full-scale operational platform to be developed in addition to the need to integrate the feedback automation loop as well as enable seamless software control of all the various components. From a chemistry perspective, several logistical issues needed to be addressed including switching to a tube-based reactor platform to allow a more modular, flexible approach to selecting the optimal configuration to carry out the desired synthetic sequences. The scope of transformations was also considered and while one- and two-step sequences (in some cases a third deprotection step could be added) were considered routine to carry out provided pre-evaluation of reagent compatibility and solubility was taken into account, longer sequences were not explored as these would likely necessitate intermediate work-up and possible purification. The outlier herein was in cases where metal-catalyzed transformations were employed in which a silica plug filter was placed in line prior to chromatography to remove polar impurities given the known propensity for metal ions to interfere with biological assays.

The initial carousel designated for reagent storage/delivery was also felt to be unsuitable for the next-generation platform given its relatively limited capacity, which was felt would not enable a sufficient degree of chemical space to be explored. Several commercially available liquid handlers were evaluated based on their simplicity, reliability, and ease of integration/control within the platform with reagent capacity also a key consideration. The potential to explore a chemical space of one

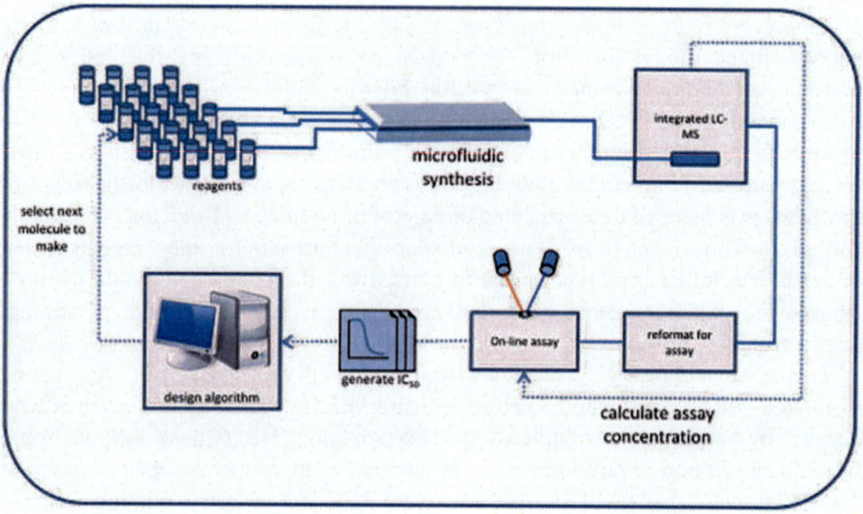

Fig. 34 Cyclofluidic optimization platform. (Reproduced by permission of ACS Publications)

million molecules (thus needing space for 300 reagents for a two-step sequence) was considered and subsequently proved to be sufficient as we will see in executing lead optimization campaigns.

While HPLC purification in tandem with ELSD quantification had proven to be highly effective especially with the implementation of a robust chromatographic method enabling the isolation of suitable amounts of test materials even from poor-yielding reactions (<5%), the ability to effectively capture the "timing" of the emergence of the "heart cut" of the peak of interest through system-controlled software proved critical. In addition, while the solution of the product emerged from the purification module in the millimolar concentration range, a significant dilution was still required (to the low micromolar range) for utilization in biological assays. A reformatting station was designed to enable a threefold dilution sequence to be achieved with judicious selection of the HPLC solvents considered for each program in order to avoid excess residual organics influencing the downstream biological assays.

The modified next-generation platform also incorporated an alternative approach to running biochemical assays with a switch away from flow-based assays to utilizing more conventional microtiter-plate-based systems (Fig. 34). With the synthesis/purification modules able to deliver an assay-ready sample, commercial vendors offered several options for dispensing of assay reagents as well as reading of either 96- or 384-well plates while a custom single probe liquid handling platform was developed. Assays typically employed fluorescence-based techniques with a versatile optical cartridge plate reader used as the detector. Continuous, automated operation of the whole platform over periods of days required standardization of the assays specifically regarding reagent stability prior to operation. In drug development, the physicochemical properties of the molecules under investigation are also a key consideration, and in the modified platform, a second HPLC system was integrated to obtain a Log D measurement as well as an estimate of concentration (UV) and QC purity.

Integrated software to ensure seamless communication, data capture, and scheduling across all components of the platform was obviously critical to the whole endeavor and a proprietary environment was developed which enabled utilization of the original vendor software with the control achieved through either application program interface or contact-closures triggering events between the system's hardware.

While robust data capture and precise control of processing parameters represent hallmarks of running operations in an automated flow-based fashion, developing an algorithm to incorporate into the platform's software and informatics infrastructure to allow compound optimization in a serial fashion while optimizing cycle times was a key challenge. While several methodologies are potentially suitable as the foundation for this algorithm, a random forest approach was selected and optimized utilizing a wide range of molecular descriptors to build a predictive model while also able to carry out a multiparameter optimization once the nature and appropriate weighting of the various parameters was defined. Regarding the implementation of the algorithm, an initial area of chemical space needed to be defined for each SAR-based exploration, which required some initial activity data to be supplied upfront in order to facilitate the subsequent optimization. For the latter, two possible strategies can be employed for screening the space chemical space available either through a broad survey of activity data across the entire collection of compounds or by identifying the best molecule through a sequential compound by compound based approach focused on incremental increases in potency. In practice though, programs are best suited to utilize a hybrid of these two approaches with initial cycles highlighting broad structural requirements that enhance activity within the lead series with these being followed by cycles that drill down to the specific compound with desired potency/properties within a subset of the overall chemical space.

5.2 *Abl Kinase Inhibitors*

To validate and confirm the potential of the platform for lead optimization purposes, a specific target-based project was undertaken [48]. BCR-Abl (breakpoint cluster region Abelson tyrosine kinase) inhibitors initially through the discovery of imatinib (**47**) have established themselves as effective medications for the treatment of chronic myeloid leukemia (CML). Despite robust efficacy, some patients relapse due to the emergence of clinically relevant mutations (as exemplified by T315I – the gatekeeper mutation), and while second- and third-generation inhibitors have been reported that demonstrate enhanced efficacy against mutated versions of the kinase, the identification of new inhibitors of BCR-Abl remains an attractive and valuable endeavor. Furthermore, BCR-Abl is an established drug target with known validated assays as well as a range of data reported for previous inhibitors, while soluble protein targets (kinases) are also commonly pursued within drug discovery.

To initiate the search for novel inhibitors of BCR-Abl, it was important to not only define the chemical space for exploration but also identify a robust flow-based sequence in order to assemble these molecules. Examination of both the previously disclosed inhibitors and utilizing structural insight derived from the binding mode exemplified by the third-generation alkyne-based inhibitor ponatinib (that does show appreciable efficacy against the T315I mutant enzyme) enabled a generic model structure to be proposed featuring a heterocyclic hinge-binding group (an imidazopyridine in ponatinib) linked through an alkyne to a structural motif that binds to the "DFG-out" conformation of Abl, and terminates with a basic amine.

Planning for the assembly of the target molecules on the integrated-platform, it was rationalized that a Sonogashira-type coupling between a (hetero)aryl alkyne and the "DFG-templated" halide would represent a robust, modular bond disconnection to the desired compounds. This reaction is mediated using Pd/Cu catalysis with numerous examples of such Cu-mediated reactions being reported in flow with the necessary exposure to the catalyst provided through employing a copper-reaction coil [49].

Further planning as to the chemical space to be investigated highlights one of the key challenges typically associated with drug discovery endeavors based around the synthesis of compound collections specifically to access to the desired building blocks required to assemble the molecules of interest. The importance of this aspect of planning cannot be over-emphasized given the strong argument that can be made that the "drug-likeness" and "potential for success" are inherently based on the judicious selection of these building blocks/monomers. This is clearly evidenced by the exponential increase in both the sheer number of pharmacophoric monomers available and the commercial sources to access them from. Usually, such offerings are organized into specific collections based on the reactive functionalities present through which these can be further elaborated.

Having said this, there are still numerous structural classes of monomers, which are under-represented in terms of commercial availability. This is exemplified in the example here with evaluation of a series of databases demonstrating that there are relatively few (hetero)aryl alkynes available. However, these can be accessed in a trivial fashion from the wide variety of either bromo- or iodo-based heterocycles that can be sourced, still this endeavor highlights an aspect of flow chemistry for library syntheses within medicinal chemistry that is often overlooked. While streamlined synthesis of libraries of drug-like compounds can be realized in flow, this exercise may rely on the need for the off-line synthesis/storage of custom-building blocks thus entailing investments both in cost and time resource. As noted here, it is somewhat ironic that the reaction chosen to synthesize these compounds is the classical Sonogashira coupling performed in batch, which also is the pivotal flow-step for the assembly of lead molecules on the platform. Triage of the heterocycles was achieved through a combined analysis involving comparison with other hinge-binding motifs and overlay with the available ponatinib-Abl co-crystal structure leading to 27 monomers being selected for purchase/synthesis.

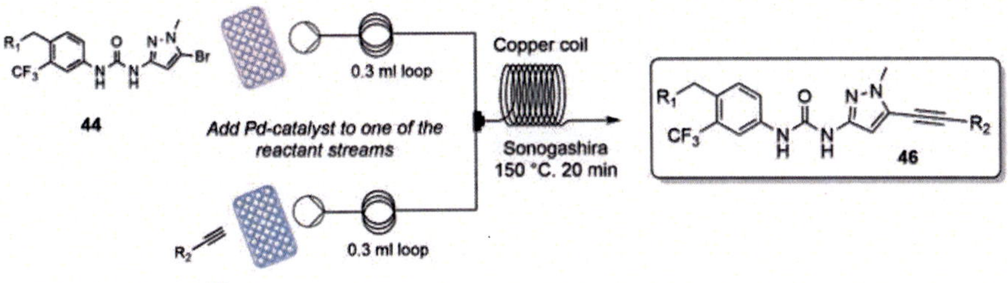

Fig. 35 Optimization of the key Sonogashira-coupling on the flow platform

Similarly, a degree of synthetic investment was required to assemble the DFG-binding templates. Key differentiators between the 10 selected motifs were the nature of both the connector (amide, reversed amide, urea) incorporated therein, the aryl moieties and the basic-amine substituent. Only a small range of these were evaluated with the main vector of structural diversity of the 270 possible new compounds intended to be the hinge-binding heterocycle.

With the necessary monomers for the platform in hand, a key requirement became identifying robust conditions in flow for the Sonogashira-coupling in order to access the compounds selected for synthesis by the design algorithm. To achieve this, a commercial Vapourtec R4 synthesizer was used as the flow instrument with 120 mM solutions of the two reactants in DMF combined and passed through a 1 mm ID copper coil (capacity 2 mL) heated to 150°C at a flow rate of 100 μL/min to provide a residence time of 20 min. The Pd-catalyst was added to one of the reactant solutions with the removal of metal residues from the exiting stream achieved by passing the crude reaction through a plug of silica. Following this at maximum concentration, 10 μL was collected in an injection loop, which was switched automatically to load the sample onto an LC/MS for purification before reformatting for the subsequent bioassay (Fig. 35).

While the synthetic chemistry carried out on the platform can be considered to be somewhat routine (a characteristic of library chemistries in general), the most significant differentiator of this approach above and beyond the integrated nature of the platform is the algorithm driving the design of the next compound from the available virtual chemical space based on the data as it is collected. This program utilizes random forest activities using as noted a blend of "chase potency" and "most active under-sampled" strategies.

Key to the use of an algorithm for a machine-learning-based exercise is the ability to break down the compounds into a combination of numerical descriptors (MW, H-bond donors, TPSA, etc.) as well as a molecule fingerprint that captures the structure of the compound in a machine-readable form. The popularity of Random Forest-based optimizations within chemistry-based endeavors is due to its ability to perform non-linear regression while being resistant to over-fitting and requiring very few tuning parameters.

Regarding the design strategies used, "chase potency" simply sorts all the possible compounds based on predicted potency and then prioritizes the "best" compounds for synthesis. The "most active under-sampled" methodology is designed to ensure fuller coverage of the virtual chemical space under consideration. To accomplish this, the program considers how often each of the available reactants is utilized within a systematic series of designs and queues compounds for synthesis to sample all the reactants as equally as possible. Ideally though in a lead discovery campaign, one wants to rapidly identify the optimal compound as quickly as possible while not missing something, and so to achieve this a combination of the two approaches is desirable running sequential loops between scanning for pockets of activity, and then drilling down to the most active compounds in each space.

Critical to the success of this endeavor as a whole is to be able to predict activity, and in order to achieve this, a "training set" of compounds with known potency against the target has to be provided to the algorithm either from within the virtual chemical space or through closely related compounds. As noted Abl1 is a well-established target, and given this, the data around ponatinib and related analogies (36 compounds in all) were used to seed the design model leading to the output of predicted activity for the 270 proposed possible compounds for synthesis (note based on the monomer selection, ponatinib itself is included within the virtual chemical space allowing data to be generated from a positive control).

In order to validate the generation of IC_{50} data on the platform, the synthesis of imatinib was carried out using the flow-based approach reported by Ley and co-workers [50–53]. Screening of the authentic sample both on the platform and in a conventional manual assay provided IC_{50} values that were consistent with those reported in the literature.

For the experimental strategy, three parts were devised in which the first one would evaluate "most active under-sampled" areas of the virtual space to ensure broad coverage while identifying areas of enhanced potency. In the second phase, drilling down into the design spaces featuring the more active compounds would be pursued through a "chase potency" approach while a third cycle would combine the two strategies to identify and subsequently optimize additional activity "hotspots."

In practice, the first part consisted of 29 design-screen-synthesis cycles taking approximately 30 h to complete. From the loops, 22 compounds (76% success rate) were assayed with six failing in the synthesis step with a further analogue failing QC. Learnings from the initial round of syntheses highlighted the pyrazole-urea-based template featured in compounds with promising activity while several compounds featuring the ponatinib heterocyclic template also stood out. In the areas of weaker potency, gratifyingly compounds with a phenyl-motif as the proposed hinge-binding component were prominent and served as a negative control given that such compounds are unable to make the necessary hydrogen-bond interactions in the pocket of the protein.

Part two enabled focus on areas identified with relatively potent compounds. A deeper evaluation around the SAR of a novel hinge-binding motif specifically investigating potency variations based on the nature of the connections in the DFG-binding templates (amide, reverse-amide, urea). 20 design/synthesis loops were carried out with 14 compounds being successfully synthesized enabling a refined heatmap of predicting activities (now incorporating the data from the 36 new compounds made into the model) to be generated for utilization in part three.

The final combined design approach utilized six alternating cycles of "most active under-sampled" and "chase potency" strategies (a total of 12 cycles). Through these, a further 41 loops were executed leading to access to 28 more compounds (68% success rate) with the main outcomes of this cycle being a more thorough exploration of two novel hinge-binding motifs predicted to have good activity.

In summary, 90 design-synthesis loops were carried out leading to 64 new compounds (71% overall success rate) being assayed against Abl1/Abl2. Whereas from a conventional library success rate, the chemistry is pretty much comparable, what is notable here is the ability for the platform to purify enough material for testing from low-yielding (~5–10%) reactions.

At this juncture, it is instructive to benchmark this endeavor with a typical medicinal-chemistry program in the hit identification space. One can hypothesize that the exercise would initially start with the goal to identify alternative lead matter to ponatinib. A similar library design could be envisioned based around the pivotal bond formation originating from a Sonogashira-coupling. As noted previously, one of the key elements of the overall design is in the selection of suitable building blocks, and this is likely to follow the same process used in a conventional approach. This would lead to a 27 × 10 matrix compound library, which is certainly within scope for many parallel-synthesis-based groups utilizing plate-based approaches though the likelihood is that a bottleneck will occur during a downstream purification step. In addition, the success of this purification may not only depend obviously on the methodology (HPLC, SFC) employed (with a common method being likely, which is analogous to the platform) but also potentially on a pre-purification QC that sets cut-offs on what level of desired product needs to be detected in a specific well for the reaction to be processed to the actual purification. In addition, further downstream processing, transfer, and reformatting will be required to prepare the samples for the relevant assays. However, once this is done, the biological assays are typically processed in a high-throughput plate-based fashion. Another interesting aspect of this comparison is in the downstream logistics and most notably in the amounts of compound made. In the conventional scenario, all the compound is purified with excess material retained after re-constitution of samples for Absorption, Distribution, Metabolism, Excretion/biological studies, and while this might in some ways be seen to be wasteful, it does provide amounts of compound for further downstream assays, enable rigorous QC analysis of the compounds, and enable samples of novel chemical matter to be added to a larger collection of compounds for future high-throughput screening campaigns. One parameter, which we did not discuss in the conventional setting, which mitigates to a large extent the issue of waste, is the scale on which a library campaign is run. With advances in liquid handling and purification capabilities there is now a tendency to run such chemistries on an analytical scale yielding 1–3 mg of the compounds of interest. In contrast, whereas the platform operates on a relatively efficient scale, only a fraction of the output from the flow reaction is segmented from the major product peak and processed through purification thus leading to the remainder being treated as waste. In this paradigm, sample transfers/logistics are effectively replaced after purification by dilution, this leaves residual organic solvents, which could in principle interfere with the assay. Finally, concentration is assayed by ELSD as opposed to arguably a more rigorous approach based on accurate weighing after a standard library prep. The biggest distinction though between the two approaches from a chemistry perspective is based on the number of and speed by which compounds get made. While with a conventional approach, the synthesis of all the compounds in the matrix will be attempted, and although ~25–30% will fail (given that the same chemistry is being employed, it is a somewhat safe assumption to say that the synthesis of the same compounds will fail in both paradigms), one is highly likely to identify the same activity trends as one would with the platform-based approach in which only a subset of compounds are synthesized (~24% in the Abl example) based on a predictive algorithm. In the latter case, these compounds in tandem with the upfront model data provided informs the heatmap for the remaining (i.e., those not synthesized) compounds potency data. One caveat to an assumption we have made herein is that use of a flow-based method of synthesis may allow access to novel ("forbidden") compound space or enable compounds to be obtained (either through superior purifications of "low-yielding" reactions or through enhanced reactivity), which would be classed as "failed" under normal circumstances. Still though, to some degree, the questions around "gaps in data" from compounds that fail synthesis (or are not made) lead to frustrations of potential missed opportunities for practicing medicinal chemists.

To conclude the study on the identification of novel Abl inhibitors, four of the compounds (**48–51**) identified from the platform were resynthesized as solids for further profiling. Three of these featured the novel pyrazole-urea while one presented an amide-linker while all possessed novel heterocycles as the hinge-binding motif. In contrast, all the compounds possessed the same aryl moiety of the DFG-binding fragment found in ponatinib. For further validation of the data obtained from the platform a further, five mildly potent compounds were resynthesized for testing. In general, for all the compounds the correlation between platform, re-test on solid material, and predicted values (Abl1 only) for Abl1/Abl2 IC50 data was excellent (Fig. 36).

The four lead compounds (**48–51**) showed excellent potency against all the clinically relevant Abl-mutants (including T315I) as well as good selectivity against closely related kinases. While the compounds had mediocre physicochemical properties (clearance, permeability), it is proposed that further loops of optimization could address this. The identification of the pyrazole-urea DFG-out motif as a replacement for the benzamide present in ponatinib is cited as the highlight of the endeavor though it is important to recognize that this decision was implicit in the compound design for the initial selection of the building blocks.

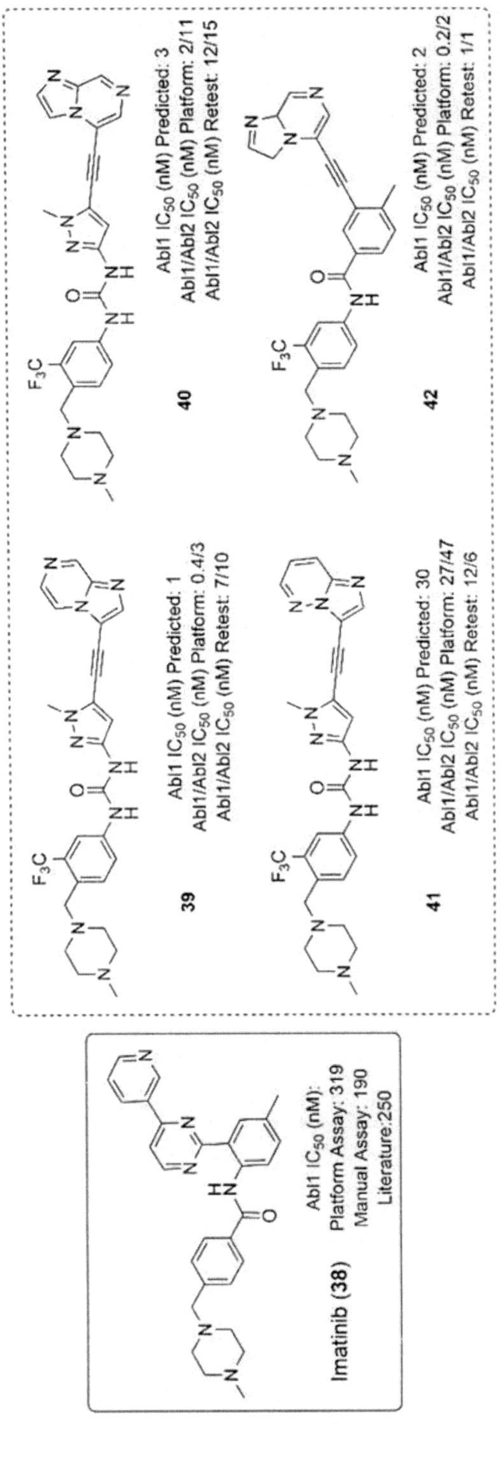

Fig. 36 Lead analogues identified through the platform-optimization process

5.3 DPP4 Inhibitors

To further validate this approach, a collaboration was undertaken with a pharma partner (Sanofi-Aventis) around a series of compounds with known DPP4 inhibitory activity [54]. The purpose of this exercise was to see whether the platform could successfully replicate the known data using the algorithm-based approach. Dipeptidyl peptidase-4 (DPP4) inhibitors (or gliptins) are a class of oral hypoglycemics used for the treatment of type-2 diabetes with the first agent in class sitagliptin (Januvia) having been approved by the FDA in 2006. DPP4 inhibitors work through increasing incretin levels inhibiting glucagon release, which leads to insulin secretion thus both decreasing gastric emptying and blood-glucose levels. For the exercise herein, Sanofi-Aventis provided a series of xanthine-based compounds with known inhibitory activity against DPP-4 for evaluation on the platform.

The first task was to develop the flow-based chemistry to access the molecules of interest, which could be derived from displacement off 8-bromo-xanthine-based scaffold **52**. Highlights of this chemistry include the need to incorporate both protected-diamines and amino-alcohols as substrates as well as develop a two-step displacement/deprotection synthetic sequence.

For the diamine-derived compounds, the initial displacement was carried out in a 2 mL stainless-steel column at 150°C and a residence time of 20 min using two equivalents of the Boc-protected diamine **53** with NMP as the reaction solvent. The output from the reaction was mixed with a 30% aqueous solution of $MeSO_3H$ from an in-line pump to affect the Boc-deprotection. The combined stream was passed through a second coil at 90°C for ~13 min with the output collected in a 20 μL injection loop at the point of maximum concentration for purification purposes (Fig. 37).

System consistency was established through the repeat synthesis of a standard compound across ten cycles with the assay being performed in duplicate on different human DPP-4 columns with the results further validated by comparison with those

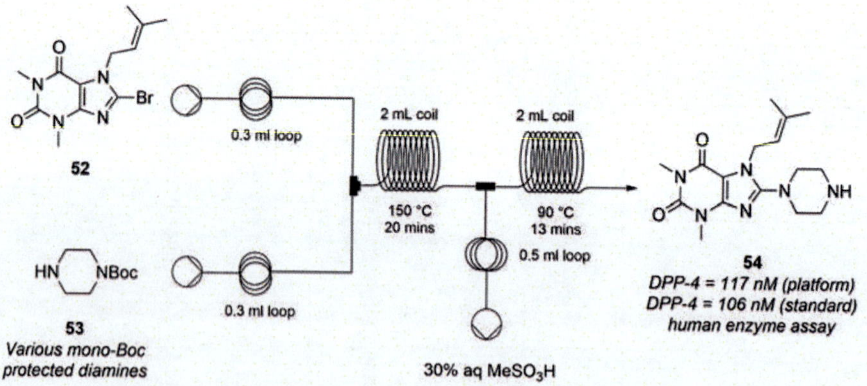

Fig. 37 Synthesis of DPP-4 inhibitors derived from mono-protected diamines

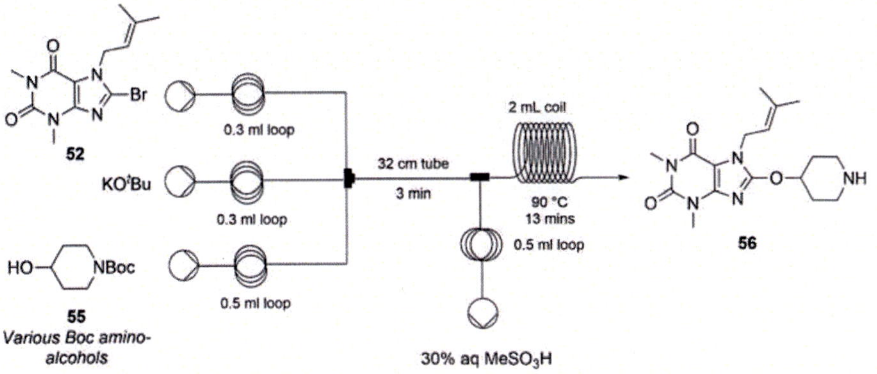

Fig. 38 Synthesis of DPP-4 inhibitors derived from protected amino-alcohols

obtained from a solid sample of the inhibitor prepared off-platform. Following this validation, all 12 of the proposed compounds were successfully synthesized with yields ranging from 3 to 38% (note again the ability to recover enough material for assay from a low-yielding reaction) with a total cycle time per sample of 120 min.

For the proposed compounds derived from amino-alcohol displacements, a modified protocol was developed to accommodate the extremely rapid reaction mediated by KO^{t}Bu. Combination of reagents **52** and **55** followed by passage through a 32 cm long 1 mm ID tube followed by an analogous Boc-deprotection step led to the desired compounds (exemplified by **56**). Eighteen of twenty-one compounds were successfully synthesized with a maximum yield of 23% (Fig. 38).

The platform assay was carried out in a plate-based format with the compounds titrated into aqueous buffer followed by addition by either porcine or human DPP-4 enzyme. Residual enzyme activity was monitored by adding H-Gly-Pro-AMC as the substrate with the data analyzed through non-linear regression to generate a slope for IC_{50} determination. Comparison with the unblinded source data revealed an excellent correlation for both the porcine- ($R^2 = 0.9253$) and human-enzyme ($R^2 = 0.9893$) assays though in the latter case only limited original data was available. This study (whilst omitting the synthesis/design algorithm) highlights the ability of the platform to efficiently synthesize and test – the whole exercise to make/test 29 compounds are reported to have taken 3 days – potentially bioactive compounds in a robust manner [55].

6 Conclusions and Outlook

The work described herein has aimed to present a somewhat different perspective on the utilization of flow chemistry with a view to its implementation in a medicinal chemistry setting. In contrast to previous reviews, here several detailed studies have been presented focusing on the implementation of flow-based technologies to provide a sustainable solution to expedite a workflow within drug discovery. Central to all the work discussed within this chapter is a significant amount of upfront resource investment required to devise, test, and validate the equipment setup, and if necessary re-work to provide a "proof-of-concept" for the proposed approach prior to the technology realizing its full utility. This is indicative of a "disruptive technology", though in the cases presented it is interesting to observe firstly the future potential uptake within the responsible departments/organizations, and whether the public disclosure of the methodologies sparks their adoption (perhaps with enhancements) within other companies/academic institutions. Whether this occurs will critically depend on an assessment of the cost vs value associated with each of the approaches disclosed.

Sections 2 and 3 of this chapter discussed alternative flow-based approaches to high-throughput experimentation (HTE), and no doubt the exponential growth in adoption of this technology within medicinal chemistry groups has fueled the drive to develop methodologies to achieve this in flow. Whereas the nanomole-screening approach disclosed by Pfizer is impressive in the sheer quantity and quality of information/data that can be rapidly acquired in a relatively-material sparing fashion while also enabling evaluation of the reaction solvent, the classic challenge remains handling heterogeneous reactions. In addition, the approach is ideal for reactions with rapid reaction kinetics with the assumption that the initial rate will be indicative of the overall reaction progress can provide misleading results specifically in cases in which catalyst deactivation may occur. Increasing the residence time does offer a solution to mitigate this problem though will lead to a commensurate decrease in throughput. The core system is relatively inexpensive to assemble, though costs escalate depending on the instrumentation employed to analyze the reactions. Whereas in-line analysis is optimal, this requires two UPLC-MS instruments to be utilized. The system can easily be adapted to run continuously to generate milligram quantities of material, and although not reported in the original disclosure, one can easily envision this setup being utilized for parallel medicinal chemistry campaigns.

The droplet-based system reported by Stephenson demonstrates a further drop in reaction scale operating on picomoles while showing the ability to handle 1000s of reactions per day. An advantage of the system is with the ESI-MS analytics employed enabling rapid screening of the droplets as they emerge from the system. Still the first read of data from this technique provides a more qualitative overview of reaction success as opposed to the precise analytics allowed by the previous methods. The system was designed and is particularly well-suited to photoredox-based chemistries with a solution to extending residence times (with again a corresponding reduction in reaction throughput) provided by developing an oscillatory system to prolong sample irradiation. Scale-up in some cases proved challenging with a different flow setup designed to access tangible amounts of material while translation to batch led to lower yields. The latter is unsurprising given the well-known advantages of carrying out photoredox reactions in flow, though this system is ideal for the discovery of new reactivity paradigms as well as rapid substrate screening in late-stage diversification campaigns.

While numerous automated synthesis platforms have been developed for accessing biologically relevant compounds through flow chemistry, the radial synthesizer described by Seeberger and Gilmore enables a more versatile approach owing to the positioning of a series of reactor positions around a central switching station. Through the employment of this technology, both single transformations or multistep sequences can be performed without the requirement for additional systems to be added downstream allowing both linear and convergent syntheses to be accommodated within the same reactor. Future development of such a modular approach holds great promise within drug discovery laboratories with the major barrier to uptake being the lack of familiarity of the current medicinal chemistry community to adoption of flow technology for the routine synthesis of analogues. The integrated design-make-test cycle platform represents an impressive achievement specifically with fully automating the drug discovery loop including the incorporation of a feedback loop with the data obtained from a series of synthetic runs informing the nature of the next set of compounds to be made. The work undertaken in this endeavor highlights the complexity of combining compound synthesis, purification with the subsequent appropriate dilution for the desired biological assay(s). The current excitement surrounding both AI/ML has in many cases focused on learning from series of training sets of compounds and legacy data to prioritize through various algorithms novel compounds for synthesis [56–60]. In addition to the sheer complexity of the task, perhaps the major drawback of the Cyclofluidics work is the lack of synthetic chemistry diversity reported. There have been numerous other reports of flow syntheses of combinatorial libraries in pharmaceutical discovery, though typically the chemistry (and in some cases in-line purification) is disconnected from the biological assays.

This chapter has attempted to describe the implementation of flow chemistry technology to address several key tasks currently of intense interest to the medicinal chemistry community. However, as can be seen, a significant amount of development is often required to provide a tangible solution to the task in hand. However, when reached, seamless integration of the technology into the conventional workflows brings major benefits regarding both throughput and resource efficiency. As the current medicinal chemistry communities gain further familiarity with flow chemistry, we will no doubt see it further exploited within a drug discovery setting.

Compliance with Ethical Standards *Funding*: No funding is applicable for this chapter.
Ethical Approval: No animal or human studies were performed in this chapter.
Informed Consent: No patients were studied in this chapter.

References

1. See for example: Gioiello A, Piccinno A, Lozza AM, Cerra B (2020) J Med Chem 63:6624–6627
2. Bogdan AR, Dombrowski AW (2019) J Med Chem 62:6422–6468
3. López E, Linares ML, Alcázar J (2020) Future Med Chem 12:1547–1563
4. Baumann M, Moody TS, Smyth M, Wharry S (2020) Org Process Res Dev 24:1802–1813

5. Bogdan AR, Organ MG (2018) Top Heterocycl Chem 56:319–342
6. Browne DL, Howard JL, Schotten C (2017) Comp Med Chem III 1:135–185
7. Farrant E (2020) ACS Med Chem Lett 11:1506–1513
8. See for example, Baranczak A, Tu NP, Marjanovic J, Searle PA, Vasudevan A, Djuric SW (2017) ACS Med Chem Lett 8:461–465
9. There have been several recent publications, which have sought to make flow chemistry more accessible to laboratory researchers. See for example: Guidi M, Seeberger PH, Gilmore K (2020) Chem Soc Rev 49:8910–8932
10. Plutschack MB, Pieber B, Gilmore K, Seeberger PH (2017) Chem Rev 117:11796–11893
11. Britton J, Jamison TF (2017) Nat Protoc 12:2423–2446
12. The H-Cube now represents standard instrumentation in most medicinal chemistry laboratories. See https://thalesnano.com/products-and-services/ (accessed June 13th, 2021). See also: Bryan MC, Wernick D, Hein CD, Peterson JV, Eschelback JW, Doherty EM (2011) Beilstein J Org Chem 7:1141–1149
13. Plowright AT, Johnstone C, Kihlberg J, Pettersson J, Robb G, Thompson RA (2012) Drug Discovery Today 17:56-62
14. Lombardino JG, Lowe JA (2004) Nat Rev Drug Discov 3:853–862
15. Murray PM, Bellany F, Benhamou L, Bučar D-K, Tabor AB, Sheppard TD (2016) Org Biomol Chem 14:2373–2384
16. Cernak T, Gesmundo NJ, Dykstra K, Yu Y, Wu Z, Shi Z-C, Vachal P, Sperbeck D, He S, Murphy BA, Sonatore L, Williams S, Madeira M, Verras A, Reiter M, Lee CH, Cuff J, Sherer EC, Kuethe J, Goble S, Perrotto N, Pinto S, Shen D-M, Nargund R, Balkovec J, DeVita RJ, Dreher SD (2017) J Med Chem 60:3594–3605
17. Cernak T, Dykstra KD, Tyagarajan S, Vachal P, Krska SW (2016) Chem Soc Rev 45:546–576
18. Zuo Z, Ahneman DT, Chu L, Terrett JA, Doyle AG, MacMillan DWC (2014) Science 345:437–440
19. Lovering F, Bikker J, Humblet C (2009) J Med Chem 52:6752–6756
20. Krska SW, DiRocco DA, Dreher SD, Shevlin M (2017) Acc Chem Res 50:2976–2985
21. Buitrago-Santanilla A, Regalado EL, Pereira T, Shevlin M, Bateman K, Campeau LC, Schneeweis J, Berritt S, Shi Z-C, Nantermet P, Liu Y, Helmy R, Welch CJ, Vachal P, Davies IW, Cernak T, Dreher SD (2014) Science 347:49–53
22. Mohamed DKB, Yu X, Li J, Wu J (2016) Tetrahedron Lett 57:3965–3977
23. Reizman BJ, Jensen KF (2015) Chem Commun 51:13290–13293
24. Perera DW, Tucker JW, Brahmbhatt S, Helal CJ, Chong A, Farrell W, Richardson P, Sach NW (2018) Science 359:429–434
25. Hawbaker N, Wittgrove E, Christensen B, Sach N, Blackmond DG (2015) Org Process Res Dev 20:465–473
26. Lennox AJJ, Lloyd-Jones GC (2014) Chem Soc Rev 43:412–443
27. Jover J, Fey N, Harvey JN, Lloyd-Jones GC, Orpen AG, Owen-Smith GJJ, Murray P, Hose DRJ, Osborne R, Purdie M (2010) Organometallics 29:6245–6258
28. Lennox AJJ, Lloyd-Jones GC (2013) Angew Chem Int Ed 52:7362–7370
29. McAtee RC, McClain EJ, Stephenson CRJ (2019) Trends Chem:111–125
30. Cambie D, Bottecchia C, Straathof NJW, Hessel V, Nöel T (2016) Chem Rev 116 (17):10276–10341
31. Tucker JW, Zhang Y, Jamison TF, Stephenson CRJ (2012) Angew Chem Int Ed 51:4144–4147
32. Corcoran EB, McMullen JP, Levesque F, Wismer MK, Naber JR (2020) Angew Chem Int Ed 59:11964–11968
33. Hwang YJ, Coley CW, Abolhasani M, Marzinzik A, Koch G, Spanka C, Lehman H, Jensen K (2017) Chem Commun 53:6649–6652
34. Beulig RJ, Warias R, Heiland JJ, Ohla S, Zeitler K, Belder D (2017) Lab Chip 17:1996–2002
35. Shang L, Cheng Y, Zhao Y (2017) Chem Rev 117:7964–8040
36. Sun AC, Steyer DJ, Allen AR, Payne EM, Kennedy RT, Stephenson CRJ (2020) Nat Commun 11:6202

37. Li J, Ballmer SG, Burke MD (2015) Science 347:1221–1226
38. Berton M, de Souza JM, Abdiaj I, Snead D, McQuade T (2020) J Flow Chem 10:73–92
39. Zhang P, Weeranoppanant N, Myerson AS, Jensen KF, Jamison TF, Adamo A (2018) Chem A Eur J 24:2776–2784
40. Adamo A, Beigessner RL, Behnam M, Chen J, Jamison TF, Jensen KF, Monbaliu JCM, Myerson AS (2016) Science 352:61–67
41. Richardson P (2014) Future Med Chem 6:845–847
42. Chatterjee S, Guidi M, Seeberger PH, Gilmore K (2020) Nature 579:379–384
43. Zhang P, Russel MG, Jamison TF (2014) Org Process Res Dev 18(11):1567–1570
44. Corcoran EB, Pirnot MT, Lin S, Dreher SD, DiRocco DA, Davies IW, Buchwald ST, MacMillan DWC (2016) Science 353:279–283
45. See: Saikin SK, Kresibeck C, Sherberla D, Becker JS, Aspuru-Guzik A (2019) Exp Opin Drug Discov 14:1–4
46. Schneder G (2017) Nat Rev Drug Discov 17:97–113
47. Parry DM (2019) ACS Med Chem Lett 10:848–856
48. Desai B, Dixon K, Farrant E, Feng Q, Gibson KR, van Hoorn WP, Mills J, Morgan T, Parry DM, Ramjee MK, Selway CN, Tarver GJ, Whitlock G, Wright AG (2013) J Med Chem 56:3033–3047
49. For a review, see: Bao J, Trammer GK (2015) Chem Commun 51:3037–3044
50. For syntheses of imatinib in flow, see: Hopkin MD, Baxendale IR, Ley SV (2010) Chem Commun 46:2450–2452
51. Hopkin MD, Baxendale IR, Ley SV (2013) Org Biomol Chem 11:1822–1839
52. Yang JC, Niu D, Karsetn BM, Lima F, Buchwald SL (2016) Angew Chem Int Ed 55:2531–2535
53. Fu WC, Jamison TF (2019) Org Lett 21:6112–6116
54. Chezchizky W, Dedio J, Desai B, Dixon K, Farrant E, Feng Q, Morgan T, Parry DM, Ramjee MK, Selway CN, Schmidt T, Tarver GJ, Wright AG (2013) ACS Med Chem Lett 4:768–772
55. For a further example of the use of this platform in the development of hepsin inhibitors, see Pant SM, Mukonoweshuro A, Desai B, Ramjee MK, Selway CN, Tarver GJ, Wright AG, Birchall K, Chapman TM, Tervonen TA, Klefström J (2018) J Med Chem 61:4335–4347
56. See for example: Schneider G, Clark DE (2019) Angew Chem Int Ed 58:10792–10803
57. Yang X, Wang Y, Byrne R, Schneider G, Yang S (2019) Chem Rev 119:10520–10594
58. Grebner C, Matter H, Plowright AT, Hessler G (2020) J Med Chem 63:8809–8823
59. Gupta R, Srivastava D, Sahu M, Tiwari S, Ambasta RK, Kumar P (2021) Mol Divers. https://doi.org/10.1007/s11030-021-10217-3
60. Ying Y, Bian Y, Hu Z, Wang L, Xie XQ (2018) AAPS 20:58